Operator Theory
Advances and Applications
Vol. 101

Editor:
I. Gohberg

Parabolic Boundary Value Problems

Samuil D. Eidelman
Nicolae V. Zhitarashu

Translated from the Russian by
Gennady Pasechnik and Andrei Iacob

Springer Basel AG

Authors:

Samuil D. Eidelman
Institute of Mathematics
Ukrainian Academy of Sciences
vul. Tereshchenkivska 3
Kiev 252601
Ukraine

Nicolae V. Zhitarashu
Moldova State University
ul. Mateevicha, 60
Kishinev 277000
Moldova

1991 Mathematics Subject Classification 35K20, 35K35, 35K50, 35B35

A CIP catalogue record for this book is available from the
Library of Congress, Washington D.C., USA

Deutsche Bibliothek Cataloging-in-Publication Data
Ejdel'man, Samuil D.:
Parabolic boundary value problems / Samuil D. Eidelman ; Nicolae V. Zhitarashu.
Übers. aus dem Russ.: Gennady Pasechnik und Andrei Jacob. – Basel ; Boston ;
Berlin : Birkhäuser, 1998
  (Operator theory ; 101)
  ISBN 978-3-0348-9765-5   ISBN 978-3-0348-8767-0 (eBook)
  DOI 10.1007/978-3-0348-8767-0

© 1998 Springer Basel AG
Originally published by Birkhäuser Verlag in 1998
Printed on acid-free paper produced from chlorine-free pulp. TCF ∞
Cover design: Heinz Hiltbrunner, Basel
ISBN 978-3-0348-9765-5

9 8 7 6 5 4 3 2 1

Contents

Chapter III
Linear Operators

Chapter IV
Parabolic Boundary Value Problems in Half-Space 79

Chapter VI
The Cauchy Problem and Parabolic Boundary Value Problems in
Spaces of Smooth Functions 181

Chapter VII
Behaviour of Solutions of Parabolic Boundary Value Problems
for Large Values of Time 233

Foreword

The present monograph is devoted to the theory of general parabolic boundary value problems. The vastness of this theory forced us to take difficult decisions in selecting the results to be presented and in determining the degree of detail needed to describe their proofs.

In the first chapter we define the basic notions at the origin of the theory of parabolic boundary value problems and give various examples of illustrative and descriptive character. The main part of the monograph (Chapters II to V) is devoted to a the detailed and systematic exposition of the L_2-theory of parabolic boundary value problems with smooth coefficients in Hilbert spaces of smooth functions and distributions of arbitrary finite order and with some natural applications of the theory.

Wishing to make the monograph more informative, we included in Chapter VI a survey of results in the theory of the Cauchy problem and boundary value problems in the traditional spaces of smooth functions. We give no proofs; rather, we attempt to compare different results and techniques. Special attention is paid to a detailed analysis of examples illustrating and complementing the results formulated. The chapter is written in such a way that the reader interested only in the results of the classical theory of the Cauchy problem and boundary value problems may concentrate on it alone, skipping the previous chapters.

Space limitations prevented us from presenting a number of interesting techniques and important results of the qualitative theory of linear parabolic system, as we desired. However, a fragment of this theory that is the closest to our research interests is presented in Chapter VI. In it we give the proofs of several stabilization theorems (the existence of a limit as time tends to infinity) of solutions of parabolic boundary value problems. Chapters II to V have been written by N.V. Zhitarashu, and Chapter I, VI and VII, by S.D. Eidel'man.

When constructing the L_2-theory of parabolic boundary value problems in the spaces of distributions, we followed the similar theory for elliptic boundary value problems elaborated by Ya.A. Roĭtberg, whose advice was very valuable to us. V.A. Solonnikov has stimulated our efforts by his work and his constant benevolence.

Our work began in the Cernovtsy University. We wish to acknowledge numerous discussions of various aspects of the theory of parabolic systems with several of its graduates: S.D. Ivasishen, Ya.S. Kushitskiĭ, M.I. Matiĭchuk, F.O. Porper and V.D. Repnikov, to all of whom we express our deep gratitude.

Chapter I
Equations and Problems

I.1. Equations

I.1.1. Introduction

Equations of parabolic type first appeared when mathematical tools were applied to phenomena of heat propagation and diffusion. These studies began with the simplest but most important representative of parabolic equations, the equation of heat conduction:

$$D_t u = a^2 \left[D_1^2 + D_2^2 + D_3^2 \right] u \equiv a^2 \Delta u. \tag{1.1}$$

The subsequent development of natural sciences and mathematics showed convincingly that this equation and its natural generalizations play a fundamental role in mathematics as well as in its applications. The aim of the present monograph is to provide a systematic account of one of the branches of the very rich theory of parabolic equations.

I.1.2. Systems parabolic in the sense of Petrovskiĭ

The basic definitions and results of the theory of highly-general parabolic equations and systems belong to I.G. Petrovskiĭ. Remarkably, it turns out that many known properties of solutions of the heat equation remain valid (with the appropriate modifications) for such systems.

Throughout the book x will stand for an n-dimensional variable and t will be a one-dimensional variable; x and t will be regarded as space and time coordinates, respectively. We will denote by Ω a set of points (x, t) of the space $\mathbb{R}^{n+1}$.

We begin by defining the notion of an equation of arbitrary order parabolic in the sense of Petrovskiĭ. Let $L(x, t, D, D_t)$ be a linear differential expression (from now on we will use the terms "differential operator" or simply "operator") of an arbitrary order with complex-valued coefficients depending on x and t and defined on Ω. For any point $(x, t) \in \Omega$ the function $L(x, t, i\xi, p)$, where ξ is an n-dimensional vector with coordinates $\xi_1, \ldots, \xi_n$ and p is a complex scalar parameter, is a polynomial in ξ_k, $k = 1, \ldots, n$ and p. Let s and r be positive integers. We will assume that the degree of the polynomial $L(x, t, i\lambda\xi, \lambda^s p)$ in λ equals sr. Denote by L_0 the principal part of the polynomial L, that is, the sum of all terms whose degree in λ equals sr:

$$L_0(x, t, i\lambda\xi, \lambda^s p) \equiv \lambda^{sr} L_0(x, t, i\xi, p).$$

Definition I.1. *The operator L is said to be parabolic in the sense of Petrovskiĭ at a point (x, t) if for any $\xi \in \mathbb{R}^n$ the p-zeroes of the polynomial $L_0(x, t, i\xi, p)$ satisfy the inequality*

$$\operatorname{Re} p(x, t, \xi) \leq -\delta_0(x, t)|\xi|^s, \qquad \delta_0(x, t) > 0. \tag{1.2}$$

The operator L is said to be *uniformly parabolic* in Ω if it is parabolic at each point (x, t) and inequality (1.2) holds at each point $(x, t) \in \Omega$ with the same positive number δ_0. Definition I.1 implies that s is an even integer, $s = 2b$. Indeed, the p-zeroes of the polynomial $L_0(x, t, i\xi, p)$ are homogeneous functions of ξ of degree s, that is, $p(x, t, \lambda\xi) = \lambda^s p(x, t, \xi)$. Therefore, letting $\lambda = +1$, $\lambda = -1$, we see that in one of these two situations inequality (1.2) is violated for odd s. For parabolic operators differentiating once with respect to time t is comparable in "strength" to differentiating $2b$ times with respect to the space coordinates $x_1, \ldots, x_n$, that is, differentiation with respect to t has weight $2b$, whereas differentiation with respect to $x_1, \ldots, x_n$ has weight 1. Consider a weighted homogeneous parabolic polynomial $R(\xi, p)$:

$$R(\lambda\xi, \lambda^{2b} p) = \lambda^q R(\xi, p). \tag{1.3}$$

Let $\xi = 0$. Then (1.3) yields $R(0, \lambda^{2b} p) = \lambda^q R(0, p)$. It follows that either $q = 2br$, or $R(0, p) \equiv 0$. The latter is impossible, since for $\operatorname{Re} p > 0$ this contradicts the assumption that $R(\xi, p)$ is a parabolic polynomial. Thus $q = 2br$ and $R(0, p) = \gamma \cdot p^r$, where γ is a nonzero constant. In a similar manner one can prove that the coefficients of ξ_k^{2br}, $k = 1, \ldots, n$, in the polynomial R are different from zero. For a parabolic operator L the numbers b and r are uniquely determined: r is the degree of the polynomial $L(x, t, i\xi, p)$ in p, and $2br$ is its degree in the variables ξ_k, $k = 1, \ldots, n$.

Definition I.2. *The matrix differential operator $\mathcal{L}(x, t, D, D_t)$ with the elements $l_{kj}(x, t, D, D_t)$, $k, j = 1, \ldots, n$, is said to be parabolic in the sense of Petrovskiĭ if*
(1) *the operator $L(x, t, D, D_t) = \det \mathcal{L}(x, t, D, D_t)$ is parabolic in the sense of Petrovskiĭ;*
(2) *the degrees of the polynomials $l_{kj}(x, t, i\lambda\xi, \lambda^{2b} p)$ in λ do not exceed $2bn_j$ and $l_{kj}(x, t, i\xi, p) = \delta_{kj} p^{n_j} + l'_{kj}(x, t, i\xi, p)$, where l'_{kj} is a polynomial that does not contain p^{n_j} and δ_{kj} is the Kronecker symbol.*

From now the principal part of the polynomial L_{kj} will be denoted by L^0_{kj}.

Thus, the systems that are $2b$-parabolic in the sense of Petrovskiĭ have the following structure:

$$D_t^{n_k} u_k = \sum_{j=1}^{n} \sum_{\substack{|\alpha| + 2b\alpha_0 \leq 2bn_j \\ \alpha_0 < n_j}} a_{\alpha\alpha_0}^{kj}(x, t) D^\alpha D_t^{\alpha_0} u_j + f_k(x, t), \qquad k = 1, \ldots, m. \tag{1.4}$$

The best studied is the case of systems parabolic in the sense of Petrovskiĭ that are of first order in the time coordinate t:

$$D_t u = \sum_{|\alpha| \leq 2b} a_\alpha(x,t) D^\alpha u + f(x,t), \tag{1.5}$$

where $a_\alpha(x,t)$ are $m \times m$ matrices, $u(x,t)$ and $f(x,t)$ are m-dimensional vectors. A very important subclass of the systems of type (1.5) that are parabolic in the sense of Petrovskiĭ is the class of strongly parabolic systems.

Definition I.3. *A system of type* (1.5) *is said to be uniformly strongly parabolic in* Ω *if there exists a positive constant* δ_0 *such that*

$$\mathrm{Re}\left(\sum_{|\alpha|=2b} a_\alpha(x,t)(i\xi)^\alpha \eta, \eta \right) \leq -\delta_0 |\xi|^{2b} \cdot |\eta|^2$$

for any $(x,t) \in \Omega$, $\eta \in \mathbb{C}^m$, $\xi \in \mathbb{R}^n$.

Strongly parabolic systems have some properties that do not hold for arbitrary systems parabolic in the sense of Petrovskiĭ.

In applications one usually encounters equations and systems of second order,

$$D_t u = \sum_{\mu,\nu=1}^{n} a_{\mu\nu}(x,t) D_\mu D_\nu u + \sum_{\mu=1}^{n} a_\mu(x,t) D_\mu u + a(x,t) u + f(x,t), \tag{1.6}$$

where $a_{\mu,\nu}$, a_μ, and a are $m \times m$ matrices, $\mu,\nu = 1,\ldots,m$.

The uniform parabolicity of the system (1.6) in Ω means that the λ-roots of the equation

$$\det\left(\sum_{\mu,\nu=1}^{n} a_{\mu\nu}(x,t)\xi_\mu\xi_\nu - \lambda I \right) = 0, \tag{1.7}$$

(where I is the $m \times m$ unit matrix) satisfy the inequality

$$\mathrm{Re}\,\lambda(x,t,\xi) \geq \delta_0|\xi|^2, \qquad \delta_0 > 0 \tag{1.8}$$

for any $(x,t) \in \Omega$ and $\xi \in \mathbb{R}^n$.

In the case of a single equation ($m = 1$) with real-valued coefficients, (1.8) becomes

$$\sum_{\mu,\nu=1}^{n} a_{\mu\nu}(x,t)\xi_\mu\xi_\nu \geq \delta_0|\xi|^2 \tag{1.9}$$

and uniform parabolicity means that the quadratic form $\sum_{\mu,\nu=1}^{n} a_{\mu\nu}(x,t)\xi_\mu\xi_\nu$, $\xi \in \mathbb{R}^n$ is positive definite uniformly in $(x,t) \in \Omega$.

Below we consider only uniformly parabolic systems, and the word "uniformly" will be omitted.

I.1.3. Systems parabolic in the sense of Solonnikov

The structure of a system that is parabolic in the sense of Petrovskiĭ allows the following generalization. For such a system the degrees of generalized homogeneity of the polynomials L_{kj}^0 equal $2bn_j$ and do not depend on k. V.A. Solonnikov suggested another definition of parabolicity, in which the degrees of generalized homogeneity of the polynomials L_{kj}^0 are allowed to depend on both j and k. For hyperbolic systems a similar generalization was introduced earlier by J. Leray, and for elliptic ones by A. Douglis and L. Nirenberg. In these definitions the order of the operator L_{kj} is defined as a sum $s_k + t_j$, where s_k, t_j, $k, j = 1, \ldots, m$ are integers.

Definition I.4. *The operator $\mathcal{L}$ is said to be parabolic in the sense of Solonnikov if*
(1) *the operator $L = \det \mathcal{L}$ is $2b$-parabolic in the sense of Petrovskiĭ;*
(2) *there exist integers s_k, t_j, $k, j = 1, \ldots, m$ such that the degree of the polynomial $l_{kj}(x, t, i\lambda\xi, \lambda^{2b}p)$ in λ does not exceed $s_k + t_j$ (if $s_k + t_j < 0$, then*

$$l_{kj} \equiv 0), \ \text{and, in addition,} \ \sum_{k=1}^m (s_k + t_k) = 2br, \ \text{where } r \text{ is the degree of the}$$

polynomial $L(x, t, i\xi, p)$ in p.

The *principal part* of the operator $\mathcal{L}$ is the operator $\mathcal{L}_0$ whose matrix has as elements the polynomials $l_{kj}^0(x, t, i\xi, p)$, the principal parts of l_{kj}. The polynomials $l_{kj}^0(x, t, i\xi, p)$ are the sums of all terms of l_{kj} for which

$$l_{kj}^0(x, t, i\lambda\xi, \lambda^{2b}p) = \lambda^{s_k + t_j} l_{kj}^0(x, t, i\xi, p).$$

If the degree of the polynomial $l_{kj}(x, t, i\lambda\xi, \lambda^{2b}p)$ in λ is less than $s_k + t_j$, then $l_{kj}^0 \equiv 0$. The polynomial $L_0 = \det \mathcal{L}_0$ is the principal part of the polynomial L.

Here a natural question arises. Let $\mathcal{L}$ be a given matrix differential operator such that $L = \det \mathcal{L}$ is parabolic in the sense of Petrovskiĭ. Can we find integers s_k, t_j such that $\mathcal{L}$ will be parabolic the sense of Solonnikov? Note that such integers must obviously satisfy the following requirements: $s_k + t_j \geq a_{kj}$, $\sum_{k=1}^m (s_k + t_k) = 2br$, where a_{kj} are the degrees of $l_{kj}(x, t, i\lambda\xi, \lambda^{2b}p)$ in λ. Of course, the numbers b and r are uniquely determined by the polynomial L.

As was shown by L.R. Volevich [11], such numbers s_k, t_j always exist. If s_k, t_j is a set of weights, then $s_k + a$, $t_j - a$ is also an admissible set. For this reason one usually subjects the set s_k, t_j to the condition $\max_k s_k = 0$.

A system parabolic in the sense of Solonnikov can be written as

$$\sum_{j=1}^n \ \sum_{|\alpha| + 2b\alpha_0 \leq s_k + t_j} a_{\alpha\alpha_0}^{kj}(x, t) D^\alpha D_t^{\alpha_0} u_j = f_k(x, t), \qquad k = 1, \ldots, m. \qquad (1.10)$$

Example I.1. Consider a system of two heat equations

$$D_t u_1 = \Delta u_1, \qquad D_t u_2 = \Delta u_2. \qquad (1.11)$$

For the system (1.11)

$$\mathcal{L}(i\xi, p) = \mathcal{L}_0(i\xi, p) = \begin{pmatrix} p + \xi^2 & 0 \\ 0 & p + \xi^2 \end{pmatrix}, \qquad \xi^2 = \xi_1^2 + \ldots + \xi_n^2.$$

The operator $\mathcal{L}$ is parabolic in the sense of Petrovskiĭ: $s_1 = s_2 = 0$, $t_1 = t_2 = 2$. But (1.11) can also be considered as a system parabolic in the sense of Solonnikov, with $s_1 = 0$, $s_2 = -c$, $t_1 = 2$, $t_2 = 2+c$, where c is an arbitrary positive integer. Thus the condition $\max_k s_k$ does not guarantee that s_k, t_j can be uniquely selected even for the simplest system (1.11). Incidentally, this arbitrariness can be used to widen the class of problems that we want to investigate and to derive weel-posedness theorems for such problems in various functional spaces.

The systems parabolic in the sense of Petrovskiĭ form a subclass of the systems parabolic in the sense of Solonnikov: for them $s_1 = \ldots = s_m = 0$, $t_j = 2bn_j$, $j = 1, \ldots, m$.

A particular case of the general parabolic systems is the class introduced by T. Shirota in 1957 [124].

Definition I.5. *A system of the form*

$$D_t u_k = \sum_{j=1}^{m} a_{kj}(x, t, D) u_j, \qquad k = 1, \ldots, m, \tag{1.12}$$

is said to be parabolic in the sense of Shirota if
(1) the polynomial $\det \left[pI - \left(a_{kj}(x, t, i\xi) \right)_{i,j=1}^{m} \right]$ *is parabolic in the sense of Petrovskiĭ;*
(2) there exist nonnegative integers s_k, $k = 1, \ldots, m$ *such that the order of the operator* $a_{kj}(x, t, D)$ *does not exceed* $s_k - s_j + 2b$.

The systems parabolic in the sense of Shirota include as a particular case the systems parabolic in the sense of Petrovskiĭ. For the system (1.5) this is obvious (all $s_k = 0$). For a system of the form of (1.4) one first reduces the system, in the standard way, to a system with first order derivatives with respect to t, and then one verifies that the latter is parabolic in the sense of Shirota.

If a system is parabolic in the sense of Shirota, then to the principal parts of its diagonal operators there correspond the polynomials $L^0(x, t, i\lambda\xi, \lambda^{2b}p) = \lambda^{2b}p + a_{kk_0}^0(x, t, i\lambda\xi) = \lambda^{s_k+t_k} l_{kk_0}^0(x, t, i\xi, p)$. Hence, $s_k + t_k = 2b$, $t_k = 2b - s_k$. Thus we have shown in what sense the general parabolic systems include the systems parabolic in the sense of Shirota.

I.2. Initial and boundary value problems

I.2.1. Introduction

Concrete physical processes cannot be described solely by differential equations. If, for example, we want to determine the temperature inside a body at an arbitrary time, we must in addition know the temperature distribution in the body at the initial moment (initial condition) and the temperature regime on the boundary ∂G of the body G (boundary condition).

The boundary conditions can vary, depending on the concrete physical or engineering problem.

In the case of the heat equation the simplest problems are formulated as follows:

(1) at any point of the surface ∂G one prescribes the temperature

$$u\big|_{\partial G} = \varphi_1, \tag{2.1}$$

where φ_1 is a known function of the current point of ∂G and time t;

(2) at any point of the surface ∂G one prescribes the heat flux $-k\frac{\partial u}{\partial \nu}$, where k is the heat conductivity coefficient and $\partial/\partial \nu$ stands for differentiation along the normal to the surface ∂G, i.e.,

$$\frac{\partial u}{\partial \nu}\bigg|_{\partial G} = \varphi_2, \tag{2.2}$$

(3) on the surface of the body there is a heat exchange with the environment, whose temperature u_0 is known. In the simplest case it is assumed that Newton's law of heat exchange is valid. According to this law, the amount of heat Q transferred per unit of time from a unit area of the surface ∂G to the environment is proportional to the temperature difference between the body surface and the environment: $Q = H(u - u_0)$, where H is the heat-exchange coefficient. By conservation of energy, this amount of heat is equal to the amount of heat transferred due to internal heat conduction across a unit area of surface per unit of time. This yields the following condition on ∂G: $H(u - u_0) = -k\partial u/\partial \nu$, or

$$\left[\frac{\partial u}{\partial \nu} + h(u - u_0)\right]_{\partial G} = 0, \qquad h = H/k. \tag{2.3}$$

Note that conditions (2.1)–(2.3) involve derivatives with respect to space variables of at most order one, that is, the order of the boundary conditions is smaller than the weighted order of the equation. However, in applications one encounters problems for the heat equation that lead to boundary conditions involving derivatives of arbitrary order. As an example we will present a problem posed by A.N. Tikhonov in 1950 [100]. It deals with heat propagation in a homogeneous semi-infinite rod with a thermally insulated lateral surface under the assumption that at any time t the temperature is constant throughout any cross-section of the rod. There are no heat sources inside the rod, and its boundary $\{x = 0\}$ is in contact to a body G_1 of heat capacity c and such high heat conductivity that one can practically assume that its temperature U depends only on t. It is also assumed that inside G_1 there are located heat sources whose density at time t equals $\varphi(t)$.

In this situation the temperature $u(x, t)$ in the rod satisfies the heat equation

$$D_t u = a^2 D^2 u, \qquad x > 0, \ t > 0. \tag{2.4}$$

To find the boundary condition at $x = 0$ at a time t one needs to take into account two circumstances: (1) for $x = 0$ the rod and the body G_1 have the same temperature; (2) the heat flux across the cross-section $x = 0$ must coincide with the amount of heat received by the body G_1 per unit of time. Thus,

$$U(t) = u(0, t),$$
$$-kD_x u(x, t)\big|_{x=0} = \varphi(t) - c\frac{dU}{dt}, \qquad t > 0,$$

Using the fact that $u(x, t)$ satisfies the heat equation (2.4) and the above equalities, we obtain the boundary condition

$$\left(ca^2 D^2 u - kDu\right)\big|_{x=0} = \varphi(t), \qquad t > 0,$$

called the condition of concentrated heat capacity, which involves a second-order derivative.

If there is a temperature jump on the boundary between the body G_1 and the rod, that is,

$$U(t) - u(x, 0) = k_1 Du(x, t)\big|_{x=0}, \qquad t > 0,$$

then the corresponding boundary condition will involve third-order derivatives. Using a number of heat capacitors with temperature drop between each pair, one can construct schemes that correspond to boundary conditions involving derivatives of arbitrary order.

The problem of finding a solution of the heat equations that satisfies the initial condition $u|_{t=0} = \psi(x)$ and some boundary condition is called the *mixed problem* for this equation (also called the *initial-boundary value* or sometimes simply the *boundary value problem*). In the case of the boundary conditions (2.1), (2.2), or (2.3) this problem is called the first, the second or the third mixed (boundary value) problem, respectively. The first problem is often called the Dirichlet problem and the second is referred to as the Neumann problem.

If we need to determine the temperature of the body and, for some reason, we are allowed to neglect the influence of the boundary, then it is natural to assume that there is no boundary at all, and we arrive at the problem of finding a solution in the whole space given the initial values. Such a problem is traditionally called the *Cauchy problem*. The practical importance of such an approach lies in the fact that usually the Cauchy problem is easier to solve and investigate than the boundary value problems. We remark that the Cauchy problem is also quite frequently encountered in problems of probabilistic origin.

I.2.2. The Cauchy problem. The initial value problem

For systems that are parabolic in the sense of Petrovskiĭ and in the sense of Shirota, solved with respect to the highest-order time derivative, it is natural to pose the Cauchy problem of finding a solution in the layer $\Pi = \mathbb{R}^n \times (0, T]$, satisfying in the case of system (1.4) the following initial conditions:

$$D_t^{\alpha_0 - 1} u_j\big|_{t=0} = \psi_j^{\alpha_0}(x), \qquad \alpha_0 = 1, \dots, n_j, \ j = 1, \dots, m. \tag{2.5}$$

Here it is assumed that $D_t^{\alpha_0 - 1} u_j(x, t)$ tends to $\psi_j^{\alpha_0}(x)$ as $t \to 0$ in the topology of the functional spaces in which the solvability of the problem is studied. In the simplest case (2.5) may be understood as the limit for any fixed x or as the uniform limit on any compact set of $\mathbb{R}^n$.

In the case of system (1.10), parabolic in the sense of Solonnikov, the Cauchy problem may be meaningless, and consequently it becomes necessary to consider a special problem, called the *initial value problem*.

Let us describe this problem. The initial conditions are given in the form

$$C(x, D, D_t) u\big|_{t=0} = \psi(x), \tag{2.6}$$

where C is a matrix differential operator with elements c_{kj}, $k = 1, \dots, m_1$, $j = 1, \dots, m$, and $\psi(x)$ is an arbitrary vector function with values in $\mathbb{R}^{m_1}$. To specify the structure of the matrix of initial conditions we proceed as follows: let γ_{kj} be the degree in λ of the polynomial $c_{kj}(x, i\lambda\xi, \lambda^{2b} p)$ (if $c_{kj} \equiv 0$, then we may take an arbitrary integer for γ_{kj}), and set $\rho_k = \max_j(\gamma_{kj} - t_j)$. The sum c_{kj}^0 of all terms of the polynomial c_{kj} that satisfy the generalized homogeneity condition

$$c_{kj}^0(x, i\lambda\xi, \lambda^{2b} p) = \lambda^{\rho_k + t_j} c_{kj}^0(x, i\xi, p) \tag{2.7}$$

is called the principal part of c_{kj}; the matrix $C_0 = (c_{kj}^0)$ is called the principal part of C.

Let us find what conditions the number m_1 and the matrix operator C should satisfy in order for us to be able to uniquely determine the value of any derivative of an arbitrary function u_j at $t = 0$ from the system

$$\mathcal{L} u = f \tag{2.8}$$

and the initial condition (2.6) by differentiating and solving linear algebraic systems (assuming that solutions, the coefficients of $\mathcal{L}$, ψ, and f are sufficiently smooth). We require that all the conditions for $\mathcal{L}$ and C be formulated in terms of $\mathcal{L}_0$ and C_0. Thus, let us consider the problem

$$\mathcal{L}_0(x, 0, D, D_t)u = f, \tag{2.9}$$

$$C_0(x, D, D_t)u\big|_{t=0} = \psi. \tag{2.10}$$

In particular, taking all possible vector-valued functions $u(t)$ that depend only on t, we derive from (2.9), (2.10) the following problem:

$$\mathcal{L}_0(x, 0, 0, D_t)u(t) = f(x, t), \tag{2.11}$$

$$C_0(x, 0, D_t)u(t)\big|_{t=0} = \psi(x). \tag{2.12}$$

Note that $\det \mathcal{L}_0(x, 0, 0, p) = \mathcal{L}_0(x, 0, 0, p) = \gamma(x)p^r$, where $2br = \sum\limits_{k=1}^{m}(t_k + s_k)$, and that $\gamma(x) \neq 0$ for all x due to the parabolicity of the system. It follows that the dimension of the set of solutions of the system of ordinary differential equations (2.11) equals r, that is, C_0 (and C) are $r \times m$ matrices.

Example I.2. Consider a system parabolic in the sense of Solonnikov with the matrix

$$\mathcal{L}_0(i\xi, p) = \begin{pmatrix} p + \xi^2 & p^{10} \\ 0 & p + \xi^2 \end{pmatrix}.$$

For this matrix $b = 1$, $r = 2$, $s_1 = 0$, $s_2 = -8$, $t_1 = 2$, $t_2 = 10$. In spite of the fact that the system involves a derivative with respect to t of the tenth order, C_0 is a 2×2 matrix. Consequently, it suffices to pose only two initial conditions. By using the method of residues to solve the homogeneous system $\mathcal{L}_0(x, 0; 0, \frac{d}{dt})u(x, t) = 0$ and a standard argument (which will be discussed in the next subsection), one can obtain a necessary condition for the initial value problem to possess all the desired properties. Denote by $\widehat{\mathcal{L}}_0$ the adjoint matrix of the matrix $\mathcal{L}_0$ ($\mathcal{L}_0\widehat{\mathcal{L}}_0 = \widehat{\mathcal{L}}_0\mathcal{L}_0 = \det \mathcal{L}_0 \cdot I$).

Definition I.6. (*Complementarity condition for the initial value problem.*) *The initial value problem* (2.6) *for a system parabolic in the sense of Solonnikov* $\mathcal{L}u = f$ *is said to satisfy the complementarity condition if the rows of the matrix* $C_0(x; 0, p)\widehat{\mathcal{L}}_0(x, 0; 0, p)$ *are linearly independent modulo the polynomial* $p^r \equiv \frac{1}{\gamma(x)}\mathcal{L}_0(x, 0; 0, p)$ *for any* x.

Recall that the linear independence of a set of polynomials $M_1(p), \ldots, M_l(p)$ modulo a polynomial $N(p)$ means that the remainders of the division of $M_k(p)$ by $N(p)$, $k = 1, \ldots, l$, are linearly independent.

Note that both the Cauchy problem and the initial value problem for parabolic systems are characteristic problems ($t = 0$ is a characteristic hyperplane). Hence, to ensure uniqueness of the solution, we must impose additional conditions on the behavior of solutions as $|x| \to \infty$.

I.2.3. Parabolic boundary value problems

Here we shall consider the solution of the general parabolic system

$$\mathcal{L}(x,t,D,D_t)u(x,t) = f(x,t) \tag{2.13}$$

in a cylindrical domain $\Omega = G \times [0,T]$, where $G \subset \mathbb{R}^n$ is a domain with smooth boundary ∂G, and $S = \partial G \times [0,T]$ is the lateral surface of Ω.

Set the initial condition for $t = 0$,

$$C(x,t,D,D_t)u\big|_{t=0} = \psi(x), \qquad x \in G, \tag{2.14}$$

and the boundary condition on S,

$$\mathcal{B}(x,t,D,D_t)u\big|_S = \varphi, \tag{2.15}$$

where $\mathcal{B}$ is a matrix differential operator and φ is an arbitrary vector-valued function on S.

Let us define the principal part $\mathcal{B}_0$ of the matrix $\mathcal{B}$. Denote by b_{qj}, $q = 1,\dots,m_2$, $j = 1,\dots,m$ the elements of $\mathcal{B}$ and by r_{qj} the degree in λ of the polynomial $b_{qj}(x,t,i\lambda\xi,\lambda^{2b}p)$ (if $b_{qj} \equiv 0$, then we can take for r_{qj} any integer). Let $\sigma_q = \max\limits_{j}(r_{qj} - t_j)$; then $\sigma_q + t_j \geq r_{qj}$.

The principal part of the polynomial b_{qj} is defined to be the sum b_{qj}^0 of all its terms that satisfy the generalized homogeneity condition

$$b_{qj}^0(x,t,i\lambda\xi,\lambda^{2b}p) = \lambda^{\sigma_q + t_j} b_{qj}^0(x,t,i\xi,p).$$

Further, denote by $\mathcal{B}_0$ the matrix with the elements b_{qj}^0. Here, of course, the choice of the principal part $\mathcal{B}_0$ depends on choice of t_j, and so, generally speaking, $\mathcal{B}_0$ is not uniquely determined.

Now fix a set of numbers s_k, t_j and σ_q. Take an arbitrary point $(x^0,t^0) \in S$, and introduce local coordinates $y_1,\dots,y_n$ in $\mathbb{R}^n$ near $x^0 \in \partial G$, such that the y_n-axis is directed along the inner normal ν. In the half-space $y_n > 0$ consider the problem

$$\mathcal{L}_0(x^0,t^0,D,D_t)u = 0, \tag{2.16}$$
$$\mathcal{B}_0(x^0,t^0,D,D_t)u\big|_{y_n=0} = \varphi. \tag{2.17}$$

Following Ya.B. Lopatinskiĭ, we will assume that the following fundamental condition is satisfied: *for any point $(x^0,t^0) \in S$ the problem (2.16), (2.17) is solvable for any smooth vector-valued function φ with compact support in the class of functions that admit a Fourier transform in the tangential variables y' (i.e., $y_1,\dots,y_{n-1}$).*

The aim of the next steps is to recast the last condition in a transparent algebraic form. They are based on a standard argument, which is very often used in the theory of boundary value problems and can be briefly described as follows.

Problem (2.16), (2.17) can be rewritten in the new coordinates:

$$\widetilde{\mathcal{L}}(D_y, D_t)u = 0$$

$$\widetilde{\mathcal{B}}(D_y, D_t)u\big|_{y_n=0} = \varphi(y', t).$$

The operator $\widetilde{\mathcal{L}}$ obviously remains parabolic in the new coordinates, with the same parabolicity constant δ_0.

Applying the Fourier transformation in the tangential variables y' and the Laplace transformation in time to the last problem (assuming that all time derivatives vanish at $t = 0$), we reduce it to the following boundary value problem for a system of ordinary equations on the half-line $y_n \geq 0$:

$$\widetilde{\mathcal{L}}\Big(i\xi, \frac{d}{dy_n}, p\Big)\widehat{u}(y_n) = 0,$$

$$\widetilde{\mathcal{B}}\Big(i\xi, \frac{d}{dy_n}, p\Big)\widehat{u}(y_n)\Big|_{y_n=0} = \widehat{\varphi}(\xi', p),$$

$$|\widehat{u}(y_n)| \to 0 \text{ as } y_n \to \infty,$$

where $\xi' = (\xi_1, \ldots, \xi_{n-1})$.

Thus, we need to consider the following problem:

$$\mathcal{P}\Big(\frac{d}{dz}\Big)W(z) = 0, \tag{2.18}$$

$$\mathcal{A}\Big(\frac{d}{dz}\Big)W(z)\Big|_{z=0} = h, \tag{2.19}$$

$$|W(z)| \to 0 \text{ as } z \to \infty. \tag{2.20}$$

The solutions of system (2.18) form a finite-dimensional space E. Denote by E_+ the subspace of E consisting of the solutions of (2.18) that satisfy condition (2.20). The dimension of E_+ is equal to the number of zeroes with positive imaginary parts of the polynomial $P(i\tau) = \det \mathcal{P}(i\tau)$ (counting multiplicities). Denote these zeroes by τ_s^+, and their number by r_+. Define the polynomial $P_+(\tau) = \prod_{s=1}^{r_+}(\tau - \tau_s^+)$. It can be directly verified that the components of any solution of (2.18) in E_+ satisfy the equation $P_+\big(\frac{d}{dz}\big)v(z) = 0$. Now let $W^{(s)}(z)$, $s = 1, \ldots, r_+$ be a basis in E_+. Then any solution of (2.18) in E_+ can be written as

$$W(z) = \sum_{s=1}^{r_+} C_s W^{(s)}(z),$$

where C_s are arbitrary constants. We can choose them so that the boundary condition (2.19) will be satisfied. To this end we must solve the system of algebraic equations

$$\sum_{s=1}^{r_+}\Big(\sum_{j=1}^{m} A_{qj}\Big(\frac{d}{dz}\Big)W_j^{(s)}(z)\big|_{z=0}\Big)C_s = h_q, \qquad q = 1, \ldots, m_2. \tag{2.21}$$

Let D be the matrix with the elements $d_{qs} = \sum\limits_{j=1}^{m} A_{qj}\left(\frac{d}{dz}\right)W_j^{(s)}(z)\Big|_{z=0}$, $q = 1,\ldots,m_2$, $s = 1,\ldots,r_+$.

From the theory of linear algebraic systems one derives the following important statement:

Lemma I.1. *The system (2.21) has a unique solution for any h_q, $q = 1,\ldots,m_2$ if and only if*

$$m_2 = r_+, \tag{2.22}$$

$$\det \mathcal{D} \neq 0. \tag{2.23}$$

Thus, we have answered the questions in which we were interested, concerning the number of boundary conditions and the solvability conditions of the posed boundary value problem. However, condition (2.23) is still not formulated directly in terms of the original problem and some extra effort is needed to obtain such a formulation. This is our next objective.

Let us use the method of residues (Cauchy's method) to solve the system (2.18). Its general solution is given by the formula

$$W(z) = \int\limits_{\gamma} \left\{\left(\widehat{P}(i\tau)Q(i\tau)e^{i\tau z}\right)/P(i\tau)\right\}d\tau,$$

where γ is an arbitrary closed loop in the complex τ-plane that encircles all the zeroes of $P(i\tau)$, and $Q(i\tau)$ is a vector whose components are arbitrary polynomials of degree less than that of $P(i\tau)$.

Any vector of E_+ can be similarly written as

$$W(z) = \int\limits_{\gamma_+} \left\{\left(\widehat{P}(i\tau)Q(i\tau)e^{i\tau z}\right)/P_+(\tau)\right\}d\tau, \tag{2.24}$$

where γ_+ is an arbitrary closed loop that encircles all the zeroes of $P_+(\tau)$, and the components of the vector $Q(i\tau)$ are arbitrary polynomials of degree $r_+ - 1$. In particular, the vector-valued functions $W^{(s)}(z)$ of the basis of E_+ can be written by means of formula (2.24) with some $Q^{(s)}(i\tau)$:

$$W^{(s)}(z) = \int\limits_{\gamma_+} \left\{\left(\widehat{P}(i\tau)Q^{(s)}(i\tau)e^{i\tau z}\right)/P_+(\tau)\right\}d\tau. \tag{2.25}$$

Consequently, the elements d_{qs} of the matrix $\mathcal{D}$ can be rewritten as

$$\begin{aligned}
d_{qs} &= \int\limits_{\gamma_+} \left\{\left(\sum_{k,j=1}^{m} A_{qk}(i\tau)\widehat{P}_{kj}(i\tau)Q_j^{(s)}(i\tau)\right)\Big/P_+(\tau)\right\}d\tau \\
&= \int\limits_{\gamma_+} \left\{\sum_{j=1}^{m} R_{qj}(i\tau)Q_j^{(s)}(i\tau)\Big/P_+(\tau)\right\}d\tau,
\end{aligned} \tag{2.26}$$

where $R_{qj}(i\tau) = \sum\limits_{k=1}^{m} A_{qk}(i\tau)\widehat{P}_{kj}(i\tau)$.

Now we proceed to the final step, which is based on the fact that condition (2.23) is equivalent to the condition that the homogeneous algebraic system

$$\sum_{q=1}^{r_+} e_q d_{qs} = 0, \qquad s = 1, \ldots, r_+ \tag{2.27}$$

have only the trivial solution.

Let us introduce the functions $R_j(i\tau) = \sum_{q=1}^{r_+} e_q R_{qj}(i\tau)$. From (2.26) and (2.27) it follows that

$$\int_{\gamma_+} \left\{ \left(\sum_{j=1}^{m} R_j(i\tau) Q_j^{(s)}(i\tau) \right) \Big/ P_+(\tau) \right\} d\tau = 0, \qquad s = 1, \ldots, r_+. \tag{2.28}$$

Since $W^{(s)}(z)$, defined by formula (2.25), $s = 1, \ldots, r_+$, form a basis in E_+, it follows that there exist constants $a_1, \ldots, a_q$ such that for any polynomial $Q_j(i\tau)$, $j = 1, \ldots, m$, of degree not higher than $r_+ - 1$ the following equality holds:

$$\int_{\gamma_+} \left\{ \left(\sum_{j=1}^{m} \widehat{P}_{kj}(i\tau) Q_j(i\tau) e^{i\tau z} \right) \Big/ P_+(\tau) \right\} d\tau$$

$$= \sum_{s=1}^{r_+} a_s \int_{\gamma_+} \left\{ \left(\sum_{j=1}^{m} \widehat{P}_{kj}(i\tau) Q_j^{(s)}(i\tau) e^{i\tau z} \right) \Big/ P_+(\tau) \right\} d\tau.$$

Now apply the operator $\sum_{q=1}^{r_+} e_q A_{qk}\left(\frac{d}{dz}\right)$ to both sides of the last equality and then sum the results over k from 1 to m. This yields

$$\int_{\gamma_+} \left\{ \left(\sum_{q=1}^{r_+} \sum_{k,j=1}^{m} e_q A_{qk}(i\tau) \widehat{P}_{kj}(i\tau) Q_j(i\tau) \right) \Big/ P_+(\tau) \right\} d\tau$$

$$= \sum_{s=1}^{r_+} a_s \int_{\gamma_+} \left\{ \left(\sum_{q=1}^{r_+} \sum_{k,j=1}^{m} e_q A_{qk}(i\tau) \widehat{P}_{kj}(i\tau) Q_j^{(s)}(i\tau) \right) \Big/ P_+(\tau) \right\} d\tau. \tag{2.29}$$

In the notations introduced above (2.29) becomes

$$\int_{\gamma_+} \left\{ \left(\sum_{j=1}^{m} R_j(i\tau) Q_j(i\tau) \right) \Big/ P_+(\tau) \right\} d\tau$$

$$= \sum_{s=1}^{r_+} a_s \int_{\gamma_+} \left\{ \left(\sum_{j=1}^{m} R_j(i\tau) Q_j^{(s)}(i\tau) \right) \Big/ P_+(\tau) \right\} d\tau,$$

and using (2.28) we obtain the equality

$$\int_{\gamma_+} \left\{ \left(\sum_{j=1}^{m} R_j(i\tau) Q_j(i\tau) \right) \Big/ P_+(\tau) \right\} d\tau = 0, \tag{2.30}$$

which holds for arbitrary polynomials $Q_j(i\tau)$ of degree $r_+ - 1$.

Since the polynomials $Q_j(i\tau)$ are arbitrary, (2.30) is equivalent to the assertion that, for any j and any polynomial $Q(i\tau)$ of degree $r_+ - 1$,

$$\int_{\gamma_+} \left\{ R_j(i\tau) Q(i\tau) / P_+(\tau) \right\} d\tau = 0.$$

The latter is evidently equivalent to the equality

$$\sum_{q=1}^{r_+} e_q R_{qj}(i\tau) = R_j(i\tau) = P_j(\tau) \cdot P_+(\tau), \qquad j = 1, \ldots, m, \tag{2.31}$$

where $P_j(\tau)$ are some polynomials. We see that condition (2.23) is equivalent to the fact that (2.31) implies $e_q = 0$.

Thus, we arrive at the following formulation of the condition of unique solvability of problem (2.18)–(2.20) directly in terms of the problem itself: *the rows of the matrix $R(i\tau) = A(i\tau)\widehat{P}(i\tau)$ are linearly independent modulo the polynomial $P_+(\tau)$.*

Now let us return to the problem (2.13)–(2.15). The number of boundary conditions is specified by the following lemma.

Lemma I.2. *Suppose that $L_0(i\xi', i\tau, p)$ is a weighted homogeneous polynomial parabolic in the sense of Petrovskiĭ, that is, $L_0(i\lambda\xi', i\lambda\tau, \lambda^{2b}p) = \lambda^{2br} L_0(i\xi', i\tau, p)$, and for any real (ξ', τ) the p-zeroes of $L_0(i\xi', i\tau, p)$ satisfy the inequality $\operatorname{Re} p(\xi', \tau) \leq -\delta_0(\xi'^2 + \tau^2)^b$. Then for any real $\xi_1, \ldots, \xi_{n-1}$ and any complex p such that the conditions $\operatorname{Re} p > -\delta_1 \xi'^{2b}$, $0 < \delta_1 < \delta_0$, $|p| + \xi'^{2b} > 0$ are satisfied, the polynomial $L_0(i\xi', i\tau, p)$ has br τ-zeroes with positive imaginary parts and br τ-zeroes with negative imaginary parts.*

Thus, for general parabolic systems we must impose $r_+ = br$ boundary conditions, that is, the matrix of boundary conditions, $\mathcal{B}$, is of size $br \times m$. Consider the problem (2.16), (2.17). Any vector $\xi \in \mathbb{R}_n$ can be uniquely represented in the form $\xi = \widetilde{\xi} + \tau\nu$, where $\widetilde{\xi}$ is a vector that lies in the tangent plane to ∂G at the point x^0, ν is the unit inner normal to ∂G at x_0, and τ is a real parameter. Denote by A^{x^0} the set of all points $(\widetilde{\xi}, p)$, where $p \in \mathbb{C}$ and $\widetilde{\xi}$ is a tangent vector to ∂G at x_0, with the property that $|p| + |\widetilde{\xi}|^{2b} > 0$, $\operatorname{Re} p \geq -\delta_1 |\widetilde{\xi}|^{2b}$, where $\delta_1 \in (0, \delta_0)$ and δ_0 is the parabolicity constant of the system. Consider the τ-polynomial $L_0(x^0, t^0, \widetilde{\xi} + \tau\nu, p)$. If $(\widetilde{\xi}, p) \in A^{x^0}$, then, by Lemma I.2, L_0 has br

τ-zeroes with positive imaginary parts (denote these zeroes by $\tau_s^+(x^0,t^0;\widetilde{\xi},p)$) and br τ-zeroes with negative imaginary parts. Let

$$L_0^+(x^0,t^0,\widetilde{\xi},\tau,p) = \prod_{s=1}^{br}\left(\tau - \tau_s^+(x^0,t^0,\widetilde{\xi},p)\right).$$

In the present case the solvability condition for a boundary value problem obtained above (Lopatinskiĭ's condition) takes the following concrete form:

Definition I.7. (*Complementarity condition of the initial value problem* (2.15) *for the system* (2.13).) *For any point* $(x^0,t^0) \in \bar{S}$ *and any* $(\widetilde{\xi},p) \in A^{x^0}$ *the rows of the matrix*

$$R(x^0,t^0,\widetilde{\xi},\tau,p) = \mathcal{B}_0\left(x^0,t^0,i(\widetilde{\xi}+\tau\nu),p\right)\widehat{\mathcal{L}}_0\left(x^0,t^0,i(\widetilde{\xi}+\tau\nu),p\right), \qquad (2.32)$$

regarded as τ-*polynomials, are linearly independent modulo the* τ-*polynomial* $L_0^+(x_0,t_0,\widetilde{\xi},\tau,p)$.

This important condition can be recast in yet another form. Let $R'(x^0,t^0,\widetilde{\xi},\tau,p)$ be the matrix whose elements are the remainders of the division of the elements of R by $L_0^|$. Then the elements R'_{qj} of R' can be written as

$$R'_{qj}(x^0,t^0,\widetilde{\xi},\tau,p) = \sum_{l=1}^{br} d_{qj}^{(l)}(x^0,t^0;\widetilde{\xi},p)\tau^{l-1}, \qquad q,j = 1,\ldots,br.$$

Now consider the $br \times (br)^2$ matrix $\mathcal{D}$ with the elements $d_{qj}^{(l)}$, $q,j,l = 1,\ldots,br$. Then the complementarity condition is equivalent to the following assumption: *for any* $(x_0,t_0) \in \bar{S}$ *and any* $(\widetilde{\xi},p) \in A^{x^0}$ *the rank of the matrix* $\mathcal{D}(x^0,t^0;\widetilde{\xi},p)$ *equals* br, *or*

$$\sum_j \left|\Delta_j(x^0,t^0;\widetilde{\xi},p)\right| > 0, \qquad (2.33)$$

where Δ_j *are the minors of order* br *of* $\mathcal{D}$.

Definition I.8. *Problem* (2.13)–(2.15) *is called a general parabolic mixed (boundary value) problem if*
(1) *the system* (2.13) *is parabolic in the sense of Solonnikov;*
(2) *the initial value problem* (2.14) *satisfies the complementarity condition;*
(3) *the boundary condition* (2.15) *satisfies the complementarity condition.*

The complementarity condition is often called the *Lopatinskiĭ condition* or the *Shapiro-Lopatinskiĭ condition*, or the *covering condition*. For systems parabolic in the sense of Petrovskiĭ the parabolicity of the mixed problem means that at $t = 0$ one has to prescribe Cauchy data and require that the complementarity condition hold for the boundary operators.

I.2.4. Particular cases. Examples

Boundary value problems for systems parabolic in the sense of Petrovskiĭ (1.5) with first-order time derivative deserve special attention. In matrix notation such problems take the form

$$D_t u = \sum_{|\alpha| \leq 2b} a_\alpha(x,t) D^\alpha u + f(x,t)$$

$$\equiv \mathcal{A}(x,t;D)u + f(x,t), \tag{2.34}$$

$$u\big|_{t=0} = \psi(x), \qquad x \in G, \tag{2.35}$$

$$\mathcal{B}(x,t;D,D_t)u\big|_S = \varphi(x',t), \qquad (x',t) \in S, \tag{2.36}$$

where $\mathcal{B}$ is a $bm \times m$ matrix operator.

In the case under consideration the complementarity condition can be written in a form that goes back to Lopatinskiĭ's fundamental work [61]: for any $(x_0,t_0) \in \bar{\tilde{S}}$ and any $(\widetilde{\xi},p) \in A^{x^0}$,

$$\mathrm{rank} \int_{\gamma_+} \mathcal{B}_0\big(x_0,t_0,i(\widetilde{\xi}+\tau\nu),p\big)$$

$$\times \Big(pI - \mathcal{A}_0\big(x_0,t_0,i(\widetilde{\xi}+\tau\nu)\big)\Big)^{-1}(I,\tau I,\ldots,\tau^{2b-1}I)d\tau = bm \tag{2.37}$$

Now let us write the complementarity condition in the case of a single equation ($m = 1$) and the half-space $x_n > 0$. To do this we must perform the following operations:

1. Write the polynomial $L_0(i\xi',i\tau,p) = (-1)^{b-1} \sum_{|\alpha|=2b} a_\alpha \xi'^{\alpha'} \tau^{\alpha_n} + p$.

2. Find its τ-zeroes $\tau_s^+(\xi',p)$, $s = 1,\ldots,b$, with positive imaginary parts and write the polynomial $L_0^+(\xi',\tau,p) = \prod_{s=1}^{b} \big(\tau - \tau_s^+(\xi',p)\big)$.

3. Write the τ-polynomials $b_q^0(i\xi',i\tau,p)$, $q = 1,\ldots,b$, corresponding to the principal parts of the boundary operators b_q.

4. Find the remainders $R_q(\xi',\tau,p) = \sum_{l=1}^{b} d_q^{(l)}(\xi',p)\tau^{l-1}$, $q = 1,\ldots,b$, of the division of the τ-polynomials $b_q^0(i\xi',i\tau,p)$ by the τ-polynomial $L_0^+(\xi',\tau,p)$.

5. Calculate the determinant $\Delta(\xi',p) = \det\left\{\big(d_q^{(l)}(\xi',p)\big)_{q,l=1}^{b}\right\}$.

6. Verify the validity of the complementarity condition: for any

$$(\xi',p) \in A^0 = \left\{(\xi',p) : \xi' \in \mathbb{R}^{n-1},\ p \in \mathbb{C};\ \mathrm{Re}\, p \geq -\delta_1|\xi'|^{2b},\ |p|+|\xi'|^{2b} > 0\right\},$$

$$0 < \delta_1 < \delta_0,$$

the determinant $\Delta(\xi',p)$ is not zero.

Example I.3. Let us check whether the complementarity condition holds for the model problems of the theory of heat conduction $D_t u = \Delta u$, $u\big|_{t=0} = \psi(x)$:

(1) The Dirichlet problem: $u\big|_{x_n=0} = \varphi_1(x',t)$.

(2) The Neumann problem: $D_n u\big|_{x_n=0} = \varphi_2(x',t)$.

(3) The oblique derivative problem: $\sum\limits_{j=1}^{n} b_j D_j u\big|_{x_n=0} = \varphi_3(x',t)$.

(4) Tikhonov's problem: $\sum\limits_{|\alpha|+2b\alpha_0 \leq r} b_{\alpha\alpha_0} D^\alpha D_t^{\alpha_0} u\big|_{x_n=0} = \varphi_4(x',t)$.

(5) The problem $D_t u\big|_{x_n=0} = \varphi_5(x',t)$.

Here $L_0(i\xi', i\tau, p) = p + a^2(\xi'^2 + \tau^2)$, $\delta_0 = a^2$, $\tau_1^+ = i\sqrt{\xi'^2 + pa^{-2}}$, where we choose the branch of the two-valued algebraic function that has positive real part; $L_0^+(\xi', \tau, p) = \tau - i\sqrt{\xi'^2 + pa^{-2}}$,

$$A^0 = \left\{ (\xi',p) \; : \; \xi' \in \mathbb{R}^{n-1}, \; \operatorname{Re} p > -(a^2 - \varepsilon)|\xi'|^2, \; 0 < \varepsilon < a^2 \right\}.$$

Further, $B_0(i\xi', i\tau, p)$ equals

$$1, \quad i\tau, \quad i\left(\sum_{i=1}^{n-1} b_j \xi_j + b_n \tau \right), \quad \sum_{|\alpha|+2b\alpha_0=r} b_{\alpha\alpha_0}(i\xi')^{\alpha'}(i\tau)^{\alpha_n} \cdot p^{\alpha_0}, \quad p,$$

respectively, and $R'(\xi', \tau, p)$ equals

$$1, \quad \sqrt{\xi'^2 + pa^{-2}}, \quad i\sum_{i=1}^{n-1} b_j \xi_j + b_n \sqrt{\xi'^2 + pa^{-2}},$$

$$R_4'(\xi', p) = \sum_{|\alpha|+2b\alpha_0=r} b_{\alpha\alpha_0}(\xi')^{\alpha'}\left(\sqrt{\xi'^2 + pa^{-2}}\right)^{\alpha_n} p^{\alpha_0}, \quad p,$$

respectively.

We see that for the Dirichlet and the Neumann problems the complementarity condition is satisfied. For the oblique derivative problem it is satisfied for $b_n \neq 0$ and is not satisfied for $b_n = 0$ (when the direction of differentiation lies in the tangent hyperplane $x_n = 0$). In the case of Tikhonov's problem the complementarity condition means that $R_4'(\xi', p) \neq 0$ in A^0. Finally, for problem 5) the condition holds for $n = 1$, since there are no restrictions on p, whereas for $n > 1$ it never holds (for $p = 0$).

Example I.4. One of the simplest systems of the theory of heat and mass exchange has the form

$$D_t u - M\Delta u = f, \tag{2.38}$$

where M is a 2×2 matrix with real constant elements and characteristic numbers λ_1 and λ_2 such that $0 < \lambda_1 < \lambda_2$. This system is parabolic in the sense of

Petrovskiĭ with the parabolicity constant $\delta_0 = \lambda_1$. By using a nonsingular linear transformation that diagonalizes the matrix M one can rewrite (2.38) as a system of two uncoupled heat equations:

$$D_t u_j - \lambda_j \Delta u_j = f_j, \qquad j = 1, 2. \tag{2.39}$$

In the domain $\mathbb{R}^{n+1}_{++} = \big\{(x', x^n, t) \; : \; x_n > 0, \; x' \in \mathbb{R}^{n-1}, t > 0\big\}$ we consider the problem of finding a solution $u(x,t) = \big(u_1(x,t), u_2(x,t)\big)$ which satisfies the initial conditions

$$u_j\big|_{t=0} = \psi_j, \qquad j = 1, 2. \tag{2.40}$$

and the boundary conditions (encountered in applications)

$$\sum_{k=1}^{2} \alpha_k D_n u_k\big|_{x_n=0} = \varphi_1, \qquad \sum_{k=1}^{2} \beta_k u_k\big|_{x_n=0} = \varphi_2, \tag{2.41}$$

where α_k, β_k, $k = 1, 2$ are given real numbers.

Let us derive the complementarity condition for problem (2.39)–(2.41). We write successively

$$\mathcal{L}_0(i\xi', i\tau, p) = \begin{pmatrix} p + \lambda_1\big(|\xi'|^2 + \tau^2\big) & 0 \\ 0 & p + \lambda_2\big(|\xi'|^2 + \tau^2\big) \end{pmatrix};$$

$$\mathcal{B}_0(i\xi', i\tau, p) = \begin{pmatrix} i\alpha_1\tau & i\alpha_2\tau \\ \beta_1 & \beta_2 \end{pmatrix};$$

$$\tau_j^+(\xi', p) = i\sqrt{(p/\lambda_j) + |\xi'|^2};$$

$$L_0^+(\xi', \tau, p) = (\tau - \tau_1^+)(\tau - \tau_2^+);$$

$$\mathcal{R}(\xi', \tau, p) = \mathcal{B}_0(i\xi', i\tau, p)\widehat{\mathcal{L}}_0(i\xi', i\tau, p)$$

$$= \begin{pmatrix} i\alpha_1\lambda_2\tau\big(\tau^2 - (\tau_1^+)^2\big) & i\alpha_2\lambda_2\tau\big(\tau^2 - (\tau_1^+)^2\big) \\ \beta_1\lambda_2\big(\tau^2 - (\tau_2^+)^2\big) & \beta_2\lambda_1\big(\tau^2 - (\tau_2^+)^2\big) \end{pmatrix}$$

The matrix $\mathcal{R}'(\xi', \tau, p)$ whose elements are the remainders of the division of the elements of $\mathcal{R}(\xi', \tau, p)$ by $L_0^+(\xi', \tau, p)$ has the form

$$\mathcal{R}'(\xi', \tau, p) = (\tau_1^+ + \tau_2^+) \begin{pmatrix} i\alpha_1\lambda_2\tau_1^+(\tau - \tau_2^+) & i\alpha_2\lambda_1\tau_2^+(\tau - \tau_1^+) \\ \beta_1\lambda_2(\tau - \tau_2^+) & \beta_2\lambda_1(\tau - \tau_1^+) \end{pmatrix}.$$

Finally, the complementarity condition reads

$$\operatorname{rank}\left((\tau_1^+ + \tau_2^+) \begin{pmatrix} i\alpha_1\lambda_2\tau_1^+ & i\alpha_2\lambda_1\tau_2^+ & -i\alpha_1\lambda_2\tau_1^+\tau_2^+ & -i\alpha_2\lambda_1\tau_1^+\tau_2^+ \\ \beta_1\lambda_2 & \beta_2\lambda_1 & -\beta_1\lambda_2\tau_2^+ & -\beta_2\lambda_1\tau_1^+ \end{pmatrix} \right) = 2,$$

that is,

$$\begin{vmatrix} \alpha_1\tau_1^+ & \alpha_2\tau_2^+ \\ \beta_1 & \beta_2 \end{vmatrix} \neq 0,$$

or

$$\alpha_1\beta_2\sqrt{(p/\lambda_1)+|\xi'|^2} - \alpha_2\beta_1\sqrt{(p/\lambda_2)+|\xi'|^2} \neq 0, \qquad (\xi',p) \in A^0. \qquad (2.42)$$

Condition (2.42) obviously holds if $\alpha_1\beta_2 = 0$ or $\alpha_2\beta_1 = 0$, but not simultaneously. An examination of (2.42) in the case $\alpha_1\alpha_2\beta_1\beta_2 \neq 0$ shows that if $(\alpha_2\beta_1/\alpha_1\beta_2) \notin \left[\sqrt{\gamma(1-\delta_1/\lambda_2)(\gamma-\delta_1/\lambda_1)}, \ \sqrt{\gamma}\right]$, where $\gamma = \lambda_2/\lambda_1$, and δ_1 is the number from the definition of A^0, then the complementarity condition holds.

Note that in this example, who at the surface may seem rather unsophisticated, one succeeds, in a very simple situation, to obtain an explicit solution of the difficult and important problem of determining conditions on the physical constants under which the boundary value problem will be well posed. In the general situation the complementarity condition, which contains in its final form the parameters $(\xi',p) \in A^{x^0}$, is, to say the least, difficult to verify.

Example I.5. Let us consider the Dirichlet problem for an equation of the form (2.34): $D_n^{q-1}u\big|_{x_n=0} = \varphi_q(x',t)$, $q = 1,\ldots,b$. In this problem $b_q(i\xi',i\tau) \equiv (i\tau)^{q-1}$, $q = 1,\ldots,b$. Since $b_q^0 = R_q = (i\tau)^{q-1}$, $q = 1,\ldots,b$, and these functions are linearly independent, it follows that the complementarity condition holds for any equation of the form (2.34) parabolic in the sense of Petrovskiĭ.

Example I.6. Now consider the Dirichlet problem for a parabolic system of the form (2.34) with one space variable:

$$D_t u = a_{2b}D^{2b}u; \qquad u\big|_{t=0} = \psi, \qquad D^{q-1}u\big|_{x=0} = \varphi_q, \quad q = 1,\ldots,b.$$

The verification of the complementarity condition reduces to proving that the problem $p\widehat{u} = a_{2b}D^{2b}\widehat{u}$; $\widehat{u}^{(\beta-1)}(0) = 0$, $q = 1,\ldots,b$, $|\widehat{u}(x,p)| \to 0$ as $x \to \infty$, has only the trivial solution. By using a linear nonsingular transformation that preserves the Dirichlet boundary conditions as well as the behavior of the solution at infinity, we can arrange that the matrix of coefficients of the new parabolic system will be in Jordan form:

$$\widehat{v}(x) = T\widehat{u}(x); \qquad p\widehat{v} = \tilde{a}_{2b}D^{2b}\widehat{v}; \qquad \widehat{v}^{(q-1)}(0) = 0, \quad q = 1,\ldots,b;$$
$$\left|\widehat{v}(x,p)\right| \to 0 \text{ as } x \to \infty.$$

Let us write the equations of the system: $p\widehat{v}_1 = \lambda_1 D_x^{2b}v_1$, $p\widehat{v}_2 = \lambda D^{2b}\widehat{v}_2 + \widehat{v}_1, \ldots$. Successive application of the complementarity condition for a single parabolic equation yields the desired result. One can show that for the Dirichlet problem the complementarity condition holds for any strongly parabolic system in an arbitrary number of space variables.

Example I.7. Consider the Dirichlet problem for a system of the form (2.34), consisting of two second-order equations in two space variables, such that its matrix

$A_0(\xi', \tau)$, $\xi' = \xi_1$, has the form

$$A_0(\xi', \tau)$$
$$= -\begin{pmatrix} a\xi'^2 - (\tau - i\sqrt{1+a}\,\xi')^2 & -4(1+a)\xi'^2 \\ (\tau - i\sqrt{1+a}\,\xi')^2 & a\xi'^2 - (\tau + i\sqrt{1+a}\,\xi')^2 - 4(1+a)\xi'^2 \end{pmatrix}.$$

It is a straightforward matter to check that $\det\left(pI - A_0(\xi', \tau)\right) = (p + \xi'^2 + \tau^2)^2$ and that for $p = a\xi'^2$ the system of ordinary differential equations

$$\left\{p - a\xi'^2 + (iD_{x_2} - i\sqrt{1+a}\,\xi')^2\right\}\widehat{u}_1 + 4(1+a)\xi'^2\widehat{u}_2 = 0,$$

$$-(iD_{x_2} - i\sqrt{1+a}\,\xi')^2)\widehat{u}_1 -$$

$$-\left\{p - a\xi'^2 + 4(1+a)\xi'^2 + (iD_{x_2} + i\sqrt{1+a}\,\xi')^2\right\}\widehat{u}_2 = 0$$

has the nontrivial solution $\widehat{u}(x_2) = \left(x_2 \exp\left\{-\sqrt{1+a}\,\xi' x_2\right\}, 0\right)$ with null Dirichlet boundary data. Thus, if the number of space variables, n, is larger than one, then there exist systems parabolic in the sense of Petrovskiĭ for which the first boundary value problem is ill posed (is not a parabolic problem).

Now suppose the Dirichlet problem for the system (2.34) parabolic in the sense of Petrovskiĭ satisfies the complementarity condition. Then, as was shown by Lopatinskiĭ [61], the matrix $\mathcal{L}(x^0, t^0, i\xi', i\tau, p) = pI - A_0\left(x^0, t^0, i(\xi' + \tau\nu)\right)$, regarded as a polynomial in τ, admits a factorization

$$\mathcal{L}_0(x^0, t^0, i\xi', i\tau, p) = \mathcal{L}_0^-(x^0, t^0, \xi', \tau, p) \cdot \mathcal{L}_0^+(x^0, t^0, \xi', \tau, p), \qquad (2.43)$$

where $\mathcal{L}_0^+$, $\mathcal{L}_0^-$ are matrices, polynomial in τ, such that the τ-zeroes of $\det\mathcal{L}_0^+$ [resp., $\det\mathcal{L}_0^-$] coincide with the τ-zeroes of $\det\mathcal{L}_0$ with $\operatorname{Im}\tau > 0$ [resp., with $\operatorname{Im}\tau < 0$].

The factorization (2.43) allows one to recast the complementarity condition (2.37) in a more concrete form: for any $(x^0, t^0) \in \bar{S}$ and any $(x', p) \in A^{x^0}$

$$\det \int_{\gamma_+} \mathcal{B}_0\left(x^0, t^0, i(\xi' + \tau\nu), p\right) \times$$
$$\times \left(\mathcal{L}_0^+(x^0, t^0, \xi', \tau, p)\right)^{-1}(I, \tau I, \ldots, \tau^{b-1}I)d\tau \neq 0. \qquad (2.44)$$

Example I.8. Consider the system parabolic in the sense of Petrovskiĭ with $n = 2$, which consists of two uncoupled heat equations:

$$\mathcal{L}_0(i\xi', i\tau, p) = \begin{pmatrix} \xi'^2 + \tau^2 + p & 0 \\ 0 & \xi'^2 + \tau^2 + p \end{pmatrix}.$$

We have already met this system in Example I.1. In the present case $b = 1$, $m = r = 2$. Our system can be also regarded as a system parabolic in the sense of Solonnikov with $s_1 = -c$, $s_2 = 0$, $t_1 = 2 + c$, $t_2 = 2$ $(c > 0)$; $s_1 = 0$, $s_2 = -c$, $t_1 = 2$, $t_2 = 2 + c$ $(c \geq 0)$.

Let $\mathcal{B}(i\xi', i\tau, p) = \left(\begin{smallmatrix} 1 & 0 \\ 0 & 1 \end{smallmatrix}\right)$. We specify for this operator numbers σ_1 and σ_2 (weights of boundary conditions): if $t_1 = 2 + c$, $t_2 = 2$ $(c > 0)$, then $\sigma_1 = -2 - c$, $\sigma_2 = -2$. If $t_1 = 2$, $t_2 = 2 + c$ $(c > 0)$, then $\sigma_1 = -2$, $\sigma_2 = -1$.

In the first case we have $\sigma_1 + t_1 = 0$, $\sigma_1 + t_2 = -c$, $\sigma_2 + t_1 = 0$, $\sigma_2 + t_2 = 0$. Therefore, $\mathcal{B}^0_{(1)}(i\xi', i\tau, p) = \left(\begin{smallmatrix} 1 & 0 \\ i\xi & 1 \end{smallmatrix}\right)$ for $c = 1$, $\mathcal{B}^0_{(2)}(i\xi', i\tau, p) = \left(\begin{smallmatrix} 1 & 0 \\ 0 & 1 \end{smallmatrix}\right)$ for $c > 1$.

In the second case $\sigma_1 + t_1 = 0$, $\sigma_1 + t_2 = 1 + c$, $\sigma_2 + t_1 = 1$, $\sigma_2 + t_2 = 2 + c$, $c \geq 0$. Hence, $\mathcal{B}^0_{(3)} = \left(\begin{smallmatrix} 1 & 0 \\ i\xi & 1 \end{smallmatrix}\right)$. The rows of this matrix are linearly dependent, and hence so are the rows of the matrix $\mathcal{B}^0_{(3)} \widehat{\mathcal{L}}_0$. We see that in this case the complementarity condition is not satisfied. In particular, it is not satisfied if we treat $\mathcal{L}_0$ as an operator parabolic in the sense of Petrovskiĭ ($s_1 = s_2 = 0$, $t_1 = t_2 = 2$). Thus, introducing the structure of the general parabolic system enables us to find new interesting well-posed boundary value problems also for the simplest cases, which are important in applications.

I.2.5. Parabolic conjugation problems

Here we shall discuss some very simple generalizations of the linear parabolic boundary value problems introduced above. One such generalization is known as the parabolic conjugation problem. For the sake of simplicity we will deal with systems parabolic in the sense of Petrovskiĭ that are of first order in the time variable t. The general case is completely similar to the case discussed here.

Consider a domain $G_0 \subset \mathbb{R}^n$ with smooth boundary ∂G_0. Assume that G_0 is divided by a smooth hypersurface ∂G_1 into two subdomains G_1 and G_2 so that $G_0 = \overline{G}_1 \cup G_2$, $G_1 \cap G_2 = \emptyset$, $\partial G_1 \cap \partial G_0 = \emptyset$, $\partial G_2 = \partial G_1 \cup \partial G_0$. Let $\Omega_\mu = G_\mu \times [0, T]$, $\mu = 1, 2$ and $S_\mu = \partial G_\mu \times [0, T]$, $\mu = 0, 1, 2$.

The conjugation problem is formulated as follows: Find a vector-valued function $u(x, t)$ defined in G_0, $u(x, t) = u_{(\mu)}(x, t)$ in Ω_μ, $\mu = 1, 2$, such that $u_{(\mu)} = \left(u_{(\mu)1}, \ldots, u_{(\mu)m_\mu}\right)$ is a solution of the system

$$\sum_{j=1}^{m_\mu} l_{(\mu)kj}(x, t, D, D_t) u_{(\mu)j}(x, t) = f_{(\mu)k}(x, t), \qquad k = 1, \ldots, m_\mu, \qquad (2.45)$$

in the domain Ω_μ, $\mu = 1, 2$, $u_{(\mu)}$ satisfies at $t = 0$ the initial conditions

$$u_{(\mu)i}(x, t)\big|_{t=0} = \psi_{(\mu)i}(x), \qquad x \in G_\mu, \ i = 1, \ldots, m_\mu, \qquad (2.46)$$

and $u_{(2)}(x, t)$ satisfies the following boundary conditions on the surface S_0:

$$\sum_{j=1}^{m_2} b_{qj}(x, t, D, D_t) u_{(2)j}(x, t)\big|_{S_0} = \varphi_{(0)q}(x', t), \qquad q = 1, \ldots, m_2. \qquad (2.47)$$

Finally, on the common boundary S_1 of the two cylinders Ω_1 and Ω_2 one imposes the *conjugation condition*

$$\left(\sum_{j=1}^{m_1} b_{(1)qj}(x,t,D,D_t)u_{(1)j} + \sum_{j=1}^{m_2} b_{(2)qj}(x,t,D,D_t)u_{(2)j} \right)\Bigg|_{S_1} = \varphi_{(1)q}(x',t), \quad (2.48)$$

$q = 1,\ldots,m_1 + m_2$. The sum $\left(v_{(1)}(x,t) + v_{(2)}(x,t)\right)\big|_{S_1}$ in (2.48) is understood as the sum of the limit values as x tends to S_1 from within Ω_1 and Ω_2, respectively.

Rewrite problem (2.45)–(2.48) in the more compact form

$$\mathcal{L}_\mu(x,t,D,D_t)u_{(\mu)}(x,t) = f_\mu(x,t),$$
$$u_{(\mu)}(x,t)\big|_{t=0} = \psi_{(\mu)}(x),$$
$$\mathcal{B}(x,t,D,D_t)u_{(2)}\big|_{S_0} = \varphi_{(2)}(x',t), \quad (2.49)$$
$$\sum_{\mu=1}^{2} B_\mu(x,t,D,D_t)u_{(\mu)}(x,t)\Big|_{S_1} = \varphi_{(1)}(x',t).$$

Definition I.9. *Problem (2.49) is called a parabolic conjugation problem if*
(1) *the systems (2.45) are systems of first order in t parabolic in the sense of Petrovskiĭ (with the same parabolicity constant δ_0), defined in Ω_1 and Ω_2, respectively;*
(2) *the boundary conditions (2.47) satisfy the complementarity condition·with respect to the operator $\mathcal{L}_2$;*
(3) *the boundary operators (B_1,B_2) satisfy the joint complementarity condition (or compatible covering condition) with respect to the operators $(\mathcal{L}_1,\mathcal{L}_2)$.*

The last definition involves a new notion, the compatible covering condition. Its precise definition is as follows.

Denote by $\mathcal{L}_{\mu0}$, $\mathcal{B}_{\mu0}$ the principal parts of the corresponding operators, and set $L_{\mu0}(x,t,i\xi,p) = \det \mathcal{L}_{\mu0}(x,t,i\xi,p)$, $\mu = 1,2$. Let (x_0,t_0) be an arbitrary point of S_1, and let ν_μ, $\mu = 1,2$, the inner (with respect to G_μ) normal to ∂G_1. Write the factorization identities

$$L_{\mu0}\big(x,t,i(\xi + \tau\nu_\mu),p\big) = L_{\mu0}^{+}(x,t,\xi,\tau,p) \cdot L_{\mu0}^{-}(x,t,\xi,\tau,p),$$

where the τ-zeroes of $L_{\mu0}^{+}$ [resp., $L_{\mu0}^{-}$] coincide with the τ-zeroes of $L_{\mu0}$ in the lower [resp., upper] half-plane of the complex τ-plane. Finally, let $A^{x_0} = \left\{ (\widetilde{\xi},p) : p \in \mathbb{C},\ \widetilde{\xi} \text{ is an arbitrary vector tangent to } \partial G_1 \text{ at the point } x_0,\ \operatorname{Re} p > -\delta_1|\widetilde{\xi}|^{2b} \right\}$.

Definition I.10. (*The complementarity condition for the conjugation problem (the compatible covering condition).*) *The operators $\{\mathcal{B}_1,\mathcal{B}_2\}$ satisfy the compatible covering condition with respect to the operators $\{\mathcal{L}_1,\mathcal{L}_2\}$ if for any $(x_0,y_0) \in S_1$ and any $(\widetilde{\xi},p) \in A^{x_0}$ the remainders $R'_\mu(x_0,t_0,\widetilde{\xi},\tau,p)$ of the division of $\mathcal{B}_{\mu0}(x_0,t_0,$*

$i(\widetilde{\xi} + \tau\nu_\mu))\widehat{\mathcal{L}}_{\mu 0}(x_0, t_0, i(\widetilde{\xi} + \tau\nu_\mu), p)$ by $L^+_{\mu 0}(x_0, t_0, \widetilde{\xi}, \tau, p)$, $\mu = 1, 2$, form a matrix $\left(\mathcal{R}'_1(x_0, t_0, \widetilde{\xi}, \tau, p),\ \mathcal{R}'_2(x_0, t_0, \widetilde{\xi}, \tau, p)\right)$, whose rows, regarded as τ-polynomials, are linearly independent.

Let us explain briefly how the compatible covering condition follows from the ordinary covering condition (the complementarity condition). Consider the model conjugation problem, where in the two domains separated by the hyperplane $x_n = 0$ there are given the systems of differential equations

$$\mathcal{L}_{(1)0}u_{(1)} = f_{(1)}, \quad x_n > 0,, \qquad \mathcal{L}_{(2)0}u_{(2)} = f_{(2)}, \quad x_n < 0, \tag{2.50}$$

together with the initial conditions

$$u_{(1)}\big|_{t=0} = 0, \quad x_n > 0, \qquad u_{(2)}\big|_{t=0} = 0, \quad x_n < 0, \tag{2.51}$$

and the conjugation conditions

$$\mathcal{B}_{10}(D', D_{x_n}, D_t)u_{(1)}\big|_{x_n=+0} + \mathcal{B}_{20}(D', D_{x_n}, D_t)u_{(2)}\big|_{x_n=-0} = \varphi_1(x', t). \tag{2.52}$$

Using the reflection $u_{(2)}(x', x_n, t) = u_{(2)}(x', -y_n, t) \equiv \widehat{u}_{(2)}(x', y_n, t)$ and preserving the original notation, we reformulate the problem (2.50)–(2.52) as a usual boundary value problem in the half-space $x_n > 0$ for the system

$$\begin{pmatrix} \mathcal{L}_{10} & 0 \\ 0 & \mathcal{L}_{20} \end{pmatrix} \begin{pmatrix} u_{(1)} \\ u_{(2)} \end{pmatrix} = \begin{pmatrix} f_{(1)} \\ f_{(2)} \end{pmatrix} \tag{2.53}$$

with a diagonal matrix and the boundary conditions

$$\left(\mathcal{B}_{10}(D', D_{x_n}, D_t)u_{(1)} + \mathcal{B}_{20}(D', -D_{x_n}, D_t)u_{(2)} \right)\Big|_{x_n=0} = \varphi_1(x', t). \tag{2.54}$$

If we write the complementarity condition for the problem (2.53), (2.54), we obtain the desired condition, that is, the compatible covering condition. For general systems the conjugation problem for two half-spaces is considered in detail in Sections IV.5 and IV.7.

Example I.9. Let us study the propagation of heat in $\mathbb{R}^3$ in a composite body G_0, consisting of parts G_μ, $\mu = 1, 2$, of densities ρ_μ, heat capacities c_μ, and heat conductivity coefficients k_μ. Let $a_\mu^2 = \rho_\mu c_\mu / k_\mu$. The mutual disposition of the parts G_1 and G_2 of G_0 is such that the conditions introduced at the beginning of this subsection are fulfilled. The sought-for temperature $u(x, t)$ of the body G_0, equal to $u_{(\mu)}(x, t)$ in Ω_μ, $\mu = 1, 2$, must satisfy the heat equations

$$D_t u_{(\mu)} = a_\mu^2 \Delta u_{(\mu)}, \qquad (x, t) \in \Omega_\mu, \ \mu = 1, 2 \tag{2.55}$$

and the initial conditions

$$u_{(\mu)}\big|_{t=0} = \psi_{(\mu)}(x), \qquad x \in \Omega_\mu, \ \mu = 1, 2. \tag{2.56}$$

On the boundary ∂G_1 separating G_1 and G_2, the temperatures and the heat fluxes must coincide:

$$u_{(1)}\big|_{\partial G_1} = u_{(2)}\big|_{\partial G_1},$$
$$k_1 \frac{\partial u_{(1)}}{\partial \nu_1}\bigg|_{\partial G_1} = k_2 \frac{\partial u_{(2)}}{\partial \nu_2}\bigg|_{\partial G_1}. \tag{2.57}$$

What we described is precisely a natural conjugation problem. Let us formulate the appropriate compatible covering condition in the model situation in which ∂G_1 is the plane $x_3 = 0$, G_1 is the half-space $x_3 > 0$, and G_2 is the half-space $x_3 < 0$. Then (2.57) becomes

$$u_{(1)}(x_1, x_2, 0) = u_{(2)}(x_1, x_2, 0),$$
$$k_1 \frac{\partial u_{(1)}}{\partial x_3}\bigg|_{x_3=0} = -k_2 \frac{\partial u_{(2)}}{\partial x_3}\bigg|_{x_3=0}.$$

In the case under consideration $L_{\mu 0}^+(\xi', \tau, p) = \tau - \tau_\mu^+$, $\tau_\mu^+ = i\sqrt{\xi_1^2 + \xi_2^2 + \frac{p}{a_\mu^2}}$; $\mathcal{B}_1 = \left(\begin{smallmatrix} 1 \\ ik_1\tau \end{smallmatrix}\right)$, $\mathcal{B}_2 = \left(\begin{smallmatrix} -1 \\ ik_2\tau \end{smallmatrix}\right)$; $\mathcal{R}' = (\mathcal{R}_1', \ \mathcal{R}_2') = \left(\begin{smallmatrix} 1 & -1 \\ k_1\tau_1^+ & k_2\tau_2^+ \end{smallmatrix}\right)_0$, $\det \mathcal{R} = (k_1\tau_1^+ + k_2\tau_2^+) \neq 0$, $k_1 > 0$, $k_2 > 0$. Thus the compatible covering condition is satisfied for all physically meaningful values of the parameters involved.

I.2.6. Nonlocal parabolic boundary value problems

Let us discuss another possible generalization of the class of boundary value problems introduced above. From a general standpoint, it is natural to call the latter problems *local* boundary value problems. Here we will consider, perhaps in the simplest version, *nonlocal* boundary value problems. We will work under the assumptions of Subsection I.2.5 concerning the structure of the domain G_0, but require in addition that the surfaces ∂G_0 and ∂G_1 be diffeomorphic, that is, that there exists a smooth one-to-one mapping of ∂G_0 onto ∂G_1: $y = \alpha(x)$, $x \in \partial G_1$, $y \in \partial G_0$. If ν_0 and ν_1 are the unit inner normals to ∂G_0 and ∂G_1, respectively, then for ε small enough the mapping $y + \varepsilon\nu_0 \mapsto x + \varepsilon\nu_1$ is a diffeomorphism of the ε-neighborhood of ∂G_0 (denoted by $G_{0,\varepsilon}$) onto $G_{1,\varepsilon}$. If in $S_{0,\varepsilon} = \partial G_{0,\varepsilon} \times (0, T]$ there are given a smooth function $u(y, t)$ and a differential operator $\mathcal{L}(y, t, D_y, D_t)$ with smooth coefficients, then the function $u(\alpha(x), t)$ and the operator $\mathcal{L}(\alpha(x), t; J \cdot D_y, D_t)$ (where J is the transposed Jacobian matrix of the mapping $\alpha^{-1}(y)$), are defined in $S_{1,\varepsilon}$.

Consider the problem of finding a vector function $u(x, t)$ that equals $u_{(\mu)}(x, t)$ in Ω_μ, $\mu = 1, 2$, and satisfies the equations (2.45), the initial conditions (2.46) and, instead of the condition (2.47), the nonlocal boundary conditions on S_1 defined below.

Let $\mathcal{B}_1(x, t; D, D_t)$ and $\mathcal{B}_2(x, t; D, D_t)$ be $r \times m_1$ and $r \times m_2$ matrix differential operators, defined in $S_{1,\varepsilon} \cap \Omega_1$ and $S_{1,\varepsilon} \cap \Omega_2$, respectively. Assume that on $S_{0,\varepsilon} \cap \Omega_0$ there is given a differential operator $\mathcal{B}_3(y, t, D_y, D_t)$, which is mapped by the diffeomorphism $y = \alpha(x)$ into the differential operator $\check{\mathcal{B}}_3(x, t, D_x, D_t) \equiv \mathcal{B}_3(\alpha(x), t, J \cdot D_x, D_t)$, given on $S_{1,\varepsilon} \cap \Omega_1$.

Under this mapping the operator $\mathcal{L}_2(y, t, D_y, D_t)$, defined in $S_{0,\varepsilon} \cap \Omega_0$, goes into the operator $\mathcal{L}_3(x, t, D_x, D_t) \equiv \mathcal{L}_2(\alpha(x), t, J \cdot D_x, D_t)$, defined on $S_{1,\varepsilon} \cap \Omega_1$. We will denote by ν_1 the inner normal with respect to $S_{1,\varepsilon} \cap \Omega_1$ at a point $x \in S_1$.

A *nonlocal boundary condition* (or *nonlocal conjugation condition*) is prescribed as follows:

$$\left(\mathcal{B}_1 u_{(1)} + \mathcal{B}_2 u_{(2)} + \mathcal{B}_3 u_{(3)}\right)\big|_{S_1} = \varphi_{(1)}, \tag{2.58}$$

where $\mathcal{B}_1, \mathcal{B}_2, \mathcal{B}_3$ are $r \times m_1$, $r \times m_2$, and $r \times m_2$ matrix operators respectively, $r = b(m_1 + 2m_2)$, and $\varphi_{(1)}(x', t)$ is an r-dimensional vector-valued function, defined on S_1.

Let us to find the compatible covering condition for the nonlocal boundary value problem (2.45), (2.58). To do this we will use gluing and reflection (for details see Section IV.5) to (locally) transform our problem into an ordinary boundary value problem.

Denote by $\mathcal{L}_{\mu 0}$, $\mathcal{B}_{\mu 0}$, $\mu = 1, 2, 3$, the principal parts of the corresponding operators. Fix a point $(x_0, t_0) \in S_1$ and consider

$$\mathcal{B}_{\mu 0}\left(x_0, t_0, i(\widetilde{\xi} + \tau \nu_\mu), p\right), \qquad \mathcal{L}_{\mu 0}\left(x_0, t_0, i(\widetilde{\xi} + \tau \nu_\mu), p\right),$$

$$L_{\mu 0}\left(x_0, t_0, i(\widetilde{\xi} + \tau \nu_\mu), p\right) = L_{\mu 0}^{+} L_{\mu 0}^{-}, \qquad \mu = 1, 2, 3.$$

Definition I.11. (*Nonlocal compatible covering condition, or nonlocal complementarity condition.*) *The operators $\{\mathcal{B}_1, \mathcal{B}_2, \mathcal{B}_3\}$ are said to satisfy the nonlocal compatible covering condition (or the nonlocal complementarity condition) with respect to the operators $\{\mathcal{L}_1, \mathcal{L}_2, \mathcal{L}_3\}$ if for any $(x_0, t_0) \in S_1$ and any $(\widetilde{\xi}, p) \in A^{x_0}$ the remainders $R'_\mu(x_0, t_0, \widetilde{\xi}, \tau, p)$ of the division of the elements of the matrix*

$$\mathcal{B}_{\mu 0}\left(x_0, t_0, i(\widetilde{\xi} + \tau \nu_\mu), p\right) \cdot \widetilde{\mathcal{L}}_{\mu 0}\left(x_0, t_0, i(\widetilde{\xi} + \tau \nu_\mu), p\right)$$

by the τ-polynomials $L_{(\mu) 0}^{+}(x_0, t_0, \widetilde{\xi}, \tau, p)$, $\mu = 1, 2, 3$, form a matrix

$$\left(\mathcal{R}_1(x_0, t_0, \widetilde{\xi}, \tau, p),\ \mathcal{R}_2(x_0, t_0, \widetilde{\xi}, \tau, p),\ \mathcal{R}_3(x_0, t_0, \widetilde{\xi}, \tau, p)\right),$$

whose rows are linearly independent as polynomials in τ.

Definition I.12. *The problem (2.45)–(2.46), (2.53) is called a nonlocal parabolic boundary value problem if*
(1) *the systems $\mathcal{L}_1$ and $\mathcal{L}_2$ are of first order in t and parabolic in the sense of Petrovskiĭ, with the same parabolicity constant δ_0, and defined in Ω_1 and Ω_2, respectively;*

(2) *the boundary operators $\{\mathcal{B}_1, \mathcal{B}_2, \mathcal{B}_3\}$ satisfy on S_1 the nonlocal compatible covering condition with respect to $\{\mathcal{L}_1, \mathcal{L}_2, \mathcal{L}_3\}$.*

Parabolic conjugation problems are formally a particular case of nonlocal parabolic boundary value problems. However, one must not forget that in conjugation problems the boundaries are not required to be diffeomorphic to one another, and consequently the outer and the inner boundary surfaces may have completely different structures.

Chapter II
Functional Spaces

II.1. Spaces of test functions and distributions

II.1.1. Definition and basic properties of distributions. Spaces $\mathcal{D}(\Omega)$ and $\mathcal{D}'(\Omega)$

Let Ω be a domain of the $(n+1)$-dimensional space $\mathbb{R}^{n+1}$ and $\varphi(x,t)$ a function continuous in Ω. The closure in Ω of the set of all points (x,t) for which $\varphi(x,t) \neq 0$ is called the *support* of the function $\varphi(x,t)$ in Ω and denoted by $\operatorname{supp}\varphi(x,t)$. For an integer $q \geq 0$ we denote by $C^q(\Omega)$ [resp., $C^q(\overline{\Omega})$] the set of all functions $\varphi(x,t)$ continuous in Ω [resp., $\overline{\Omega}$] together with their derivatives of order less than or equal to q. $C_0^q(\Omega)$ [resp., $C_0^\infty(\Omega)$] will designate the set of functions $\varphi(x,t) \in C^q(\Omega)$ [resp., $C^\infty(\Omega)$] with supports in Ω.

Let $\Omega' \subset \Omega$ be a compact set. A sequence of functions $\left\{\varphi_j(x,t) \in C_0^\infty(\Omega)\right\}_{j=1}^{\infty}$ with supports in Ω' is said to converge to zero as $j \to \infty$ if $D^\alpha D_t^\beta \varphi(x,t) \to 0$ as $j \to \infty$ uniformly on Ω' for any α and β. The set $C_0^\infty(\Omega)$ with the topology of uniform convergence introduced in this manner is called the *space of test functions* and is denoted by $\mathcal{D}(\Omega)$.

A linear form (linear functional) (u,φ) on $C_0^\infty(\Omega)$ that satisfies the estimate

$$|(u,\varphi)| < c(\Omega') \sum_{|\alpha|+\beta < q} \sup_{\Omega'} \left|D_x^\alpha D_t^\beta \varphi(x,t)\right| \tag{1.1}$$

(for all $\varphi \in C_0^\infty(\Omega)$) with some q and some constant $c(\Omega')$ and any $\Omega' \subset \Omega$, is called a *distribution* (or *generalized function*) $u(x,t)$ in Ω. The smallest number q for which estimate (1.1) holds is called the *order* of the distribution $u(x,t)$ in Ω.

The space $\mathcal{D}(\Omega)$ is a locally convex topological vector space. The space consisting of all distributions $u(x,t)$ on $\mathcal{D}(\Omega)$ is the dual space to $\mathcal{D}(\Omega)$ and is denoted by $\mathcal{D}'(\Omega)$. The linearity of the functional $u(x,t)$ means that $(u, a_1\varphi_1 + a_2\varphi_2) = \bar{a}_1(u,\varphi_1) + \bar{a}_2(u,\varphi_2)$ for any $\varphi_1, \varphi_2 \in \mathcal{D}$. The sum of distributions $u_1(x,t)$ and $u_2(x,t)$ and multiplication by scalars a_1, a_2 are defined by the equality $(a_1 u_1 + a_2 u_2, \varphi) = a_1(u_1,\varphi) + a_2(u_2,\varphi)$ for any $u_1, u_2 \in \mathcal{D}'(\Omega)$, $\varphi \in \mathcal{D}(\Omega)$. The functional $u(x,t)$ considered only on $C_0^\infty(Q)$ is called the *restriction* of the distribution $u(x,t) \in \mathcal{D}'(\Omega)$ to $Q \subset \Omega$. Two distributions $u_1(x,t)$ and $u_2(x,t)$ in

$\mathcal{D}'(\Omega)$ are considered to be equal in a neighborhood U of a point $(x_0, t_0) \in Q$ if their restrictions to U coincide, that is, $u_1 = u_2$ in U if $(u_1 - u_2, \varphi) = 0$ for all $\varphi(x, t)$ with support in U. The set of points that have no neighborhood in which $u(x, t) = 0$ will be called the *support* (denoted $\operatorname{supp} u(x, t)$) of the distribution $u(x, t) \in \mathcal{D}'(\Omega)$. As in the case of ordinary functions, the support of the distribution $u(x, t)$ is a closed set and the complement of $\operatorname{supp} u$ in Ω is an open set in which $u \equiv 0$.

II.1.2. Differentiation of distributions. Multiplication of distributions by smooth functions

For sufficiently smooth functions $u(x, t)$ and $a(x, t) \in C^\infty(\Omega)$, $\varphi(x, t) \in C_0^\infty(\Omega)$ the equalities

$$\int (au)\overline{\varphi}\, dx\, dt = \int u\overline{(\bar{a}\varphi)}\, dx\, dt,$$

$$\int D_t^\beta D^\alpha u \overline{\varphi}\, dx\, dt = (-1)^{|\alpha|+\beta} \int u D_t^\beta D^\alpha \overline{\varphi}\, dx\, dt$$

hold. Using these equalities, we define for a distribution $u(x, t) \in \mathcal{D}'(\Omega)$ the operations of differentiation and multiplication by a smooth function $a(x, t)$:

$$(au, \varphi) = (u, \bar{a}\varphi),$$

$$\left(D^\alpha D_t^\beta u, \varphi\right) = (-1)^{|\alpha|+\beta}\left(u, D_t^\beta D^\alpha \varphi\right), \qquad \forall \varphi(x, t) \in C_0^\infty(\Omega).$$

It is known [10, 15] that every distribution $u(x, t) \in \mathcal{D}'(\Omega)$ has the following structure: for any compact set $\Omega' \subset \Omega$ there exists a function $f(x, t) \in L_\infty(\Omega')$ such that $u(x, t) = D_t^m D_1^m \cdots D_n^m f(x, t)$, with some $m = m(\Omega)$.

II.1.3. Distributions with compact support. The spaces $\mathcal{E}(\Omega)$ and $\mathcal{E}'(\Omega)$, $\mathcal{S}(\Omega)$ and $\mathcal{S}'(\Omega)$

1. Consider the set $C^\infty(\Omega)$ with the above-introduced topology of uniform convergence. The resulting function space will be denoted by $\mathcal{E}(\Omega)$. The space dual to $\mathcal{E}(Q)$ will be denoted by $\mathcal{E}'(\Omega)$. As is shown in [10], $\mathcal{E}(\Omega)$ consists of the distributions $u(x, t)$ in Ω with compact support, and $\mathcal{E}'(\Omega)$ can be identified with the set of distributions $u(x, t)$ supported in Ω.

2. It is often more convenient to think of distributions $u(x, t) \in \mathcal{D}'(\Omega)$ as being defined on wider spaces of test functions and study them as distributions on these spaces. As a representative of such spaces one takes the space $\mathcal{S}(\mathbb{R}^{n+1})$ of all infinitely differentiable functions that decay at infinity together with all their derivatives faster than any polynomial: $\sup_{(x,t)} \left| t^\beta x^\alpha D^\alpha D_t^\beta \varphi(x, t) \right| < \infty$ for any α and β. Convergence in $\mathcal{S}(\mathbb{R}^{n+1})$ is defined as follows. A sequence $\{\varphi_j(x, t)\}_{j=1}^\infty \subset \mathcal{S}(\mathbb{R}^{n+1})$ converges to $\varphi(x, t)$ if in any bounded domain Ω and for any pair α, β the function

$D^\alpha D_t^\beta \varphi_j$ converges to $D^\alpha D_t^\beta \varphi(x,t)$ uniformly and $\left|x^\alpha t^\beta D^\alpha D_t^\beta \varphi_j\right| < C_{\alpha\beta}$, where $C_{\alpha\beta}$ does not depend on j. The set of all continuous functionals on $\mathcal{S}(\mathbb{R}^{n+1})$ is denoted by $\mathcal{S}'(\mathbb{R}^{n+1})$ and is called the space of *temperate distributions* (or *distributions of temperate growth*). Convergence in $\mathcal{S}'(\mathbb{R}^{n+1})$ is defined as follows. A sequence $\left\{u_j(x,t)\right\}_{j=1}^\infty \subset \mathcal{S}'(\mathbb{R}^{n+1})$ converges to $u(x,t)$ in $\mathcal{S}'(\mathbb{R}^{n+1})$ if for any $\varphi(x,t) \in \mathcal{S}(\mathbb{R}^{n+1})$ the sequence (u_j, φ) converges to (u, φ) as $j \to \infty$. The set $C_0^\infty(\mathbb{R}^{n+1})$ is dense in $\mathcal{S}(\mathbb{R}^{n+1})$, hence a distribution $u(x,t)$ can be extended by continuity from $\mathcal{D}(\mathbb{R}^{n+1})$ to $\mathcal{S}(\mathbb{R}^{n+1})$. Note, in particular, that all compactly-supported distributions and all regular functionals corresponding to functions that grow slower than a polynomial at infinity can be extended from $\mathcal{D}(\mathbb{R}^{n+1})$ to $\mathcal{S}(\mathbb{R}^{n+1})$.

It follows from the embedding of function spaces $\mathcal{D}(\mathbb{R}^{n+1}) \subset \mathcal{S}(\mathbb{R}^{n+1})$ that $\mathcal{S}'(\mathbb{R}^{n+1}) \subset \mathcal{D}'(\mathbb{R}^{n+1})$ and that the space $\mathcal{S}'(\mathbb{R}^{n+1})$ has the following structure: every distribution $u(x,t) \in \mathcal{S}'(\mathbb{R}^{n+1})$ can be represented as

$$u(x,t) = \sum_{|\alpha|+\beta<q} D^\alpha D_t^\beta u_k(x,t),$$

where $u_k(x,t)$ are regular functionals corresponding to continuous functions of polynomial growth: $|u_k(x,t)| < C(1 + |t| + |x|)^N$ with some C and N.

II.1.4. Convolution and direct product of distributions

Let $f(x,t),\ \varphi(x,t) \in \mathcal{S}(\mathbb{R}^{n+1})$. In classical calculus the *convolution* $f * \varphi$ of $f(x,t)$ and $\varphi(x,t)$ is defined as

$$(f * \varphi)(x,t) = \int_{\mathbb{R}^{n+1}} f(x-y, t-\tau)\varphi(y,\tau)\, dy\, d\tau.$$

The convolution product is commutative: $f * \varphi = \varphi * f$ and $D^\alpha D_t^\beta (f * \varphi) = D^\alpha D_t^\beta f * \varphi = f * D^\alpha D_t^\beta \varphi$. In fact, convolution is defined for wider classes of functions. For instance, if $|f(x,t)| < C(1 + |x| + |t|)^q$ and $\varphi(x,t) \in \mathcal{S}(\mathbb{R}^{n+1})$, then $f * \varphi$ exists and $|D^\alpha D_t^\beta(f * \varphi)| < C(\varphi)(1 + |x| + |t|)^q$. In particular, if $f(x,t)$ and $\varphi(x,t)$ belong to $\mathcal{S}(\mathbb{R}^{n+1})$, then $(f * \varphi) \in \mathcal{S}(\mathbb{R}^{n+1})$.

For an arbitrary vector $h = (h_1, \ldots, h_n, h_0) = (\tilde{h}, h_0)$, we define the translation $u_h(x,t)$ by the formula $(u_h, \varphi) = \left(u, \varphi(x + \tilde{h}, t + h_0)\right)$. In accordance with the definition of convolution for ordinary functions, we define the convolution of distributions $u(x,t) \in \mathcal{S}'(\mathbb{R}^{n+1})$ and $\varphi(x,t) \in \mathcal{S}(\mathbb{R}^{n+1})$ by the rule $(u * \varphi)(x,t) = \left(u(y,\tau), \varphi(x - y, t - \tau)\right)$, where (x,t) is fixed and the distribution $u(x,t)$ acts on $\varphi(x - y, t - \tau)$ with respect to (y,τ). It easily verified that $u * \varphi$ is an infinitely differentiable function defined everywhere in $\mathbb{R}^{n+1}$ and which, together with all its derivatives, grows no faster than a polynomial at infinity [10, 15].

Let $\mathbb{R}^{n+1} = \mathbb{R}^m \times \mathbb{R}^k$, $(x,t) = (x', y'')$, where $x' \in \mathbb{R}^m$, $y'' \in \mathbb{R}^k$ and $f(x')$, $g(y'')$ are distributions defined in $\mathbb{R}^m$ and $\mathbb{R}^k$, respectively; let $\varphi(x', y'')$ be a test function. We call the distribution $\big(f(x'), (g(y''), \varphi(x', y''))\big)$ the *direct product* $h(x', y'') = f(x') \times g(y'')$ of the distributions $f(x')$ and $g(y'')$. In particular, if $\varphi(x', x'') = \varphi_1(x') \cdot \varphi_2(x'')$, then $(f \times g, \varphi) = (f, \varphi_1) \cdot (g, \varphi_2)$. If $F = \operatorname{supp} f(x')$, $G = \operatorname{supp} g(x'')$, then $\operatorname{supp}(f \times g) = F \times G$ is the direct product of the sets F and G. The direct product of an arbitrary number of distributions $f_1, \ldots, f_r$, each of which belongs to the corresponding space $\mathcal{S}'$, is defined in the same fashion.

II.1.5. The Fourier and Laplace transformations of distributions

(a) Fourier transformation of functions in $\mathcal{S}(\mathbb{R}^{n+1})$. For $\varphi(x,t) \in \mathcal{S}(\mathbb{R}^{n+1})$ we define the Fourier transform of φ by the formula

$$\widetilde{\varphi}(\xi, \xi_0) = F(\varphi) = \int\limits_{\mathbb{R}^{n+1}} e^{-ix\cdot\xi - it\xi_0} \varphi(x,t)\, dx\, dt.$$

The inverse Fourier transform is given by

$$\varphi(x,t) = F^{-1}(\widetilde{\varphi}) = (2\pi)^{-n-1} \int\limits_{\mathbb{R}^{n+1}} e^{ix\cdot\xi + i\xi_0 t} \widetilde{\varphi}(\xi, \xi_0)\, d\xi\, d\xi_0.$$

Here is a list of main familiar properties of the Fourier transformation which will be often used below [10, 15].

1) Fourier transform of derivatives: $F(D^\alpha D_t^\beta \varphi) = (i\xi)^\alpha \cdot (i\xi_0)^\beta \widetilde{\varphi}(\xi, \xi_0)$.

2) Fourier transform of convolution: $F(\varphi * \psi) = F(\varphi) \cdot F(\psi) = \widetilde{\varphi}(\xi, \xi_0) \cdot \widetilde{\psi}(\xi, \xi_0)$.

3) Fourier transform of translates: $F(u_h) = e^{ih\cdot x + ih_0 t} \widetilde{u}(\xi, \xi_0)$.

4) Parseval identity: for all $\varphi(x,t), \psi(x,t) \in \mathcal{S}(\mathbb{R}^{n+1})$,

$$(\varphi, \psi) = \int \varphi(x,t) \cdot \overline{\psi}(x,t)\, dx\, dt$$

$$= (2\pi)^{-n-1} \int \widetilde{\varphi}(\xi, \xi_0) \cdot \overline{\widetilde{\varphi}(\xi, \xi_0)}\, d\xi\, d\xi_0 = (2\pi)^{-n-1}(\widetilde{\varphi}, \widetilde{\psi}). \quad (1.2)$$

5) The Fourier transformation is a one-to-one continuous mapping of $\mathcal{S}(\mathbb{R}^{n+1})$ onto itself. This means that if $\varphi(x,t) \in \mathcal{S}(\mathbb{R}^{n+1})$, then $\widetilde{\varphi}(\xi, \xi_0) \in \mathcal{S}(\mathbb{R}^{n+1})$ and the convergence $\varphi_\nu(x,t) \to \varphi(x,t)$ in $\mathcal{S}(\mathbb{R}^{n+1})$ implies that $\widetilde{\varphi}_\nu(\xi, \xi_0) \to \widetilde{\varphi}(\xi, \xi_0)$ in $\mathcal{S}(\mathbb{R}^{n+1})$ and vice versa.

    Denote the subspace of functions $\varphi(x,t) \in \mathcal{S}(\mathbb{R}^{n+1})$ with supports in $\overline{\mathbb{R}}^{n+1}_{++} = \{(x,t) \in \mathbb{R}^{n+1}: x_n \geq 0,\, t \geq 0\}$ by $\mathcal{S}_{++}(\mathbb{R}^{n+1})$.

6) The Fourier image $\widetilde{\mathcal{S}}_{++}(\mathbb{R}^{n+1}) = F\mathcal{S}(\mathbb{R}^{n+1})$ of $\mathcal{S}_{++}(\mathbb{R}^{n+1})$ consists of the functions $\widetilde{\varphi}(\xi', \zeta_n, \zeta_0)$ that can be analytically continued with respect to the complex variables $\zeta_n = \xi_n + i\eta_n$, $\zeta_0 = \xi_0 + i\eta_0$ to the half-planes $\eta_n < 0$,

$\eta_0 < 0$, are infinitely differentiable with respect to (ξ, ξ_0), and satisfy the estimates

$$\left(1 + |\xi'| + |\zeta_n| + |\zeta_0|\right)^m \cdot \left| D_{\xi'}^{k'} D_{\xi_n}^{k_n} D_{\xi_0}^{k_0} \widetilde{\varphi}(\xi', \zeta_n, \zeta_0) \right| < C_{m,k,k_0} \quad \forall k, k_0, \ m < \infty. \tag{1.3}$$

(b) The Fourier and Laplace transformations of growing functions. Let $\varphi(x, t)$ be a function such that $\psi(x, t) = e^{-\eta_0 t} \varphi(x, t) \in \mathcal{S}$ for a certain $\eta_0 > 0$. Then the Fourier transform $\widetilde{\psi}(\xi, \xi_0)$ of the function $\psi(x, t)$, is defined by

$$\widetilde{\psi}(\xi, \xi_0) = \int_{\mathbb{R}^{n+1}} e^{-ix\cdot\xi - \eta_0 t - it\xi_0} \varphi(x, t) \, dx \, dt$$

$$= \int_{\mathbb{R}^{n+1}} e^{-ix\cdot\xi - (\eta_0 + i\xi_0)t} \varphi(x, t) \, dx \, dt. \tag{1.4}$$

If $\widetilde{\phi}(\xi, \xi_0)$ is defined, (1.4) can be rewritten as $\widetilde{\psi}(\xi, \xi_0) = \widetilde{\varphi}\big(\xi, -i(\eta_0 + i\xi_0)\big) = \widetilde{\varphi}(\xi, -ip)$, $p = \eta_0 + i\xi_0$.

In classical calculus the function $\widehat{\varphi}(\xi, p) = \widetilde{\varphi}(\xi, -ip)$ is called the *Fourier transform* in x and the *bilateral* (or *two-sided*) *Laplace transform* of the function $\varphi(x, t)$. If $\operatorname{supp} \varphi(x, t) \subseteq \overline{E}_+^{n+1} = \{(x, t) \in \mathbb{R}^{n+1} : t \geq 0\}$ and $\varphi(x, t) \cdot e^{\eta_0 t} \in \mathcal{S}_+(E^{n+1})$, then the integration in (1.4) is carried out over E_+^{n+1}, and in this case formula (1.4) defines the Fourier transform in x and the *unilateral* (or *one-sided*) Laplace transform of the function $\varphi(x, t)$, which is usually referred to simply as the Laplace transform of $\varphi(x, t)$. Since the mapping $\zeta_0 \mapsto p = i\zeta_0$ takes the lower half-plane $\operatorname{Im} \zeta_0 < 0$ into the right half-plane $\operatorname{Re} p > 0$, it is obvious that if $\widetilde{\varphi}(\xi, \xi_0)$ can be analytically continued with respect to ξ_0 to the lower half-plane $\operatorname{Im} \zeta_0 < 0$ of the complex ζ_0-plane, then $\widetilde{\varphi}(\xi, -ip)$ can be analytically continued with respect to p to the right half-plane $\operatorname{Re} p > 0$ of the complex p-plane. Let us agree to denote by $\widehat{\varphi}(\xi, p) = F_{x\to\xi} L_{t\to p}(\varphi(x, t))$ the Fourier transform in x and the Laplace transform in t of the function $\varphi(x, t)$. It follows from the properties of the Fourier transformation that

$$FL(D^\alpha D_t^\beta \varphi) = (i\xi)^\alpha p^\beta F \cdot L(\varphi) = (i\xi)^\alpha p^\beta \widehat{\varphi}(\xi, p), \tag{1.5}$$

$$FL\big(\varphi(x - h, t - \tau)\big) = e^{i\xi\cdot h + \tau\xi_0} \widehat{\varphi}(\xi, p), \tag{1.6}$$

$$\int_{\mathbb{R}^n} dx \int_0^\infty |\varphi(x, t)|^2 e^{-2\eta_0 t} \, dt = (2\pi)^{-n-1} \int_{\mathbb{R}^n} d\xi \int_{-\infty}^\infty |\widehat{\varphi}(\xi, p)|^2 \, d\xi_0. \tag{1.7}$$

In particular, if $\operatorname{supp} \varphi(x, t) \subseteq \overline{\mathbb{R}}_{++}^{n+1}$, then $\widehat{\varphi}(\xi', \zeta_n, p)$ can be analytically continued in the complex variables ζ_n and p for $\operatorname{Im} \zeta_n < 0$, $\operatorname{Re} p > \eta_0$, and the estimates (1.3) with ζ_0 replaced by p hold for $\widehat{\varphi}(\xi', \zeta_n, p)$.

(c) The Fourier and Laplace transformations of distributions. Consider equality
(1.2), which is valid for all $\varphi, \psi \in \mathcal{S}(\mathbb{R}^{n+1})$. Since the left-hand side of the Parseval
identity (1.2) can be extended by continuity to all possible pairs $\varphi \in \mathcal{S}'(\mathbb{R}^{n+1})$,
$\psi \in \mathcal{S}(\mathbb{R}^{n+1})$, and the Fourier transformation maps $\mathcal{S}(\mathbb{R}^{n+1})$ onto $\mathcal{S}(\mathbb{R}^{n+1})$, the
right-hand side of (1.2) can also be extended by continuity to all possible couples
$\widetilde{\varphi}(\xi, \xi_0) \in \mathcal{S}'(\mathbb{R}^{n+1})$ and $\widetilde{\psi}(\xi, \xi_0) \in \mathcal{S}(\mathbb{R}^{n+1})$. It is this equality that uniquely defines the Fourier transform of a distribution $u(x, t) \in \mathcal{S}'(\mathbb{R}^{n+1})$ as the distribution
$\widetilde{u}(\xi, \xi_0)$ that satisfies the equality

$$(u, \varphi) = (2\pi)^{-n-1}(\widetilde{u}, \widetilde{\varphi}), \quad \forall \varphi(x, t) \in \mathcal{S}(\mathbb{R}^{n+1}). \tag{1.8}$$

The distribution $\widetilde{u}(\xi, \xi_0)$ is defined on the whole space $\mathcal{S}(\mathbb{R}^{n+1})$ and is continuous;
F is a linear continuous operator in $\mathcal{S}'(\mathbb{R}^{n+1})$ and equalities (1.5)–(1.6) remain
valid, by continuity, for distributions $\varphi(x, t) \in \mathcal{S}'(\mathbb{R}^{n+1})$. If $f * \varphi$ is the convolution
of $f(x, t) \in \mathcal{S}'(\mathbb{R}^{n+1})$ and $\varphi(x, t) \in \mathcal{S}(\mathbb{R}^{n+1})$, then $F(f * \varphi) = \widetilde{f}(\xi, \xi_0) \cdot \widetilde{\varphi}(\xi, \xi_0)$. The
equality $F(f_1 \times f_2) = \widetilde{f}(\xi') \times \widetilde{f}_2(\xi'')$ holds for the direct product $f_1(x') \times f_2(x'')$,
where $f_1(x') \in \mathcal{S}'(\mathbb{R}^m), f_2(x'') \in \mathcal{S}'(\mathbb{R}^q)$ $(\mathbb{R}^{n+1} = \mathbb{R}^m \times \mathbb{R}^q)$. Here $\widetilde{f}_1(\xi') \in$
$\mathcal{S}'(\mathbb{R}^m), \widetilde{f}_2(\xi'') \in \mathcal{S}'(\mathbb{R}^q)$ are the Fourier transforms of $f_1(x')$ and $f_2(x'')$, respectively.

Suppose that $\operatorname{supp} u(x, t) \subseteq \overline{E}_+^{n+1}$ and that there exists $\gamma > 0$ such that
$u(x, t) \cdot e^{-\gamma t} \in \mathcal{S}'(E^{n+1})$. The distribution $\widehat{u}(\xi, p) \in \mathcal{S}'(\mathbb{R}^{n+1})$ with the property
that $\left(e^{-\gamma t} u(x, t), \varphi(x, t)\right) = (2\pi)^{-n-1} i(\widehat{u}(\xi, p), \widetilde{\varphi}(\xi, \xi_0))$, for $\varphi(x, t) \in \mathcal{S}_+(E^{n+1})$ is
called the Fourier transform in x and the Laplace transform in t of the distribution
$u(x, t)$. For the Laplace and Fourier transforms defined in this manner, properties
(1.5)–(1.7) remain valid. Moreover, if $\operatorname{supp} u(x, t) \subseteq \overline{\mathbb{R}}_{++}^{n+1}$, then for almost every
$\xi' \in \mathbb{R}^{n-1}$ the function $\widehat{u}(\xi', \zeta_n, p)$ can be continued analytically in ζ_n and p for
$\operatorname{Im} \zeta_n < 0, \operatorname{Re} p > \gamma$.

II.2. The Hilbert spaces H^s and $\mathcal{H}^s$

II.2.1. The isotropic spaces $H^s(\mathbb{R}^n)$ and $H_+^s(\mathbb{R}^n)$

For each $s \in \mathbb{R}^1$, the set of distributions $v(x)$ whose Fourier transforms are locally
Lebesgue-integrable functions $\widetilde{v}(\xi)$ for which the norm

$$\left\|[v(x), \mathbb{R}^n]\right\|_s^2 = \int\limits_{\mathbb{R}^n} \left(1 + |\xi|^2\right)^s |\widetilde{v}(\xi)|^2 d\xi \tag{2.1}$$

is finite is called the *isotropic Sobolev-Slobodetskiĭ space* $H^s(\mathbb{R}^n)$.

If we equip the space $H^s(\mathbb{R}^n)$ with the inner product

$$\left[v(x), \psi\right]_s = \int\limits_{\mathbb{R}^n} \widetilde{v}(\xi) \cdot \overline{\widetilde{\psi}(\xi)} \left(1 + |\xi|^2\right)^s d\xi, \tag{2.2}$$

then it becomes a Hilbert space, i.e., is complete. Obviously, if $v(x) \in \mathcal{S}(\mathbb{R}^n)$, then $v(x) \in H^s(\mathbb{R}^n)$ for any s, hence convergence $v_n \to v$ in $\mathcal{S}$ implies convergence $v_n \to v$ in $H^s(\mathbb{R}^n)$.

By the Cauchy-Bunyakovskiĭ inequality,

$$\left| [v, \psi] \right| = (2\pi)^{-n} \left| \int_{\mathbb{R}^n} \widetilde{v}(\xi) \cdot \overline{\widetilde{\psi}(\xi)} d\xi \right|$$

$$\leq (2\pi)^{-n} \int_{\mathbb{R}^n} (1 + |\xi|^2)^{s/2} |\widetilde{v}(\xi)| \cdot (1 + |\xi|^2)^{-s/2} |\widetilde{\psi}(\xi)| \, d\xi \tag{2.3}$$

$$\leq (2\pi)^{-n} \left\| [v, \mathbb{R}^n] \right\|_s \left\| [\psi, \mathbb{R}^n] \right\|_{-s}.$$

These inequalities show that if $v_n \to v$ in $H^s(\mathbb{R}^n)$, then $[v_n, \psi] \to [v, \psi]$ for all $\psi \in \mathcal{S}$. Thus, by definition, $v_n \to v$ in $\mathcal{S}'(\mathbb{R}^n)$. Hence, $\mathcal{S}(\mathbb{R}^n) \subset H^s(\mathbb{R}^n) \subset \mathcal{S}'(\mathbb{R}^n)$ for any $s \in \mathbb{R}^1$, the embeddings being compatible with the respective topologies. For $s = 0$ we have $H^0(\mathbb{R}^n) = L_2(\mathbb{R}^n)$, and the space $H^s(\mathbb{R}^n)$ is isometrically isomorphic to the space $L_2(\mathbb{R}^n)$ with the weight $(1 + |\xi|^2)^{s/2}$. It can be shown [14, 92, 113] that $C_0^\infty(\mathbb{R}^n)$ (and hence $\mathcal{S}$) is dense in $H^s(\mathbb{R}^n)$ with respect to the norm $\left\| [v, \mathbb{R}^n] \right\|_s$.

It follows from (2.2) and (2.3) that the spaces $H^s(\mathbb{R}^n)$ and $H^{-s}(\mathbb{R}^n)$ may be considered as being in duality with respect to the extension of the inner product $[\cdot, \cdot]$ in $L_2(\mathbb{R}^n)$. For any $s_1, s_2 \in \mathbb{R}^1$, $s_1 < s_2$, we get $H^{s_2}(\mathbb{R}^n) \subset H^{s_1}(\mathbb{R}^n)$, the embedding being compatible with the topologies.

For $0 \leq s = m + \lambda$, where m is an integer, $0 \leq \lambda < 1$, along with the norm (2.1) defined by means of the Fourier transform, one also uses the "natural" equivalent norm $\left\| [v, \mathbb{R}^n] \right\|_s'$, defined by

$$\left\| [v, \mathbb{R}^n] \right\|_s'^2 = \sum_{|\alpha| \leq m} \int_{\mathbb{R}^n} |D^\alpha u|^2 dx + \sum_{|\alpha| = m} \iint_{\mathbb{R}^n \times \mathbb{R}^n} \frac{|D^\alpha v(x) - D^\alpha v(y)|^2}{|x - y|^{n+2\lambda}} \, dx \, dy,$$

where the second sum is absent if $\lambda = 0$. The equivalence of these two norms can be proved by making use of the Parseval identity [14, 92, 106].

Let $H_+^s(\mathbb{R}^n)$ [resp., $H_-^s(\mathbb{R}^n)$] be the subspace of $H^s(\mathbb{R}^n)$ consisting of the distributions $v(x)$ with support in $\overline{\mathbb{R}_+^n}$ [resp., $\overline{\mathbb{R}_-^n}$]. By definition, $v(x) \in H_+^s(\mathbb{R}^n)$ means that $(v(x), \varphi(x))) = 0$ for all $\varphi(x) \in C_0^\infty(\mathbb{R}_-^n)$. The subspace $H_+^s(\mathbb{R}^n)$ is closed in $H^s(\mathbb{R}^n)$, and $C_0^\infty(\mathbb{R}_+^n)$ is dense in $H_+^s(\mathbb{R}^n)$. An analogous statement holds true for $H_-^s(\mathbb{R}^n)$. For detailed proofs of these results see [113].

II.2.2. The isotropic spaces $H^s(\mathbb{R}_+^n)$ and their duals

Let $V(x) \in H^s(\mathbb{R}^n)$ and let $v(x)$ denote the restriction of the distribution $V(x)$ to $\mathbb{R}_+^n$. The space $H^s(\mathbb{R}_+^n)$ is defined as the set of all such distributions $v(x)$ with domains in $\mathbb{R}_+^n$, equipped with the quotient norm

$$\left\| [v(x), \mathbb{R}_+^n] \right\|_s = \inf_V \left\| [V(x), \mathbb{R}^n] \right\|_s, \tag{2.4}$$

where the infimum is taken over all $V(x)$ such that $V(x) = v(x)$ on $\mathbb{R}^n_+$. Obviously, if $V_1(x)$ and $V_2(x)$ are two extensions of $v(x)$ from $\mathbb{R}^n_+$ to $\mathbb{R}^n$, then $V_2 - V_1 = 0$ on $\mathbb{R}^n_+$ and $(V_2 - V_1) \in H^s_-(\mathbb{R}^n)$. The norm defined by (2.4) is the norm in the quotient space of $H^s(\mathbb{R}^n)$ by $H^s_-(\mathbb{R}^n)$, i.e., $H^s(\mathbb{R}^n)/H^s_-(\mathbb{R}^n)$. This implies the completeness of $H^s(\mathbb{R}^n_+)$. A similar statement holds true for $H^s(\mathbb{R}^n_-)$.

Let $s \in \mathbb{R}^1$. We denote the spaces dual to $H^s_+(\mathbb{R}^n)$ and $H^s(\mathbb{R}^n_+)$, with respect to the extension of the inner product in $L_2(\mathbb{R}^n)$ by $\left(H^s_+(\mathbb{R}^n)\right)^*$ and $\left(H^s(\mathbb{R}^n_+)\right)^*$, respectively. The space $\left(H^s_+(\mathbb{R}^n)\right)^*$ is isomorphic to $H^{-s}(\mathbb{R}^n_+)$, while $\left(H^{-s}(\mathbb{R}^n_+)\right)^*$ is isomorphic to $H^s_+(\mathbb{R}^n)$. Thus, $H^s_+(\mathbb{R}^n)$ and $H^{-s}(\mathbb{R}^n_+)$ are dual spaces with respect to the extension of the inner product

$$\left[u_+(x), V(x)\right]_0 = \int\limits_{\mathbb{R}^n} \widetilde{u}_+(\xi) \cdot \overline{\widetilde{V}(\xi)} \, d\xi = \overline{\left[V, u_+\right]}_0, \tag{2.5}$$

which does not depend on the extension $V(x)$. Identity (2.5) may be considered as a formula that defines a functional in the dual space.

II.2.3. Restriction to a hyperplane

Let $x' = (x_1, \ldots, x_{n-1})$, $v(x', x_n) \in H^s(\mathbb{R}^n)$ and let $\langle\!\langle \cdot, \mathbb{R}^{n-1} \rangle\!\rangle'_s$ be the norm in $H^s(\mathbb{R}^{n-1})$. It is known [3, 14, 92] that if $s > \frac{1}{2}$, then $v(x', x_n)$ is a continuous function of $x_n \in \mathbb{R}^1$ with values in $H^{s-\frac{1}{2}}(\mathbb{R}^{n-1})$ and the estimate

$$\langle\!\langle v(x' x_n), \mathbb{R}^{n-1} \rangle\!\rangle'_{s-\frac{1}{2}} < C_s \big\lvert [v(x), \mathbb{R}^n] \big\rvert_s \tag{2.6}$$

holds, where the constant C_s depends neither on v nor on x_n. From this it follows, in particular, that if $v(x) \in H^s(\mathbb{R}^n)$ and $s > \frac{1}{2}$, then there exist the traces of the derivatives, $D_n^{k-1} v(x', 0) \in H^{s-k+1/2}(\mathbb{R}^{n-1})$, for any k such that $k < s + \frac{1}{2}$, and

$$\langle\!\langle D_n^{k-1} v(x', 0), \mathbb{R}^{n-1} \rangle\!\rangle'_{s-k+\frac{1}{2}} < C_s \langle\!\langle v(x), \mathbb{R}^n \rangle\!\rangle_s. \tag{2.7}$$

The converse assertion that there exists a function $v(x)$ with given traces on the hyperplane $x_n = 0$ is also true: for any collection of functions $\psi_k(x') \in H^{s-k+\frac{1}{2}}(\mathbb{R}^{n-1})$, $k < s + \frac{1}{2}$, there exists a function $v(x) \in H^s(\mathbb{R}^n)$ such that $D_n^{k-1} v(x', 0) = \psi_k(x')$ and the estimate

$$\big\lvert [v(x), \mathbb{R}^n] \big\rvert_s < \sum_k \langle\!\langle \psi_k(x'), \mathbb{R}^{n-1} \rangle\!\rangle'_{s-k+\frac{1}{2}} \tag{2.8}$$

holds, where C is a constant that does not depend on $\psi_k(x')$. In particular, if $v(x) \in H^s_+(\mathbb{R}^n)$, then $D_n^{k-1} v(x', 0) = 0$ for all k such that $1 \leq k < s - \frac{1}{2}$.

II.2.4. The anisotropic Sobolev-Slobodetskiĭ spaces $\mathcal{H}^s$

Let $0 < b$ be a fixed integer. Let $p = \gamma + i\xi_0$. Set

$$\rho(1,\xi,p) = \left(1 + |\xi|^2 + |\gamma|^{1/b} + |\xi_0|^{1/b}\right)^{1/2}, \quad \rho_1(1,\xi',p) = \rho(1,\xi',0,p),$$

$\rho_0(1,\xi) = \rho(1,\xi,0)$, $\rho_2(1,\xi') = \rho(1,\xi',0,0)$.

Given any real s, r and σ, let $\mathcal{H}^{s,r}(\mathbb{R}^{n+1})$, $\mathcal{H}^{s,\sigma}(E^{n+1})$, and $\mathcal{H}^s(E^n)$ be the sets of distributions belonging to $\mathcal{S}'(\mathbb{R}^{n+1})$ and $\mathcal{S}'(E^n)$, respectively, whose Fourier transforms are locally Lebesgue-integrable functions for which the following norms are finite ($\mathbb{R}^{n+1} = E^{n+1}$):

$$\left\|u(x,t), \mathbb{R}^{n+1}\right\|_{s,r}^2 = \int_{\mathbb{R}^{n+1}} \rho^{2s}(1,\xi,p) \cdot \rho_1^{2r}(1,\xi',p) \cdot \left|\widetilde{u}(\xi,\xi_0)\right|^2 d\xi\, d\xi_0, \quad (2.9)$$

$$\left\{\!\left\{u(x,t), E^{n+1}\right\}\!\right\}_{s,\sigma}^2 = \int_{E^{n+1}} \rho^{2s}(1,\xi,p) \cdot \rho_0^{2\sigma}(1,\xi) \cdot \left|\widetilde{u}(\xi,\xi_0)\right|^2 d\xi\, d\xi_0, \quad (2.10)$$

$$\left\langle\!\left\langle\varphi(x',t), E^n\right\rangle\!\right\rangle_s^2 = \int_{E^n} \rho_1^{2s}(1,\xi',p) \cdot \left|\widetilde{\varphi}(\xi',\xi_0)\right|^2 d\xi'\, d\xi_0. \quad (2.11)$$

As in the case of $H^s(\mathbb{R}^n)$, if we equip $\mathcal{H}^{s,r}(\mathbb{R}^{n+1})$ and $\mathcal{H}^{s,\sigma}(E^{n+1})$ with the inner products

$$(u,v)_{s,r} = \int_{\mathbb{R}^{n+1}} \widetilde{u}(\xi,\xi_0) \cdot \overline{\widetilde{v}(\xi,\xi_0)} \rho^{2s} \rho_1^{2r}\, d\xi\, d\xi_0,$$

$$\{u,v\}_{s,\sigma} = \int_{E^{n+1}} \widetilde{u}(\xi,\xi_0) \cdot \overline{\widetilde{v}(\xi,\xi_0)} \rho^{2s} \rho_0^{2\sigma}\, d\xi\, d\xi_0,$$

they become Hilbert spaces. One can regard $\mathcal{H}^{s,r}(\mathbb{R}^{n+1})$ and $\mathcal{H}^{s,\sigma}(E^{n+1})$ as spaces of distributions $u(x,t)$ that are smooth of weighted order s in x and t and are "additionally" smooth in the "tangent" variables (x',t) and x, respectively. By the Cauchy-Bunyakovskiĭ inequality,

$$\left|(u,v)_0\right| < (2\pi)^{-n-1}\left\|u, \mathbb{R}^{n+1}\right\|_{s,r} \cdot \left\|v, \mathbb{R}^{n+1}\right\|_{-s,-r}, \quad (2.12)$$

$$\left|\{u,v\}_0\right| < (2\pi)^{-n-1}\left\{\!\left\{u, \mathbb{R}^{n+1}\right\}\!\right\}_{s,\sigma} \cdot \left\{\!\left\{v, E^{n+1}\right\}\!\right\}_{-s,-\sigma}. \quad (2.13)$$

It follows that $\mathcal{S}(\mathbb{R}^{n+1}) \subset \mathcal{H}^{s,r}(\mathbb{R}^{n+1}) \subset \mathcal{S}'(\mathbb{R}^{n+1})$, $\mathcal{S}(\mathbb{R}^{n+1}) \subset \mathcal{H}^{s,\sigma}(E^{n+1}) \subset \mathcal{S}'(\mathbb{R}^{n+1})$, the embeddings being compatible with the topologies. Obviously, $\mathcal{H}^{s,r}(\mathbb{R}^{n+1})$ and $\mathcal{H}^{s,\sigma}(E^{n+1})$ are isometrically isomorphic to the spaces $L_2(\mathbb{R}^{n+1})$ with the weights $\rho^s \rho_1^r$ and $\rho^s \rho_0^\sigma$, respectively. In particular, if $\sigma = 0$, $r = 0$, the norms defined by (2.9) and (2.10) are equal and the corresponding spaces coincide. In that case we write $\mathcal{H}^{s,0}(\mathbb{R}^{n+1}) = \mathcal{H}^s(\mathbb{R}^{n+1})$ $(\equiv \mathcal{H}_x^s \,{}_t^{s/2b}(\mathbb{R}^{n+1}))$ and denote the norm by $\|u, \mathbb{R}^{n+1}\|_s$; $\mathcal{H}^0(\mathbb{R}^{n+1}) \equiv L_2(\mathbb{R}^{n+1})$. The sets $C_0^\infty(\mathbb{R}^{n+1})$ and $\mathcal{S}(\mathbb{R}^{n+1})$ are dense in these spaces. Formulas (2.12) and (2.13) immediately imply that the pairs $\mathcal{H}^{s,r}(\mathbb{R}^{n+1})$ and $\mathcal{H}^{-s,-r}(\mathbb{R}^{n+1})$, as well as $\mathcal{H}^{s,\sigma}(E^{n+1})$ and $\mathcal{H}^{-s,-\sigma}(E^{n+1})$, are dual with respect to the extensions by continuity of the inner product in $L_2(\mathbb{R}^{n+1})$. The definitions of the norms (2.9), (2.10) immediately imply that for any $s_1 < s_2$, $r_1 \le r_2$ and $\sigma_1 < \sigma_2$, we have $\mathcal{H}^{s_2,r_2}(\mathbb{R}^{n+1}) \subset \mathcal{H}^{s_1,r_1}(\mathbb{R}^{n+1})$ and $\mathcal{H}^{s_2,\sigma_2}(E^{n+1}) \subset \mathcal{H}^{s_1,\sigma_1}(E^{n+1})$, the embeddings being compatible with the topologies. These embeddings will be often used below.

Denote by $\mathcal{H}_+^{s,r}(\mathbb{R}^{n+1})$, $\mathcal{H}_+^{s,\sigma}(E^{n+1})$, $\mathcal{H}_{++}^{s,r}(\mathbb{R}^{n+1})$ and $\mathcal{H}_+^s(E^n)$ the subspaces of $\mathcal{H}^{s,r}(\mathbb{R}^{n+1})$, $\mathcal{H}^{s,\sigma}(E^{n+1})$, $\mathcal{H}^{s,r}(\mathbb{R}^{n+1})$ and $\mathcal{H}^s(E^n)$ that consist of the distributions with support in $\overline{\mathbb{R}}_+^{n+1}$, $\overline{E}_+^{n+1}$, $\overline{\mathbb{R}}_{++}^{n+1}$ and $\overline{E}_+^n$, respectively. The sets $C_0^\infty(\mathbb{R}_+^{n+1})$, $C_0^\infty(E_+^{n+1})$, $C_0^\infty(\mathbb{R}_{++}^{n+1})$ and $C_0^\infty(E_+^n)$ are dense in the spaces $\mathcal{H}_+^{s,r}(\mathbb{R}^{n+1})$, $\mathcal{H}_+^{s,\sigma}(E^{n+1})$, $\mathcal{H}_+^{s,r}(\mathbb{R}^{n+1})$, $\mathcal{H}_{++}^{s,r}(\mathbb{R}^{n+1})$ and $\mathcal{H}_+^s(E^n)$, respectively.

In the spaces $\mathcal{H}^s(\mathbb{R}^{n+1})$, with $s \ge 0$, together with the norm $\|u, \mathbb{R}^{n+1}\|_s$, one often uses the equivalent "natural" norm defined directly through $u(x,t)$ in the (x,t)-representation.

Let $s = s' + \alpha = s'' + \beta$, where $0 \le \alpha < 1$, $0 \le \beta < 2b$, $s'' = 2bq$, q and s' are integers.

The norm $\|u, \mathbb{R}^{n+1}\|_s'$, equivalent to $\|u, \mathbb{R}^{n+1}\|_s$, is defined as

$$
\|u, \mathbb{R}^{n+1}\|_s'^2 = \sum_{|k|+2bk_0 \le s'} \int_{\mathbb{R}^{n+1}} \left| D^k D_t^{k_0} u \right|^2 dx\, dt
$$

$$
+ \sum_{|k|=s'} \int_{\mathbb{R}^1} dt \iint_{\mathbb{R}^n \times \mathbb{R}^n} \frac{\left| D^k u(x,t) - D^k u(y,t) \right|^2}{|x-y|^{n+2\alpha}} dx\, dy \qquad (2.14)
$$

$$
+ \int_{\mathbb{R}^n} dx \iint_{\mathbb{R}^1 \times \mathbb{R}^1} \frac{\left| D_t^{s''} u(x,t) - D_\tau^{s''}(x,\tau) \right|^2}{|t-\tau|^{1+2\beta}} dt\, d\tau,
$$

where the second sum is absent for $\alpha = 0$, while if $\beta = 0$, the last sum is also absent. In particular, a similar formula defines an equivalent norm in $\mathcal{H}^s(E^n)$.

II.2.5. The anisotropic spaces $\mathcal{H}^s$ on $\mathbb{R}_+^{n+1}$, E_+^{n+1}, $\mathbb{R}_{++}^{n+1}$ and E_+^n

Let $U(x,t)$, $\Phi(x',t)$ be distributions defined in $\mathbb{R}^{n+1}$ and E^n, respectively. Let Ω be one of the following domains: $\mathbb{R}_+^{n+1}$, E_+^{n+1} or $\mathbb{R}_{++}^{n+1}$; let $C\Omega$ denote the complement of Ω in $\mathbb{R}^{n+1}$. For any s,r, define the space $\mathcal{H}^{s,r}(\Omega)$ [resp., $\mathcal{H}(E_+^n)$] as the set consisting of the restrictions $u(x,t)$ [resp., $\varphi(x',t)$] of the distributions $U(x,t) \in \mathcal{H}^s(\mathbb{R}^{n+1})$ [resp., $\Phi(x',t) \in \mathcal{H}^s(E^n)$] to Ω [resp., E_+^n], equipped with the norm

$$
\left\| u(x,t), \Omega \right\|_{s,r} = \inf_U \left\| U(x,t), \mathbb{R}^{n+1} \right\|_{s,r},
$$

$$
[\text{resp.,} \quad \left\langle\!\left\langle \varphi(x',t), E_+^n \right\rangle\!\right\rangle_s = \inf_\Phi \left\langle\!\left\langle \Phi(x',t), E^n \right\rangle\!\right\rangle_s],
$$

where the infimum is taken over all $U(x,t)$ [resp., $\overset{\bullet}{\Phi}(x',t)$] that are equal to $u(x,t)$ in Ω [resp., $\varphi(x't)$ in E_+^n]. The spaces so defined are the quotient spaces of $\mathcal{H}^{s,r}(\mathbb{R}^{n+1})$ and $\mathcal{H}^s(E^n)$ by $\mathcal{H}^{s,r}_{\overline{C\Omega}}(\mathbb{R}^{n+1})$ and $\mathcal{H}^s_-(E^n)$, respectively:

$$\mathcal{H}^{s,r}(\Omega) = \mathcal{H}^{s,r}(\mathbb{R}^{n+1})/\mathcal{H}^{s,r}_{\overline{C\Omega}}(\mathbb{R}^{n+1}), \quad \mathcal{H}^s(E_+^n) = \dot{\mathcal{H}}^s(E^n)/\mathcal{H}^s_-(E^n),$$

where $\mathcal{H}^{s,r}_{\overline{C\Omega}}(\mathbb{R}^{n+1})$ is the subspace of $\mathcal{H}^{s,r}(\mathbb{R}^{n+1})$ consisting of the distributions with support in $\overline{C\Omega}$. In the same way we define the space $\mathcal{H}^{s,\sigma}(E_+^{n+1})$ and its norm, $\left\{\left\{u, E^{n+1}\right\}\right\}_{s,\sigma}$.

II.2.6. The dual spaces $\mathcal{H}^s$

From the definitions of the spaces $\mathcal{H}^{s,r}(\Omega)$, $\mathcal{H}^s(E_+^n)$ and the duality of $\mathcal{H}^{s,r}(\mathbb{R}^{n+1})$ and $\mathcal{H}^{-s,-r}(\mathbb{R}^{n+1})$, $\mathcal{H}^{s,\sigma}(E^{n+1})$ and $\mathcal{H}^{-s,-\sigma}(E^{n+1})$ with respect to the extension of the inner product in $L_2(\mathbb{R}^{n+1})$,

$$(u, w)_0 = \int \widetilde{u}(\xi, \xi_0)\overline{\widetilde{w}(\xi, \xi_0)} \, d\xi \, d\xi_0, \tag{2.15}$$

it follows that the pairs of spaces $\mathcal{H}^{s,r}_+(\mathbb{R}^{n+1})$ and $\mathcal{H}^{-s,-r}(\mathbb{R}^{n+1}_+)$, $\mathcal{H}^{s,r}_{++}(\mathbb{R}^{n+1})$ and $\mathcal{H}^{-s,-r}(\mathbb{R}^{n+1}_{++})$, $\mathcal{H}^{s,\sigma}_+(E^{n+1})$ and $\mathcal{H}^{-s,-\sigma}(E_+^{n+1})$ are also dual with respect to the extended inner product in $L_2(\mathbb{R}^{n+1})$. For instance, the value of any functional $u(x,t) \in \mathcal{H}^{-s,-r}_{++}(\mathbb{R}^{n+1})$ on an element $w(x,t) \in \mathcal{H}^{s,r}(\mathbb{R}^{n+1}_{++})$ is given by formula (2.15) and does not depend on the extension of $w(x,t)$ to the whole $\mathbb{R}^{n+1}$. Similar assertions hold for other pairs of dual spaces, in particular, for the pair $\mathcal{H}^s_+(E^n)$, $\mathcal{H}^{-s}(E_+^n)$.

II.2.7. Traces and continuation of functions in the anisotropic spaces $\mathcal{H}^s$

Here we consider the problem of the existence of traces of distributions $u(x,t)$ belonging to $\mathcal{H}^{s,r}(\mathbb{R}^{n+1})$ and $\mathcal{H}^{s,\sigma}(E^{n+1})$ on the hyperplanes $x_n = \text{const}$, $t = \text{const}$.

Theorem II.1 [14, 92]. 1) *Let $u(x,t) \in \mathcal{H}^{s,r}(\mathbb{R}^{n+1})$, $s > \frac{1}{2}$, $r \in \mathbb{R}^1$. Then for every k such that $1 \le k < s+\frac{1}{2}$, the mapping $x_n \mapsto D_n^{k-1}u(\cdot, x_n)$ is a continuous function on $\mathbb{R}^1$ with values in $\mathcal{H}^{s+r-k+\frac{1}{2}}(E^n)$ and one has the estimate*

$$\left\langle\!\left\langle D_n^{k-1}u(\cdot, x_n), E^n \right\rangle\!\right\rangle_{s+r-k+\frac{1}{2}} \le C_s\|u(x,t), \mathbb{R}^{n+1}\|_{s,r}, \tag{2.16}$$

with a constant C_s that does not depend on $u(x,t)$.

2) *Let $u(x,t) \in \mathcal{H}^{s,\sigma}(E^{n+1})$, $s > b$, $\sigma \in \mathbb{R}^1$. Then for every integer λ such that $1 \le \lambda < (\frac{s}{2b}) + (\frac{1}{2})$ the mapping $t \mapsto D_t^{\lambda-1}u(\cdot, t)$ is a continuous function on $\mathbb{R}^1$ with values in $H^{s+\sigma-2b\lambda+b}(\mathbb{R}^n)$ and one has the estimate*

$$\left\|D_t^{\lambda-1}u(\cdot, t), \mathbb{R}^n\right\|_{s+\sigma-2b\lambda+b} \le C\{\{u(x,t), E^{n+1}\}\}_{s,\sigma}, \tag{2.17}$$

with a constant C that does not depend on $u(x,t)$.

Proof: Since the proofs of the two assertions of the theorem are similar, we will prove here only assertion 2).

Suppose for the moment that $u(x,t) \in C_0^\infty(E^{n+1})$ and let $\widetilde{u}(\xi,t)$ be its Fourier transform in x. Since

$$D_t^{\lambda-1}\widetilde{u}(\xi,t) = \frac{1}{2\pi}\int (i\xi_0)^{\lambda-1}\widetilde{u}(\xi,\xi_0)e^{it\xi_0}\,d\xi_0,$$

the Cauchy-Bunyakovskiĭ inequality implies that

$$\left|D_t^{\lambda-1}\widetilde{u}(\xi,t)\right|^2 \le (2\pi)^{-2}\int \left(1+|\xi|^2+|\xi_0|^{1/b}\right)^s\left|\widetilde{u}(\xi,\xi_0)\right|^2 d\xi_0$$

$$\times \int |\xi_0|^{2(\lambda-1)}\left(1+|\xi|^2+|\xi_0|^{1/b}\right)^{-s} d\xi_0. \qquad (2.18)$$

The change of variable $\xi_0 = \left(1+|\xi|^2\right)^b\eta_0$ yields

$$\int \left(1+|\xi|^2+|\xi_0|^{1/b}\right)^{-s}|\xi_0|^{2(\lambda-1)}d\xi_0 = C_s\left(1+|\xi|^2\right)^{-s+b+2b(\lambda-1)}.$$

Thus, multiplying (2.18) by $\left(1+|\xi|^2\right)^{s+\sigma-2b\lambda+b}$ and integrating with respect to ξ, we obtain (2.17) for $u(x,t) \in C_0^\infty(E^{n+1})$. Since $C_0^\infty(\mathbb{R}^{n+1})$ is dense in $\mathcal{H}^{s,\sigma}(E^{n+1})$, estimate (2.17) holds also for all $u(x,t) \in \mathcal{H}^{s,\sigma}(E^{n+1})$. The continuity of the mapping $t \mapsto D_t^{\lambda-1}u(\cdot,t)$ in t follows from the continuity of the translation in the space $L_2(E^{n+1})$.

Corollary II.1. *Let* $u(x,t) \in \mathcal{H}^s(\mathbb{R}^{n+1})$, $s > b + \frac{1}{2}$. *Then for every* k,λ *such that* $k + 2b\lambda < s + b + \frac{1}{2}$, *the mapping* $(x_n,t) \mapsto D_n^{k-1}D_t^{\lambda-1}u(\cdot,x_n,t)$ *is a continuous function on* $\mathbb{R}^2$ *with values in* $H^{s-k-2b\lambda+b+\frac{1}{2}}(\mathbb{R}^{n-1})$ *and one has the estimate*

$$\left\langle\!\left\langle D_n^{k-1}D_t^{\lambda-1}u(\cdot,x_n,t),\mathbb{R}^{n-1}\right\rangle\!\right\rangle'_{s-k-2b\lambda+b+\frac{1}{2}} \le C_s\|u,\mathbb{R}^{n+1}\|_s, \qquad (2.19)$$

with a constant C_s *that does not depend on* x_n, t *and* $u(x,t)$.

It follows from Theorem II.1 and the definition of the spaces $\mathcal{H}^s(\mathbb{R}^{n+1}_{++})$ that, if $u(x,t) \in \mathcal{H}^s(\mathbb{R}^{n+1}_{++})$, then for any pair (k,λ) such that $k + 2b\lambda < s + b + \frac{1}{2}$ the traces of the derivatives $D_n^{k-1}u(x',0,t)$ and $D_t^{\lambda-1}u(x,0)$ satisfy on $\mathbb{R}^{n-1}$ the compatibility conditions

$$D_t^{\lambda-1}D_n^{k-1}u(x',0,+0) = D_n^{k-1}D_t^{\lambda-1}u(x',+0,0).$$

In particular, if $u \in \mathcal{H}^s_{++}(\mathbb{R}^{n+1})$, then $D_n^{k-1}u(x',0,t) = 0$ for all $k < s + \frac{1}{2}$; $D_t^{\lambda-1}u(x,0) = 0$ for all $\lambda < s + b$; $D_n^{k-1}D_t^{\lambda-1}u(x',0,0) = 0$, $k + 2b\lambda < s + b + \frac{1}{2}$.

Now let us address the problem of the existence of functions $u(x,t)$ with given traces on a fixed hyperplane. The following assertion holds.

Theorem II.2. 1) *Let $s > \frac{1}{2}$. Then for any choice of $\varphi_k(x',t) \in \mathcal{H}^{s+r-k+\frac{1}{2}}(E^n)$, $1 \le k < s+\frac{1}{2}$, there exists a function $u(x,t) \in \mathcal{H}^{s,r}(\mathbb{R}^{n+1})$ for which $D_n^{k-1}u(x',0,t) = \varphi_k(x',t)$ and*

$$\left\|u(x,t),\mathbb{R}^{n+1}\right\|_{s,r} \le C_s \sum_k \left\langle\!\!\left\langle \varphi_k(x',t), E^n \right\rangle\!\!\right\rangle_{r+s-k+\frac{1}{2}}, \tag{2.20}$$

with a constant C_s that does not depend on $\varphi_k(x',t)$.

2) *Let $s > b$. Then for any choice of $\Psi_\lambda(x) \in H^{s+\sigma-2b\lambda+b}(\mathbb{R}^n)$, $1 \le \lambda < s+b$, there exists a function $u(x,t) \in \mathcal{H}^{s,\sigma}(E^{n+1})$ such that $D_t^{\lambda-1}u(x,0) = \Psi_\lambda(x)$, and*

$$\left\{\!\left\{u(x,t), E^{n+1}\right\}\!\right\}_{s,\sigma} \le C_s \sum_\lambda \left\|\left[\Psi_\lambda(x),\mathbb{R}^n\right]\right\|_{s+\sigma-2b\lambda+b}, \tag{2.21}$$

with a constant C_s that does not dependent on $\Psi_\lambda(x)$.

The proof of this theorem is similar to that of Theorem 2.5.7 of [106].

Corollary II.2. *Let $s > b+\frac{1}{2}$. Then for any $\Psi_{\lambda k}(x') \in H^{s-k-2b\lambda+b+\frac{1}{2}}(\mathbb{R}^{n-1})$ there exists a function $u(x,t) \in \mathcal{H}^s(\mathbb{R}^{n+1})$ such that $D_n^{k-1}D_t^{\lambda-1}u(x',0,0) - \Psi_{\lambda k}(x')$, for (λ,k) satisfying $k + 2b\lambda < s + b + \frac{1}{2}$. Moreover,*

$$\left\|u(x,t),\mathbb{R}^{n+1}\right\|_s \le C_s \sum_{\lambda,k} \left\langle\!\!\left\langle \Psi_{\lambda,k}(x'), \mathbb{R}^{n-1} \right\rangle\!\!\right\rangle'_{s-k-2b\lambda+b+\frac{1}{2}}, \tag{2.22}$$

where the constant C_s does not depend on $\Psi_{\lambda k}(x')$.

II.2.8. Weighted anisotropic spaces; basic properties

For any real s, r, σ, γ, we define the weighted anisotropic spaces $\mathcal{H}^{s,r}(\mathbb{R}^{n+1},\gamma)$, $\mathcal{H}^{s,\sigma}(E^{n+1},\gamma)$ and $\mathcal{H}^s(E^n,\gamma)$ as the sets of the distributions $u(x,t)$ and $\varphi(x',t)$ with the property that after being multiplied by $e^{-\gamma t}$ they become elements of $\mathcal{H}^{s,r}(\mathbb{R}^{n+1})$, $\mathcal{H}^{s,\sigma}(E^{n+1})$ and $\mathcal{H}^s(E^n)$, respectively. The norms in these spaces are defined by the following formulas:

$$\left\|u(x,t),\mathbb{R}^{n+1},\gamma\right\|_{s,r}'^{\,2} = \left\|u \cdot e^{-\gamma t}, \mathbb{R}^{n+1}\right\|_{s,r}^2$$

$$= \int (1 + |\xi|^2 + |\xi_0|^{1/b})^s (1 + |\xi'|^2 + |\xi_0|^{1/b})^r |\widehat{u}(\xi,p)|^2 d\xi\, d\xi_0, \tag{2.23}$$

$$\left\{\!\left\{u(x,t), E^{n+1}, \gamma\right\}\!\right\}_{s,\sigma}'^{\,2} = \left\{\!\left\{u \cdot e^{-\gamma t}, E^{n+1}\right\}\!\right\}_{s,\sigma}^2$$

$$= \int (1 + |\xi|^2 + |\xi_0|^{1/b})^s (1 + |\xi|^2)^\sigma |\widehat{u}(\xi,p)|^2 d\xi\, d\xi_0, \tag{2.24}$$

$$\left\langle\!\!\left\langle \varphi(x',t), E^n, \gamma \right\rangle\!\!\right\rangle_s'^{\,2} = \left\langle\!\!\left\langle \varphi \cdot e^{-\gamma t}, E^n \right\rangle\!\!\right\rangle_s^2$$

$$= \int (1 + |\xi'|^2 + |\xi_0|^{1/b})^s \, |\widehat{\varphi}(\xi', p)|^2 d\xi' \, d\xi_0. \tag{2.25}$$

The following pairs of functional spaces are in duality with respect to the corresponding extensions of the L_2-inner products: $\big(\mathcal{H}^{s,r}(\mathbb{R}^{n+1}, \gamma), \mathcal{H}^{-s,-r}(\mathbb{R}^{n+1}, -\gamma)\big)$, $\big(\mathcal{H}^{s,\sigma}(E^{n+1}, \gamma), \mathcal{H}^{-s,-\sigma}(E^{n+1}, -\gamma)\big)$, and $\big(\mathcal{H}^s(E^n, \gamma), \mathcal{H}^{-s}(E^n, -\gamma)\big)$.

Denote by $\mathcal{H}^{s,r}_+(\mathbb{R}^{n+1}, \gamma)$, $\mathcal{H}^{s,r}_{++}(\mathbb{R}^{n+1}, \gamma)$, $\mathcal{H}^{s,\sigma}_+(E^{n+1}, \gamma)$ and $\mathcal{H}^s_+(E^n, \gamma)$ the subspaces of $\mathcal{H}^{s,r}(\mathbb{R}^{n+1}, \gamma)$, $\mathcal{H}^{s,\sigma}(E^{n+1}, \gamma)$, and $\mathcal{H}^s(E^n, \gamma)$ consisting of the distributions with support in $\overline{\mathbb{R}}^{n+1}_+$, $\overline{\mathbb{R}}^{n+1}_{++}$, $\overline{E}^{n+1}_+$ and $\overline{E}^n_+$, respectively. Also, denote by $\mathcal{H}^{s,r}(\mathbb{R}^{n+1}_+, \gamma)$, $\mathcal{H}^{s,r}(\mathbb{R}^{n+1}_{++}, \gamma)$, $\mathcal{H}^{s,\sigma}(E^{n+1}_+, \gamma)$, and $\mathcal{H}^s(E^n_+, \gamma)$ the subspaces consisting of the restrictions to $\mathbb{R}^{n+1}_+$, $\mathbb{R}^{n+1}_{++}$, E^{n+1}_+ and E^n_+, respectively, of the distributions that belong to the corresponding spaces with the quotient-space norms.

For any s, r, γ, the three pairs of spaces $\big(\mathcal{H}^{s,r}(\mathbb{R}^{n+1}_+, \gamma), \mathcal{H}^{-s,-r}_+(\mathbb{R}^{n+1}, -\gamma)\big)$, $\big(\mathcal{H}^{s,r}(\mathbb{R}^{n+1}_{++}, \gamma), \mathcal{H}^{-s,-r}_{++}(\mathbb{R}^{n+1}, -\gamma)\big)$, and $\big(\mathcal{H}^{s,\sigma}(E^n_+, \gamma), \mathcal{H}^{-s}_+(E^n, -\gamma)\big)$ are dual with respect to the corresponding extensions of the inner products in L_2.

Note that for $r = 0$ and $s \geq 0$, the definition of $\mathcal{H}^s(\mathbb{R}^{n+1}, \gamma)$ and (2.9) imply that the norm defined by (2.23) in $\mathcal{H}^s(\mathbb{R}^{n+1}, \gamma)$ is equivalent to the norm defined by (2.14), provided that $u(x, t)$ is replaced by $u(x, t)e^{-\gamma t}$.

II.2.9. Embeddings, traces and continuation of functions in weighted anisotropic spaces

The following complete analogues of Theorems II.1, II.2 and their corollaries are valid for functions $u(x, t) \in \mathcal{H}^{s,r}(\mathbb{R}^{n+1}, \gamma)$.

Theorem II.3. 1) *Let* $u(x, t) \in \mathcal{H}^{s,r}(\mathbb{R}^{n+1}, \gamma)$, $s > \frac{1}{2}$, $r \in \mathbb{R}^1$. *Then for every* k *with* $1 \leq k < s + \frac{1}{2}$, *the mapping* $x_n \mapsto D_n^{k-1} u(\cdot, x_n)$ *is a continuous function on* $\mathbb{R}^1$ *with values in* $\mathcal{H}^{s+r-k+\frac{1}{2}}(E^n, \gamma)$ *and one has the estimate*

$$\big\langle\!\big\langle D_n^{k-1} u(\cdot, x_n), E^n, \gamma \big\rangle\!\big\rangle_{s+r-k+\frac{1}{2}} \leq C_s \big\| u(x, t), \mathbb{R}^{n+1} \big\|_{s,r},$$

with a constant C_s *that does not depend on* $u(x, t), \gamma$.

2) *Let* $u(x, t) \in \mathcal{H}^{s,\sigma}(E^{n+1})$, $s > b$, $\sigma \in \mathbb{R}^1$. *Then for every integer* λ *with* $1 \leq \lambda < (\frac{s}{2b}) + (\frac{1}{2})$ *the mapping* $t \mapsto D_t^{\lambda-1} u(\cdot, t)$ *is a continuous function on* $\mathbb{R}^1$ *with values in* $H^{s+\sigma-2b\lambda+b}(\mathbb{R}^n)$ *and*

$$\big[D_t^{\lambda-1} u(\cdot, t), \mathbb{R}^n \big]_{s+\sigma-2b\lambda+b} \leq C\{\{u(x, t), E^{n+1}\}\}_{s,\sigma},$$

where the constant C *does not depend on* $u(x, t)$.

In addition, the assertions of Theorem II.1 and Corollary II.1 remain valid when the spaces $\mathcal{H}^{s,r}(\mathbb{R}^{n+1})$ and $\mathcal{H}^{s,\sigma}(E^{n+1})$ (and their norms) are replaced by the spaces $\mathcal{H}^{s,r}(\mathbb{R}^{n+1}, \gamma)$ and $\mathcal{H}^{s,\sigma}(E^{n+1}, \gamma)$, respectively.

Let Ω_1 be a domain $\mathbb{R}^{n+1}$, $\mathbb{R}^{n+1}_+$ or $\mathbb{R}^{n+1}_{++}$, and let Ω_2 be E^{n+1} or E^{n+1}_+. It follows from the definitions of the spaces $\mathcal{H}^{s,r}(\Omega_1, \gamma)$ and $\mathcal{H}^{s,\sigma}(\Omega_2, \gamma)$ that for any $r, s_1, s_2 \in \mathbb{R}^1$ with $s_1 < s_2$, $s_1 + r_1 \le s_2 + r_2$, the space $\mathcal{H}^{s_2,r_2}(\Omega_1, \gamma)$ is continuously embedded in $\mathcal{H}^{s_1,r_1}(\Omega_1, \gamma)$, and if $s_1 + \sigma_1 \le s_2 + \sigma_2$, $\sigma_1 < \sigma_2$, the space $H^{s_2,\sigma_2}(\Omega_2, \gamma)$ is continuously embedded in $H^{s_1,\sigma_1}(\Omega_2, \gamma)$ and the following inequalities hold:

$$\|u, \Omega_1, \gamma\|_{s_1,r_1} \le C \|u, \Omega_1, \gamma\|_{s_2,r_2},$$

$$\{\{u, \Omega_2, \gamma\}\}_{s_1,\sigma_1} \le \{\{u, \Omega_2, \gamma\}\}_{s_2,\sigma_2},$$

where the constant C does not depend on u, γ.

II.2.10. Equivalent norms in $\mathcal{H}^{s,r}(\mathbb{R}^{n+1}, \gamma)$, $\mathcal{H}^{s,\sigma}(E^{n+1}, \gamma)$, $\mathcal{H}^s(E^n, \gamma)$

Together with the norms defined by (2.23)–(2.25), we will use some other equivalent norms. Specifically, the elementary inequalities

$$\min\left(1, |\gamma|^{1/b}\right) \cdot \left(1 + |\xi|^2 + |\xi_0|^{1/b}\right) \le |\gamma|^{1/b} + |\xi|^2 + |\xi_0|^{1/b}$$

$$\le \max\left(1, |\gamma|^{1/b}\right) \cdot \left(1 + |\xi|^2 + |\xi_0|^{1/b}\right),$$

which hold for all $\gamma \ne 0$, imply that the norms $\|u, \mathbb{R}^{n+1}, \gamma\|'_{s,r}$, $\{\{u, E^{n+1}, \gamma\}\}'_{s,\sigma}$ and $\langle\langle \varphi, E^n, \gamma \rangle\rangle'_s$ are equivalent to the norms defined by the equalities

$$\|u, \mathbb{R}^{n+1}, \gamma\|_{s,r}^2 = \int \left(|\gamma|^{1/b} + |\xi|^2 + |\xi_0|^{1/b}\right)^s$$

$$\times \left(|\gamma|^{1/b} + |\xi'|^2 + |\xi_0|^{1/b}\right)^r |\widehat{u}(\xi, p)|^2 d\xi\, d\xi_0, \tag{2.26}$$

$$\{\{u, E^{n+1}, \gamma\}\}_{s,\sigma}^2 = \int \left(|\gamma|^{1/b} + |\xi|^2 + |\xi_0|^{1/b}\right)^s \left(1 + |\xi|^2\right)^\sigma |\widehat{u}(\xi, p)|^2 d\xi\, d\xi_0, \tag{2.27}$$

$$\langle\langle \varphi, E^n, \gamma \rangle\rangle_s^2 = \int \left(|\gamma|^{1/b} + |\xi'|^2 + |\xi_0|^{1/b}\right)^s |\widehat{\varphi}(\xi', p)|^2 d\xi'\, d\xi_0. \tag{2.28}$$

Thanks to the quasihomogeneity of the weight functions, the norms defined by (2.26)–(2.28) have the following important properties, which will be often used below. For any s, r, σ and $\alpha > 0, \beta > 0$, the following estimates hold:

$$\|u, \mathbb{R}^{n+1}, \gamma\|_{s,r} \le C |\gamma|^{-(\alpha+\beta)/2b} \|u, \mathbb{R}^{n+1}, \gamma\|_{s+\alpha,r+\beta}, \tag{2.29}$$

$$\{\{u, E^{n+1}, \gamma\}\}_{s,\sigma} \le C |\gamma|^{-\alpha/2b} \{\{u, \mathbb{R}^{n+1}, \gamma\}\}_{s+\alpha,\sigma}, \tag{2.30}$$

$$\langle\langle \varphi, E^n, \gamma \rangle\rangle_s \le C |\gamma|^{-\alpha/2b} \langle\langle \varphi, E^n, \gamma \rangle\rangle_{s+\alpha}. \tag{2.31}$$

where the constants C do not depend on u, φ and γ. In addition, for $s > 0$, the norm $\|u, \mathbb{R}^{n+1}, \gamma\|_s$ is equivalent to the norm defined by the formula

$$\|u, \mathbb{R}^{n+1}, \gamma\|_s''^2 = \int \left(|\gamma|^{s/b} + |\xi|^{2s} + |\xi_0|^{s/b}\right) |\widehat{u}(\xi, p)|^2 d\xi\, d\xi_0. \tag{2.31'}$$

II.2.11. The Spaces $H^s(G)$ and $H^s(\Gamma)$

Let G be a connected finite or infinite domain in $\mathbb{R}^n$ whose boundary Γ has the following structure. At each point $M \in \Gamma$ there exist the tangent plane and a ball K centered at M, with some fixed radius $d_0 > 0$, such that every straight line parallel to $\nu(M)$ (the unit inner normal to Γ at M) intersects the part of Γ lying inside K at at most one point. By local coordinate system at a point $M \in \Gamma$ we will understand a coordinate system $y_1, \ldots, y_n$ for which the axes $y_1, \ldots, y_{n-1}$ lie in the tangent plane to Γ at M and the axis y_n is directed along $\nu(M)$.

Let $\{g_i\}$ be a locally-finite covering of G by neighbourhoods (subsets g_i, $G = \bigcup_i g_i$) with the following property: Any subset g_i that does not intersect Γ is contained inside a ball K_i of radius d centered at a point M_i which lies at the distance $d > 0$ from Γ. Any subset g_i abutting on Γ (i.e., such that $\overline{g}_i \cap \Gamma \neq \emptyset$) is mapped by a local diffeomorphism (local straightening of Γ) into a subset $\breve{g}_i$ of a semi-ball $K_i^+ \subset \mathbb{R}_+^n$, while $\sigma_i = \overline{g}_i \cap \Gamma$ is mapped into a subset $\breve{\sigma}_i$ of the hyperplane $z_n = 0$: $\breve{\sigma}_i \subset \overline{K}_i \cap \mathbb{R}^{n-1}$.

Denote by $C^\infty(\Gamma)$ the set of infinitely differentiable functions defined on Γ. For each set g_i abutting on Γ, let $z = \alpha_i(x)$ be a diffeomorphism that maps g_i into $\breve{g}_i \subset K_i^+$ and maps σ_i into $\breve{\sigma}_i$. For a function $\varphi(x)$ let $\breve{\varphi}(z)$ be the expression of $\varphi(x)$ in variables $\{z\}$ in which σ_i is given by the equation $z_n = 0$. We assume that if $g_i \cap g_j \neq \emptyset$, then the Jacobian of the transition from the i-th to j-th local coordinate system is nonsingular. The above conditions imposed on Γ guarantee that in a neighbourhood of any point $M \in \Gamma$ the piece of Γ lying inside K is given by the equation $y_n = F(y')$, where $F(y')$ is a sufficiently smooth function. Obviously, if $x = \mathcal{F}(y)$ is the mapping effecting the transition from the coordinate system $\{x\}$ to a local coordinate system at $M \in \Gamma$ and $z' = y'$, $z_n = F(y') + y_n$ is the transformation of coordinates that straightens σ_i, then the composition of these two transformations is precisely α, the diffeomorphism described above.

A distribution $w(x)$ on Γ is defined to be a continuous linear functional on $C^\infty(\Gamma)$. Further, a set of functions $\{\varphi_i(x)\}$ in $C^\infty(\Gamma)$ such that $\operatorname{supp} \varphi_i(x) \subset g_i$ and $\sum_i \varphi_i(x) \equiv 1$ in $\overline{G}$ is called a *partition of unity* in $\overline{G}$ subordinate to the covering $\{g_i\}$.

The space $H^s(G)$ is defined as the set of all $v(x)$ that are restrictions of distributions $V(x) \in H^s(\mathbb{R}^n)$ to G, and is equipped with the norm

$$\big\|[v(x), G]\big\|_s' = \inf_V \big\|[V(x), \mathbb{R}^n]\big\|_s ,$$

where the infimum is taken over all $V(x) \in H^s(\mathbb{R}^n)$ that are equal to $v(x)$ on G.

Denote by $H_+^s(G)$ the subspace of $H^s(\mathbb{R}^n)$ consisting of the distributions with support contained in $\overline{G}$. The spaces $H^s(G)$ and $H_+^{-s}(G)$ are dual with respect to extension of the inner product in $L_2(G)$ [106].

Now let $\omega(x')$ be a distribution defined in Γ , and let $\breve{\omega}(z')$ be its expression in terms of the coordinates $\{z\}$. By definition, the space $H^s(\Gamma)$ is the set of all distributions $\omega(x')$ on Γ for which the following norm is finite:

$$\langle\!\langle \omega(x'), \Gamma \rangle\!\rangle_s'^2 = \sum_i \langle\!\langle \check{\varphi}_i(z', 0)\check{\omega}(z'), \mathbb{R}^{n-1} \rangle\!\rangle_s'^2, \tag{2.32}$$

where the sum is taken over all i such that $\sum_i \varphi_i(x') \equiv 1$ on Γ. Norms defined by different partitions of unity are equivalent. The space $H^s(\Gamma)$ may also be defined as the closure of the set $C^\infty(\Gamma)$ with respect to the norm (2.32).

II.2.12. The spaces $\mathcal{H}^s(S_+, \gamma)$ and $\mathcal{H}^s(\Omega_+, \gamma)$

Set $\Omega^\infty = G \times \mathbb{R}^1$, $S^\infty = \Gamma \times \mathbb{R}^1$, $\Omega_+ = G \times \mathbb{R}^1_+$, $S_+ = \Gamma \times \mathbb{R}^1_+$ and $\Omega = G \times (0, T)$, where $0 < T < \infty$.

For any $s, \gamma \in \mathbb{R}^1$ denote by $\mathcal{H}^s(\Omega_+, \gamma)$ and $\mathcal{H}^s(\Omega)$ the spaces consisting of the restrictions to Ω_+ and Ω of distributions $u(x, t)$ belonging to $\mathcal{H}^s(\mathbb{R}^{n+1}, \gamma)$ and $\mathcal{H}^s(\mathbb{R}^{n+1})$, respectively, equipped with the norms

$$\left\| u(x, t), \Omega_+, \gamma \right\|_s = \inf_U \left\| U, \mathbb{R}^{n+1}, \gamma \right\|_s, \qquad \left\| u, \Omega \right\|_s = \inf_U \left\| U, \mathbb{R}^{n+1} \right\|_s,$$

where the infimum is taken over all U that are equal to $u(x, t)$ in Ω_+ and Ω, respectively.

If $z = \alpha_i(x)$ is the diffeomorphism introduced above, then $(x, t) \mapsto (z, t)$: $z = \alpha_i(x)$, $t = t$ is a diffeomorphism of the cylinder $\Omega_i = g_i \times \mathbb{R}^1$ onto the cylinder $\check{\Omega}_i$ lying in $\mathbb{R}^{n+1}_+$; the image of $S_i = \Gamma_i \times \mathbb{R}^1$ lies in the hyperplane $z_n = 0$ of $\mathbb{R}^{n+1}$. Let $\check{v}(z, t)$ be the representation of the distribution $v(x, t)$ defined on Ω_i in the coordinates (z, t).

By definition, the space $\mathcal{H}^s(S^\infty, \gamma)$ is the set of all distributions $v(z, t)$ on S^∞ for which the norm

$$\langle\!\langle v(z, t), S^\infty, \gamma \rangle\!\rangle_s^2 = \sum_i \langle\!\langle \check{\varphi}_i(z', 0)\check{v}(z', 0, t), E^n, \gamma \rangle\!\rangle_s^2,$$

is finite. The spaces $\mathcal{H}^s(S_+, \gamma)$ and $\mathcal{H}^s(S)$ are defined as the sets of restrictions $v(x, t)$ to S_+ and S of the distributions $V(x, t) \in \mathcal{H}^s(S^\infty, \gamma)$ and $V(x, t) \in \mathcal{H}^s(S^\infty)$, respectively, equipped with the norms

$$\langle\!\langle v(x, t), S_+, \gamma \rangle\!\rangle_s = \inf_V \langle\!\langle V, S^\infty, \gamma \rangle\!\rangle_s, \qquad \langle\!\langle v(x, t), S \rangle\!\rangle_s = \inf_V \langle\!\langle V, S^\infty \rangle\!\rangle_s.$$

Denote by $\mathcal{H}^s_{++}(\Omega^\infty, \gamma)$ and $\mathcal{H}^s_+(S^\infty, \gamma)$ the subspaces of $\mathcal{H}^s(\Omega^\infty, \gamma)$ and $\mathcal{H}^s(S^\infty, \gamma)$ consisting of the distributions with support in $\overline{\Omega}_+$ and $\overline{S}_+$, respectively. For any γ and s, the pairs of spaces $\left(\mathcal{H}^s_{++}(\Omega^\infty, \gamma), \mathcal{H}^{-s}(\Omega_+, -\gamma) \right)$ and $\left(\mathcal{H}^s_+(S^\infty, \gamma), \mathcal{H}^{-s}(S_+, -\gamma) \right)$ are in duality with respect to the corresponding extensions of the inner products in L_2. By definition, $\mathcal{H}^s(\Omega^\infty, 0)$ coincides with $\mathcal{H}^s(\Omega^\infty)$ and $\mathcal{H}^s(S^\infty, 0)$ with $\mathcal{H}^s(S^\infty)$. Theorem II.2 has the following consequence.

Theorem II.4. 1) *Let $u(x,t) \in \mathcal{H}^s(\Omega^\infty, \gamma)$, $s > \frac{1}{2}$ and ν be the unit inner normal to S. Then for $1 \le k < s + \frac{1}{2}$ there exist the traces $D_\nu^{k-1} u(x,t)|_{S^\infty} \in \mathcal{H}^{s-k+\frac{1}{2}}(S^\infty, \gamma)$, and one has the estimate*

$$\left\langle\!\!\left\langle D_\nu^{k-1} u, S^\infty, \gamma \right\rangle\!\!\right\rangle_{s-k+\frac{1}{2}} \le C_s \| u(x,t), \Omega_\infty, \gamma \|_s,$$

with a constant C_s that does not depend on γ and $u(x,t)$.

2) *Let $u(x,t) \in \mathcal{H}^s(\Omega^\infty, \gamma)$ and $s > b$. Then for $1 \le \lambda < s + b$, there exist the traces $D_t^{\lambda-1} u(x,0) \in \mathcal{H}^{s-2b\lambda+b}(G)$, and one has the estimate*

$$\left\| [D_t^{\lambda-1} u(x,0), G] \right\|_{s-2b\lambda+b} \le C_s \| u(x,t), \Omega^\infty, \gamma \|_s,$$

with a constant C_s that does not depend on $u(x,t)$.

Corollary II.3. *Let $s > b + \frac{1}{2}$, $u(x,t) \in \mathcal{H}^s(\Omega^\infty, \gamma)$. Then for any pair (k, λ) such that $k + 2b\lambda < s + b + \frac{1}{2}$ there exist the traces $D_\nu^{k-1} D_t^{\lambda-1} u(x,t)\big|_\Gamma \in \mathcal{H}^{s-k-2b\lambda+b+\frac{1}{2}}(\Gamma)$, and one has the estimate*

$$\left\langle\!\!\left\langle D_\nu^{k-1} D_t^{\lambda-1} u\big|_\Gamma, \Gamma \right\rangle\!\!\right\rangle'_{s-2b\lambda-k+b+\frac{1}{2}} \le C_s \| u(x,t), \Omega^\infty, \gamma \|_s.$$

Note that Theorem II.4 implies similar statements about traces of functions $u(x,t)$ belonging to $\mathcal{H}^s(\Omega)$ and $\mathcal{H}^s(\Omega_+, \gamma)$ on S and G, respectively.

II.3. Banach spaces of Hölder functions

II.3.1. The spaces $C^s(G)$ and $C^s(\Omega)$

Let $s \ge 0$ be a real number, $v(x)$ and $u(x,t)$ functions defined in domains $G \subset \mathbb{R}^n$ and $\Omega \subset \mathbb{R}^{n+1}$, respectively. Let $C^s(G)$ and $C^s(\Omega) \equiv C_x^s\,{}_t^{s/2b}(\Omega)$ be the linear spaces of functions $v(x)$ and $u(x,t)$ that have continuous derivatives $D^\alpha v(x)$, where $|\alpha| \le s$, and $D^\alpha D_t^\mu u(x,t)$, where $|\alpha| + 2b\mu \le [s]$, and for which the following norms are finite:

$$\left| v(x), G \right|_s = \sum_{|\alpha| \le [s]} \sup_{x \in G} \left| D^\alpha v(x) \right| + \sum_{\substack{|\alpha| = [s]}} \sup_{\substack{x,y \in G \\ x \ne y}} \frac{\left| D^\alpha v(x) - D^\alpha v(y) \right|}{|x-y|^{s-[s]}}, \qquad (3.1)$$

$$\left| u(x,t), \Omega \right|_s = \sum_{|\alpha| + 2b\mu \le [s]} \sup_{(x,t) \in \Omega} \left| D^\alpha D_t^\mu u(x,t) \right|$$

$$+ \sum_{|\alpha| + 2b\mu = [s]} \sup_{(x,t),\,(y,t) \in \Omega} \frac{\left| D^\alpha D_t^\mu u(x,t) - D^\alpha D_t^\mu u(y,t) \right|}{|x-y|^{s-[s]}} \qquad (3.2)$$

$$+ \sum_{0 < s - 2b\mu - |\alpha| < 2b} \sup_{(x,t),\,(y,t) \in \Omega} \frac{\left| D^\alpha D_t^\mu u(x,t) - D^\alpha D_t^\mu u(x,\tau) \right|}{|t-\tau|^{(s-2b\mu-|\alpha|)/2b}}.$$

When s is an integer, the second sums in (3.1) and (3.2) are absent, and when s is a multiple of $2b$ the third sum in (3.2) is absent, too. $C^s(G)$ and $C^s(\Omega)$, equipped with the norms (3.1) and (3.2), are Banach spaces. We denote by $\overset{\circ}{C}{}^s(\Omega)$ the subspace of $C^s(\Omega)$ consisting of the functions $u(x,t)$ that satisfy the conditions $D_t^\mu u(x,0) = 0$, $\mu = 0, 1, \ldots, [s/2b]$.

II.3.2. The spaces $C^s(\Gamma)$ and $C^s(S)$

Now let us introduce the spaces $C^s(\Gamma)$ and $C^s(S) \equiv C_{x'\,t}^{s\,s/2b}(S)$ of functions $\omega(x')$ and $v(x',t)$ defined on Γ and $S = \Gamma \times (0,T]$, respectively. Suppose the surface Γ satisfies the conditions listed in Subsection II.2.11. Let K be a standard neighbourhood of the point $M \in \Gamma$ in which the piece of Γ is given, in the corresponding local coordinate system, by the equation $y_n = F(y')$; let σ be the piece of the hyperplane $z_n = 0$ into which $\Gamma_k = \Gamma \cap K$ is mapped by the transformation from the local coordinates $\{y\}$ to the coordinates z: $z' = y'$, $z_n = y_n - F(y')$.

We will say that $\Gamma \in C^s$ if $F(y') \in C^s(\sigma)$ and $|F(y'), \sigma|_s < C$, where C is a constant independent of $M \in \Gamma$.

Let $\Gamma \in C^s$. When we pass from the coordinates (x,t) to the coordinates (z,t), functions $\omega(x)$ and $u(x,t)$, defined on Γ and $S = \Gamma \times (0,T)$, respectively, go over into functions $\breve{\omega}_y(z')$ and $\breve{u}_y(z',t)$, defined on σ and $S_\sigma = \sigma \times (0,T)$, respectively. The spaces $C^s(\Gamma)$ and $C^s(S) \equiv C_{x'\,t}^{s\,s/2b}(S)$ are defined to be the sets of the functions $\omega(x')$ and $u(x',t)$ for which the norms

$$\left|\omega(x'),\Gamma\right|_s = \sup_{y\in\Gamma}\left|\breve{\omega}_y(z'),\sigma\right|_s, \qquad \left|u(x,t),S\right|_s = \sup_{y\in\Gamma}\left|\breve{u}_y(z',t),\sigma\times(0,T)\right|_s$$

are finite. $C^s(\Gamma)$ and $C^s(G)$ are Banach spaces. For every $s_1, s_2 \in \mathbb{R}^1$, $s_1 < s_2$, and $\Gamma \in C^{s_2}$ the space $C^{s_2}(\Gamma)$ is embedded in $C^{s_1}(\Gamma)$, and $C^{s_2}(S)$ in $C^{s_1}(S)$, the emebddings being compatible with the topologies.

As above, we denote by $\overset{\circ}{C}{}^s(S)$ the subspace of $C^s(S)$ consisting of the functions $u(x',t)$ for which $D_t^\mu u(x',0) = 0$, for $\mu = 0, \ldots, [s/2b]$. For a more detailed study and references on Hölder spaces the reader may consult the monograph [94].

Chapter III
Linear Operators

III.1. Operators of potential type

From now on we will denote by $\mathbb{C}$ the complex p-plane, $p = \gamma + i\xi_0$, and use $\mathbb{C}_\gamma$ for the half-plane $\operatorname{Re} p > \gamma$, $\gamma \in \mathbb{R}^1$, and $\overline{\mathbb{C}}_\gamma$ for its closure. Let $f(x,t)$, $\varphi(x',t)$, $\psi(x)$, and $\omega(x')$ be distributions defined in $\mathbb{R}^{n+1}$, E^n, $\mathbb{R}^n$, and $\mathbb{R}^{n-1}$, respectively. Denote by $\hat{f}(\xi,p)$ [resp., $\hat{\varphi}(\xi',p)$] the Fourier transform in x [resp., x'] and the Laplace transform in t of $f(x,t)$ [resp., $\varphi(x',t)$]. Similarly, denote by $\tilde{\psi}(\xi)$ [resp., $\tilde{\omega}(\xi')$] the Fourier transform in x [resp., x'] of the distribution $\psi(x)$ [resp., $\omega(x')$]. We will assume that $\widehat{G}(\xi,p)$ is an ordinary function of (ξ,p) defined in $A_0 = \mathbb{R}^n \times \mathbb{C}_0$ and locally integrable with respect to (ξ,ξ_0) in $\mathbb{R}^{n+1}$.

The inverse Fourier transforms in x and the inverse Laplace transforms in t of the products $\widehat{G}(\xi,p) \cdot \hat{f}(\xi,p)$, $\widehat{G}(\xi,p) \cdot \hat{\varphi}(\xi',p)$, $\widehat{G}(\xi,p) \cdot \tilde{\psi}(\xi)$, and $\widehat{G}(\xi,p) \cdot \tilde{\omega}(\xi')$ will be called *convolutions* (or *potentials*) and will be denoted by $G(x,t) * f(x,t)$, $G(x,t) *_1 \varphi(\xi',t)$, $G(x,t) *_2 \psi(x)$, and $G(x,t) *_3 \omega(x')$, respectively. Needless to say, these convolutions exist only under certain conditions imposed on $\widehat{G}$, f, φ, ψ, and ω and, generally speaking, are distributions. We are interested in the boundedness of these potentials in the spaces $\mathcal{H}^{s,r}(\mathbb{R}^{n+1},\gamma)$ and $\mathcal{H}^{s,\sigma}(E^{n+1},\gamma)$.

The following fundamental lemma will play an important role in many of our considerations.

Lemma III.1. *Let $\widehat{G}(\xi,p)$ be a function, locally integrable with respect to $(\xi,\xi_0) \in \mathbb{R}^{n+1}$ and quasihomogeneous of degree β: $\widehat{G}(a\xi, a^{2b}p) = a^\beta \widehat{G}(\xi,p)$ for all $a > 0$. Suppose that $\widehat{G}(\xi,p)$ obeys the estimate*

$$\left|\widehat{G}(\xi,p)\right| \le C\left(|\xi| + |p|^{1/2b}\right)^\beta. \tag{1.1}$$

Then

$$\left\|G(x,t) * f(x,t),\ \mathbb{R}^{n+1},\ \gamma\right\|_{s,r} \le C\left\|f(x,t),\ \mathbb{R}^{n+1},\ \gamma\right\|_{s+\beta,r},\ \text{for all } s,r \in \mathbb{R}^1, \tag{1.2}$$

$$\left\|G(x,t) *_1 \varphi(\xi',t),\ \mathbb{R}^{n+1},\ \gamma\right\|_{s,r} \le C\left\langle\!\left\langle\varphi(x',t),\ E^n,\ \gamma\right\rangle\!\right\rangle_{s+r+\beta+1/2} \tag{1.3}$$

for all $r \in \mathbb{R}^1$, $s < -\beta + \frac{1}{2}$,

$$\left\| G(x,t) *_2 \psi(x), \ \mathbb{R}^{n+1}, \ \gamma \right\|_{s,r} \leq C \left\| [\psi(x), \mathbb{R}^n] \right\|_{s+r+\beta+b}, \tag{1.4}$$

for all $r \in \mathbb{R}^1_+$, $s + r + \beta + b < 0$,

$$\left\| G(x,t) *_3 \omega(x'), \ \mathbb{R}^{n+1}, \ \gamma \right\|_{s,r} \leq C \langle\!\langle \omega(x'), \ \mathbb{R}^{n-1} \rangle\!\rangle_{s+r+\beta+b+1/2}, \tag{1.5}$$

for $s < -\beta + \frac{1}{2}$, $s + r + \beta + b + \frac{1}{2} < 0$, and

$$\left\{\!\left\{ G(x,t) *_2 \psi(x), \ E^{n+1}, \ \gamma \right\}\!\right\}_{s,\sigma} \leq C \left\| [\psi(x), \mathbb{R}^n] \right\|_{s+\beta+\sigma+b}, \tag{1.6}$$

for all $\sigma \in \mathbb{R}^1$, $s + \beta + b < 0$, with constants C that depend on γ and the indices of the norms of the right-hand sides of (1.2)–(1.6).

Proof: We will prove here only the estimate (1.5); the proof of the remaining estimates is similar. Using the elementary inequalities

$$C_1 \left(|\xi|^2 + |\xi_0|^{1/b} + |\gamma|^{1/b} \right)^{1/2} < \left(|\xi|^2 + |p|^{1/2b} \right) < C_2 \left(|\xi|^2 + |\xi_0|^{1/b} + |\gamma|^{1/b} \right)^{1/2},$$

estimate (1.1) and the definitions of the norms involved, we obtain

$$\left\| G *_3 \omega, \ \mathbb{R}^{n+1}, \ \gamma \right\|_{s,r}^2 \leq C \int\limits_{\mathbb{R}^n} \left(|\xi'|^2 + |\xi_0|^{1/b} + |\gamma|^{1/b} \right)^r \left| \widetilde{\omega}(\xi') \right|^2 d\xi_0 d\xi' \cdot$$
$$\times \int\limits_{\mathbb{R}^1} \left(|\xi|^2 + |\xi_0|^{1/b} + |\gamma|^{1/b} \right)^{s+\beta} d\xi_n.$$

Since for $s + \beta < -1/2$ we have

$$\int\limits_{\mathbb{R}^1} \left(|\xi|^2 + |\xi_0|^{1/b} + |\gamma|^{1/b} \right)^{s+\beta} d\xi_n$$
$$= \left(|\xi'|^2 + |\gamma|^{1/b} + |\xi_0|^{1/b} \right)^{s+\beta+1/2} \int\limits_{\mathbb{R}^1} \left(1 + \zeta_n^2 \right)^{s+\beta} d\zeta_n$$
$$\equiv J(s+\beta) \left(|\xi'|^2 + |\xi_0|^{1/b} + |\gamma|^{1/b} \right)^{s+\beta+1/2},$$

it follows that, for $s + r + \beta + 1/2 < -b$,

$$\left\| G *_3 \omega, \ \mathbb{R}^{n+1}, \ \gamma \right\|_{s,r}^2$$

$$\leq C J(s+\beta) \int\limits_{\mathbb{R}^{n-1}} \left| \widetilde{\omega}(\xi') \right|^2 d\xi' \cdot \int\limits_{\mathbb{R}^1} \left(|\xi'|^2 + |\xi_0|^{1/b} + |\gamma|^{1/b} \right)^{s+r+\beta+1/2} d\xi_0 =$$

$$= CJ(s+\beta) \int_{\mathbb{R}^1} \left(1+|\zeta_0|^{1/b}\right)^{s+r+\beta+1/2} d\zeta_0 \cdot \int_{\mathbb{R}^{n-1}} \left(|\xi'|^2+|\gamma|^{1/b}\right)^{s+r+\beta+b+1/2} |\widetilde{\omega}(\xi')|^2 \, d\xi'$$

$$\leq CJ(s+\beta)J(s+r+\beta+\tfrac{1}{2})$$

$$\times \max\left(1,|\gamma|^{(s+r+\beta+b+1/2)/b}\right) \langle\!\langle \omega(x'), \mathbb{R}^{n-1}\rangle\!\rangle'^2_{s+r+\beta+b+1/2}, \tag{1.7}$$

where $J(\cdot)$ denotes the corresponding integrals figuring in the the right-hand side of (1.7). This shows that for $\gamma \geq \gamma_0$, where γ_0 is fixed, the constant C in (1.5) depends only on γ_0 and s, r, β. Since in what follows we will assume that $\gamma > 1$, the constants C in (1.2)–(1.6) are actually independent of γ.

Remark III.1. Let us mention that in our considerations $\widehat{G}(\xi, p)$ is additionally an analytic function of p for $\operatorname{Re} p > -c_0|\operatorname{Im} p|$. It follows that $G(x, t) = F^{-1}_{\xi\to x} L^{-1}_{p\to t} \widehat{G}(\xi, p)$ is an ordinary function, equal zero for $t < 0$, which satisfies an estimate similar to that for the fundamental solution of a parabolic equation, or to the estimates of the Poisson kernels for the model parabolic boundary value problem [94, 108]. Therefore, the convolutions defined above in a formal manner do exist and can be expressed by the corresponding integrals. To write these integrals, we observe that $\widehat{\varphi}(\xi', p) = \widehat{\varphi}(\xi', p)1(\xi_n)$, $\widetilde{\psi}(\xi) = 1(p)\widetilde{\psi}(\xi)$, and $\widetilde{\omega}(\xi') = 1(p)1(\xi_n)\widetilde{\omega}(\xi')$, and their inverse Fourier transforms in x and Laplace transforms in t equal $\varphi(x', t) \times \delta(x_n)$, $\psi(x) \times \delta(t)$ and $\omega(x') \times \delta(x_n, t)$, respectively, where δ is the Dirac delta-function.

For suitable $f(x, t)$ and $\varphi(x', t)$ that vanish for $t < 0$, and $\psi(x)$ and $\omega(x')$, we obtain, by the properties of convolution and of the δ-function,

$$G(x, t) * f(x, t) = \int_0^t d\tau \int_{\mathbb{R}^n} G(x - y, t - \tau) f(y, \tau) \, dy, \tag{1.8}$$

$$G(x, t) *_1 \varphi(x', t) = \int_0^t d\tau \int_{\mathbb{R}^{n-1}} G(x - y', t - \tau) \varphi(y', \tau) \, dy', \tag{1.9}$$

$$G(x, t) *_2 \psi(x) = \int_{\mathbb{R}^n} G(x - y, t) \psi(y) \, dy, \tag{1.10}$$

$$G(x, t) *_3 \omega(x') = \int_{\mathbb{R}^{n-1}} G(x - y', t) \omega(y') \, dy', \tag{1.11}$$

The integral (1.8) is the volume potential with kernel $G(x - y, t - \tau)$ and density function $f(x, t)$, (1.9) is the integral of Poisson type for the heat equation, (1.10) is the surface potential with surface density $\varphi(x', t)$, and finally (1.11) is the potential with density concentrated in $\mathbb{R}^{n-1}$. Note that each of the integrals (1.9)–(1.11) can be regarded as the volume potential with the respective concentrated density $\varphi(x', t) \times \delta(x_n)$, $\psi(x) \times \delta(t)$, or $\omega(x') \times \delta(x_n, t)$.

III.2. Operators of multiplication by a function

III.2.1. Boundedness of truncation operators

Denote by $\theta(y)$ the characteristic function of the half-line $y > 0$: $\theta(y) = 1$ for $y > 0$ and $\theta(y) = 0$ for $y < 0$. Consider the operators of multiplication of functions belonging to the spaces H^s and $\mathcal{H}^s$ by $\theta(x_n)$ and $\theta(t)$. They will be referred to as *truncation operators* or *cutoff operators*.

Lemma III.2. *Let $|s| < 1/2$. Then for any $\gamma, r \in \mathbb{R}^1$ the operator of multiplication by $\theta(x_n)$ is bounded in $\mathcal{H}^{s,r}(\mathbb{R}^{n+1}, \gamma)$:*

$$\left\| \theta(x_n)u(x,t), \ \mathbb{R}^{n+1}, \ \gamma \right\|_{s,r} < C_s \left\| u(x,t), \ \mathbb{R}^{n+1}, \ \gamma \right\|_{s,r}, \tag{2.1}$$

with a constant C_s that does not depend on $u(x,t)$.

Proof: By definition,

$$\left\| \theta(x_n)u, \ \mathbb{R}^{n+1}, \ \gamma \right\|_{s,r}^2$$

$$= \int\limits_{\mathbb{R}^n} d\xi_0 d\xi' \int\limits_{\mathbb{R}^1} \left(1 + |\xi_0|^{1/b} + |\xi'|^2 + |\xi_n|^2\right)^s \left(1 + |\xi_0|^{1/b} + |\xi'|^2\right)^r \left|\widehat{\theta(x_n)u}\right|^2 d\xi_n$$

$$\sim \int\limits_{\mathbb{R}^n} d\xi_0 d\xi' \int\limits_{\mathbb{R}^1} \left(1 + |\xi'| + |\xi_0|^{1/2b} + |\xi_n|\right)^{2s} \left(1 + |\xi_0|^{1/2b} + |\xi'|\right)^{2r} \left|\widehat{\theta(x_n)u}\right|^2 d\xi_n.$$

Assume first that $u(x,t)e^{-\gamma t}$ belongs to $\mathcal{S}(\mathbb{R}^{n+1})$. It is known [10, 15, 113] that if $\widehat{u}(\xi', x_n, p)$ is the Fourier transform in x' and the bilateral Laplace transform in t of the function $u(x,t)$, then

$$F_n\big(\theta(x_n) \cdot \widehat{u}(\xi', x_n, p)\big) = \int\limits_{\mathbb{R}^1} \frac{\widehat{u}(\xi', \eta_n, p)}{\xi_n - i0 - \eta_n}\, d\eta_n = \lim_{\tau \to +0} \int\limits_{\mathbb{R}^1} \frac{\widehat{u}(\xi', \eta_n, p)}{\xi_n - i\tau - \eta_n}\, d\eta_n.$$

Hence,

$$\left\| \theta(x_n)u, \ \mathbb{R}^{n+1}, \ \gamma \right\|_{s,r}^2$$

$$\sim \int\limits_{\mathbb{R}^n} d\xi_0 d\xi' \int\limits_{\mathbb{R}^1} \left| \int\limits_{\mathbb{R}^1} \left(\frac{1 + |\xi'| + |\xi_n| + |\xi_0|^{1/2b}}{1 + |\xi'| + |\eta_n| + |\xi_0|^{1/2b}} \right)^s \cdot \frac{\widehat{u}_0(\xi', \eta_n, p)}{\xi_n - i0 - \eta_n}\, d\eta_n \right|^2 d\xi_n,$$

where $\widehat{u}_0(\xi', \eta_n, p) = \left(1 + |\xi_0|^{1/2b} + |\xi'| + \eta_n\right)^s \left(1 + |\xi_0|^{1/2b} + |\xi'|\right)^r \widehat{u}(\xi', \eta_n, p)$.

Obviously, if $u(x,t) \in \mathcal{H}^{s,r}(\mathbb{R}^{n+1}, \gamma)$, then $\widehat{u}_0(\xi', \eta_n, p) \in L_2(\mathbb{R}^{n+1})$. Therefore, denoting by Π_s the operator acting as

$$(\Pi_s \widehat{u}_0)(\xi', \eta_n, p) = \frac{i}{2\pi} \int\limits_{\mathbb{R}^1} \left(\frac{1 + |\xi_0|^{1/2b} + |\xi'| + |\xi_n|}{1 + |\xi_0|^{1/2b} + |\xi'| + |\eta_n|} \right)^s \frac{\widehat{u}_0(\xi', \eta_n, p)}{\xi_n - i0 - \eta_n}\, d\eta_n,$$

we obtain $\left\|\theta(x_n)u,\ \mathbb{R}^{n+1},\ \gamma\right\|_{s,r} \sim \left\|\Pi_s\widehat{u}_0,\ \mathbb{R}^{n+1}\right\|_0$ and it remains to estimate the norm $\left\|\Pi_s\widehat{u}_0,\ \mathbb{R}^{n+1}\right\|_0$. Since the detailed estimation of such an integral operator was carried out in [113], we only sketch it here.

Let us write $\Pi_s\widehat{u}_0$ as a sum of four terms

$$
\begin{aligned}
\Pi_s\widehat{u}_0 = {} & \theta(\xi_n)\Pi_s\theta(\eta_n)\widehat{u}_0(\xi',\eta_n,p) \\
& + \theta(\xi_n)\Pi_s\big(1 - \theta(\eta_n)\big)\widehat{u}_0(\xi',\eta_n,p) \\
& + \big(1 - \theta(\eta_n)\big)\Pi_s\theta(\eta_n)\widehat{u}_0(\xi',\eta_n,p) \\
& + \big(1 - \theta(\xi_n)\big)\Pi_s\big(1 - \theta(\eta_n)\big)\widehat{u}_0(\xi',\eta_n,p)
\end{aligned}
\tag{2.2}
$$

and estimate the L_2-norm of each of these terms. Since all the terms have the same structure, it suffices to estimate the L_2-norm of the first, say. Thus, let us show that

$$
\int_{\mathbb{R}^n} d\xi'd\xi_0 \int_0^\infty \big|\theta(\xi_n)\Pi_s\theta(\eta_n)\widehat{u}_0\big|^2\,d\xi_n \leq C_s \int_{\mathbb{R}^n} d\xi'd\xi_0 \int_0^\infty \theta(\eta_n)\big|\widehat{u}_0(\xi',\eta_n,p)\big|^2\,d\eta_n.
\tag{2.3}
$$

Indeed, the change of variables

$$
\xi_n = \big(1 + |\xi_0|^{1/2b} + |\xi'|\big)(\widetilde{t} - 1), \quad \eta_n = \big(1 + |\xi_0|^{1/2b} + |\xi'|\big)(\tau - 1)
$$

yields

$$
\theta(\xi_n)\Pi_s\theta(\eta_n)\widehat{u}_0 = \theta(\widetilde{t} - 1)\int_1^\infty \left(\frac{\widetilde{t}}{\tau}\right)^s \widehat{u}_0\Big(\xi',\big(1 + |\xi'| + |\xi_0|^{1/2b}\big)(\tau - 1),p\Big)\frac{d\tau}{\widetilde{t} - i0 - \tau}.
$$

The integral

$$
(Jg)(\widetilde{t}) = \int_0^\infty \left(\frac{\widetilde{t}}{\tau}\right)^s \frac{g(\tau)}{\widetilde{t} - i0 - \tau}\,d\tau
$$

obeys the following estimate (see [113]):

$$
\int_0^\infty \big|(Jg)(\widetilde{t})\big|^2\,dt \leq C \int_0^\infty \big|g(\tau)\big|^2\,d\tau.
\tag{2.4}
$$

Therefore, upon setting $g(\tau) = \widehat{u}_0\big(\xi',\big(1 + |\xi'| + |\xi_0|^{1/2b}\big)(\tau - 1),p\big)$ for $\tau \geq 1$ and $g(\tau) \equiv 0$ for $0 \leq \tau < 1$, multiplying (2.4) by $\big(1 + |\xi'| + |\xi_0|^{1/2b}\big)$, and integrating with respect to ξ_0 and ξ', we obtain (2.3).

The following lemma is proved in a similar way.

Lemma III.3. *Let $|s| < b$. Then for any $\gamma, \sigma \in \mathbb{R}^1$ the operator of multiplication by $\theta(t)$ is bounded in $\mathcal{H}^{s,\sigma}(E^{n+1}, \gamma)$:*

$$\{\{\theta(t)u(x,t),\ E^{n+1},\ \gamma\}\}_{s,\sigma} < C_s\{\{u(x,t),\ E^{n+1},\ \gamma\}\}_{s,\sigma}, \tag{2.5}$$

with a constant C_s that does not depend on $u(x,t)$.

Remark III.2. The arguments used in the proof of Lemma III.1 remain valid for any $b > 0$, in particular, for $b = 1/2$, $\gamma = 0$, $r = 0$, which corresponds to the case of the isotropic non-weighted spaces $H^s(\mathbb{R}^n)$. Therefore, Lemma III.1 yields the estimate

$$\left|\left[\theta(x_n)v(x),\ \mathbb{R}^n\right]\right|_s < C_s\left|\left[v(x),\ \mathbb{R}^n\right]\right|_s, \tag{2.6}$$

where the constant C_s does not depend on $v(x)$.

Lemma III.3 for the case $\sigma = 0$, $|s| < b$ gives estimate

$$\left\langle\!\left\langle\theta(t)\varphi(x',t),\ E^n,\ \gamma\right\rangle\!\right\rangle_s < C_s\left\langle\!\left\langle\varphi(x',t),\ E^n,\ \gamma\right\rangle\!\right\rangle_s, \tag{2.7}$$

where the constant C_s does not depend on $\varphi(x',t)$.

III.2.2. Boundedness of the operators of multiplication by smooth functions

It is known [3, 14, 92] that the operators of multiplication of the functions belonging to the spaces H^s, $\mathcal{H}^s$ with $s \geq 0$ by sufficiently smooth Hölder functions are bounded. From this fact and the definition of the spaces H^s, $\mathcal{H}^s$ with $s < 0$ one concludes that these operators are bounded for all $s \in \mathbb{R}^1$, i.e.,

$$\left\|a(x,t)u(x,t),\ \Omega\right\|_s < C\left|a(x,t),\ \Omega\right|_{|s|+\varepsilon}\left\|u(x,t),\ \Omega\right\|_s, \quad s \in \mathbb{R}^1, \tag{2.8}$$

$$\left\langle\!\left\langle b(x',t)\cdot\varphi(x',t),\ S\right\rangle\!\right\rangle_s < C\left|b(x',t),\ S\right|_{|s|+\varepsilon}\left\langle\!\left\langle\varphi(x',t),\ S\right\rangle\!\right\rangle_s, \tag{2.9}$$

$$\left|\left[c(x)\cdot v(x),\ G\right]\right|_s < C\left|c(x),\ G\right|_{|s|+\varepsilon}\left|\left[v(x),\ G\right]\right|_s, \tag{2.10}$$

$$\left\langle\!\left\langle d(x')\omega(x'),\ \Gamma\right\rangle\!\right\rangle'_s < C\left|d(x'),\ \Gamma\right|_{|s|+\varepsilon}\left\langle\!\left\langle\omega(x'),\ \Gamma\right\rangle\!\right\rangle'_s, \tag{2.11}$$

where $\varepsilon > 0$ is an arbitrary number and the constants C do not depend on u, φ, v, ω.

However, in the case of the weighted spaces $\mathcal{H}^s(\Omega_+, \gamma)$ and $\mathcal{H}^s(S_+, \gamma)$ the estimates (2.8) and (2.9) can be sharpened. These new estimates, which play an essential role in the study of parabolic boundary value problems with variable coefficients in the spaces $\mathcal{H}^s(\Omega_+, \gamma)$, follow from the definitions of the norms (2.26), (2.28) and from inequalities (2.29), (2.30) of Chapter II, .

Lemma III.4. *Assume that $a(x,t) \in C^{|s|+\varepsilon}(\Omega_+)$, $b(x',t) \in C^{|s|+\varepsilon}(S_+)$, $u(x,t) \in \mathcal{H}^s(\Omega_+, \gamma)$, $\varphi(x',t) \in \mathcal{H}^s(S_+, \gamma)$, where $\varepsilon > 0$ is an arbitrary number and $\gamma \gg 1$. Then*

$$\left\|a\cdot u(x,t),\ \Omega_+,\ \gamma\right\|_s \leq C\left(|a, \Omega_+|_0 + \gamma^{-1/2b}|a, \Omega_+|_{|s|+\varepsilon}\right)\left\|u,\ \Omega_+,\ \gamma\right\|_s, \tag{2.12}$$

$$\left\langle\!\left\langle b\cdot\varphi,\ S_+,\ \gamma\right\rangle\!\right\rangle_s \leq C\left(|b, S_+|_0 + \gamma^{-1/2b}|b, S_+|_{|s|+\varepsilon}\right)\left\langle\!\left\langle\varphi,\ S_+\right\rangle\!\right\rangle_s, \tag{2.13}$$

with a constant C that does not depend on u, φ, γ.

Proof: Set $\alpha = s - [s]$, $\beta = \frac{s}{2b} - [s/2b]$, $\delta = [s/2b]$. We will prove (2.12) considering separately the cases 1) $s \geq 0$ and 2) $s < 0$.

1) Let $\bar{a}(x,t) \in C^{|s|+\varepsilon}$ and $U(x,t) \in \mathcal{H}^s(\mathbb{R}^{n+1}, \gamma)$ be smooth extensions of $a(x,t)$ and $u(x,t)$, respectively, from Ω_+ to all $\mathbb{R}^{n+1}$. Put $\overline{U}(x,t,\gamma) = U(x,t)e^{-\gamma t}$. The squared norm $\left\| \bar{a}U,\ \mathbb{R}^{n+1},\ \gamma \right\|_s^2$ is equivalent to the integral

$$\int\limits_{\mathbb{R}^{n+1}} \left(\gamma^{s/b} + |\xi_0|^{s/b} + |\xi|^{2s} \right) |\widehat{\bar{a} \cdot \overline{U}}|^2 \, d\xi \, d\xi_0.$$

By the properties of the Fourier transformation and the Parseval equality, this integral is equal (up to a constant factor) to the following expression:

$$I = \gamma^{s/b} \int\limits_{\mathbb{R}^{n+1}} \left| \bar{a} \cdot \overline{U}(x,t,\gamma) \right|^2 dx \, dt$$

$$+ \sum_{|k|=[s]} \int\limits_{\mathbb{R}^1} dt \int\limits_{\mathbb{R}^{2n}} \left| \Delta_x^y D^k (\bar{a} \cdot \overline{U}) \right|^2 \frac{dx \, dy}{|x - y|^{n+2\alpha}} \qquad (2.14)$$

$$+ \int\limits_{\mathbb{R}^n} dx \int\limits_{\mathbb{R}^2} \left| \Delta_t^\tau D_t^\delta (\bar{a} \cdot \overline{U}) \right|^2 \frac{d\tau \, dt}{|t - \tau|^{1+2\beta}} \equiv I_1 + I_2 + I_3,$$

where in the case when s is an integer the integrals over $\mathbb{R}^{2n}$ are replaced by the integrals over $\mathbb{R}^n$ of $D^k(\bar{a}\overline{U})$, and where in the case when s is a multiple of $2b$ the integral over $\mathbb{R}^2$ with respect to τ and t is replaced by the integral over $\mathbb{R}^1$ of $D_t^\delta(\bar{a}\overline{U})$. Here and in what follows one denotes $\Delta_x^y v(x,t) = v(x + y, t) - v(x,t)$, $\Delta_t^\tau v = v(x, t + \tau) - v(x,t)$.

First consider the case when s or $s/2b$ is an integer. Then applying the Leibniz formula to calculate the derivatives $D^k(\bar{a}\overline{U})$, $D_t^\delta(\bar{a}\overline{U})$ and the estimate (2.29) of Chapter II we conclude that for $\gamma \geq \gamma_0 > 0$, the terms I_1 and I_2 or I_1 and I_3 are bounded by $C\big(|a, \Omega_+|_0 + \gamma^{-1/2b}|a, \Omega_+|_s\big)\|u,\ \Omega_+,\ \gamma\|_s$. This implies the assertion of Lemma III.4 for this case.

Now suppose that both s and $s/2b$ are not integers. We will estimate here I_3; the estimation of I_2 when s is not an integer is carried out in a similar manner. Using the Leibniz formula we obtain, after elementary calculations,

$$I_3 \leq C \sum_{m+q=\delta} \left(\int\limits_{\mathbb{R}^n} dx \int\limits_{\mathbb{R}^2} \left| \Delta_t^\tau D_t^m \bar{a} \right|^2 \frac{|D_t^q \overline{U}|^2}{|t - \tau|^{1+2\beta}} \, d\tau \, dt \right.$$

$$\left. + \int\limits_{\mathbb{R}^n} dx \int\limits_{\mathbb{R}^2} |D_t^m \bar{a}|^2 \frac{|\Delta_t^\tau D_t^q \overline{U}|^2}{|t - \tau|^{1+2\beta}} \, d\tau \, dt \right). \qquad (2.15)$$

It is easily seen that

$$\int\limits_{\mathbb{R}^1} |\Delta_t^\tau D_t^m \bar{a}|^2 \frac{d\tau}{|t-\tau|^{1+2\beta}} \leq C_m^2(\beta,\varepsilon)|\bar{a},\mathbb{R}^{n+1}|_{2b(m+\varepsilon)},$$

$$\int\limits_{\mathbb{R}^n} dx \int\limits_{\mathbb{R}^2} |D_t^m \bar{a}|^2 \frac{|\Delta_t^\tau D_t^q \overline{U}|^2}{|t-\tau|^{1+2\beta}}\, d\tau\, dt \leq C \cdot |\bar{a},\mathbb{R}^{n+1}|_{2bm} \|\overline{U},\ \mathbb{R}^{n+1}\|^2_{2b(\delta-m)}.$$

Therefore (2.15) yields the estimate

$$I_3 \leq C \sum_{m=0}^{\delta} \left(C_m^2 |\bar{a},\mathbb{R}^{n+1}|^2_{s+\varepsilon} \|u,\ \mathbb{R}^{n+1},\ \gamma\|^2_{2b(m-\delta)} \right.$$
$$\left. + \left|D_t^m a,\ \mathbb{R}^{n+1}\right|^2_0 \|U,\ \mathbb{R}^{n+1},\ \gamma\|^2_{s-2bm} \right).$$

Hence, by the estimate (2.29) of Chapter II,

$$I_3 \leq C \sum_{m=0}^{\delta} \left(C_m^2 |\bar{a},\mathbb{R}^{n+1}|_{s+\varepsilon} \gamma^{-2(\beta+m)} + |\bar{a},\mathbb{R}^{n+1}|_{2bm} \gamma^{-m/b} \right) \|U,\ \mathbb{R}^{n+1},\ \gamma\|^2_s.$$

We conclude that

$$I_3 \leq C\big(|a,\Omega|_0 + C_1(s)|a,\Omega|_{s+\varepsilon}\gamma^{-1/2b}\big)\|U,\ \mathbb{R}^{n+1},\ \gamma\|^2_s. \tag{2.16}$$

A similar argument yields the bounds

$$I_1,\ I_2 \leq C\big(|a,\Omega|_0 + C_1(s)|a,\Omega|_{s+\varepsilon}\gamma^{-1/2b}\big)\|U,\ \mathbb{R}^{n+1},\ \gamma\|^2_s. \tag{2.17}$$

If in (2.16), (2.17) we take the infimum over U, we obtain (2.12) for $s \geq 0$.

2) Now let $s < 0$. Regarding $u(x,t) \in \mathcal{H}^s(\Omega_+,\gamma)$ as a functional on the space $\mathcal{H}^{-s}(\Omega_+,-\gamma)$ and making use of duality and the estimate (2.12) for $s \geq 0$, we get

$$\|au,\ \Omega_+,\ \gamma\|_s = \sup_{v\in\mathcal{H}^{-s}(\Omega_+,-\gamma)} \frac{|(au,v)|}{\|v,\ \Omega_+,\ -\gamma\|_{-s}}$$

$$= \sup_{v\in\mathcal{H}^{-s}} \frac{|(u,\bar{a}v)|}{\|v,\ \Omega_+,\ -\gamma\|_{-s}}$$

$$\leq \|u,\ \Omega_+,\ \gamma\|_s C_s\big(|a,\Omega_+|_0 + |a,\Omega_+|_{|s|+\varepsilon}\gamma^{-1/2b}\big);$$

here $\bar{a}$ denotes the complex conjugate of a. Estimate (2.12) is proved. Estimate (2.13) is established in a similar manner, using a partition of unity on Γ and the local straightening of Γ by means of local coordinate systems.

III.3. Commutators. Green formulas

III.3.1. The operators J_n, J_0

Recall that above we have called the operators of multiplication of functions $u(x,t)$, $v(x)$ or $\varphi(x',t)$ by the characteristic functions $\theta(x_n)$ or $\theta(t)$ the truncation (or cutoff) operators with respect to x_n or t, respectively. The functions $u_{++}(x,t) = \theta(x_n)\theta(t)u(x,t)$, $v^+(x) = \theta(x_n)v(x)$, and $\varphi^+(x',t) = \theta(t)\varphi(x',t)$ will be called the *truncations* of the functions u, v, φ respectively. In further considerations an essential role will be played by the formulas giving the action of differential operators on truncations in the sense of distributions, $\mathcal{P}(x,D)v^+(x)$, $b(x',t,D',D_t)\varphi^+(x',t)$ and $l(x,t,D,D_t)u_{++}$, and, in particular, the commutation formulas for the operators $\mathcal{P}(x,D)$ and $\theta(x_n)$, $b(x',t,D',D_t)$ and $\theta(t)$, and $l(x,t,D,D_t)$ and $\theta(x_n)\theta(t)$. We begin by introducing some notations that will be repeatedly used below.

Let

$$\mathcal{P}(x,D) = \sum_\sigma \mathcal{P}_\sigma(x,D')D_n^\sigma,$$

$$b(\cdot,D',D_t) = \sum_\mu b_\mu(\cdot,D')D_t^\mu,$$

$$l(\cdot,D,D_t) = \sum_\sigma l_\sigma(\cdot,D',D_t)D_n^\sigma \equiv \sum_{\sigma,\mu} l_{\sigma,\mu}(\cdot,D')D_n^\sigma D_t^\mu$$

be linear differential operators with sufficiently smooth coefficients of weighted orders α, q, and m, respectively. The dot indicates the dependence of these operators on x and t.

On the set of polynomials $l(\cdot,\zeta_n,p)$, $b(\cdot,p)$, and $\mathcal{P}(\cdot,\zeta_n)$ in (ζ_n,p), p, and ζ_n, respectively, consider the linear operators J_n and J_0 that act on $l(\cdot,\zeta_n,p)$, $b(\cdot,p)$, and $\mathcal{P}(\cdot,\zeta_n)$ according to the formulas

$$J_0^\lambda J_n^k l \equiv \left[p^{-\lambda}\zeta_n^{-k}l(\cdot,\zeta_n,p)\right] \equiv l^{(k,\lambda)}(\cdot,\zeta_n,p),$$

$$J_0^\lambda b \equiv \left[p^{-\lambda}b(\cdot,p)\right] \equiv b^{(\lambda)}(\cdot,p),$$

$$J_n^k \mathcal{P} \equiv \left[\zeta_n^{-k}\mathcal{P}(\cdot,\zeta_n)\right] \equiv \mathcal{P}^{(k)}(\cdot,\zeta_n), \qquad k,\lambda \in \mathbb{Z},$$

where $[p^{-\lambda}\zeta_n^{-k}l]$, $[p^{-\lambda}b]$, and $[\zeta_n^{-k}\mathcal{P}]$ are the integral parts (polynomials) of the division of $l(\cdot,\zeta_n,p)$, $b(\cdot,p)$, and $\mathcal{P}(\cdot,\zeta_n)$ by $\zeta_n^k p^\lambda$, p^λ, and ζ_n^k, respectively.

From now on $[s]$ will designate the integral part of s for any $s \in \mathbb{R}^1$; also, $s' = [s+1/2]$, $s'' = [(s+b)/2b]$. Assume that $u(x,t) \in \mathcal{H}^s(\mathbb{R}^{n+1},\gamma)$, $s \geq 0$. Set $u_k(x',t) = D_n^{k-1}u(x',0,t)$, $k = 1,\ldots,s'$, $v_\lambda(x) = D_t^{\lambda-1}u(x,0)$, $\lambda = 1,\ldots,s''$, and $\omega_{\lambda k}(x') = D_n^{k-1}D_t^{\lambda-1}u(x',0,0)$, for any pair (k,λ) such that $k + 2b\lambda \leq s' + b$. Similarly, for $v(x) \in H^s(\mathbb{R}^n)$ and $\Phi(x',t) \in \mathcal{H}^s(E^n,\gamma)$, $s \geq 0$, set $\omega_k(x') = D_n^{k-1}v(x',0)$, $k = 1,\ldots,s'$, $\Phi_\lambda(x') = D_t^{\lambda-1}\Phi(x',0)$, $\lambda = 1,\ldots,s''$.

III.3.2. Formulas for calculating $\mathcal{P}(x, D)v^+(x)$ and $b(x', t, D', D_t)\Phi^+(x', t)$

Lemma III.5. *Let $v(x) \in H^s(\mathbb{R}^n)$, where $s > \frac{1}{2}$, and $s + \frac{1}{2}$ is not an integer for $s < \alpha$. Then*

$$\mathcal{P}(\cdot, D)v^+(x) = \mathcal{P}^{(s')}(\cdot, D)\left(D_n^{s'} v\right)^+ + \sum_{k=1}^{s'} \mathcal{P}^{(k)}(\cdot, D)\omega_k(x') \times \delta(x_n)$$

$$+ \sum_{\sigma=0}^{s'-1} \mathcal{P}_\sigma(\cdot, D')\left(D_n^\sigma v\right)^+, \tag{3.1}$$

where $\delta(x_n)$ is the Dirac delta-function and $\omega_k(x') \times \delta(x_n)$ is the direct product of the functions $\omega_k(x')$ and $\delta(x_n)$.

Proof: By Theorem II.3 on traces, for any $k \le s'$ there exist the traces $\omega_k(x') = D_n^{k-1}v(x', +0)$. Consequently, for any $\sigma \le s'$, the equality

$$D_n^\sigma v^+ = \left(D_n^\sigma v\right)^+ + \sum_{k=1}^{\sigma} D_n^{\sigma-k} \cdot \left(D_n^{k-1}v(x', 0) \times \delta(x_n)\right)$$

$$= \left(D_n^\sigma v\right)^+ + \sum_{k=1}^{\sigma} D_n^{\sigma-k}\omega_k(x') \times \delta(x_n) \tag{3.2}$$

holds in the sense of distributions.

Let us rewrite $\mathcal{P}(\cdot, D)v^+$ in the form

$$\mathcal{P}(\cdot, D)v^+(x) = \sum_{\sigma=0}^{s'} \mathcal{P}_\sigma(\cdot, D')D_n^\sigma v^+(x) + \sum_{\sigma=s'+1}^{\alpha} \mathcal{P}_\sigma(\cdot, D')D_n^{\sigma-s'} \cdot D_n^{s'} v^+(x),$$

and then replace $D_n^\sigma v^+$ and $D_n^{s'} v^+$ in the formula for $\mathcal{P}v^+$ by their expressions, given by (3.2). Interchanging the order of summation with respect to σ and k and using the above-introduced notations, we obtain (3.1).

Lemma III.6. *Let $\Phi(x', t) \in \mathcal{H}^s(E^n, \gamma)$, where $s > b$, and $(s + b)/2b$ is not an integer for $b < s < q$. Then*

$$b(\cdot, D', D_t)\Phi^+(x', t) = b^{(s'')}(\cdot, D', D_t)\left(D_t^{s''} \Phi\right)^+$$

$$+ \sum_{\lambda=1}^{s''} b^{(\lambda)}(\cdot, D', D_t)\Phi_\lambda(x') \times \delta(t) + \sum_{\mu=0}^{s''-1} b_\mu(\cdot, D')\left(D_t^\mu \Phi\right)^+. \tag{3.3}$$

The proof of this lemma is similar to that of Lemma III.5.

III.3.3. Formulas for calculating $l(x, t, D, D_t)u_{++}(x, t)$

Set $s_\lambda = s' - 2b\lambda + b$, $s_0 = \big[(s+b-1/2)/2b\big]$, $\sigma_s = \big[(s+b-\sigma)/2b\big]$, $\alpha_\lambda = [s] - 2b\lambda + b$ and define $\mathbb{Z}_1 = \{s \in \mathbb{R}^1 : s \text{ is half-integer}\}$, $\mathbb{Z}_{2b} = \{s \in \mathbb{R}^1 : s/2b \text{ is half-integer}\}$, $\mathbb{Z}_{1,2b} = \{s \in \mathbb{R}^1 : s \text{ or } s/2b \text{ or } (s+1/2)/2b \text{ is half-integer}\}$.

Lemma III.7. *Let* $u(x,t) \in \mathcal{H}^s(\mathbb{R}^{n+1}, \gamma)$, *where* $s > \frac{1}{2}$, *and* $s \notin \mathbb{Z}_{1,2b}$ *for* $s < m$. *Then*

$$lu_{++}(x, t) = l^{(s',0)} \cdot \big(D_n^{s'} u\big)_{++} + \sum_{k=1}^{s'} l^{(k,0)} u_k^+ \times \delta(x_n) + \sum_{\lambda=1}^{s''} l^{(0,\lambda)} v_\lambda^+ \times \delta(t)$$

$$- \sum_{\lambda=1}^{s_0} \sum_{k=1}^{s_\lambda} l^{(k,\lambda)} \omega_{\lambda k}(x') \times \delta(x_n, t) + \big[l, \theta(x_n, t)\big]_s u(x, t), \qquad (3.4)$$

where

$$\big[l, \theta(x_n, t)\big]_s u = \sum_{\sigma=[s]-b+1}^{s'-1} l_\sigma \big(D_n^\sigma u\big)_{++} + \sum_{\sigma=0}^{[s]-b} l_\sigma^{(\sigma_s)} \cdot \big(D_t^{\sigma_s} D_n^\sigma u\big)_{++}$$

$$+ \sum_{\sigma+2b\mu \leq [s]-b} l_{\sigma\mu} \cdot \big(D_t^\mu D_n^\sigma u\big)_{++}$$

$$- \sum_{\lambda=1}^{s_0} l^{(\alpha_\lambda+1,\lambda)} D_n^{[s+1]-s'} \cdot \big(D_n^{s_\lambda} v_\lambda^+(x) \times \delta(t)\big)$$

$$- l^{(\alpha_{s''}+1, s'')} \cdot D_n^{\alpha_{s''}+1} v_{s''}^+(x) \times \delta(t).$$

(3.5)

Here the convention is that sums with negative limits of summation are identically equal to zero.

Proof: Since the form of the right-hand side of (3.4) depends on s, we will consider three cases.

1. Let $\frac{1}{2} < s < b$. Then because D_n and $\theta(t)$ commute, repeating the argument of the proof of Lemma 3.5 we obtain

$$lu_{++} = l^{(s',0)} \big(D_n^{s'} u\big)_{++} + \sum_{k=1}^{s'} l^{(k,0)} u_k^+ \times \delta(x_n) + \sum_{\sigma=0}^{s'-1} l_\sigma \big(D_n^\sigma u\big)_{++}, \qquad (3.6)$$

which proves (3.4) for $\frac{1}{2} < s < b$.

2. Let $b < s < b + \frac{1}{2}$. In this case we transform the last sum in the right-hand side of (3.6) further, using Lemma III.5 to isolate the terms containing $v_\lambda^+ \times \delta(t)$. However, since the order of l_σ in t is at most $\big[(m - \sigma)/2b\big]$, we can isolate the indicated terms only from those terms of $l_\sigma(D_n^\sigma u)_{++}$, for which $(s - \sigma)/2b > \frac{1}{2}$,

that is, $\sigma \leq [s] - b$. As suggested by these considerations, we break the sum with respect to σ into two sums,

$$\sum_{\sigma=0}^{s'-1} l_\sigma (D_n^\sigma u)_{++} = \sum_{\sigma=0}^{[s]-b} l_\sigma (D_n^\sigma u)_{++} + \sum_{\sigma=[s]+1-b}^{s'-1} l_\sigma (D_n^\sigma u)_{++},$$

and then, using the fact that D_n and $\theta(t)$ commute, we apply Lemma III.5 to each term $l_\sigma \theta(t) D_n^\sigma u$. This yields

$$\sum_{\sigma=0}^{s'-1} l_\sigma (D_n^\sigma u)_{++} = \sum_{\sigma=[s]+1-b}^{s'-1} l_\sigma (D_n^\sigma u)_{++} + \sum_{\sigma=0}^{[s]-b} l_\sigma^{(\sigma_s)} (D_t^{\sigma_s} D_n^\sigma u)_{++}$$

$$+ \sum_{\sigma=0}^{[s]-b} \sum_{\mu=0}^{\sigma_s-1} l_{\sigma\mu} (D_n^\sigma D_t^\mu u)_{++} + \sum_{\sigma=0}^{[s]-b} \sum_{\lambda=1}^{\sigma_s} l_\sigma^{(\lambda)} \cdot (D_n^\sigma v_\lambda)^+ \times \delta(t). \qquad (3.7)$$

Substituting this expression in (3.6), we obtain (3.4).

3. Let $b + \frac{1}{2} < s < m - \frac{1}{2}$. Then we transform the last double sum (with respect to σ and λ) in the right-hand side of (3.7) further, isolating the terms that contain $\omega_{\lambda k}(x') \times \delta(x_n, t)$. To do this, we first isolate the term with $\sigma = 0$ and interchange the order of summation with respect to σ and λ in the remaining double sum. This yields

$$\sum_{\sigma=0}^{[s]-b} \sum_{\lambda=1}^{\sigma_s} l_\sigma^{(\lambda)} (D_n^\sigma v_\lambda)^+ \times \delta(t) = \sum_{\lambda=1}^{s''} l_0^{(\lambda)} v_\lambda^+ \times \delta(t) + \sum_{\lambda=1}^{\lambda_0} \sum_{\sigma=1}^{\alpha_\lambda} l_\sigma^{(\lambda)} \cdot (D_n^\sigma v_\lambda)^+ \times \delta(t), \quad (3.8)$$

where $\lambda_0 = [(s + b - 1)/2b]$. Next, in the double sum with respect to σ and λ in the right-hand side of (3.8) we transform $(D_n^\sigma v_\lambda)^+$ by means of formula (3.2). This formula can be applied to those summands of $D_n^\sigma v_\lambda^+$ for which $s - 2b\lambda - \sigma + b + \frac{1}{2} > 0$, that is, $\sigma + 2b\lambda < s + b + \frac{1}{2}$. But since σ and λ in the double sum with respect to σ and λ of (3.8) are such that $s - \sigma - 2b\lambda + b > 0$, i.e., $\sigma + 2b\lambda < s + b$, the more so $\sigma + 2b\lambda < s + b + \frac{1}{2}$. Consequently,

$$\sum_{\lambda=1}^{\lambda_0} \sum_{\sigma=1}^{\alpha_\lambda} l_\sigma^{(\lambda)} (D_n^\sigma v_\lambda)^+ \times \delta(t)$$

$$= \sum_{\lambda=1}^{\lambda_0} \sum_{\sigma=1}^{\alpha_\lambda} l_\sigma^{(\lambda)} \left[D_n^\sigma v_\lambda^+ - \sum_{k=1}^{\sigma} D_n^{\sigma-k} \omega_{\lambda k} \times \delta(x_n) \right] \times \delta(t) \qquad (3.9)$$

$$= \sum_{\lambda=1}^{\lambda_0} \sum_{\sigma=1}^{\alpha_\lambda} l_\sigma^{(\lambda)} D_n^\sigma v_\lambda^+ \times \delta(t) - \sum_{\lambda=1}^{\lambda_0} \sum_{\sigma=1}^{\alpha_\lambda} \sum_{k=1}^{\sigma} l_\sigma^{(\lambda)} D_n^{\sigma-k} \omega_{\lambda k}(x') \times \delta(x_n, t).$$

Consider now the inner double sum over k and σ in the triple sum appearing in the right-hand side of (3.9). Let J_n and J_0 be the operators introduced above. Since

$$\sum_{\sigma=1}^{\alpha_\lambda}\sum_{k=1}^{\sigma} l_\sigma^{(\lambda)} D_n^{\sigma-k}\omega_{\lambda k}(x') \times \delta(x_n,t) = \sum_{k=1}^{\alpha_\lambda}\left(\sum_{\sigma=k}^{\alpha_\lambda} J_0^\lambda l_\sigma D_n^{\sigma-k}\right)\omega_{\lambda k}(x') \times \delta(x_n,t)$$

$$= \sum_{k=1}^{\alpha_\lambda} J_0^\lambda J_n^k\left(\sum_{\sigma=0}^{\alpha_\lambda} l_\sigma D_n^\sigma\right)\omega_{\lambda k}(x') \times \delta(x_n,t);$$

$$\sum_{\sigma=0}^{\alpha_\lambda} J_0^\lambda l_\sigma D_n^\sigma = J_0^\lambda\left(\sum_{\sigma=0}^{\alpha_\lambda} l_\sigma D_n^\sigma\right);$$

$$\sum_{\lambda=1}^{s''} J_0^\lambda l_0 v_\lambda^+(x) \times \delta(t) + \sum_{\lambda=1}^{\lambda_0}\sum_{\sigma=1}^{\alpha_\lambda} J_0^\lambda l_\sigma \cdot D_n^\sigma v_\lambda^+(x) \times \delta(t)$$

$$= \sum_{\lambda=1}^{s''}\sum_{\sigma=0}^{\alpha_\lambda} J_0^\lambda l_\sigma \cdot D_n^\sigma v_\lambda^+(x) \times \delta(t),$$

it follows that for the double sum over σ and λ in the right-hand side of (3.9) we get

$$\sum_{\sigma=0}^{[s]-b}\sum_{\lambda=1}^{\sigma_s} J_0^\lambda l_\sigma\left(D_n^\sigma v_\lambda\right)^+ \times \delta(t) = \sum_{\lambda=1}^{s''} J_0^\lambda\left(\sum_{\sigma=0}^{\alpha_\lambda} l_\sigma D_n^\sigma\right)v_\lambda^+(x) \times \delta(t)$$

$$- \sum_{\lambda=1}^{\lambda_0}\sum_{k=1}^{\alpha_\lambda} J_0^\lambda J_n^k\left(\sum_{\sigma=0}^{\alpha_\lambda} l_\sigma D_n^\sigma\right)\omega_{\lambda k}(x') \times \delta(x_n,t) \equiv I_1 + I_2. \qquad (3.10)$$

Consider the expressions for I_1 and I_2 separately. If in the inner sum over σ appearing in I_1 we add and subtract the terms that are necessary to complete the expression of $l(\cdot, D, D_t)$, we obtain

$$I_1 = \sum_{\lambda=1}^{s''} l^{(0,\lambda)} \cdot v_\lambda^+(x) \times \delta(t) - \sum_{\lambda=1}^{s''} J_0^\lambda\left(\sum_{\sigma=\alpha_\lambda+1}^{m} l_\sigma \cdot D_n^\sigma\right)v_\lambda^+(x) \times \delta(t) = I_{11} + I_{12}.$$

Now take I_{12} and break the sum over λ into two sums:

$$I_{12} = -\sum_{s_0<\lambda\le s''} J_0^\lambda\left(\sum_{\sigma=\alpha_\lambda+1}^{m} l_\sigma D_n^\sigma\right)v_\lambda^+(x) \times \delta(t)$$

$$-\sum_{\lambda=1}^{s_0}\sum_{\sigma=\alpha_\lambda+1}^{m} J_0^\lambda l_\sigma D_n^\sigma v_\lambda^+(x) \times \delta(t) = I_{12}^{(1)} + I_{12}^{(2)}.$$

Note that if $s_0 = s''$, then $I_{12}^{(1)} \equiv 0$.

Further, using (3.2) rewrite the expression for $I_{12}^{(2)}$ as

$$I_{12}^{(2)} = -\sum_{\lambda=1}^{s_0}\sum_{\sigma=\alpha_\lambda+1}^{m} J_0^\lambda\big(l_\sigma(\cdot,D',D_t)\big)\cdot D_n^{\sigma-s_\lambda}\cdot D_n^{s_\lambda}v_\lambda^+(x)\times\delta(t)$$

$$-\sum_{\lambda=1}^{s_0}\sum_{\sigma=\alpha_\lambda+1}^{m} J_0^\lambda\big(l_\sigma(\cdot,D',D_t)\big)\cdot D_n^{\sigma-s_\lambda}\cdot \big(D_n^{s_\lambda}v_\lambda\big)^+\times\delta(t)$$

$$-\sum_{\lambda=1}^{s_0}\sum_{k=1}^{s_\lambda} J_0^\lambda J_n^k\bigg(\sum_{\sigma=\alpha_\lambda+1}^{m} l_\sigma D_n^\sigma\bigg)\omega_{\lambda k}(x')\times\delta(x_n,t)=\Sigma_1+\Sigma_2.$$

Adding and subtracting the sum over σ from 0 to α_λ in the expression for Σ_2, we obtain

$$\Sigma_2 = -\sum_{\lambda=1}^{s_0}\sum_{k=1}^{s_\lambda} l^{(k,\lambda)}\omega_{\lambda k}(x')\times\delta(x_n,t)$$

$$+\sum_{\lambda=1}^{s_0}\sum_{k=1}^{s_\lambda} J_0^\lambda J_n^k\bigg(\sum_{\sigma=0}^{\alpha_\lambda} l_\sigma D_n^\sigma\bigg)\omega_{\lambda k}(x')\times\delta(x_n,t)\equiv\Sigma_{21}+\Sigma_{22}.$$

Now rewrite the expression for Σ_{22} in the form

$$\Sigma_{22} = \sum_{k+2b\lambda<[s]+b} J_0^\lambda J_n^k\bigg(\sum_{\sigma=0}^{\alpha_\lambda} l_\sigma D_n^\sigma\bigg)\omega_{\lambda k}(x')\times\delta(x_n,t)$$
$$+ \sum_{[s]+b<k+2b\lambda<s'+b} J_0^\lambda J_n^k\bigg(\sum_{\sigma=0}^{\alpha_\lambda} l_\sigma D_n^\sigma\bigg)\omega_{\lambda k}(x')\times\delta(x_n,t). \tag{3.11}$$

In the second sum in the right-hand side of (3.11) we have $k+2b\lambda>[s]+b$, i.e., $k>[s]-2b\lambda+b=\alpha_\lambda$. Hence, $J_n^k\big(\sum_{\sigma=0}^{\alpha_\lambda} l_\sigma D_n^\sigma\big)=0$ and the second sum in (3.11) is identically equal to zero. Next, it is readily seen that the first sum in the right-hand side of (3.11) equals exactly $-I_2$. Consequently,

$$I = \sum_{\lambda=1}^{s''} l^{(0,\lambda)}v_\lambda^+(x)\times\delta(t) - \sum_{\lambda=1}^{s_0}\sum_{k=1}^{s_\lambda} l^{(k,\lambda)}\omega_{\lambda k}(x')\times\delta(x_n,t)$$

$$- \sum_{s_0<\lambda\le s''} J_0^\lambda\bigg(\sum_{\sigma=\sigma_\lambda+1}^{m} l_\sigma D_n^\sigma\bigg)v_\lambda^+(x)\times\delta(t) \tag{3.12}$$

$$- \sum_{\lambda=1}^{s_0}\sum_{\sigma=\alpha_\lambda+1}^{m} J_0^\lambda l_\sigma D_n^{\sigma-s_\lambda}\big(D_n^{s_\lambda}v_\lambda\big)^+\times\delta(t)$$

Finally, let us transform the two remaining sums in the right-hand side of (3.12) that contain the operators J_0^λ. Since

$$J_0^\lambda \left(\sum_{\sigma=\alpha_\lambda+1}^{m} l_\sigma D_n^\sigma \right) = J_0^\lambda \left(\sum_{\sigma=\alpha_\lambda+1}^{m} l_\sigma D_n^{\sigma-\alpha_\lambda-1} \right) D_n^{\alpha_\lambda+1}$$

$$= J_0^\lambda J_n^{\alpha_\lambda+1} l \cdot D_n^{\alpha_\lambda+1} \equiv l^{(\alpha_\lambda+1,\lambda)} \cdot D_n^{\alpha_\lambda+1},$$

$$J_0^\lambda \left(\sum_{\sigma=\alpha_\lambda+1}^{m} l_\sigma \cdot D_n^{\sigma-s_\lambda} \right) = J_0^\lambda \left(\sum_{\sigma=\alpha_\lambda+1}^{m} l_\sigma D_n^{\sigma-\alpha_\lambda-1} \right) D_n^{\alpha_\lambda+1-s_\lambda}$$

$$= J_0^\lambda J_n^{\alpha_\lambda+1} l \cdot D_n^{\alpha_\lambda+1-s_\lambda} \equiv l^{(\alpha_\lambda+1,\lambda)}(\cdot, D, D_t) D_n^{\alpha_\lambda+1-s_\lambda},$$

and since $\alpha_\lambda + 1 - s_\lambda = 1 + [s] - s'$, it follows that

$$I = \sum_{\lambda=1}^{s''} l^{(0,\lambda)} v_\lambda^+(x) \times \delta(t) - \sum_{\lambda=1}^{s_0} \sum_{k=1}^{s_\lambda} l^{(k,\lambda)} \omega_{\lambda k}(x') \times \delta(x_n, t)$$

$$- l^{(\alpha_{s''}+1,s'')} \cdot D_n^{\alpha_{s''}+1} v_{s''}^+(x) \times \delta(t) \tag{3.13}$$

$$- \sum_{\lambda=1}^{s_0} l^{(\alpha_\lambda+1,\lambda)} D_n^{1+[s]-s'} \left(D_n^{s_\lambda} v_\lambda(x) \right)^{\,|} \times \delta(t).$$

Equalities (3.8) and (3.11)–(3.13) yield (3.4). If, in addition, $s_0 = s''$, then the last term in (3.5) is absent.

III.3.4. Corollaries. Commutation formulas for differential and truncation operators. Green formulas

The following remark plays an essential role in the sequel: in the case of sufficiently smooth functions $v(x) \in H^s(\mathbb{R}^n)$, $s > \alpha$, $\Phi(x',t) \in \mathcal{H}^s(E^n,\gamma)$, $s > q$, $u(x,t) \in \mathcal{H}^s(\mathbb{R}^{n+1},\gamma)$, $s > m$, formulas (3.1), (3.3) and (3.4) become, respectively,

$$\mathcal{P}(x,D)v^+(x) - \sum_{k=1}^{\alpha} \mathcal{P}^{(k)} \omega_k(x') \times \delta(x_n) = \left(\mathcal{P}(x,D)v \right)^+, \tag{3.14}$$

$$b(x',t,D',D_t)\Phi^+(x',t) - \sum_{\lambda=1}^{[q/2b]} b^{(\lambda)} \Phi_\lambda(x') \times \delta(t) = (b(x',t,D',D_t)\Phi)^+, \tag{3.15}$$

$$l(\cdot,D,D_t)u_{++}(x,t) - \sum_{k=1}^{m} l^{(k,0)}(\cdot,D,D_t)u_k^+(x',t) \times \delta(x_n)$$

$$- \sum_{\lambda=1}^{[m/2b]} l^{(0,\lambda)}(\cdot,D,D_t)v_\lambda^+(x) \times \delta(t) \tag{3.16}$$

$$+ \sum_{k+2b\lambda<m} l^{(k,\lambda)}(\cdot,D,D_t)\omega_{\lambda k}(x') \times \delta(x_n,t) = \left(l(\cdot,D,D_t)u \right)_{++}.$$

In a similar way one can prove the following formula:

$$l(\cdot, D, D_t)u_{++}(x,t) + \sum_{k=1}^{m} l^{(k,0)}(\cdot, D, D_t)u_k^+(x',t) \times \delta(x_n)$$

$$- \sum_{\lambda=1}^{[m/2b]} l^{(0,\lambda)}(\cdot, D, D_t)v_\lambda(x) \times \delta(t)$$

$$- \sum_{k+2b\lambda\leq m} l^{(k,\lambda)}(\cdot, D, D_t)\omega_{\lambda k}(x') \times \delta(x_n,t) = \theta(-x_n)\theta(t)l(\cdot, D, D_t)u. \qquad (3.16')$$

Formulas (3.14)–(3.16) are explicit commutation formulas for the operators $\mathcal{P}$, b, and l and the operators of multiplication by the characteristic functions $\theta(x_n)$, $\theta(t)$, and $\theta(x_n,t)$, respectively.

Let us also point out, for the sake of clarity, that each of the formulas (3.14), (3.15), and (3.16) (in the case of operators with constant coefficients) is an alterntive form of the generalized classical Leibniz formula for calculating $\mathcal{P}\big(\theta(x_n)v(x)\big)$, $b(\cdot, D', D_t)\big(\theta(t)\Phi(x',t)\big)$, $l\big(\theta(x_n,t)u(x,t)\big)$ in the sense of distributions (see [106, p.19]).

For an arbitrary operator $\mathcal{P}(D)$ with constant coefficients the Leibniz formula has the form

$$\mathcal{P}(D) = \sum_{\alpha} \frac{1}{\alpha!}(D^\alpha a)\big(\mathcal{P}^{(\alpha)}(D)v\big), \qquad a \in C^\infty,\ v \in \mathcal{D}',\ D = -i\partial x, \qquad (3.17)$$

where

$$\mathcal{P}^{(\alpha)}(\eta) = \frac{\partial^\alpha \mathcal{P}(\eta)}{\partial\eta_1^{\alpha_1}\ldots\partial\eta_n^{\alpha_n}}, \qquad \eta = (\eta_1,\ldots,\eta_n).$$

In point of fact this formula can be extended to all functions a and distributions v for which it remains meaningful.

The usefulness of equalities (3.14)–(3.16) resides in the fact that, if one regards them as expressions for the corresponding Green formulas in terms of the theory of distributions, they lie at the basis of the definition of generalized (weak) solutions of the corresponding equation in the closure of the underlying domain. How this can be done will be shown in Section III.6. As concerns formulas (3.1), (3.3), and (3.4), it is natural to regard them as formulas of "partial commutation" of differential and truncation operators in the case of insufficiently smooth functions $v(x)$, $\Phi(x',t)$, and $u(x,t)$.

Note that if $l(x,t,D,D_t)$ does not contain the mixed derivatives $D_t^\lambda D_n^k$, then (3.4) does not contain terms with $\omega_{\lambda k}(x')$. Also, if $u(x,t)$ and $\Phi(x',t)$ vanish at $t=0$ together with all their t-derivatives up to a certain order, then (3.3) [resp., (3.4)] does not involve the terms with $\Phi_\lambda(x')$ [resp., $v_\lambda(x)$, $\omega_{\lambda k}(x')$].

III.4. On equivalent norms in $\mathcal{H}^s(\mathbb{R}_+^{n+1},\gamma)$, $\mathcal{H}^s(E_+^{n+1},\gamma)$, and $H^s(\mathbb{R}_+^n)$, $s \geq 0$

III.4.1. Equivalents norms defined by truncations

The spaces $\mathcal{H}^s(\Omega,\gamma)$ (where Ω is either $\mathbb{R}_+^{n+1} = \{(x,t) : x_n \geq 0\}$ or $E_+^{n+1} = \{(x,t) : t > 0\}$) were equipped above with the natural quotient norm $\|u,\Omega,\gamma\|_s = \inf_U \|U,\mathbb{R}^{n+1},\gamma\|_s$, $U \in \mathcal{H}^s(\mathbb{R}^{n+1},\gamma)$, $U = u$ in Ω. Defined by using smooth extensions U, this norm is not always convenient in practice, since we will deal, as a rule, with non-smooth extensions, for example, with continuation by zero (truncation). For this reason we introduce below, by means of truncations, other norms that are more convenient for applications.

Let $\mathcal{A}(i\xi,p)$ be a given function, sufficiently smooth in $\xi \in \mathbb{R}^n$ and $p = \gamma+i\xi_0$, which grows in (ξ,p) not faster than a polynomial. The equality

$$\mathcal{A}(D,D_t)u(x,t) = (2\pi)^{-n-1} \int_{\mathbb{R}^{n+1}} e^{ix\cdot\xi+pt}\hat{u}(\xi,p)\mathcal{A}(i\xi,p)\,d\xi d\xi_0$$

defines a pseudodifferential operator (PDO) $\mathcal{A}(D,D_t)$ (with *symbol* $\mathcal{A}(i\xi,p)$). Let $\Lambda(D,D_t)$, $\Lambda_1(D',D_t)$, and $\Lambda_{++}(D,D_t)$ be the PDOs with the symbols $\Lambda(i\xi,p) = \big(|\xi|^{2b}+p\big)^{1/2b}$, $\Lambda_1(i\xi',p) = \big(|\xi'|^{2b}+p\big)^{1/2b}$, and $\Lambda_{++}(i\xi,p) = \xi_n-i|\xi'|-(ip)^{1/2b}e^{i\pi/4}$, respectively, where $z^\lambda = \exp(\lambda \ln z)$, $-\pi < \arg z < \pi$ (see [9]).

Denote by $\widehat{\mathcal{H}}_+^s(\mathbb{R}^{n+1},\gamma)$, $\widehat{\mathcal{H}}_+^s(E^{n+1},\gamma)$, and $\widehat{\mathcal{H}}_{++}^s(\mathbb{R}^{n+1},\gamma)$ the images under the Fourier transformation in x and the Laplace transformation in t of the spaces $\mathcal{H}_+^s(\mathbb{R}^{n+1},\gamma)$, $\mathcal{H}_+^s(E^{n+1},\gamma)$, and $\mathcal{H}_{++}^s(\mathbb{R}^{n+1},\gamma)$ ($\gamma > 0$). By the definition of the spaces involved and by the properties of the Fourier transform of distributions with support in cones [10], $\widehat{\mathcal{H}}_+^s(\mathbb{R}^{n+1},\gamma)$, $\widehat{\mathcal{H}}_+^s(E^{n+1},\gamma)$, and $\widehat{\mathcal{H}}_{++}^s(\mathbb{R}^{n+1},\gamma)$ consist of the ordinary functions $\hat{u}_+(\xi,p)$, $\hat{u}^+(\xi,p)$, and $\hat{u}_{++}(\xi,p)$ with the property that for almost all values of the remaining arguments they are analytic functions of ξ_n, p, and (ξ_n,p) in the domains $\operatorname{Im}\xi_n < 0$, $\operatorname{Re}p > \gamma$ and $A_{\gamma,0} = \big\{(\xi_n,p) : \operatorname{Im}\xi_n < 0,\ \operatorname{Re}p > \gamma\big\}$, respectively. It can be easily seen that the operators $\Lambda^\alpha \cdot \Lambda_1^\beta$ and $\Lambda_{++}^\alpha \cdot \Lambda_1^\beta$ effect an isomorphism of the spaces $\mathcal{H}^{s,r}(\mathbb{R}^{n+1},\gamma)$ and $\mathcal{H}^{s-\alpha,r-\beta}(\mathbb{R}^{n+1},\gamma)$. Moreover (and this is the main reason why we consider these operators), thanks to the analyticity of $\Lambda(i\xi,p)$ in the variable p for $\operatorname{Re}p > 0$, the operator $\Lambda^\alpha(D,D_t)$ effects an isomorphism of the subspaces $\mathcal{H}_+^s(E^{n+1},\gamma)$ and $\mathcal{H}_+^{s-\alpha}(E^{n+1},\gamma)$, and, thanks to the analyticity of $\Lambda_{++}^\alpha(i\xi',i\xi_n,p)\Lambda_1^\beta(i\xi',p)$ in the variables $(\xi_n,p) \in A_{\gamma,0}$, the operator $\Lambda_{++}^\alpha(D,D_t) \cdot \Lambda_1^\beta(D',D_t)$ effects an isomorphism of the subspaces $\mathcal{H}_+^{s,r}(\mathbb{R}^{n+1},\gamma)$ and $\mathcal{H}_+^{s-\alpha,r-\beta}(\mathbb{R}^{n+1},\gamma)$, $\mathcal{H}_{++}^{s,r}(\mathbb{R}^{n+1},\gamma)$ and $\mathcal{H}_{++}^{s-\alpha,r-\beta}(\mathbb{R}^{n+1},\gamma)$ for any $\gamma > 0$.

Hence, if $U(x,t) \in \mathcal{H}^s(\mathbb{R}^{n+1},\gamma)$ is a smooth extension of $u_0(x,t)$, then for any $s \geq 0$, $s \notin \mathbb{Z}_1$ the expression $\Lambda_{++}^{s-s'}(D,D_t)\theta(x_n)\Lambda_{++}^{s'}(D,D_t)U(x,t)$ does not depend $U(x,t)$, and $\Lambda_{++}^{s-s'}\theta(x_n)\Lambda_{++}^{s'}$ is a bounded operator that maps the space $\mathcal{H}^s(\mathbb{R}^{n+1},\gamma)$ into the subspace $\mathcal{H}_+^0(\mathbb{R}^{n+1},\gamma)$.

The following assertion holds true.

Lemma III.8. *For any positive non-half-integer s the norms $\left\|u_0(x,t),\mathbb{R}_+^{n+1},\gamma\right\|_s$ and $\left\|\Lambda_{++}^{s-s'}\theta(x_n)\Lambda_{++}^{s'}U,\mathbb{R}^{n+1},\gamma\right\|_0$ are equivalent. Here $\theta(x_n)\Lambda_{++}^{s'}u_0(x,t)$ is understood as the continuation of the function $\Lambda_{++}^{s'}u_0$ by zero for $x_n < 0$.*

Proof: By the non-degeneracy of the symbol $\Lambda_{++}(i\xi,p)$ for $\xi \in \mathbb{R}^n$ and $\operatorname{Re}p \geq \gamma > 0$, the norm in $\mathcal{H}^s(\mathbb{R}^{n+1},\gamma)$ is defined by $\left\|U(x,t),\mathbb{R}^{n+1},\gamma\right\|_s = \left\|\Lambda_{++}^s U, \mathbb{R}^{n+1},\gamma\right\|_0$ and the norm in $\mathcal{H}^s(\mathbb{R}_+^{n+1},\gamma)$ as the quotient norm $\left\|u_0(x,t),\mathbb{R}_+^{n+1},\gamma\right\|_s = \inf_U \left\|U,\mathbb{R}^{n+1},\gamma\right\|_s$, where the infimum is taken over all $U(x,t)$ that are equal to $u_0(x,t)$ in $\mathbb{R}_+^{n+1}$. Therefore,

$$\left\|u_0(x,t),\mathbb{R}_+^{n+1},\gamma\right\|_s = \inf_U \left\|U,\mathbb{R}^{n+1},\gamma\right\|_s = \inf_U \left\|\Lambda_{++}^{s-s'}\Lambda_{++}^{s'}U,\mathbb{R}^{n+1},\gamma\right\|_0$$

$$= \inf_U \left\|\Lambda_{++}^{s'}U,\mathbb{R}^{n+1},\gamma\right\|_{s-s'}. \tag{4.1}$$

Since the operator $\Lambda_{++}^{s'}$ maps the space $\mathcal{H}^s(\mathbb{R}^{n+1},\gamma)$ isomorphically onto the space $\mathcal{H}^{s-s'}(\mathbb{R}^{n+1},\gamma)$, it follows that the subset of functions $E_s = \left\{U \in \mathcal{H}^s(\mathbb{R}^{n+1},\gamma), U = u_0 \text{ on } \mathbb{R}_+^{n+1}\right\}$ is isomorphically mapped onto $E_{s-s'} = \left\{\Lambda_{++}^{s'}U \in \mathcal{H}^{s-s'}(\mathbb{R}^{n+1},\gamma), \Lambda_{++}^{s'}U = \Lambda_{++}^{s'}u_0 \text{ on } \mathbb{R}_+^{n+1}\right\}$. Consequently,

$$\inf_{U\in\mathcal{H}^s} \left\|\Lambda_{++}^{s'}U,\mathbb{R}^{n+1},\gamma\right\|_{s-s'} = \inf_{\Lambda_{++}^{s'}U} \left\|\Lambda_{++}^{s'}U,\mathbb{R}^{n+1},\gamma\right\|_{s-s'}$$

$$\leq \left\|\theta(x_n)\Lambda_{++}^{s'}U,\mathbb{R}^{n+1},\gamma\right\|_{s-s'} \tag{4.2}$$

$$= \left\|\Lambda_{++}^{s-s'}\theta(x_n)\Lambda_{++}^{s'}u_0(x,t),\mathbb{R}^{n+1},\gamma\right\|_0.$$

Relations (4.1) and (4.2) yield

$$\left\|u_0(x,t),\mathbb{R}_+^{n+1},\gamma\right\|_s \leq \left\|\Lambda_{++}^{s-s'}\theta(x_n)\Lambda_{++}^{s'}u_0(x,t),\mathbb{R}^{n+1},\gamma\right\|_0. \tag{4.3}$$

On the other hand, by Lemma III.2,

$$\left\|\Lambda_{++}^{s-s'}\theta(x_n)\Lambda_{++}^{s'}u_0,\mathbb{R}^{n+1},\gamma\right\|_0 = \left\|\theta(x_n)\Lambda_{++}^{s'}u_0,\mathbb{R}^{n+1},\gamma\right\|_{s-s'}$$

$$= \left\|\theta(x_n)\Lambda_{++}^{s'}U,\mathbb{R}^{n+1},\gamma\right\|_{s-s'} \leq \left\|\theta(x_n)\right\|_{s-s'} \cdot \left\|U,\mathbb{R}^{n+1},\gamma\right\|_s, \tag{4.4}$$

where $\left\|\theta(x_n)\right\|_{s-s'}$ is the norm of the operator of multiplication by $\theta(x_n)$ in $\mathcal{H}^{s-s'}$. Taking in (4.4.) the infimum over $U \in E_s$, we obtain

$$\left\|\Lambda_{++}^{s-s'}\theta(x_n)\Lambda_{++}^{s'}u_0,\mathbb{R}^{n+1},\gamma\right\|_0 \leq \left\|\theta(x_n)\right\|_{s-s'} \cdot \left\|u_0(x,t),\mathbb{R}_+^{n+1},\gamma\right\|_s. \tag{4.5}$$

Inequalities (4.3) and (4.5) prove the assertion of Lemma III.8.

Proceeding in much the same manner one can prove

Lemma III.9. *For any positive s such that $s/2b$ is not a half-integer, the norms* $\left\|\Lambda^{s-2bs''}\theta(t)\Lambda^{2bs''}u_0(x,t), \mathbb{R}^{n+1}, \gamma\right\|_0$ *and* $\left\|u_0(x,t), E^{n+1}_+, \gamma\right\|_s$ *are equivalent.*

In the analysis of model parabolic problems the typical situation is that one deals with a distribution with support in some closed half-space, for example, $u_+(x,t)$ with $\operatorname{supp} u_+(x,t) \subseteq \overline{\mathbb{R}}^{n+1}_+$ [resp., $u^+(x,t)$ with $\operatorname{supp} u^+(x,t) \subseteq \overline{E}^{n+1}_+$] and it is known a priori that its restriction to the corresponding open half-space, $u^0(x,t)$ [resp., $u_0(x,t)$], is an ordinary function belonging to the space $\mathcal{H}^s(\mathbb{R}^{n+1}_+, \gamma)$ [resp., $\mathcal{H}^s(E^{n+1}_+, \gamma)$] for some $s \geq 0$. Using Lemmas III.8 and III.9 and the commutation formulas for the differentiation operators and the operators of multiplication by $\theta(x_n)$ and $\theta(t)$, we can write the formulas for calculating the norms of the restrictions $u^0(x,t)$ [resp., $u_0(x,t)$] in $\mathcal{H}^s(\mathbb{R}^{n+1}_+, \gamma)$ [resp., $\mathcal{H}^s(E^{n+1}_+, \gamma)$].

Thus, let P_n^+ and P_0^+ be the operators whose action is to remove the terms (distributions) that are supported in the hyperplanes $x_n = 0$ and $t = 0$, respectively (in the Fourier and the Laplace transform picture—to remove the terms that are polynomial in ξ_n and $\dot{p}$). Let $u_+(x,t)$, $u^+(x,t) \in \mathcal{H}^\sigma(\mathbb{R}^{n+1}, \gamma)$ $(|\sigma| < \infty)$ be the continuations of $u^0(x,t)$ and $u_0(x,t)$ from $\mathbb{R}^{n+1}_+$ and respectively E^{n+1}_+ by zero for $x_n < 0$, $t < 0$. Clearly, $u_+(x,t) = u^0_+(x,t) + Q\bigl(x',t,\delta(x_n)\bigr)$, $u^+(x,t) = u^+_0(x,t) + R\bigl(x, \delta(t)\bigr)$, where Q and R are finite linear combinations of the corresponding δ-functions and their derivatives. Then

$$\Lambda^{s'}_{++}u^0(x,t) = P_n^+ \Lambda^{s'}_{++}(D, D_t)u_+(x,t), \tag{4.6}$$

$$\theta(t)\Lambda^{2bs''}u_0(x,t) = P_0^+ \Lambda^{2bs''}(D, D_t)u^+(x,t), \tag{4.7}$$

Lemmas III.8, III.9 and equalities (4.6), (4.7) yield the following assertion.

Lemma III.10. *Let $u_+(x,t)$ and $u^+(x,t)$ be distributions of finite order supported in $\overline{\mathbb{R}}^{n+1}_+$ and $\overline{E}^{n+1}_+$, and let their restrictions $u^0(x,t)$ and $u_0(x,t)$ to $\mathbb{R}^{n+1}_+$ and E^{n+1}_+ belong to the spaces $\mathcal{H}^s(\mathbb{R}^{n+1}_+, \gamma)$ and $\mathcal{H}^s(E^{n+1}_+, \gamma)$, respectively, $\gamma \geq \gamma_0 > 0$. Then the following norms are equivalent:*

$$\left\|u^0(x,t), \mathbb{R}^{n+1}_+, \gamma)\right\|_s \sim \left\|\Lambda^{s-s'}_{++} P_n^+ \Lambda^{s'}_{++} u_+(x,t), \mathbb{R}^{n+1}, \gamma)\right\|_0, \tag{4.8}$$

for any $s \geq 0$, $s \notin \mathbb{Z}_1$, and

$$\left\|u_0(x,t), E^{n+1}_+, \gamma)\right\|_s \sim \left\|\Lambda^{s-2bs''} P_0^+ \Lambda^{2bs''} u^+(x,t), \mathbb{R}^{n+1}, \gamma)\right\|_0, \quad s \notin \mathbb{Z}_{2b}. \tag{4.9}$$

Similar formulas hold for the distributions $v_+(x)$ supported in $\overline{\mathbb{R}}^n_+$, where $v_+(x) \in \mathcal{H}^\sigma(\mathbb{R}^n)$ and σ is an arbitrary finite number. The norm $\left\|v(x), \mathbb{R}^n_+\right\|_s$ is equivalent to the norm $\left\|\Lambda^{s-s'}_0(D)P_n^+\Lambda^{s'}_0(D)v_+(x), \mathbb{R}^n\right\|_\sigma$, with $s \geq 0$, $s \notin \mathbb{Z}_1$, where $\Lambda_0(D)$ is the pseudodifferential operator with the symbol $\Lambda_0(i\xi) = \xi_n - i\bigl(1 + |\xi'|^2\bigr)^{1/2}$, $i = \sqrt{-1}$.

III.4.2. Restriction of distributions to an open half-space

Let $u_+(x,t)$ and $u^+(x,t)$ be distributions supported in $\overline{\mathbb{R}}_+^{n+1}$ and $\overline{E}_+^{n+1}$, respectively. Let us clarify when their restrictions to the open half-spaces $\mathbb{R}_+^{n+1}$ and E_+^{n+1} belong to the spaces $\mathcal{H}^s(\mathbb{R}_+^{n+1},\gamma)$ and $\mathcal{H}^s(E_+^{n+1},\gamma)$, respectively.

Lemma III.11. *Let $u_+(x,t)$ be a distribution of finite order and assume that* $\operatorname{supp} u_+(x,t) \subseteq \overline{\mathbb{R}}_+^{n+1}$. *Then $u^0(x,t) \in \mathcal{H}^s(\mathbb{R}_+^{n+1},\gamma)$, with $s \geq 0$, $s \notin \mathbb{Z}_1$, if and only if*

$$
\Lambda_{++}^{s'}(D,D_t)u_+(x,t) = f_+(x,t) + \sum_{\nu=0}^{s'-1} a_{s'-\nu-1}(x',t)D_n^{\nu}\delta(x_n)
$$
$$
+ \Lambda_{++}^{s'}(D,D_t)Q\big(\delta(x_n)\big),
$$
(4.10)

where $f_+(x,t) \in \mathcal{H}_+^{s-s'}(\mathbb{R}_+^{n+1},\gamma)$, $a_{s'-\nu-1}(x',t) \in \mathcal{H}^{s-s'+\nu+1/2}(E^n,\gamma)$, $Q\big(\delta(x_n)\big)$ $= \sum_\sigma \beta_\sigma(x',t) \times D_n^{\sigma-1}\delta(x_n)$. Moreover, $f_+(x,t) = \big(\Lambda_{++}^{s'}(D,D_t)u^0(x,t)\big)^+$.

Proof: The necessity of (4.10) is a straightforward consequence of the commutation formula for the operators $\Lambda_{++}(D,D_t)$ and $\theta(x_n)$, while its sufficiency follows from (4.8).

A similar assertion holds in the case of the half-space E_+^{n+1}.

Lemma III.12. *Let $u^+(x,t)$ be a distribution of finite order such that $\operatorname{supp} u^+(x,t)$ $\subseteq \overline{E}_+^{n+1}$. Then $u_0(x,t) \in \mathcal{H}^s(E_+^{n+1},\gamma)$, $s > 0$, $s \notin \mathbb{Z}_{2b}$ if and only if*

$$
\Lambda^{2bs''}(D,D_t)u^+(x,t) = f^+(x,t) + \sum_{\nu=0}^{s''-1} c_{s''-\nu-1}(x)D_t^{\nu}\delta(t) + \Lambda^{2bs''}(D,D_t)Q_1\big(\delta(t)\big),
$$

where $f^+(x,t) \in \mathcal{H}_+^{s-2bs''}(E^{n+1},\gamma)$, $c_{s''-\nu-1}(x) \in H^{s-2bs''+2b\nu+b}(\mathbb{R}^n)$, and the expression $Q_1\big(\delta(t)\big)$ is similar to $Q\big(\delta(x_n)\big)$.

A similar assertion is valid for distributions $v_+(x)$ supported in $\overline{\mathbb{R}}_+^n$.

III.5. The spaces $\widetilde{H}^s$ and $\widetilde{\mathcal{H}}^s$

In this section we present the definition and the principal properties of the spaces $\widetilde{H}_{(k)}^s(G)$, which are used in the theory of generalized solutions of elliptic boundary value problems. Then we introduce and study similar functional spaces $\widetilde{\mathcal{H}}^s$, which play an essential role in building a theory dealing with well-posedness of linear parabolic boundary value problems in a family of functional spaces that contain distributions of any finite order.

III.5.1. The spaces $\widetilde{H}^s_{(K)}(G)$ and $\widetilde{\mathcal{H}}^{s,r}_{(Q)}(E^{n+1}_+,\gamma)$

Let G be a domain of $\mathbb{R}^n$ with smooth boundary $\Gamma = \partial G$. Let ν be the inner normal to Γ at the point $x' \in \Gamma$ and let K be a given set of positive integers k, $k_0 = \max k \in K$. For any $v(x) \in C^\infty(\overline{G})$ we denote $v_0(x) = v(x)\big|_{\overline{G}}$, $\omega_k(x') = D_\nu^{k-1} v(x)\big|_\Gamma$.

Definition III.1. *Let $s \in \mathbb{R}^1$ be an arbitrary number, assumed not to be a half-integer when $s \in (0, k_0)$. Then the space $\widetilde{H}^s_{(K)}(G)$ is the completion of $C^\infty(\overline{G})$ with respect to the norm (see [83–85])*

$$\big|\{v(x), G\}\big|^2_{s,(K)} = \big\|[v_0(x), G]\big\|^2_s + \sum_{k\in K} \big\langle\!\big\langle \omega_k(x'), \Gamma \big\rangle\!\big\rangle'^2_{s-k+1/2}. \tag{5.1}$$

For the remaining values of s the space $\widetilde{H}^s_{(K)}(G)$ is the interpolation space of the couple $\widetilde{H}^{s-\varepsilon}_{(K)}(G)$, $\widetilde{H}^{s+\varepsilon}_{(K)}(G)$, $|\varepsilon| < 1/2$, constructed by the complex interpolation method [4, 102].

Obviously, if $s > k_0 - 1/2$, then $\widetilde{H}^s_{(K)}(G)$ and $H^s(G)$ coincide and their norms are equivalent.

To each function $v(x) \in C^\infty(\overline{G})$ we assign the vector $V(x) = \big(v_0(x), \omega_k(x'),$ $k \in K\big)$ and we consider the mapping $\chi_0 : v(x) \mapsto V(x) = \chi_0 v(x)$ from $C^\infty(\overline{G})$ into the space $K^s(G) = H^s(G) \times \prod_{k\in K} H^{s-k+1/2}(\Gamma)$.

Denote by $\widetilde{K}^s(G)$ the subspace of $K^s(G)$ consisting of the vectors $V(x)$ whose coordinates satisfy on Γ the compatibility conditions $D_\nu^{k-1} v_0(x)\big|_\Gamma = \omega_k(x')$, for all $k \le s'$.

Lemma III.13. *For any $s \in \mathbb{R}^1$ the closure of the mapping χ_0 with respect to the $\widetilde{H}^s_{(K)}(G)$-norm effects an isomorphism of the spaces $\widetilde{H}^s_{(K)}(G)$ and $\widetilde{K}^s(G)$.*

Proof: We carry out the proof for the case $s \le k_0$; for $s > k_0$ it is similar and in fact easier. Let $s \le k_0$. Since $C^\infty(\overline{G})$ is dense in $H^s(G)$, there exists a sequence $v_0^{(n)}(x) \in C^\infty(\overline{G})$ that converges to $v_0(x)$ in $H^s(G)$. Then the sequence $\big\{D_\nu^{k-1} v_0^{(n)}(x)\big|_\Gamma\big\}_{n=1}^\infty$ converges to $\omega_k(x')$ in $H^{s-k+1/2}(\Gamma)$ for $k < s+\frac{1}{2}$; if $k \ge s+\frac{1}{2}$, this sequence, generally speaking, does not converge.

Let $\big\{\omega_k^{(n)}(x')\big\}_{n=1}^\infty$ $\big(\omega_k^{(n)}(x') \equiv D_\nu^{k-1} v_0^{(n)}(x)\big|_\Gamma$ for $s - k + 1/2 > 0$ and $\omega_k^{(n)}(x') \equiv 0$ for $k \notin K\big)$ be a sequence of infinitely differentiable functions that converges to $\omega_k(x')$ in $H^{s-k+1/2}(\Gamma)$ for $k = 1, \ldots, k_0$. Assume that $v_1^{(n)}(x)$ is a solution of the Dirichlet problem $\Delta^{2k_0} v_1^{(n)}(x) = 0$ in G, such that $D_\nu^{k-1} v_1^{(n)}(x)\big|_\Gamma = \omega_k^{(n)}(x')$, $k = 1, \ldots, k_0$. Here Δ is the Laplace operator. Thanks to the properties of solutions of elliptic equations [12, 85, 93, 114], $v_1^{(n)}(x) \in H^s(G) \cap C^\infty(G)$. Therefore, $v_0^{(n)}(x) - v_1^{(n)}(x) \in H^s_+(G)$. Since $C_0^\infty(G)$ is dense in $H^s_+(G)$, it follows that for every n there one can find a function $\psi^{(n)}(x) \in C_0^\infty(G)$ such that

$\big\|[v_0^{(n)}(x) - v_1^{(n)}(x) - \psi^{(n)}(x), G]\big\|_s < 1/n$, $n = 1, \dots$. Hence, the sequence $\{v^{(n)}(x) = v_1^{(n)}(x) + \psi^{(n)}(x)\}_{n=1}^{\infty}$ converges to $v_0(x)$ in $\widetilde{H}^s_{(K)}(G)$.

Now let us define a similar space in the anisotropic case. Assume that $u(x,t) \in C_0^{\infty}(E^{n+1})$, $v_\lambda(x) = D_t^{\lambda-1} u(x,0)$, $u_0(x,t) = u(x,t)\big|_{\overline{E}_+^{n+1}}$, and let Q be a given set of positive integers λ, $\lambda_0 = \max \lambda \in Q$.

Definition III.2. *Let $s \in \mathbb{R}^1$ be an arbitrary number, and let $r \geq 0$ be such that $(s+r)/2b$ is not an integer for $s \in (0, 2b\lambda_0)$. The space $\widetilde{\mathcal{H}}^{s,r}_{(Q)}(E_+^{n+1}, \gamma)$ is the completion of the set of functions $\{u(x,t) \in C_0^{\infty}(E_+^{n+1})\}$ with respect to the norm*

$$\big\|\{u(x,t), E_+^{n+1}, \gamma\}\big\|_{s,r,(Q)}^2 = \big\{\{u_0(x,t), E_+^{n+1}, \gamma\}\big\}_{s,r}^2 + \sum_{\lambda \in Q} \big\|[v_\lambda(x), \mathbb{R}^n]\big\|_{s+r-2b\lambda+b}^2.$$

$$(5.2)$$

For the remaining values of s the space $\widetilde{\mathcal{H}}^{s,r}_{(Q)}(E_+^{n+1}, \gamma)$ is the interpolation space of the couple $\widetilde{\mathcal{H}}^{s-\varepsilon,r}_{(Q)}(E_+^{n+1}, \gamma)$, $\widetilde{\mathcal{H}}^{s+\varepsilon,r}_{(Q)}(E_+^{n+1}, \gamma)$, $0 < \varepsilon < \frac{1}{2}$.

As above, to every function $u(x,t) \in C_{(0)}^{\infty}(E_+^{n+1})$ we assign the vector $U_0(x,t) = \big(u_0(x,t), v_\lambda(x), \lambda \in Q\big)$ and we consider the mapping $\chi_1 : u(x,t) \mapsto U_0(x,t) = \chi_1 u(x,t)$, acting from $C_0^{\infty}(E_+^{n+1})$ to the space

$$\mathcal{M}^{s,r} = \mathcal{H}^{s,r}(E_+^{n+1}, \gamma) \times \prod_{\lambda \in Q} H^{s+r-2b\lambda+b}(\mathbb{R}^n).$$

Denote by $\widetilde{\mathcal{M}}^{s,r}$ the subspace of $\mathcal{M}^{s,r}$ consisting of the vector-valued functions $U_0(x,t)$ whose components satisfy at $t = 0$ the compatibility conditions that follow from the theorems on traces: $D_t^{\lambda-1} u_0(x,0) = v_\lambda(x)$ for all λ with $2b\lambda < s+b$. The spaces $\widetilde{\mathcal{H}}^{s,0}_{(Q)}$ and $\mathcal{M}^{s,0}$ will be denoted simply by $\widetilde{\mathcal{H}}^s$ and $\mathcal{M}^s$.

Lemma III.14. *For any $s \in \mathbb{R}^1$ the closure of the mapping χ_1 with respect to the $\widetilde{H}^s_{(Q)}(E_+^{n+1}, \gamma)$-norm effects an isomorphism of the spaces $\mathcal{H}^s_{(Q)}(E_+^{n+1}, \gamma)$ and $\widetilde{\mathcal{M}}^s(E_+^{n+1}, \gamma)$.*

The proof of this lemma is analogous to that of Lemma III.13.

III.5.2.   The spaces $\widetilde{\mathcal{H}}^s_{(\kappa,\tau,P)}(\Omega)$ and $\widetilde{\mathcal{H}}^s_{(\kappa,\tau,P)}(\Omega_+, \gamma)$

Let κ, τ be positive integers. Let P be a given set of pairs of positive integers (k, λ) $(k \geq 1, \lambda \geq 1)$ and for each fixed λ denote by P_λ the set of all k such that $(k, \lambda) \in P$. For $u(x,t) \in C^{\infty}(\overline{\Omega}_+)$ we denote $u_0(x,t) = u(x,t)\big|_{\overline{\Omega}_+}$, $u_k(x',t) = D_\nu^{k-1} u(x,t)\big|_{\overline{S}_+}$, $v_\lambda(x) = D_t^{\lambda-1} u(x,0)\big|_{\overline{G}}$, $\omega_{\lambda k}(x') = D_\nu^{k-1} D_t^{\lambda-1} u(x,t)\big|_{\Gamma}$, $M = \max(2b\kappa, \tau, k + 2b\lambda)$ for $(k, \lambda) \in P$.

Definition III.3. *Let $s \in \mathbb{R}^1$ be an arbitrary number such that $s \overline{\in} \mathbb{Z}_{1,2b} \cap (0, M)$. The space $\widetilde{\mathcal{H}}^s_{(\kappa,\tau,P)}(\Omega_+, \gamma)$ is the completion of $C_0^\infty(\Omega_+)$ with respect to the norm*

$$\left\| u(x,t), \Omega_+, \gamma \right\|^2_{s,(\kappa,\tau,P)} = \left\| u_0, \Omega_+, \gamma \right\|^2_s + \sum_{k=1}^{\tau} \left\langle\!\left\langle u_k(x',t), S_+, \gamma \right\rangle\!\right\rangle^2_{s-k+1/2}$$

$$+ \sum_{\lambda=1}^{\kappa} \left\| [v_\lambda(x), G] \right\|^2_{s-2b\lambda+b} + \sum_{(k,\lambda)\in P} \left\langle\!\left\langle \omega_{\lambda k}(x'), \Gamma \right\rangle\!\right\rangle'^2_{s-k-2b\lambda+b+1/2}, \qquad (5.3)$$

where the sum corresponding to the choice $\kappa = 0$, $\tau = 0$, $P = \{\emptyset\}$ vanishes. For the remaining values of s the space $\widetilde{\mathcal{H}}^s_{(\kappa,\tau,P)}(\Omega_+, \gamma)$ is the interpolation space of the couple $\widetilde{\mathcal{H}}^{s-\varepsilon}_{(\kappa,\tau,P)}(\Omega_+, \gamma)$, $\widetilde{\mathcal{H}}^{s+\varepsilon}_{(\kappa,\tau,P)}(\Omega_+, \gamma)$ with $s \in \mathbb{Z}_{1,2b}$, $0 < |\varepsilon| < \frac{1}{2}$.

To each function $u(x,t) \in C_0^\infty(\overline{\Omega}_+)$, we assign the vector-valued function

$$U(x,t) = \big(u_0(x,t), u_1(x',t), \ldots, u_\tau(x',t), v_1(x), \ldots, v_\kappa(x), \omega_{\lambda k}(x'),$$

$$(k,\lambda) \in P, x' \in \Gamma\big)$$

and we consider the mapping $\chi : u(x,t) \mapsto U(x,t) = \chi u(x,t)$ from $C_0^\infty(\Omega_+)$ to the direct product

$$\mathcal{K}^s(\Omega_+, \gamma) = \mathcal{H}^s(\Omega_+, \gamma) \times \prod_{k=1}^{\tau} \mathcal{H}^{s-k+1/2}(S_+, \gamma)$$

$$\times \prod_{\lambda=1}^{\kappa} \mathcal{H}^{s-2b\lambda+b}(G) \times \prod_{(k,\lambda)\in P} \mathcal{H}^{s-2b\lambda-k+b+1/2}(\Gamma).$$

Denote by $\widetilde{\mathcal{K}}^s(\Omega_+, \gamma)$ the subspace of $\mathcal{K}^s(\Omega_+, \gamma)$ consisting of the vector-valued functions $U(x,t)$ whose components satisfy the compatibility conditions that follow from the Theorem II.4 on traces:

$$u_k(x',t) = D_\nu^{k-1} u_0(x,t)\big|_{S_+}, \quad \text{for any } k < s + \tfrac{1}{2},$$

$$v_\lambda(x) = D_t^{\lambda-1} u_0(x,0), \quad \text{for any } \lambda < s + b,$$

$$\omega_{\lambda k}(x') = D_\nu^{k-1} D_t^{\lambda-1} u_0(x,t)\big|_\Gamma, \quad \text{for any } (k,\lambda) \in P \text{ with } k + 2b\lambda < s + b + \tfrac{1}{2}.$$

Clearly, the number of compatibility conditions and as well as the number of compatible components of the vector-valued function $U(x,t)$ depends on s; for $s < 1/2$, $\widetilde{\mathcal{K}}^s(\Omega_+, \gamma) = \mathcal{K}^s(\Omega_+, \gamma)$.

Lemma III.15. *The closure χ of the mapping $U(x,t) = \chi u(x,t)$ with respect to the norm (5.3) is an isometric isomorphism of the spaces $\widetilde{\mathcal{H}}^s_{(\kappa,\tau,P)}(\Omega_+,\gamma)$ and $\widetilde{\mathcal{K}}^s(\Omega_+,\gamma)$.*

Proof: We will prove the lemma only for $s < M$. For any other s the proof is similar and follows from the argument for the case $s < M$, given below. It is clear that the closure of the mapping χ is isometric and to each $u(x,t) \in \widetilde{\mathcal{H}}^s_{(\kappa,\tau,P)}(\Omega_+,\gamma)$ there corresponds a unique $U(x,t) = \chi u(x,t) \in \widetilde{\mathcal{K}}^s(\Omega_+,\gamma)$.

Now let us show that to each vector-valued function $U(x,t)$ there corresponds a unique element $u(x,t) \in \widetilde{\mathcal{H}}^s_{(\kappa,\tau,P)}(\Omega_+,\gamma)$, which is the limit of a sequence $\{u_0^{(n)}(x,t)\} \in C^M(\Omega_+) \cap \mathcal{H}^s(\Omega_+,\gamma)$; moreover, $\chi u_0^{(n)}(x,t) = U^{(n)}(x,t) \to U(x,t)$ in $\widetilde{\mathcal{K}}^s(\Omega_+,\gamma)$ as $n \to \infty$.

Let $U(x,t) \in \widetilde{\mathcal{K}}^s(\Omega_+,\gamma)$ and $u_0(x,t) \in \mathcal{H}^s(\Omega_+,\gamma)$ be the first component of $U(x,t)$. By the completeness of $\mathcal{H}^s(\Omega_+,\gamma)$, there exists a sequence $\{u_0^{(n)}(x,t)\} \subset \mathcal{H}^s(\Omega_+,\gamma) \cap C^M(\overline{\Omega}_+)$ that converges to $u_0(x,t)$ in $\mathcal{H}^s(\Omega_+,\gamma)$. Then, by Theorem II.4 on traces,

$$u_k^{(n)}(x',t) = D_\nu^{k-1} u_0^{(n)}(x,t)\big|_{S_+} \longrightarrow u_k(x',t) \ \text{ in } \ \mathcal{H}^{s-k+1/2}(S_+,\gamma),$$
$$\text{for any } \ k < s + \tfrac{1}{2};$$

$$v_\lambda^{(n)}(x) = D_t^{\lambda-1} u_0^{(n)}(x,t)\big|_{t=0} \longrightarrow v_\lambda(x) \ \text{ in } \ H^{s-2b\lambda+b}(G),$$
$$\text{for any } \ \lambda \text{ with } 2b\lambda < s + b;$$

$$\omega_{\lambda k}^{(n)}(x') = D_\nu^{k-1} D_t^{\lambda-1} u_0^{(n)}(x,t) \longrightarrow \omega_{\lambda k}(x') \ \text{ in } \ H^{s-2b\lambda-k+b+1/2}(\Gamma),$$
$$\text{for any } \ (\lambda,k) \text{ with } 2b\lambda + k < s + b + \tfrac{1}{2}.$$

Note that, in general, the sequences $\{u_k^{(n)}(x',t)\}$, $\{v_\lambda^{(n)}(x)\}$, $\{\omega_{\lambda k}^{(n)}(x')\}$ of the traces of the other derivatives of $u_0^{(n)}(x,t)$ do not converge in the spaces under consideration to the given functions $u_k(x',t)$, $v_\lambda(x)$, $\omega_{\lambda k}(x')$.

Now let us modify the sequence $\{u_0^{(n)}(x,t)\}$ in order to obtain a sequence $u^{(n)}(x,t)$ with the property that $\chi u^{(n)}(x,t) = U^{(n)}(x,t) \to U(x,t)$ in $\widetilde{\mathcal{K}}^s(\Omega_+,\gamma)$. We proceed in four steps.

1. For each $1 \le \lambda \le [M/2b]$ consider the vector-valued function $V_\lambda(x) = \big(v_\lambda(x), \omega_{\lambda k}(x'), \ k = 1,\ldots,M - 2b\lambda\big)$, where $v_\lambda(x)$ are the given components of $U(x,t)$ for $\lambda = 1,\ldots,\kappa$, $v_\lambda(x) = D_t^{\lambda-1} u_0(x,0)$ for $s - 2b\lambda + b > 0$ and $v_\lambda(x) \equiv 0$ for all the other λ, and where $\omega_{\lambda k}(x')$ are the given components of $U(x,t)$ for $(\lambda,k) \in P$, $\omega_{\lambda k}(x') = D_\nu^{k-1} D_t^{\lambda-1} u_0(x',0,0)$ for $s - k - 2b\lambda + b + \tfrac{1}{2} > 0$, and $\omega_{\lambda k}(x') \equiv 0$ for all the other pairs (λ,k).

By Lemma III.13, for every $\lambda = 1,\ldots,\lambda_0 = \max\big(\kappa, [M/2b]\big)$ there exists a sequence of functions $\{\bar{v}_\lambda^{(n)}(x)\} \subset C^M(\overline{G})$ such that $\bar{v}_\lambda^{(n)}(x) \to v_\lambda(x)$ in the space $\widetilde{H}^{s-2b\lambda+b}_{(M-2b\lambda)}(G)$ as $n \to \infty$, that is, $\bar{v}^{(n)}_{\lambda 0}(x) \to v_\lambda(x)$ in $H^{s-2b\lambda+b}(G)$, $D^{k-1}_\nu \bar{v}^{(n)}_\lambda(x)\big|_\Gamma \to \omega_{\lambda k}(x')$ in $H^{s-2b\lambda-k+b+1/2}(\Gamma)$ for $k = 1, \ldots, M - 2b\lambda$.

2. For each $k = 1, \ldots, M$ consider the vector-valued function $U_k(x',t) = \big(u_k(x',t), \omega_{\lambda k}(x'), \lambda = 1, \ldots, [(M-k)/2b]\big)$, where $u_k(x',t)$ are the given components of $U(x,t)$ for $k = 1, \ldots, \tau$, or $u_k(x',t) = D^{k-1}_\nu u_0(x,t)\big|_{S_+}$ for $s - k + \frac{1}{2} > 0$ and $u_k(x',t) \equiv 0$ for all the other k; $\omega_{\lambda k}(x')$ are the above-defined functions.

Arguing as in the proof of Lemma III.13, for every k we construct a sequence $\{\bar{u}^{(n)}_k(x',t)\}$ of infinitely differentiable functions that converges to $u_k(x',t)$ in $\mathcal{H}^{s-k+1/2}(S_+, \gamma)$ and such that $D^{\lambda-1}_t \bar{u}^{(n)}_k(x',0) = \omega^{(n)}_{\lambda k}(x')$ for $\lambda = 1, \ldots, [(M-k)/2b]$.

3. Let $r > 0$ be an integer such that $br \geq \max(M, b\kappa)$. For every $n = 1, 2, \ldots$ consider the mixed parabolic boundary value problem .

$$\big(D_t + (-\Delta)^b\big)^r \bar{u}^{(n)}_0(x,t) = 0 \ \text{ in } \Omega_+, \tag{5.4}$$

$$D^{k-1}_\nu \bar{u}^{(n)}_0(x,t)\big|_{S_+} = \bar{u}^{(n)}_k(x',t), \quad k = 1, \ldots, br,$$

$$D^{\lambda-1}_t \bar{u}^{(n)}_0(x,0) = \bar{v}^{(n)}_\lambda(x), \qquad \lambda = 1, \ldots, r, \tag{5.5}$$

where Δ is the Laplace operator and $\bar{u}^{(n)}_k(x',t)$ $(k = 1, \ldots, M - 2b\lambda)$, $\bar{v}^{(n)}_\lambda(x)$, $(\lambda = 1, \ldots, \lambda_0)$ are the functions introduced above or the functions identically equal zero for the remaining values of k and λ. By construction, the functions $\bar{u}^{(n)}_k(x',t)$ and $\bar{v}^{(n)}_\lambda(x)$ satisfy in Γ the compatibility condition, and (5.4), (5.5) is a problem of parabolic type. Hence, by Theorem 10.1 of [94] (see also Section V.1 of Chapter V), problem (5.4), (5.5) can be uniquely solved in $\mathcal{H}^{2br}(\Omega_+, \gamma)$. Taking r sufficiently large and applying the theorems on the embedding of $\mathcal{H}^s(\Omega_+, \gamma)$ in the space C^M_{loc}, we conclude that $\bar{u}^{(n)}_0(x,t) \in C^M(\Omega_+) \cap \mathcal{H}^{2br}(\Omega_+, \gamma)$.

4. Finally, consider the functions $w^{(n)}_0(x,t) = u^{(n)}_0(x,t) - \bar{u}^{(n)}_0(x,t)$. It is readily verified that $w^{(n)}_0(x,t) \in \mathcal{H}^s_{++}(\Omega_+, \gamma)$, and since $C^\infty_0(\Omega_+)$ is dense in $\mathcal{H}^s_{++}(\Omega_+, \gamma)$, for each $n = 1, 2, \ldots$ there exists a function $\Phi^{(n)}(x',t) \in C^\infty_0(\Omega_+)$ such that $\big\|\bar{u}^{(n)}_0(x,t) - \bar{u}_0(x,t) - \Phi^{(n)}(x,t), \Omega_+, \gamma\big\|_S < \frac{1}{n}$, $n = 1, 2, \ldots$.

By the construction of $\bar{u}^{(n)}_0(x,t)$ and the choice of $\Phi^{(n)}(x,t)$, the sequence $\{u^{(n)}_0(x,t) + \Phi^{(n)}(x,t)\}$ converges to $u_0(x,t)$ in $\mathcal{H}^s(\Omega_+, \gamma)$ and $\chi\big(\bar{u}^{(n)}_0(x,t) + \Phi^{(n)}(x,t)\big) = U^{(n)}(x,t) \to U(x,t)$ in $\widetilde{\mathcal{K}}^s(\Omega_+, \gamma)$ as $n \to \infty$.

III.6. Differential operators in the space $\widetilde{\mathcal{H}}^s$

In this section we establish the rules for applying the linear differential operators $l(x,t,D,D_t)$ in $\overline{\Omega}_+$, on $\overline{S}_+$, $\overline{G}$, and Γ, to elements $u(x,t) \in \widetilde{\mathcal{H}}^s(\Omega_+, \gamma)$, and we prove the boundedness of these operators in the corresponding functional spaces.

III.6.1. Definition of the operator lu in $\overline{\Omega}_+$; its boundedness

Let $l = \sum\limits_{\alpha,\beta} l_{\alpha\beta} D^\alpha D_t^\beta$ be an operator of order q. Suppose that in the neighborhood of S_+, G, and Γ the operator l can be written as

$$l(x,t,D,D_t) \equiv \sum_{k=1}^{q+1} l_k D_\nu^{k-1} \equiv \sum_{\lambda=1}^{q_0+1} \mathcal{L}_\lambda D_t^{\lambda-1} \equiv \sum_{k,\lambda} \widetilde{l}_{k\lambda} D_\nu^{k-1} D_t^{\lambda-1}, \qquad (6.1)$$

respectively, where ν is the unit inner normal to Γ and $q_0 = [q/2b]$; the operators $\widetilde{l}_{k\lambda}$ are expressions in the operators D' of differentiation in the directions tangent to Γ, and are not related to the operators $l^{(k,\lambda)}(\cdot, D_\nu, D_t)$, defined in Subsection III.3.1. For the operator l denote by L the set of the pairs of positive integers (k,λ) for which $l^{(k,\lambda)}(\cdot, D_\nu, D_t) \not\equiv 0$ on Γ. Suppose that $u(x,t) \in \widetilde{\mathcal{H}}^s_{(\kappa,\tau,P)}(\Omega_+,\gamma)$, where $\kappa \geq q_0$, $\tau \geq q+1$, $L \subseteq P$. Based on formulas (3.16), we define (in $\overline{\Omega}_+$) the action of the operator l on the element $u(x,t) \in \widetilde{\mathcal{H}}^s_{(\kappa,\tau,P)}(\Omega_+,\gamma)$ by the rule

$$lu\big|_{\overline{\Omega}_+} \equiv lu_{0++} - l^{(k,0)}u_k^+(x',t) \times \delta(S) - l^{(0,\lambda)}v_\lambda^+(x) \times \delta(t) + l^{(k,\lambda)}\omega_{\lambda k}(x') \times \delta(\Gamma),$$

where the sign "+" stands for the corresponding continuation by zero in x outside $\overline{G}$ and in t for $t < 0$, and repetition of k and λ indicates summation over all their values. Let us study the mapping $u \mapsto lu\big|_{\overline{\Omega}_+}$.

Lemma III.16. *Let $l_{\alpha\beta}(x,t)$ belong to $C^{|s-|\alpha|-2b\beta|+\varepsilon}(\Omega_+)$, where $\varepsilon > 0$ is an arbitrary number. Then for any $s \in \mathbb{R}^1$, $s \notin \mathbb{Z}_{1,2b} \cap (0,q)$, $l(x,t,D,D_t)$ is a bounded operator from $\widetilde{\mathcal{H}}^s_{(\kappa,\tau,P)}(\Omega_+,\gamma)$ to $\mathcal{H}^{s-q}(\Omega_+,\gamma)$, i.e.,*

$$\big\|lu,\Omega_+,\gamma\big\|_{s-q} < C_s \big\|\!\big|u(x,t),\Omega_+,\gamma\big\|\!\big|_{s,(\kappa,\tau,P)}, \qquad (6.2)$$

with a constant C_s that does not depend on $u(x,t)$.

Proof: By using a partition of unity and a local straightening of S_+, the general case can be reduced to the case when $\Omega_+ = \mathbb{R}^{n+1}_{++}$, and hence it suffices to prove (6.2) in that particular case.

1. Thus, let $\Omega_+ = \mathbb{R}^{n+1}_{++}$. Then the expression $lu\big|_{\overline{\Omega}_+}$ coincides with the left-hand side of (3.16), the form of which depends on s. For this reasons we shall examine separately three cases: (a) $s < \frac{1}{2}$; (b) $\frac{1}{2} < s < q - \frac{1}{2}$; (c) $s > q - \frac{1}{2}$.

(a) To estimate the norm $\big\|\cdot, \mathbb{R}^{n+1}, \gamma\big\|_{s-q}$ of each term in the left-hand side of (3.16) we resort to Lemmas III.1 and III.4. Thanks to the smoothness of $l_{\alpha\beta}(x,t)$ each term $l_{\alpha\beta} D^\alpha D_t^\beta u_{0++}$ belongs to the space $\mathcal{H}^{s-2b\beta-|\alpha|}$, which is continuously embedded in $\mathcal{H}^{s-q}$. Consequently,

$$\big\|l_{\alpha\beta} D^\alpha D_t^\beta u_{0++}, \mathbb{R}^{n+1}, \gamma\big\|_{s-q} \leq \big\|l_{\alpha\beta} D^\alpha D_t^\beta u_{0++}, \mathbb{R}^{n+1}, \gamma\big\|_{s-|\alpha|-2b\beta}$$

$$\leq C \big| l_{\alpha\beta}, \mathbb{R}^{n+1} \big|_{s-|\alpha|-2b|\beta|+\varepsilon} \big\| u_{0++}, \mathbb{R}^{n+1}, \gamma \big\|_s ,$$

whence

$$\big\| l u_{0++}, \mathbb{R}^{n+1}, \gamma \big\|_{s-q} \leq C \big\| u_{0++}, \mathbb{R}^{n+1}, \gamma \big\|_s . \tag{6.3}$$

Further, if $s < \frac{1}{2}$, then

$$u_k^+(x', t) \times \delta(x_n) \in \mathcal{H}^{s-k}(\mathbb{R}^{n+1}, \gamma),$$

$$v_\lambda^+(x', t) \times \delta(t) \in \mathcal{H}^{s-2b\lambda}(\mathbb{R}^{n+1}, \gamma),$$

$$\omega_{\lambda k}(x') \times \delta(x_n, t) \in \mathcal{H}^{s-k-2b\lambda}(\mathbb{R}^{n+1}, \gamma).$$

Hence, repeating the arguments used above to derive (6.3), we obtain

$$\big\| l^{(k,0)} u_k^+(x', t) \times \delta(x_n), \mathbb{R}^{n+1}, \gamma \big\|_{s-q} \leq C \big\langle\!\big\langle u_k^+(x', t) \overline{E}_+^n, \gamma \big\rangle\!\big\rangle_{s-k+1/2}, \tag{6.4}$$

$$\big\| l^{(0,\lambda)} v_\lambda^+(x) \times \delta(t), \mathbb{R}^{n+1}, \gamma \big\|_{s-q} \leq C \big[v_\lambda^+(x), \overline{\mathbb{R}}_+^n \big]_{s-2b\lambda+b}, \tag{6.5}$$

$$\big\| l^{(k,\lambda)} \omega_{\lambda k}(x') \times \delta(x_n, t), \mathbb{R}^{n+1}, \gamma \big\|_{s-q} \leq C \big\langle\!\big\langle \omega_{\lambda k}(x'), \mathbb{R}_+^{n-1} \big\rangle\!\big\rangle'_{s-k-2b\lambda+b+1/2}. \tag{6.6}$$

Formulas (6.3) and (6.4)–(6.6) yield estimate (6.2).

(b) Let $\frac{1}{2} < s < q - \frac{1}{2}$. By Lemma III.7, the expression for lu becomes

$$\begin{aligned}
lu = &- \sum_{k=s'+1}^{q} l^{(k,0)} u_k^+ \times \delta(x_n) - \sum_{\lambda=s''+1}^{q_0+1} l^{(0,\lambda)} v_\lambda^+ \times \delta(t) \\
&+ \sum_{k+2b\lambda \geq s'+b+1} l^{(k,\lambda)} \omega_{\lambda k}(x') \times \delta(x_n, t) + l^{(s',0)} \big(D_n^{s'} u_0 \big)_{++} + [l, \theta]_s u_0 ,
\end{aligned} \tag{6.7}$$

where $[l, \theta]_s u_0$ is defined by (3.5). By the estimates (1.2)–(1.5) and Lemma III.4, the estimates (6.4)–(6.6) remain valid for the norms $\big\| \cdot, \mathbb{R}^{n+1}, \gamma \big\|_{s-q}$ of the terms in the right-hand side of (6.7) that contain δ-functions. Thus, it remains to estimate the norms $\big\| \cdot, \mathbb{R}^{n+1}, \gamma \big\|_{s-q}$ of the last two terms in the right-hand side of (6.7). The norms of all the terms that appear in the expressions for $l^{(s',0)} \big(D_n^{s'} u_0 \big)_{++}$ or $[l, \theta]_s u_0$ and contain no δ-functions can be estimated in the same way. Therefore we will consider in detail only the norms of the terms $l_\sigma^{(\sigma_s)}(\cdot, D', D_t) \cdot \big(D_t^{\sigma_s} D_n^\sigma u_0 \big)_{++}$. Since $0 \leq \sigma \leq [s] - b$, it follows that $s - \sigma \geq b$ and $D_n^\sigma u_0$ belongs to the space $\mathcal{H}^{s-\sigma}(\mathbb{R}^{n+1}, \gamma)$, which in turn is continuously embedded in $\mathcal{H}^{0,s-\sigma}(\mathbb{R}^{n+1}, \gamma)$. Hence, $\theta(x_n) D_n^\sigma u_0 \in \mathcal{H}_+^{0,s-\sigma}(\mathbb{R}^{n+1}, \gamma)$, and so if $s - \sigma \neq 2b\lambda + b$, $\big(D_t^{\sigma_s} D_n^\sigma u_0 \big)_{++} \in \mathcal{H}_{++}^{0,s-\sigma-2b\sigma_s}(\mathbb{R}^{n+1}, \gamma)$. The operator $l_\sigma^{(\sigma_s)}(x, t, D', D_t)$ is the sum of terms of the form $l_{\alpha_1 \sigma \beta}(x, t) D_{x'}^{\alpha_1} D_t^{\beta-\sigma_s}$ with $\beta \geq \sigma_s$. It follows that $D_{x'}^{\alpha_1} D_t^{\beta-\sigma_s} \big(D_t^{\sigma_s} D_n^\sigma u_0 \big)_{++}$

belongs to the space $\mathcal{H}_{++}^{0,s-\sigma-|\alpha_1|-2b\beta}(\mathbb{R}^{n+1},\gamma)$, which in turn is continuously embedded in $\mathcal{H}_{++}^{0;\sigma-q}(\mathbb{R}^{n+1},\gamma)$. By what has been said above, for $s-\sigma \neq 2b\lambda+b$,

$$\left\|l_{\alpha_1\sigma\beta}(x,t)D_{x'}^{\alpha_1}D_t^{\beta-\sigma_s}\left(D_t^{\sigma_s}D_n^{\sigma}u_0\right)_{++},\mathbb{R}^{n+1},\gamma\right\|_{s-q}$$

$$\leq \left\|l_{\alpha_1\sigma\beta}(x,t)D_{x'}^{\alpha_1}D_t^{\beta-\sigma_s}\left(D_t^{\sigma_s}D_n^{\sigma}u_0\right)_{++},\mathbb{R}^{n+1},\gamma\right\|_{0,s-\sigma-|\alpha_1|-2b\beta}$$

$$\leq C\left|l_{\alpha_1\sigma\beta}(x,t),\mathbb{R}^{n+1}\right|_{|s-|\alpha_1|-\sigma-2b\beta|+\varepsilon}$$

$$\times \left\|D_{x'}^{\alpha_1}D_t^{\beta-\sigma_s}\left(D_t^{\sigma_s}D_n^{\sigma}u_0\right)_{++},\mathbb{R}^{n+1},\gamma\right\|_{0,s-\sigma-|\alpha_1|-2b\beta} \tag{6.8}$$

$$\leq C\left\|\left(D_t^{\sigma_s}D_n^{\sigma}u_0\right)_{++},\mathbb{R}^{n+1},\gamma\right\|_{0,s-\sigma-2b\sigma_s} \leq C\left\|\theta(x_n)D_n^{\sigma}u_0,\mathbb{R}^{n+1},\gamma\right\|_{0,s-\sigma}$$

$$\leq C\left\|D_n^{\sigma}u_0,\mathbb{R}^{n+1},\gamma\right\|_{0,s-\sigma} \leq C_s\left\|u_0(x,t),\mathbb{R}_{++}^{n+1},\gamma\right\|_s.$$

If $s-\sigma = 2b\lambda+b$, then, since $s \notin \mathbb{Z}_{1,2b}$, we have $\sigma \neq 0$. Therefore, assuming that $q = 2bq_1$, $q_1 \in \mathbb{Z}$, we conclude that each term $l_{\alpha_1\sigma\beta}(x,t)D_{x'}^{\alpha_1}D_t^{\beta-\sigma_s}$ either contains at least one differentiation with respect to x' ($|\alpha_1| \geq 1$), or is an operator of order not higher than $q-\sigma-1-2b\sigma_s$. In the former case we first commute one differentiation with respect to x' with the operator $\theta(t)$, while in the latter case we use the above argument to get a direct estimate. This yields the estimate (6.8) in the case $s-\sigma = 2b\lambda+b$. Estimates (2.1), (2.5) imply that

$$\left\|l_{\sigma}^{(\sigma_s)}(x,t,D',D_t)\left(D_t^{\sigma_s}D_n^{\sigma}u_0\right)_{++},\mathbb{R}^{n+1},\gamma\right\|_{s-q} \leq C_s\left\|u_0,\mathbb{R}_{++}^{n+1},\gamma\right\|_s. \tag{6.9}$$

Proceeding in the same manner, we obtain

$$\left\|l_{\sigma}(\cdot,D',D_t)\left(D_n^{\sigma}u_0\right)_{++},\mathbb{R}^{n+1},\gamma\right\|_{s-q} \leq C_s\left\|u_0,\mathbb{R}_{++}^{n+1},\gamma\right\|_s, \tag{6.10}$$

$$\left\|l_{\sigma\mu}(\cdot,D')\left(D_t^{\mu}D_n^{\sigma}u_0\right)_{++},\mathbb{R}^{n+1},\gamma\right\|_{s-q} \leq C_s\left\|u_0,\mathbb{R}_{++}^{n+1},\gamma\right\|_s, \tag{6.11}$$

$$\left\|l^{(s',0)}(\cdot,D',D_t)\left(D_n^{s'}u_0\right)_{++},\mathbb{R}^{n+1},\gamma\right\|_{s-q} \leq C_s\left\|u_0,\mathbb{R}_{++}^{n+1},\gamma\right\|_s. \tag{6.12}$$

Further, the terms of the expression $\left[l,\theta(x_n,t)\right]_s u_0$ that contain the delta-function $\delta(t)$ obey estimates similar to (6.5):

$$\left\|l^{(k_\lambda,\lambda)}D_n^{1+[s]-s'}\left(D_n^{\alpha_\lambda}v_\lambda\right)^{+} \times \delta(t),\mathbb{R}^{n+1},\gamma\right\|_{s-q} \leq C_s\left[v_\lambda(x),\mathbb{R}_+^n\right]_{s-2b\lambda+b},$$
$$\lambda = 1,2,\ldots, \tag{6.13}$$

$$\left\|l^{(k_{s''},s'')}D_n^{k_{s''}}v_{k_{s''}}^{+}(x) \times \delta(t),\mathbb{R}^{n+1},\gamma\right\|_{s-q} \leq C_s\left[v_{s''}^{+}(x),\mathbb{R}_+^n\right]_{s-2bs''+b}. \tag{6.14}$$

Estimates (6.4)–(6.6) (for $k \geq s'+1$, $\lambda \geq s''+1$, $k+2b\lambda \geq s'+b+1$, respectively) and (6.9)–(6.14) imply estimate (6.2) in the case $\frac{1}{2} < s < q-\frac{1}{2}$.

(c) If $s > q-\frac{1}{2}$, then by (3.16) we get $lu = (lu_0)_{++}$. Since $l_{\alpha\beta} \in C^{s-q}$, we obtain $lu_0 \in \mathcal{H}^{s-q}(\mathbb{R}_{++}^{n+1},\gamma)$ and so estimate (6.2) is valid.

2. In the general case $\Omega_+ = \overline{G} \times \overline{\mathbb{R}}^1_+$, where G is a domain of $\mathbb{R}^n$ with compact boundary Γ, the assertion of the lemma follows from the above proof for $\Omega_+ = \overline{\mathbb{R}}^{n+1}_{++}$. Indeed, if $\sum_i \varphi_i(x) \equiv 1$ is the partition of unity in $\overline{G}$ constructed in Section II.11 of Chapter II, then, writing each term of the expression $lu\big|_{\Omega_+} \equiv \sum_i l(\varphi_i u)\big|_{\Omega_+}$ in local coordinates and using the equivalence of norms (4.8), we obtain (6.2).

III.6.2. Definition of the operator $lu\big|_{\overline{S}_+}$; its boundedness

Let $l(\cdot, D, D_t)$ be the operator defined above and denote by B the set of pairs (k, λ) such that $l^{(k,\lambda)}(\cdot, D, D_t) \not\equiv 0$ on $\overline{S}_+$. For each $u(x,t) \in \widetilde{\mathcal{H}}^s_{(\kappa,\tau,P)}(\Omega_+, \gamma)$, where $\kappa \geq \max_k \operatorname{ord}_t l_k$, $\tau \geq q+1$ and $P \supseteq B$ we define $lu\big|_{\overline{S}_+}$ by the rule

$$lu\big|_{\overline{S}_+} \equiv \sum_{k=1}^{q+1} \left[l_k u_k^+ - \sum_{\lambda=1}^{\kappa_k} l_k^{(\lambda)}(\cdot, D', D_t) \cdot \omega_{\lambda k}(x') \times \delta(t) \right] \equiv \sum_k l_k u_k\big|_{\overline{S}_+}, \qquad (6.15)$$

where $\kappa_k = \big[(q-k)/2b\big] + 1$. Let us analyze the mapping $u \mapsto lu\big|_{\overline{S}_+}$.

Lemma III.17. *Let $l_{\alpha\beta}(x',t) \in C^{a+\varepsilon}(S_+)$, where $a = \min\big(|s-q-\tfrac{1}{2}|, |s-|\alpha|-2b\beta|\big)$ and $\varepsilon > 0$ is an arbitrary number. Then for any $s \in \mathbb{R}^1$, $s \notin \mathbb{Z}_{1,2b} \cap (0,q)$*

$$\big\langle\!\big\langle l(\cdot, D, D_t)u\big|_{\overline{S}_+}, S_+, \gamma \big\rangle\!\big\rangle_{s-q-1/2} \leq C_s \big\||u, \Omega_+, \gamma\big\|_{s,(\kappa,\tau,P)}, \qquad (6.16)$$

with a constant C_s that does not depend on $u(x,t)$.

Proof: 1. First let $S_+ = E_+^n$. In this case we can rewrite each of the expressions $l_k u_k\big|_{\overline{S}_+}$ by using formula (3.3). The form of each term $l_k u_k\big|_{\overline{E}_+^n}$ depends on k and s, therefore, as in Lemma III.16, we will estimate the norm $\|\cdot, E_+^n, \gamma\|_{s-q-1/2}$ of each term separately for the two cases (a) $s - k + \tfrac{1}{2} < b$ and (b) $s - k + \tfrac{1}{2} > b$.

(a) Let $s - k + \tfrac{1}{2} < b$. Then $u_k^+(x',t) \in \mathcal{H}^{s-k+1/2}(E_n, \gamma)$ and since

$$l_k(\cdot, D', D_t) = \sum_{|\alpha_1|+2b\beta \leq q-k+1} l_{\alpha_1 k-1, \beta}(x',t) D_{x'}^{\alpha_1} D_t^{\beta},$$

it follows that $D_{x'}^{\alpha_1} D_t^{\beta} u_k^+(x',t) \in \mathcal{H}^{s-k-|\alpha_1|-2b\beta+1/2}(E^n, \gamma)$. Hence, since $|\alpha_1| + 2b\beta + k - 1 \leq q$, we obtain

$$\big\langle\!\big\langle l_k(\cdot, D', D_t)u_k^+, E^n, \gamma \big\rangle\!\big\rangle_{s-q-1/2}$$

$$\leq C_s \sum_{\alpha,\beta} \big\langle\!\big\langle l_{\alpha_1 k-1\beta} \cdot D_{x'}^{\alpha_1} D_t^{\beta} u_k^+, E^n, \gamma \big\rangle\!\big\rangle_{s-|\alpha_1|-2b\beta-k+1/2}$$

$$\leq C_s \sum_{\alpha,\beta} \big|l_{\alpha_1 k-1\,\beta}, E_+^n\big|_{|s-|\alpha_1|-2b\beta-k+1/2|+\varepsilon} \big\langle\!\big\langle u_k^+, E^n, \gamma \big\rangle\!\big\rangle_{s-k+1/2}$$

$$\leq C_s \big\langle\!\big\langle u_k^+, E^n, \gamma \big\rangle\!\big\rangle_{s-k+1/2}. \tag{6.17}$$

Next, since $\omega_{\lambda k}(x') \times \delta(t) \in \mathcal{H}^{s-k-2b\lambda+1/2}(E^n, \gamma)$ for $s - k + \frac{1}{2} < b$ and

$$l_k^{(\lambda)}(\cdot, D', D_t) = \sum_{\beta \geq \lambda} l_{\alpha_1 k - 1\beta}(x', t) D_{x'}^{\alpha_1} D_t^{\beta-\lambda}, \tag{6.18}$$

it follows that $D_{x'}^{\alpha_1} D_t^{\beta-\lambda} \omega_{\lambda k}(x') \times \delta(t) \in \mathcal{H}^{s-|\alpha_1|-2b\beta-k+1/2}(E^n, \gamma)$. Therefore,

$$\big\langle\!\big\langle l_k^{(\lambda)} \omega_{\lambda k}(x') \times \delta(t), E^n, \gamma \big\rangle\!\big\rangle_{s-q-1/2}$$

$$\leq \sum_{\alpha,\beta} \big|l_{\alpha_1 k-1\beta}, E^n\big|_{|s-|\alpha_1|-2b\beta-k+1/2|+\varepsilon} \big\langle\!\big\langle \omega_{\lambda k}, \mathbb{R}^{n-1} \big\rangle\!\big\rangle'_{s-k-2b\lambda+b+1/2} \tag{6.19}$$

$$\leq C_s \big\langle\!\big\langle \omega_{\lambda k}(x'), \mathbb{R}^{n-1} \big\rangle\!\big\rangle'_{s-k-2b\lambda+b+1/2}.$$

Estimates (6.17) and (6.19) yield the estimate

$$\big\langle\!\big\langle l_k u_k\big|_{\overline{E}_+^n}, E^n, \gamma \big\rangle\!\big\rangle_{s-q-1/2} < C_s \big\|u(x,t), \mathbb{R}_{++}^{n+1}, \gamma \big\|_{s,(\kappa,\tau,P)} \tag{6.20}$$

for all k such that $k > s + b + \frac{1}{2}$. Finally, estimates (6.19) and (6.20) yield the estimate (6.16).

(b) Let $s - k + \frac{1}{2} > b$. Then, by Lemma III.6,

$$l_k u_k\big|_{\overline{E}_+^n} = \sum_{\mu=0}^{a_k-1} l_\mu(\cdot, D')\big(D_t^\mu u_k(x', t)\big)^+ + l_k^{(a_k)}(\cdot, D', D_t)(D_t^{a_k} u_k)^+ -$$

$$- \sum_{\lambda \geq a_k+1} l_k^{(\lambda)}(\cdot, D', D_t)\omega_{\lambda k}(x') \times \delta(t), \quad a_k = \big[(s-k+b+1/2)/2b\big]. \tag{6.21}$$

For the terms of (6.21) that contain $\delta(t)$ estimate (6.19) remains valid. Consider the first term in the right-hand side of (6.21). By (2.7), $(D_t^{a_k} u_k)^+$ belongs to the space $\mathcal{H}^{s-k-2ba_k+1/2}(E^n, \gamma)$. Consequently,

$$\big\langle\!\big\langle l_{\alpha_1 k-1\beta} D_{x'}^{\alpha_1} D_t^{\beta-a_k}(D_t^{a_k} u_k)^+, E^n, \gamma \big\rangle\!\big\rangle_{s-q-1/2} \leq C_s \big\langle\!\big\langle u_k, E_+^n, \gamma \big\rangle\!\big\rangle_{s-k+1/2}. \tag{6.22}$$

Further, (6.18) implies

$$\big\langle\!\big\langle l_k^{(a_k)}(D_t^{a_k} u_k)^+, E^n, \gamma \big\rangle\!\big\rangle_{s-q-1/2} \leq C_s \big\langle\!\big\langle u_k, E_+^n, \gamma \big\rangle\!\big\rangle_{s-k+1/2}. \tag{6.23}$$

Proceeding in much the same manner, we obtain

$$\big\langle\!\big\langle l_\mu(D_t^\mu u_k)^+, E^n, \gamma \big\rangle\!\big\rangle_{s-q-1/2} \leq C_s \big\langle\!\big\langle u_k, E_+^n, \gamma \big\rangle\!\big\rangle_{s-k+1/2}. \tag{6.24}$$

Note that if $s > q - \frac{1}{2}$, then $lu\big|_{\overline{E}_+^n} = \big(lu\big|_{E^n}\big)^+$. Hence, the membership $l_{\alpha\beta} \in C^{s-q-1/2+\varepsilon}$ implies

$$\big\langle\!\big\langle lu_0\big|_{\overline{E}_+^n}, E_+^n, \gamma \big\rangle\!\big\rangle_{s-q-1/2} \leq C_s \big\|u_0(x,t), \mathbb{R}_{++}^{n+1}, \gamma \big\|_s, \tag{6.25}$$

where C_s is a constant that does not depend on $u(x,t)$. Estimate (6.16) for $s - k + \frac{1}{2} > b$ follows from (6.23)–(6.25). Thus, estimate (6.25) holds for all s and gives us (6.16) in the case $S_+ = E_+^n$.

2. The general case $S_+ = \Gamma \times \overline{\mathbb{R}}_+^1$ is dealt with in a standard way by using a partition of unity and a local straightening of S_+.

III.6.3. Definition of the operators $lu\big|_{\overline{G}}$ and $lu\big|_\Gamma$; their boundedness

Let $\mathcal{C}$ be the set of all pairs (k,λ) such that $\mathcal{L}_\lambda^{(k)}(x,0,D) \not\equiv 0$ in $\overline{G}$ and $\mathcal{D}$ be the set of all pairs (k,λ) such that $\widetilde{l}_{k,\lambda}(x') \not\equiv 0$ on Γ. For $u(x,t) \in \widetilde{\mathcal{H}}^s_{(\kappa,\tau,P)}(\Omega_+,\gamma)$, where $P \supseteq \mathcal{C}$ [resp., $P \supseteq \mathcal{D}$] we define $lu\big|_{\overline{G}}$ [resp., $lu\big|_\Gamma$ by the rule

$$lu\big|_{\overline{G}} \equiv \sum_{\lambda=1}^{\kappa_0+1} \left[\mathcal{L}_\lambda(x,0,D)v_\lambda^+(x) - \sum_{k=1}^{k_\lambda} \mathcal{L}_\lambda^{(k)}(x,0,D)\omega_{\lambda k}(x') \times \delta(\Gamma) \right] \equiv \sum_\lambda \mathcal{L}_\lambda v_\lambda\big|_{\overline{G}},$$

where $k_\lambda = \operatorname{ord}\mathcal{L}_\lambda$

$$[\text{resp.,}\quad lu\big|_\Gamma = \sum_{k,\lambda} l_{k,\lambda}(x',0,D')\omega_{\lambda k}(x')].$$

Lemma III.18. *Let* $l_{\alpha\beta}(x,0) \in C^{d+\varepsilon}(\overline{G})$, $d = \min\left(\big|s-|\alpha|-b\big|+\varepsilon, |s-q|\right)$, *where* $\varepsilon > 0$ *is an arbitrary number. Then for any* $s \in \mathbb{R}^1$, $s \notin \mathbb{Z}_{1,2b} \cap (0,q)$

$$\big\|\big[l(x,0,D,D_t)u\big|_{\overline{G}},\overline{G}\big]\big\|_{s,q,b} \leq C_s\big\||u,\Omega_+,\gamma\big\|\|_{s,(\kappa,\tau,P)}, \tag{6.26}$$

with a constant C_s *that does not depend on* $u(x,t)$.

The proof is similar to that of Lemma III.17. By using a partition of unity in $\overline{G}$ and a local straightening of Γ the derivation of estimate (6.26) is reduced to the case $G = \mathbb{R}^n_+$. In the latter case estimate (6.26) is obtained by using Lemma 3.5 and the argument presented above.

From the boundedness of the operator of multiplication by a smooth function in the spaces $H^s(\Gamma)$ and the definition of $lu\big|_\Gamma$ we deduce the following assertion.

Lemma III.19. *Let* $l_{\alpha\beta}(x',0) \in C^{d+\varepsilon}(\Gamma)$. *Then for any* $s \in \mathbb{R}^1$, $s \notin \mathbb{Z}_{1,2b} \cap (0,q)$,

$$\big\langle\!\big\langle lu\big|_\Gamma,\Gamma\big\rangle\!\big\rangle'_{s-q-b-\frac{1}{2}} \leq C_s\big\||u,\Omega_+,\gamma\big\|\|_{s,(\kappa,\tau,P)}, \tag{6.27}$$

with a constant C_s *that does not depend on* $u(x,t)$.

Chapter IV
Parabolic Boundary
Value Problems in Half-Space

This chapter plays the central role in the book. Here we study in detail model parabolic problems in families of spaces $\widetilde{\mathcal{H}}^s$, which form the basis for the study of general parabolic boundary value problems in these spaces. The model parabolic problems are analyzed by the technique of the integral Fourier transformation. This technique works when both the sought solutions and the right-hand sides of the problem are defined in the whole space. To obtain such problems it is first necessary to continue the equations of the problem in question (in the case of equations with constant coefficients one extends the sought solution and the given right-hand sides) from the original domain to the whole space, and then the integral Fourier transformation is applied to the extended equations. Since, as a rule, we are interested in obtaining the solution of a problem only in the closure of the underlying domain, one continues the sought solution outside the domain of definition by zero. Accordingly, the right-hand sides are also continued by zero outside the domain. Thus in a natural way we are led to solving model parabolic problems in classes of functions supported in closed domains.

Continuation by zero leads to specific properties of the Fourier transformation of distributions supported in cones which are described by the generalized Paley-Wiener theorem. Using these properties and the theory of analytic functions, one can show that the parabolicity of a problem guarantees the well-posedness of a whole set of model parabolic problems in suitable functional spaces $\widetilde{\mathcal{H}}^s$. From this point of view the analysis of model parabolic boundary value problems given below can be regarded as an investigation of concrete systems of equations in cones by using the theory of the Fourier transformation of distributions with supports in cones and of the corresponding spaces of holomorphic functions in tube domains.

IV.1. Non-homogeneous systems in the space $\mathcal{H}^s_{++}(\mathbb{R}^{n+1}, \gamma)$

Here we consider the general model parabolic system of linear differential equations in $\mathbb{R}^{n+1}$:

$$\sum_{j=1}^m l_{ij}(D, D_t)u_j(x, t) = \mathcal{F}_i(x, t), \qquad i = 1, \ldots, m, \tag{1.1}$$

under the assumption that $\mathcal{F}_i(x,t)$ are distributions with support in some closed convex cone K. We investigate the problem of solvability of system (1.1) in the cone K, that is, we seek conditions on $\mathcal{F}_i(x,t)$ which guarantee the existence of a solution $u(x,t)$ that is also supported in K.

The problem of the solvability of convolution equations (in particular, differential equations) in cones plays an important role in theoretical and mathematical physics. It is closely connected with the study of integral transformations of distributions with supports in cones and the corresponding spaces of holomorphic functions in tube domains. In our case this problem essentially reduces to the problem of division by a polynomial in a suitable space of holomorphic functions. This problem is the subject of a large number of works (see [10] and references therein).

The question of the solvability of system (1.1) in cones arises naturally in studying general linear parabolic boundary value problems by reducing them to operator equations using a regularizer. Here a major role is played by theorems on the solvability of model parabolic systems (1.1) in anisotropic spaces $\mathcal{H}^s$ in two particular cases:

1. $K = \overline{E}_+^{n+1}$, which corresponds to the "generalized Cauchy problem" in the terminology of [10];

2. $K = \overline{\mathbb{R}}_{++}^{n+1}$, which corresponds to the boundary value problem for system (1.1) in the closed quadrant $\overline{\mathbb{R}}_{++}^{n+1}$.

IV.1.1.   Non-homogeneous systems in $\overline{E}_+^{n+1}$

Consider the system (1.1) and assume that $\mathcal{F}_i(x,t) \equiv \mathcal{F}_i^+(x,t)$ are distributions supported in $\overline{E}_+^{n+1}$.

Theorem IV.1. *Let $\mathcal{F}_i^+(x,t) \in \mathcal{H}_+^{s-s_i,\sigma}(E^{n+1},\gamma)$, $\gamma \geq \gamma_0 > 0$. Then the system (1.1) has a unique solution $u_j^+(x,t) \in \mathcal{H}_+^{s+t_j,\sigma}(E^{n+1},\gamma)$, which satisfies the following estimate:*

$$\{\{u_j^+(x,t), E^{n+1}, \gamma\}\}_{s+t_j,\sigma} < C \sum_{i=1}^{m} \{\{\mathcal{F}_i^+(x,t), E^{n+1}, \gamma\}\}_{s-s_i,\sigma}. \qquad (1.2)$$

Here and below, in Sections IV.1 and IV.2, the constant C depends on the parabolicity constant δ_0 of system (1.1) and on n, N, m, and does not depend on $u_j^+(x,t)$, where $2N = \sum_k (s_k + t_k)$, $\operatorname{ord} l_{ij} = s_i + t_j$.

Proof: The membership $\mathcal{F}_i^+(x,t) \in \mathcal{H}_+^{s-s_i,\sigma}(E^{n+1},\gamma)$ means that $\widehat{\mathcal{F}}_i^+(x,t)$, the Fourier transform in x and the Laplace transform in t of $\mathcal{F}_i^+(x,t)$, is a function locally integrable with respect to $\xi_0, \xi_1, \ldots, \xi_n$ ($\xi_0 = \operatorname{Im} p$) and analytic with respect to p in the half-plane $\operatorname{Re} p > \gamma$ of the complex p-plane, and that the following integrals converge:

$$\int\limits_{\mathbb{R}^{n+1}} \left(|\xi|^2 + |\gamma|^{1/b} + |\xi_0|^{1/b}\right)^{s-s_i} \left(1 + |\xi|^2\right)^\sigma \left|\widehat{\mathcal{F}}_i^+(\xi, p)\right|^2 d\xi \, d\xi_0.$$

If we seek a solution of system (1.1) in the form $u_j^+(x, t) = L_{p\to t}^{-1} F_{\xi\to x}^{-1} \widehat{u}_j^+$, we obtain the system of algebraic equations $\mathcal{L}(i\xi, p)\widehat{u}^+(\xi, p) = \widehat{\mathcal{F}}^+(\xi, p)$ in $\widehat{u}^+ = (\widehat{u}_1^+, \ldots, \widehat{u}_m^+)$. By the parabolicity of the system (1.1), $\mathcal{L}(i\xi, p)$ is a non-singular matrix for any $\xi \in \mathbb{R}^n$ and any p with $\operatorname{Re} p \geq \gamma$. This assertion follows from the fact that for any $\xi \in \mathbb{R}^n$ the p-zeroes of $\det \mathcal{L}(i\xi, p)$ lie in the half-plane $\operatorname{Re} p \leq -\delta_0|\xi|^{2b}$. Consequently, $\widehat{u}^+(\xi, p)$ is given by the formula

$$\widehat{u}^+(\xi, p) = \mathcal{L}^{-1}(i\xi, p)\widehat{\mathcal{F}}^+(\xi, p), \quad \text{or}$$

$$\widehat{u}_j^+(\xi, p) = L^{-1}(i\xi, p) \sum_{i=1}^m L_{ij}(i\xi, p)\widehat{\mathcal{F}}_i^+(\xi, p), \tag{1.3}$$

where $L_{ij}(i\xi, p)$ is the cofactor of the element $l_{ij}(i\xi, p)$ of the matrix $\mathcal{L}(i\xi, p)$, $L = \det \mathcal{L}$. Let us verify that $\operatorname{supp} u_j^+(x, p) \subseteq E_+^{n+1}$, $j = 1, \ldots, m$, and estimate (1.2) holds. Indeed, since $L^{-1}(i\xi, p)$ and $\widehat{\mathcal{F}}_i^+(\xi, p)$ are analytic functions of p for any $\xi \in \mathbb{R}^n$ and $\operatorname{Re} p > \gamma$, it follows that for any $\xi \in \mathbb{R}^n$ the functions $\widehat{u}_j^+(\xi, p)$ are analytic in p for $\operatorname{Re} p > \gamma$. Hence, by the properties of the Laplace transformation, $\operatorname{supp} u_j^+(x, t) \subseteq \overline{E}_+^{n+1}$. Further, the homogeneity of $L(i\xi, p)$ and $L_{ij}(i\xi, p)$ and the non-singularity of $L(i\xi, p)$ for the indicated ξ and p imply

$$\left|L^{-1}(i\xi, p)L_{ij}(i\xi, p)\right| < C\left(|\xi|^2 + |p|^{1/b}\right)^{-s_i - t_j}.$$

Therefore,

$$\left\{\left\{u_j^+(x, t), E^{n+1}, \gamma\right\}\right\}^2_{s+t_j, \sigma}$$

$$= \int\limits_{\mathbb{R}^{n+1}} \left(|\xi|^2 + |\gamma|^{1/b} + |\xi_0|^{1/b}\right)^{s+t_j} \left(1 + |\xi|^2\right)^\sigma \left|\widehat{u}_j^+(\xi, p)\right|^2 d\xi \, d\xi_0$$

$$\leq C \sum_{i=1}^m \int\limits_{\mathbb{R}^{n+1}} \left(|\xi|^2 + |\gamma|^{1/b} + |\xi_0|^{1/b}\right)^{s-s_i} \left(1 + |\xi|^2\right)^\sigma \left|\widehat{\mathcal{F}}_i^+(\xi, p)\right|^2 d\xi \, d\xi_0,$$

which yields the membership $u_j^+(x, t) \in \mathcal{H}_+^{s+t_j, \sigma}(E^{n+1}, \gamma)$ and the estimate (1.2).

IV.1.2.   Non-homogeneous systems in $\overline{\mathbb{R}}_{++}^{n+1}$

Suppose that in system (1.1) $\mathcal{F}_i(x, t) \equiv \mathcal{F}_{i++}(x, t) \in \mathcal{H}_{++}^{s-s_i, r}(\mathbb{R}^{n+1}, \gamma)$ $(r \in \mathbb{R}^1)$ are distributions supported in $\overline{\mathbb{R}}_{++}^{n+1}$. Then it follows from the above arguments and (1.3) that the system (1.1) has a unique solution $u_{j++}(x, t)$, supported in $\overline{E}_+^{n+1}$ and belonging to $\mathcal{H}^{s+t_j, r}(\mathbb{R}^{n+1}, \gamma)$, $\gamma > 0$. Let us clarify under what conditions on $\mathcal{F}_{i++}(x, t)$ one has that $\operatorname{supp} u_{j++} \subseteq \overline{\mathbb{R}}_{++}^{n+1}$, i.e., $u_{j++}(x, t) \in \mathcal{H}_{++}^{s+t_j, r}(\mathbb{R}^{n+1}, \gamma)$.

Theorem IV.2. *Let $\mathcal{F}_i(x,t) \equiv \mathcal{F}_i^{++}(x,t) \in \mathcal{H}_{++}^{s-s_i,r}(\mathbb{R}^{n+1},\gamma)$.*

1. The solution $u(x,t) = (u_1,\ldots,u_m)$ of the system (1.1), defined by formula (1.3), belongs to the space $\prod_{i=1}^{m} \mathcal{H}_{++}^{s+t_j,r}(\mathbb{R}^{n+1},\gamma)$ if and only if

$$\int_{-\infty}^{\infty} \Lambda_{++}^{s+t_j-N}(i\xi,p)L^{-1}(i\xi,p)\sum_{i=1}^{m} L_{ij}(i\xi,p)\widehat{\mathcal{F}}_i^{++}(\xi,p)\xi_n^{q-1}\,d\xi_n = 0,$$

$$(q = 1,\ldots,N, \quad j = 1,\ldots,m) \qquad (1.4)$$

for almost all $\xi' \in \mathbb{R}^{n-1}$ and for all p with $\operatorname{Re}p \geq \gamma$.

2. For any fixed $\xi' \in \mathbb{R}^{n-1}$ and any p with $\operatorname{Re}p \geq \gamma$ among the $N \cdot m$ conditions (1.4) there are exactly N linearly independent conditions.

Proof: Without loss of generality we can assume that $r = 0$; indeed, if $\mathcal{F}_i^{++}(x,t)\in \mathcal{H}_{++}^{s-s_i,r}(\mathbb{R}^{n+1},\gamma)$, then thanks to the properties of the operator $\Lambda_1^r(D',D_t)$ the distributions $\Lambda_1^r(D',D_t)\mathcal{F}_i^{++}(\dot{x},t)$ belong to $\mathcal{H}_{++}^{s-s_i}(\mathbb{R}^{n+1},\gamma)$ and (1.1) is equivalent to the system $l_{ij}(D,D_t)\Lambda_1^r(D',D_t)u_j(x,t) = \Lambda_1^r(D',D_t)\,\mathcal{F}_i^{++}(x,t)$ (in the unknowns $\Lambda_1^r u_j$), for which the conditions (1.4) do not depend on Λ_1^r.

1. Let $\mathcal{F}_i^{++}(x,t) \in \mathcal{H}_{++}^{s-s_i}(\mathbb{R}^{n+1},\gamma)$ and $u_j^{++}(x,t) = L_{p\to t}^{-1}F_{\xi\to x}^{-1}\,\widehat{u}_j^{++}(\xi,p)$, where $\widehat{u}_j^{++}(\xi,p) = L^{-1}(i\xi,p)\sum_{i=1}^{m}L_{ij}(i\xi,p)\widehat{\mathcal{F}}_i^{++}(\xi,p) \equiv L^{-1}(i\xi,p)\,\widehat{\Phi}_j^{++}(\xi,p)$. It follows from the properties of the Fourier and Laplace transformations that the inclusion $\operatorname{supp}u_j^{++} \subseteq \overline{\mathbb{R}}_{++}^{n+1}$ holds if and only if, for almost any fixed $\xi' \in \mathbb{R}^{n-1}$, $\widehat{u}_j^{++}(\xi,p)$ are analytic functions of (ζ_n,p) in the tube domain

$$A_{0,\gamma} = \{(\zeta_n = \xi_n + i\eta_n,p) : \operatorname{Im}\zeta_n < 0, \ \operatorname{Re}p > \gamma\}.$$

As we already pointed out, the parabolicity of system (1.1) implies that the roots ξ_n of the equation $L(i\xi',i\xi_n,p) \equiv \det \mathcal{L}(i\xi,p) = 0$, are distributed in equal numbers in the upper and the lower half-planes of the complex ξ_n-plane.

Let $L(i\xi,p) = L_+(\xi,p)L_-(\xi,p)$, where L_+ [resp., L_-] is a polynomial in ξ_n whose ξ_n-zeroes have only positive [resp., negative] imaginary parts, and write $L_-(\xi,p) = \prod_{\nu=1}^{l}\left(\xi_n - \xi_{n\nu}^-(\xi',p)\right)^{k_\nu}$, where $k_1 + \ldots + k_l = N$ and $k_\nu = k_\nu(\xi',p)$ is the multiplicity of the root $\xi_{n\nu}^-(\xi',p)$. Since $\widehat{\mathcal{F}}_i^{++}(\xi,p)$ are analytic functions of $(\xi_n,p) \in A_{0,\gamma}$ for almost all $\xi' \in \mathbb{R}^{n+1}$, it follows that $\widehat{u}_j^{++}(\xi,p)$ are meromorphic functions of $(\xi_n,p) \in A_{0,\gamma}$. Therefore, $\widehat{u}_j^{++}(\xi,p)$ are analytic functions of $(\xi_n,p) \in A_{0,\gamma}$ if and only if, for a given $\xi' \in \mathbb{R}^{n-1}$ and p with $\operatorname{Re}p \geq 0$, the points $\xi_n = \xi_{n\nu}^-(\xi',p)$ are removable singular points of $\widehat{u}_j^{++}(\xi,p)$. This means that $\xi_{n\nu}^-(\xi',p)$ are zeroes of the same multiplicity of the function $\widehat{\Phi}_j^{++}(\xi,p)$:

$$\frac{\partial^{\sigma-1}}{\partial\xi_n^{\sigma-1}}\,\widehat{\Phi}_j^{++}(\xi',\xi_n,p)\big|_{\xi_n=\xi_{n\nu}^-} = 0, \quad \sigma = 1,\ldots,k_\nu \qquad (\nu = 1,\ldots,l, \ j = 1,\ldots,m).$$

$$(1.5)$$

To complete the proof of the first part of the theorem we must establish the equivalence of conditions (1.5) and (1.4). To do this, we first show that conditions (1.5) are equivalent to the conditions

$$\int_{\Gamma^-} \Lambda_{++}^{s+t_j-N}(i\xi,p)L^{-1}(i\xi,p)\widehat{\Phi}_j^{++}(\xi,p)\zeta_n^{q-1}\,d\zeta_n = 0,$$

$$(\xi',\zeta_n) = \xi \qquad (q=1,\ldots,N,\ j=1,\ldots,m). \tag{1.6}$$

The latter, in turn, are obviously equivalent to the conditions

$$J_j = \int_{\Gamma^-} \Lambda_{++}^{s+t_j-N}L^{-1}\widehat{\Phi}_j^{++}P_N(\zeta_n)\,d\zeta_n = 0, \qquad j=1,\ldots,m, \tag{1.7}$$

where Γ^- is a contour in the half-plane $\operatorname{Im}\zeta_n < 0$ that encircles all the ζ_n-zeroes of $L_-(\xi,p)$, and $P_N(\zeta_n)$ is an arbitrary polynomial of degree $N-1$. Conditions (1.7) obviously follow from (1.5). We claim that conditions (1.7) imply (1.5). Indeed, using the arbitrariness of $P_N(\zeta_n)$ we select it so that all the zeroes $\xi_{n\nu}^-(\xi',p)$ except for an arbitrarily fixed one $\xi_{n\lambda}^-(\xi',p)$ will be zeroes of $P_N(\zeta_n)$ of the same multiplicity as the zeroes $\xi_{n\nu}^-(\xi',p)$ of the polynomial $L_-(\xi',\xi_n,p)$, that is,

$$P_N(\zeta_n) = P_{N\lambda}(\zeta_n)\prod_{\nu\neq\lambda}\left(\zeta_n - \xi_{n\nu}^-(\xi',p)\right)^{k_\nu},$$

where $P_{N\lambda}(\zeta_n)$ is an arbitrary polynomial of the degree k_λ. Then, by the Cauchy theorem, (1.7) can be rewritten as

$$J_{j\lambda} = \int_{\Gamma_\lambda^-} \Lambda_{++}^{s+t_j-N}P_{N\lambda}\widehat{\Phi}_j^{++}\left(\zeta_n - \xi_{n\lambda}^-(\xi',p)\right)^{-k_\lambda}\,d\zeta_n = 0, \qquad j=1,\ldots,m, \tag{1.8}$$

where Γ_λ^- is any contour that encircles only the root $\xi_{n\lambda}^-(\xi',p)$. Calculating the integral $J_{j\lambda}$ by means of the residue formula at a pole of order k_λ we obtain

$$\frac{d^{k_\lambda-1}}{d\zeta_n^{k_\lambda-1}}\left[\Lambda_{++}^{s+t_j-N}\Phi_j^{++}P_{N\lambda}(\zeta_n)\right]_{\zeta_n=\xi_{n\lambda}^-} = 0, \qquad j=1,\ldots,m,. \tag{1.9}$$

Next, choosing $P_{N\lambda}(\zeta_n)$ successively so that the derivative $P_{N\lambda}^{(k_\lambda-\sigma)}(\xi_{n\nu}^-)$ be different from zero, $\sigma=1,\ldots,k_\nu$, and all the other derivatives be zero, we infer (1.5) from (1.9).

Now let us show that conditions (1.6) can be rewritten in the form (1.4). Indeed, since the functions $\widehat{V}_j(\xi',\xi_n,p) = \Lambda_{++}^{s+t_j}\widehat{u}_j^{++}(\xi,p)$ belong to $L_2(\mathbb{R}^1)$ as functions of ξ_n for fixed (ξ',p), and $\widehat{V}_j(\xi',\zeta_n,p)$ is the analytic continuation of the Fourier transform $\widehat{V}_j(\xi', \xi_n, p)$ to the complex domain $\operatorname{Im}\zeta_n < 0$, it follows that $\left|\widehat{V}_j(\xi', \zeta_n, p)\right|$ is a bounded function for $\operatorname{Im}\zeta_n < -\varepsilon$ for any $\varepsilon > 0$. Further, $\left|\Lambda_{++}^{-N}(\xi', \zeta_n, p) \cdot \zeta_n^{q-1}\right| \to 0$ as $|\zeta_n| \to \infty$, $\operatorname{Im}\zeta_n < -\varepsilon$ and

$$\int_{-\infty-i\eta_n}^{\infty-i\eta_n} \widehat{V}_j(\xi', \zeta_n, p)\, d\zeta_n \longrightarrow 0 \quad \text{as } \eta_n \to \infty, \ \zeta_n = \xi_n + i\eta_n.$$

Hence, applying the Cauchy theorem to the integrals (1.6) we can pass from the integral over Γ^- to an integral over a line, namely

$$\int_{-\infty-i\varepsilon}^{\infty-i\varepsilon} \Lambda_{++}^{s+t_j-N}(i\xi', i\zeta_n, p) L^{-1}\widehat{\Phi}_j^{++}\zeta_n^{q-1}\, d\zeta_n = 0 \quad (q = 1, \ldots, N, \ j = 1, \ldots, m).$$

Now letting $\varepsilon \to 0$, we obtain (1.4). Furthermore, inequalities (1.2) obviously yield the estimate

$$\sum_{j=1}^{m} \left\| u_j^{++}(x,t), \mathbb{R}^{n+1}, \gamma \right\|_{s+t_j,r} < C \sum_{i=1}^{m} \left\| \mathcal{F}_i^{++}(x,t), \mathbb{R}^{n+1}, \gamma \right\|_{s-s_i,r}. \tag{1.10}$$

2. Since $\mathcal{H}_{++}^{s-s_i,r}(\mathbb{R}^{n+1}, \gamma)$ is a closed subspace of $\mathcal{H}^{s-s_i,r}(\mathbb{R}^{n+1}, \gamma)$ and the set of infinitely differentiable functions with compact support in $\overline{\mathbb{R}}_{++}^{n+1}$ is dense in $\mathcal{H}_{++}^{s-s_i}(\mathbb{R}^{n+1}, \gamma)$, it suffices to prove the second assertion of the Theorem IV.2 for the functions $\mathcal{F}_i^{++}(x,t) \in C_0^\infty(\mathbb{R}^{n+1})$ with $\operatorname{supp}\mathcal{F}_i^{++}(x,t) \subseteq \overline{\mathbb{R}}_{++}^{n+1}$.

Thus, assume that $\mathcal{F}_i^{++}(x,t) \in C_0^\infty(\mathbb{R}^{n+1})$ and let

$$\widehat{\mathcal{F}}_i^{++}(\xi', \zeta_n, p) = \int_0^\infty \widehat{\mathcal{F}}_i^{++}(\xi', p, x_n) e^{-i\zeta_n x_n}\, dx_n,$$

where $\widehat{\mathcal{F}}_i^{++}(\xi', p, x_n)$ is the Fourier transform in x' and the Laplace transform in t of the function $\mathcal{F}_i^{++}(x,t)$. For convenience, let us write conditions (1.4) as the matrix equalities

$$\int_{-\infty}^{\infty} \mathcal{L}^{-1}(i\xi', i\xi_n, p)\Lambda_{++}(i\xi', i\xi_n, p)\xi_n^{q-1} E \cdot \widehat{\mathcal{F}}^{++}(\xi, p)\, d\xi_n = 0,$$

or, equivalently (by the foregoing discussion), as

$$\int_{\Gamma^-} \mathcal{L}^{-1}(i\xi', i\zeta_n, p)\Lambda_{++}(i\xi', i\zeta_n, p)\zeta_n^{q-1} E \cdot \widehat{\mathcal{F}}^{++}(\xi', \zeta_n, p)\, d\zeta_n = 0, \tag{1.11}$$

where Λ_{++} is the diagonal matrix with the elements $\left(\zeta_n - ia(\xi', p)\right)^{s+t_j-N}$, $a(\xi', p)$ $= |\xi'| + (ip)^{1/2b}e^{i\pi/4}$, and E is the unit matrix. Express $\widehat{\mathcal{F}}^{++}(\xi, p)$ in terms of its partial Fourier transform in x' and the Laplace transform in t, $\widehat{\mathcal{F}}^{++}(\cdot, x_n)$, and then change the order of integration with respect to ζ_n and x_n. Then equalities (1.11) take the form

$$\int\limits_0^\infty \left(\int\limits_{\Gamma^-} \mathcal{L}^{-1}(i\xi', i\zeta_n, p)\Lambda_{++}(i\xi', i\zeta_n, p)\zeta_n^{q-1} E e^{-i\zeta_n x_n}\, d\zeta_n \right) \widehat{\mathcal{F}}^{++}(\cdot, x_n)\, dx_n = 0,$$

(1.12)

where Γ^- is a contour in the half-plane $\operatorname{Im}\zeta_n < 0$ that encircles all the ζ_n-roots of the equation $L_-(\xi', \zeta_n, p) = 0$.

Now combine the equalities (1.12) into one matrix equality and transform the latter as follows. First in the integrals with respect to ζ_n change ζ_n to $-\zeta_n$, and then regard each integral as a twice conjugate complex number. This yields

$$\int\limits_0^\infty \left(\overline{\int\limits_{\Gamma^+} e^{i\zeta_n x_n} \overline{\mathcal{L}^{-1}(i\xi', -i\zeta_n, p)}\ \overline{\Lambda_{++}}(i\xi', -i\zeta_n, p)} \right.$$

(1.13)

$$\left. \times \overline{\left(E, -\bar\zeta_n E, \ldots, (-\bar\zeta_n)^{N-1} E\right)}\, d\bar\zeta_n \right) \widehat{\mathcal{F}}^{++}(\xi', p, x_n)\, dx_n = 0,$$

where Γ^+ is a closed contour in the half-plane $\operatorname{Im}\zeta_n > 0$ that encircles all the ζ_n-roots of the equation $L_-(\xi', p, -\zeta_n) = 0$, and the bar denotes complex conjugation.

Let $\mathcal{L}^*(i\xi', D_n, p) = \left(l_{ji}(\cdot, -D_n)\right)_{i,j=1}^m$ be the operator formally adjoint (in the sense of Lagrange) to the ordinary matrix differential operator $\mathcal{L}(i\xi', D_n, p)$ $= \left(l_{ij}(\cdot, D_n)\right)_{i,j=1}^m$. It can be easily seen that $\left(\mathcal{L}^*(i\xi', -i\zeta_n, p)\right)^{-1}$, the inverse of the matrix $\mathcal{L}^*(i\xi', -i\zeta_n, p)$, has the form

$$\left(\mathcal{L}^*\right)^{-1} = \overline{L^{-1}(i\xi', -i\zeta_n, p)}\ \overline{\left(L_{ij}(i\xi', -i\zeta_n, p)\right)_{i,j=1}^m}.$$

Hence, the meaning of relations (1.13) is that the vector $\widehat{\mathcal{F}}^{++}(\xi', p, x_n)$ is orthogonal in the space $\prod\limits_1^m L_2(\mathbb{R}_+^1)$ to all $w_r(\xi', p, x_n)$, which are the columns of the matrix

$$w(\xi', p, x_n) = \int\limits_{\Gamma^+} e^{i\bar\zeta_n x_n} \left(\mathcal{L}^*(i\xi', i\zeta_n, p)\right)^{-1} \overline{\Lambda}_{++}(i\xi', -i\zeta_n, p)$$

(1.14)

$$\times \left(E, -\zeta_n E, \ldots, (-\zeta_n)^{N-1} E\right) d\bar\zeta_n,$$

The relations

$$\mathcal{L}^*(i\xi', D_n, p) \int\limits_{\Gamma^+} e^{i\bar\zeta_n x_n} \left(\mathcal{L}^*\right)^{-1} \overline{\Lambda}_{++}(-\bar\zeta_n)^r\, d\bar\zeta_n =$$

$$= \int_{\Gamma^+} e^{i\bar{\zeta}_n x_n} \overline{\Lambda}_{++}(i\xi', -i\zeta_n, p)(-\bar{\zeta}_n)^r \, d\bar{\zeta}_n$$

$$= \int_{\Gamma^-} e^{-i\zeta_n x_n} \Lambda_{++}(i\xi', i\zeta_n, p)\zeta_n^r \, d\zeta_n = 0, \quad (r = 0, 1, \ldots, N-1)$$

mean that the columns of the matrix $\widehat{w}(\xi', p, x_n)$ are solutions of the homogeneous adjoint system $\mathcal{L}^*(i\xi', D_n, p)\widehat{w}(\xi', p, x_n) = 0$ on the half-line $\mathbb{R}^1_+$. Since the elements of the matrices $\mathcal{L}^{-1}(i\xi', i\zeta_n, p)$ and $\Lambda_{++}(i\xi', i\xi_n, p)$ are homogeneous and since the contour Γ^+ can be chosen in such a way that for $\zeta_n \in \Gamma^+$ we have $\operatorname{Im}\zeta_n > c(|\xi'| + |p|^{1/2b})$, $c > 0$, one can readily show that the columns of the matrix $\widehat{w}(\xi', p, x_n)$ decay exponentially as $x_n \to \infty$ for any $(\xi', p) \neq 0$. Thus, each column $\widehat{w}_r(\xi', p, x_n)$ of the matrix $\widehat{w}(\xi', p, x_n)$ is an exponentially decaying solution of the system $\mathcal{L}^*\widehat{w}_r = 0$, and consequently is a linear combination of the columns that form a basis in the subspace of the solutions of the system $\mathcal{L}^*(i\xi', D_n, p)\widehat{w}_r(\xi', p, x_n) = 0$ that decay exponentially as $x_n \to \infty$. Since the dimension of the basis of stable (i.e., exponentially decreasing as $x_n \to \infty$) solutions equals exactly N due to the parabolicity of $\mathcal{L}(D, D_t)$, it follows that among the $m \cdot N$ columns of the matrix (1.14) there are at least N linearly independent ones. On the other hand, since $(\mathcal{L}\widehat{u}, \widehat{v}) = (\widehat{u}, \mathcal{L}^*\widehat{v})$ for all $\widehat{v} \in C_0^\infty(\mathbb{R}^1)$, for the system $\mathcal{L}(i\xi', D_n, p)\widehat{u}(\xi', p, x_n) = \widehat{\mathcal{F}}^{++}(\xi', p, x_n)$ to be solvable in $L_2 \times \cdots \times L_2$ it is necessary that the vector $\mathcal{F}^{++}(\xi', p, x_n)$ be orthogonal to all linearly independent solutions of the homogeneous adjoint system $\mathcal{L}^*(i\xi', D_n, p)\widehat{v}(\xi', p, x_n) = 0$. Thus, if one assumes that the number of linearly independent columns of the matrix (1.14) is strictly less than N, one concludes that less than N orthogonality conditions are necessary for the solvability of the non-homogeneous system $\mathcal{L}\widehat{u} = \mathcal{F}^{++}$, which is impossible. Hence, for any choice of the pair $(\xi', p) \neq 0$, among the $m \cdot N$ columns of the matrix (1.14) there are exactly N linearly independent ones. This means that among the $m \cdot N$ equations (1.4) there are exactly N linearly independent ones, and all the others can be linearly expressed through them.

IV.2. Initial value and Cauchy problems for parabolic systems in spaces $\mathcal{H}^s$

IV.2.1. Formulation of the initial value problem in the spaces of distributions $\mathcal{H}^s$

In this section we pose the initial boundary value problem for general parabolic systems in the spaces of distributions $\mathcal{H}^s$. Since considering arbitrary parabolic systems complicates the analysis (which is extremely tedious even for simpler cases) and does not bring into the picture any new ideas, we restrict ourselves to the case of a parabolic system for which the numbers s_i and t_j are multiples of $2b$: $s_i = 2bs'_i$, $t_j = 2bt'_j$, where s'_i and t'_j are integers.

Let $\mathcal{L}(D, D_t) = \big(l_{ij}(D, D_t)\big)_{i,j=1}^{m}$ be a linear quasihomogeneous matrix operator with constant coefficients, $\operatorname{ord} l_{ij} = s_i + t_j$. Assume that $u(x) = (u_{10}, \ldots, u_{m0})$ is a sufficiently smooth vector-valued function that solves in $\overline{E}_+^{n+1}$ the system

$$\mathcal{L}_i u \equiv l_{ij}(D, D_t) u_{j0}(x, t) = f_{i0}(x, t), \qquad i = 1, \ldots, m. \tag{2.1}$$

Continue the functions $u_{j0}(x, t)$ by zero in E_-^{n+1} (for $t < 0$), denote by $u_{j0}^+(x, t)$, $f_{i0}^+(x, t)$ the continuations by zero of $u_{j0}(x, t)$, $f_{i0}(x, t)$ for $t < 0$, and set $v_{j\lambda}(x) = D_t^{\lambda-1} u_{j0}(x, +0)$, $\lambda = 1, \ldots, t_j'$. Then, by the Green formula (3.15),

$$\mathcal{L}_i u \big|_{\overline{E}_+^{n+1}} \equiv l_{ij}(D, D_t) u_{j0}^+(x, t) - l_{ij}^{(0,\lambda)}(D, D_t) v_{j\lambda}(x) \times \delta(t) = f_{i0}^+(x, t),$$

$$i = 1, \ldots, m. \tag{2.2}$$

Let $C(D, D_t) = \big(C_{\alpha j}(D, D_t)\big)$, $\alpha = 1, \ldots, r$ be a matrix linear differential operator $\big(C_{\alpha j}(D, D_t) = C_{\alpha j\lambda}(D) D_t^{\lambda-1}$, $\operatorname{ord} C_{\alpha j} = \rho_\alpha + t_j\big)$ of initial conditions

$$C_\alpha u \big|_{\mathbb{R}^n} \equiv C_{\alpha j\lambda}(D) D_t^{\lambda-1} u_{j0}(x, 0) = C_{\alpha j\lambda}(D) v_{j\lambda}(x) = \psi_\alpha(x), \quad \alpha = 1, \ldots, r, \tag{2.3}$$

which satisfies the complementarity condition [94].

We regard equations (2.2), (2.3) as a system of differential equations in the unknown vector-valued functions $U_j = \big(u_{j0}^+(x, t), v_{j1}(x), \ldots, v_{jt_j'}(x)\big)$.

Clearly, by formulas (1.3), from system (2.2) one can uniquely express the functions $u_{j0}^+(x, t)$ in terms of $f_{i0}^+(x, t)$ and $v_{j\lambda}(x)$. Therefore, to find all the above functions it remains to find $v_{j\lambda}(x)$. Since $\sum(s_i' + t_j') = r$, it is clear that if among all s_i' there is at least one negative number, then the system (2.3) for determining $v_{j\lambda}(x)$ is not complete. To find all the functions $v_{j\lambda}(x)$ that figure in (2.2), it is necessary to complement the system of initial conditions (2.3) by additional conditions in order to get a uniquely solvable system. Let us describe a natural procedure for completing (2.3) to a uniquely solvable system. This procedure is presented with detailed proofs in [94, Section 2].

Thus, suppose, for definiteness, that $s_1' \le s_2' \le \ldots \le s_m' = 0$, $t_1' \le t_2' \le \ldots \le t_m'$, $\sum s_i' < 0$. The parabolicity of the operator $\mathcal{L}$ and the complementarity condition for C imply that if $s_i = 2bs_i'$, $t_j = 2bt_j'$, then the numbers ρ_α are also multiples of $2b$: $\rho_\alpha = 2b\rho_\alpha'$ and $-t_m' \le \rho_\alpha' \le -1$. For every $\rho' \ge -t_m'$ we define the numbers $\mu = \mu(\rho')$ and $\nu = \nu(\rho')$ such that $s_1' \le \ldots \le s_\mu' \le \rho' < s_{\mu+1}' \le \ldots \le s_m'$, $-t_n' \le \ldots \le \rho' \le -t_\nu' \le \ldots \le -t_1'$, $\mu(\rho') = m$ for $\rho' \ge 0$.

Since $l_{ij} \equiv 0$ for $s_i + t_j < 0$ and $C_{\alpha j} \equiv 0$ for $\rho_\alpha + t_j < 0$, the first $\mu(\rho')$ equations of the system $\mathcal{L}u = f$ as well as the initial conditions with those α, for which $\rho_\alpha' = \rho'$ do not contain $u_1, \ldots, u_\nu$, that is, they contain only $u_{\nu+1}, \ldots, u_m$. Let $l_{ij}(D, D_t) = l_{ij\lambda}(D) D_t^{\lambda-1}$.

For a fixed ρ' consider the system of equations consisting of the initial conditions with those α for which $\rho_\alpha' = \rho'$ and the first $\mu(\rho')$ equations of the system

$\mathcal{L}u = f$, all differentiated with respect to t for $t = 0$:

$$C_\alpha(D, D_t)u\big|_{t=0} = \sum_{\lambda=1}^{\rho'_\alpha+t'_j+1} C_{\alpha j \lambda}(D)v_{j\lambda}(x) = \psi_\alpha(x), \qquad \forall \alpha : \rho'_\alpha = \rho', \quad (2.4)$$

$$D_t^{\rho'-s'_i}\mathcal{L}_i u\big|_{t=0} = \sum_{\lambda=1}^{s'_i+t'_j} l_{ij\lambda}(D)D_t^{\rho'-s'_i+\lambda-1}u_{j0}(x,0)$$

$$= l_{ij\lambda}(D)v_{j\,\rho'-s'_i+\lambda}(x) = f_{i\,\rho'-s'_i+1}(x) \qquad (2.5)$$

$$\left(\equiv D_t^{\rho'-s'_i}f_{i0}(x,0)\right), \quad \forall i : i = 1,\dots,\mu(\rho').$$

As shown in [94], for each fixed ρ' the structure of the system (2.4), (2.5) is such that the matrix composed of the coefficients of $v_{j\,\rho'-t'_j+1}(x)$ has a non-zero determinant. Therefore, from the system (2.4), (2.5) one can uniquely express the functions $v_{j\,\rho'+t'_j+1}(x)$ in terms of $\psi_\alpha(x)$, $f_{i\,\rho'-s'_i+1}(x)$ and $v_{j\lambda}(x)$ with $\lambda < \rho' + t'_j + 1$. Since for $\rho' = -t'_m$ the functions $v_{j1}(x)$ with j such that $t'_j = t'_m$ can be expressed solely in terms of $\psi_\alpha(x)$ with α such that $\rho'_\alpha = -t'_m$ and $f_{j1}(x)$ with $1 \le i < \mu(t'_m)$, it follows that by assigning to ρ' the values $-t'_m,\dots,-1$ we can successively express all the $v_{j\lambda}(x)$ with $1 \le \lambda \le t'_j$ from the system (2.4), (2.5) solely in terms of $\psi_\alpha(x)$ and $f_{i\lambda}(x)$. In particular, to express the functions $v_{j\lambda}(x)$ with $\lambda > t'_j$ we need only equations (2.5). Expressing all the $v_{j\lambda}(x)$ solely in terms of $\psi_\alpha(x)$ and $f_{i\lambda}(x)$ and substituting them into (2.2), we can express $u_{j0}^+(x,t)$ solely in terms of the right-hand sides of equations (2.2)–(2.5). Thus, by adding conditions (2.5) with $\rho' = -t'_m,\dots,-1$ to the initial conditions (2.3), we obtain the uniquely solvable system (2.2)–(2.5) in the (sufficiently smooth) components of the vector-valued functions $U_j(x,t) = \left(u_{j0}^+(x,t), v_{j1}(x),\dots,v_{jt'_j}(x)\right)$.

To pass to the case of nonsmooth functions we note that each of equations (2.2)–(2.5) admits a continuous extension from the set of sufficiently smooth vector-valued functions $u = (u_1,\dots,u_m)$ to the vector-valued functions $u(x,t)$ with $u_j(x,t) \in \widetilde{\mathcal{H}}_{(t'_j)}^{s+t_j}\left(\overline{E}_+^{n+1}, \gamma\right)$. Indeed, by Lemmas III.13 and III.15, the system of equations (2.2)–(2.5) defines a bounded operator $\mathfrak{U}_0$ that acts from the product space $\prod_{j=1}^{m}\widetilde{\mathcal{H}}_{(t'_j)}^{s+t_j}\left(\overline{E}_+^{n+1},\gamma\right)$ into the space

$$\prod_{i=1}^{m}\mathcal{H}^{s-s_i}\left(\overline{E}_+^{n+1},\gamma\right) \times \prod_{\lambda=1}^{-s'_i} H^{s-s_i-2b\lambda+b}(\mathbb{R}^n) \times \prod_{\alpha=1}^{m} H^{s-\rho_\alpha-b}(\mathbb{R}^n)$$

of vector-valued functions $\left(f_{10}^+(x,t), f_{11}(x),\dots,f_{1,-s'_1}(x),\dots,f_{m0}^+(x,t),\dots,f_{m,-s'_m}(x),\right.$ $\left.\psi_1(x),\dots,\psi_m(x)\right)$. The components of the vector-valued functions

$$\mathcal{F}_i(x,t) = \left(f_{i0}^+(x,t), f_{i1}(x),\dots,f_{i,-s'_i}(x)\right)$$

satisfy the compatibility conditions that follow from the theorem on traces:

$$f_{i\lambda}(x) = D_t^{\lambda-1} \mathcal{L}_i u\big|_{t=0} = D_t^{\lambda-1} f_i(x,0), \qquad \forall \lambda : s - s_i - 2b\lambda + b > 0,$$

that is, $\mathcal{F}_i(x,t) = \chi_0 f_i(x,t)$, where $f_i(x,t) \in \widetilde{\mathcal{H}}^{s-s_i}_{(-s_i')}\big(\overline{E}^{n+1}_+, \gamma\big)$. Thus the natural formulation of the initial value problem is as follows.

Given functions $f_i(x,t) \in \widetilde{\mathcal{H}}^{s-s_i}_{(-s_i')}\big(\overline{E}^{n+1}_+, \gamma\big)$, $\chi_0 f_i(x,t) = \big(f_{i0}^+(x,t), f_{i1}(x), \ldots,$ $f_{i,-s_i'}(x)\big)$, and $\psi_\alpha(x) \in H^{s-\rho_\alpha-b}(\mathbb{R}^n)$, find a solution $u \in \prod_j \widetilde{\mathcal{H}}^{s+t_j}_{(t_j')}\big(\overline{E}^{n+1}_+, \gamma\big)$ of the problem

$$\mathcal{L}_i u\big|_{\overline{E}^{n+1}_+} \equiv l_{ij}(D, D_t) u_{j0}^+(x,t) - l_{ij}^{(0,\lambda)}(D, D_t) v_{j\lambda}(x) \times \delta(t) = f_{i0}^+(x,t)$$

$$i = 1, \ldots, m, \tag{2.6}$$

$$D_t^{\rho'-s_i'} \mathcal{L}_i u\big|_{t=0} \equiv l_{ij\lambda}(D) v_{j\,\rho'-s_i'+\lambda}(x) = f_{i\,\rho'-s_i'+1}(x),$$

$$\rho' = -s_i', \ldots, 1, \quad \forall i : -s_i' \geq 1, \tag{2.7}$$

$$C_\alpha(D, D_t) u\big|_{t=0} = C_{\alpha j\lambda}(D) v_{j\lambda}(x) = \psi_\alpha(x), \qquad \alpha = 1, \ldots, m. \tag{2.8}$$

Prescribing the initial conditions by means of the matrix C (which satisfies the complementarity condition) in the case of general parabolic systems differs, generally speaking, from prescribing the usual Cauchy data at $t = 0$. The formulation of the initial problem becomes simpler in the case of systems parabolic in the sense of Petrovskiĭ, for which it is natural to pose the Cauchy problem. To make all the derivations less tedious, here we will study in detail the Cauchy problem for systems parabolic in the sense of Petrovskiĭ; the analysis of the initial value problem (2.6)–(2.8) involves no essentially new ideas.

IV.2.2. The Cauchy problem in $\widetilde{\mathcal{H}}^s$ for a system parabolic in the sense of Petrovskiĭ

Let $\mathcal{L} = (l_{ij})_{i,j=1}^m$ be a parabolic quasihomogeneous linear operator with constant coefficients, $\operatorname{ord} l_{ij} = t_j = 2bt_j'$, $\operatorname{ord} L = \operatorname{ord} \det \|l_{ij}\| = 2N = 2b\kappa$. For the system $\mathcal{L}u = f$ in $\widetilde{\mathcal{H}}^s$ we consider the Cauchy problem: find a solution $u = (u_1, \ldots, u_m) \in \widetilde{\mathcal{H}}^s$ of *)

$$\mathcal{L}_i u\big|_{\overline{E}^{n+1}_+} \equiv l_{ij}(D, D_t) u_{j0}^+(x,t) - l_{ij}^{(0,\lambda)}(D, D_t) v_{j\lambda}(x) \times \delta(t)$$

$$= f_{i0}^+(x,t) \in \mathcal{H}^s\big(\overline{E}^{n+1}_+, \gamma\big), \qquad i = 1, \ldots, m, \tag{2.9}$$

$$\big(D_t^{\lambda-1} u_{j0}(x,t) = \big) \, v_{j\lambda}(x) = \psi_{j\lambda}(x), \qquad \lambda = 1, \ldots, t_j', \; j = 1, \ldots, m, \tag{2.10}$$

---

*) From now on, the spaces $\mathcal{H}^s(\overline{\Omega}, \gamma)$, $H^s(\overline{G})$, (where Ω and G are domains of $\mathbb{R}^{n+1}$ and $\mathbb{R}^n$) are understood as the spaces $\mathcal{H}^s_+(\Omega, \gamma)$, $H^s_+(G)$ for $s < 0$ and $\mathcal{H}^s(\Omega, \gamma)$, $H^s(G)$ for $s \geq 0$. See also Section V.5.

where repetition of j, λ indicates summation with respect to $\lambda = 1, \ldots, t'_j$ and $j = 1, \ldots, m$. Obviously, the Cauchy problem (2.9), (2.10) is a particular case of the initial value problem (2.6)–(2.8). The Cauchy data (2.10) correspond to a matrix $C(D, D_t)$, consisting of blocks $C_j(D_t)$, in each of which all the elements are zero except for the elements of the j-th row, which equal $1, D_t, D_t^2, \ldots, D_t^{t'_j - 1}$, respectively.

We will analyze the Cauchy problem (2.9), (2.10) in two stages. First we consider the case $s < -t_m + b$ and then $s > -t_m + b$. The reason for this is that for $s < -t_m + b$ the functions $u_{j0}(x, t)$, generally speaking, do not have traces on the hyperplane $t = 0$, while for $s > -t_m + b$ the traces $D_t^{\lambda-1} u_{j0}(x, t) = v_{j\lambda}(x)$ exist for j, λ such that $s + t_j - 2b\lambda + b > 0$, therefore the components $u_{j0}(x)$ and $v_{j\lambda}(x)$ are compatible.

(a) Let $s < -t_m + b$. Then the following assertion holds true.

Theorem IV.3. *Let* $f_{i0}^+(x, t) \in \mathcal{H}^s(\overline{E}_+^{n+1}, \gamma)$, $\gamma > 0$, $\psi_{j\lambda}(x) \in H^{s+t_j-2b\lambda+b}(\mathbb{R}^n)$. *Then for any* $s < -t_m + b$ *the problem* (2.9)–(2.10) *has a unique solution* $u = (u_1, \ldots, u_m) \in \widetilde{\mathcal{H}}^s$, *which obeys the estimate*

$$\{\{u_{j0}^+(x, t), \overline{E}_+^{n+1}, \gamma\}\}_{s+t_j-r,r} + \left\|[v_{j\lambda}(x), \mathbb{R}^n]\right\|_{s+t_j-2b\lambda+b}$$

$$< C\left(\sum_{i=1}^m \{\{f_{i0}^+(x, t), \overline{E}_+^{n+1}, \gamma\}\}_{s-r,r} + \left\|[\psi_{j\lambda}(x), \mathbb{R}^n]\right\|_{s+t_j-2b\lambda+b} \right), \quad (2.11)$$

where $r \geq 0$ *is an arbitrary number and* C *is a constant that does not depend on* $u(x, t)$.

Proof: If in (2.9) we replace $v_{j\lambda}(x)$ by their values $\psi_{j\lambda}(x)$ and then take the Fourier transformation in x and the Laplace transformation in t of (2.9), we obtain

$$\widehat{u}_{j0}^+(\xi, p) = L^{-1}(i\xi, p) L_{ij}(i\xi, p) \widehat{f}_{i0}^+(\xi, p)$$

$$+ L^{-1}(i\xi, p) L_{ij}(i\xi, p) l_{i\nu}^{(0,\lambda)}(i\xi, p) \widetilde{\psi}_{\nu\lambda}(\xi), \quad (2.12)$$

where $L_{ij}(i\xi, p)$ is the cofactor of the element l_{ij} of the matrix $\mathcal{L}$. From the parabolicity of the operator $\mathcal{L}$ and the quasihomogeneity of the functions $L_{ij}(i\xi, p)$ and $l_{ij}^{(0,\lambda)}(i\xi, p)$ it follows that the functions $\Lambda^{s+t_j}(i\xi, p) L^{-1}(i\xi, p)$, $L_{ij}(i\xi, p)$ and $\Lambda^{s+t_j}(i\xi, p) L_{ij}(i\xi, p) l_{i\nu}^{(0,\lambda)}(i\xi, p) L^{-1}(i\xi, p)$ satisfy for $\operatorname{Re} p \geq \gamma_0 > 0$ the assumptions of Lemma III.1 with $\beta = s$ and $\beta = s + t_\nu - 2b\lambda$, respectively. Applying that lemma to estimate the norm $\{\{\cdot, \overline{E}_+^{n+1}, \gamma\}\}_{s+t_j-r,r}$ of every term in the right-hand side of (2.12), we get

$$\{\{u_{j0}^+(x, t), \overline{E}_+^{n+1}, \gamma\}\}_{s+t_j-r,r}$$

$$< C\left(\sum_{i=1}^m \{\{f_{i0}^+(x, t), \overline{E}_+^{n+1}, \gamma\}\}_{s-r,r} + \left\|[\psi_{j\lambda}(x), \mathbb{R}^n]\right\|_{s+t_j-2b\lambda+b} \right),$$

which in conjunction with (2.10) gives (2.11).

Before we proceed to the analysis of the Cauchy problem (2.9), (2.10) for $s > -t_m + b$ we will prove an auxiliary lemma that plays an essential role below.

Lemma IV.1. *Let* $u = (u_1, \ldots, u_m)$, *where* $u_j(x,t) \in \widetilde{\mathcal{H}}^{s+t_j}_{(t'_j)}\left(\overline{E}^{n+1}_+, \gamma\right)$, $s < -t_m + b$ *be the solution of the system* $\mathcal{L}(D, D_t)u = f^+_0(x,t)$, *where one assumes that* $f^+_{i0}(x,t) \in \mathcal{H}^q_+\left(E^{n+1}, \gamma\right)$, $q > -b$. *If* $u_{j0}(x,t) \in \mathcal{H}^{q+t_j}\left(E^{n+1}_+, \gamma\right)$, *where* $u_{j0}(x,t)$ *is the restriction of* $u^+_{j0}(x,t)$ *to* E^{n+1}_+, *then* $u^+_{j0}(x,t) = \left(u_{j0}(x,t)\right)^+$ *and* $v_{j\lambda}(x) = D^{\lambda-1}_t u_{j0}(x,t)$, $\lambda = 1, \ldots, t'_j$, *for any* $j = 1, \ldots, m$.

Proof: The fact that $u(x,t)$ is a solution of the system $\mathcal{L}u = f^+(x,t)$ in $\overline{E}^{n+1}_+$ means, by definition, that the components $u^+_{j0}, v_{j1}, \ldots, v_{jt'_j}$ of the vector-valued functions $u_j(x,t) \in \widetilde{\mathcal{H}}^{s+t_j}_{(t'_j)}\left(\overline{E}^{n+1}_+, \gamma\right)$ solve the system (2.9). Since $u_{j0}(x,t) \in \mathcal{H}^{q+t_j}\left(E^{n+1}_+, \gamma\right)$ and $q > -b$, Theorem II.1 implies that there exist the traces $D^{\lambda-1}_t u_{j0}(x,0) = v'_{j\lambda}(x) \in H^{q+t_j-2b\lambda+b}\left(\mathbb{R}^n\right)$, $\lambda = 1, \ldots, t'_j$, and $w^+_{j0}(x,t) = u^+_{j0}(x,t) - \left(u_{j0}(x,t)\right)^+$ are distributions supported on the hyperplane $t = 0$. Further, by the Green formula (3.15) of Chapter III,

$$l_{ij}(D, D_t)\left(u_{j0}(x,t)\right)^+ - l^{(0,\lambda)}_{ij}(D, D_t)v'_{j\lambda}(x) \times \delta(t) = \left(f_{i0}(x,t)\right)^+, \quad i = 1, \ldots, m.$$
$$(2.13)$$

Since $f^+_{i0}(x,t) - \left(f_{i0}(x,t)\right)^+ \in \mathcal{H}^q_+\left(E^{n+1}, \gamma\right)$ and $f^+_{i0}(x,t) - \left(f_{i0}(x,t)\right)^+ = 0$ in E^{n+1}_+, we obtain from (2.9), (2.13) the relations

$$l_{ij}(D, D_t)w^+_j(x,t) - l^{(0,\lambda)}_{ij}(D, D_t)v''_{j\lambda}(x) \times \delta(t) = 0, \quad v''_{j\lambda}(x) = v_{j\lambda} - v'_{j\lambda}. \quad (2.14)$$

From the foregoing analysis and the inclusions $w^+_j(x,t) \in \mathcal{H}^{s+t_j}\left(\overline{E}^{n+1}_+, \gamma\right)$, $s < -t_m + b$ it follows that $w^+_j(x,t) = \sum_{q=0}^{t'_m - t'_j} \beta_{jq}(x)D^{t'_m - t'_j - q}_t \delta(t)$, $j = 1, \ldots, m$. If we write l_{ij} and $l^{(0,\lambda)}_{ij}$ in the respective forms

$$l_{ij}(D, D_t) = \sum_{r=0}^{t'_j} l_{ijr}(D)D^{t'_j - r}_t,$$

$$(2.15)$$

$$l^{(0,\lambda)}_{ij}(D, D_t) = \sum_{r=0}^{t'_j - \lambda} l_{ijr}(D)D^{t'_j - r - \lambda}_t,$$

and then apply the Fourier transformation in x and the Laplace transformation in t to the equations (2.15), replace $\widehat{w}^+_j(\xi, p)$, $l_{ij}(i\xi, p)$, and $l^{(0,\lambda)}_{ij}(i\xi, p)$ by their expressions, and switch the order of summation with respect to r, j, q, we obtain

$$\sum_{\sigma=0}^{t'_m}\left(\sum_{j:t'_j\leq t'_m-\sigma}\sum_{r+q=\sigma} l_{ijr}(i\xi)\widetilde{\beta}_{jq}(\xi)\right)p^{t'_m-\sigma}$$

$$=\sum_{j:t'_j>t'_m-\sigma}\sum_{\eta=1}^{t'_j}\left(\sum_{r+\lambda=\eta} l_{ijr}(i\xi)\widetilde{v}''_{j\lambda}(\xi)\right)p^{t'_j-\eta}, \qquad i=1,\ldots,m. \quad (2.16)$$

To determine $\widetilde{\beta}_{jq}(\xi)$, $\widetilde{v}''_{j\lambda}(\xi)$, we equate the coefficients of the terms with the same exponent of p in each equation of system (2.16) and then put successively $\sigma=0,\ldots,t'_m$. For $\sigma=0$ we have $\sum_{j=1}^{m} l_{ij0}\widetilde{\beta}_{j0}=0$, $i=1,\ldots,m$. Hence, by the parabolicity of $\mathcal{L}$, $\det\|l_{ij0}\| = \mathrm{const}\neq 0$ and, therefore $\widetilde{\beta}_{j0}=0$, $i=1,\ldots,m$. Then, equating the coefficients of $p^{t'_m-1}$ to determine $(\widetilde{\beta}_{j1}, -\widetilde{v}''_{j1})$, we obtain the system

$$\sum_{j:t'_j\leq t'_m-1} l_{ij0}\widetilde{\beta}_{j1} - \sum_{j:t'_j=t'_m} l_{ij0}\widetilde{v}''_{j1} = 0, \qquad i=1,\ldots,m,$$

which gives $\widetilde{\beta}_{j1}=0$, $\widetilde{v}''_{j1}=0$. To find the remaining $\widetilde{\beta}_{jq}$ and $\widetilde{v}''_{j\lambda}$ we use induction.

Assume that $\widetilde{\beta}_{jq}=0$ for any j such that $t'_j\leq t'_m-q$, $q=1,\ldots,\sigma-1$ and $\widetilde{v}''_{j\lambda}=0$ for any j,λ such that $t'_j-\lambda=t'_m-q$, $q=1,\ldots,\sigma-1$. We want to show that $\widetilde{\beta}_{j\sigma}=0$ for any j such that $t'_j\leq t'_m-\sigma$, $\widetilde{v}''_{j\,t'_j-t'_m+\sigma}=0$ for any j such that $t'_j>t'_m-\sigma$. Indeed, taking into account that $\widetilde{\beta}_{jq}=0$ and $\widetilde{v}''_{j\,t'_j-t'_m+q}=0$ for the indicated values of q, j, and λ, the equality between the coefficients of $p^{t'_m-\sigma}$ yields the relations

$$\sum_{j:t'_j\leq t'_m-\sigma} l_{ij0}\widetilde{\beta}_{j\sigma} - \sum_{j:t'_j>t'_m-\sigma} l_{ij0}\widetilde{v}''_{j\,t'_j-t'_m+\sigma} = 0, \qquad i=1,\ldots,m.$$

This shows that $\widetilde{\beta}_{j\sigma}=0$ for any j such that $t'_j\leq t'_m-\sigma$ and $\widetilde{v}''_{j\,t'_j-t'_m+\sigma}=0$ for any j such that $t'_j>t'_m-\sigma$.

(b) Now we proceed to the Cauchy problem in the case $s>-t_m+b$.

Theorem IV.4. *Let* $f^+_{i0}(x,t)\in\mathcal{H}^s\big(\overline{E}^{n+1}_+,\gamma\big)$, $\psi_{j\lambda}(x)\in H^{s+t_j-2b\lambda+b}(\mathbb{R}^n)$. *Then for any* $s>-t_m+b$ *the problem* (2.9)–(2.10) *has a unique solution* $u=(u_1,\ldots,u_m)\in\widetilde{\mathcal{H}}^s$, *which obeys the estimate*

$$\big|\{\{u_j(x,t),\overline{E}^{n+1}_+,\gamma\}\}\big|_{s+t_j,(t'_j)}$$

$$< C_s\left(\sum_{i=1}^{m}\{\{f_{i0}(x,t),\overline{E}^{n+1}_+,\gamma\}\}_s + [[\Psi_{j\lambda}(x),\mathbb{R}^n]]_{s+t_j-2b\lambda+b}\right),$$

$$s+t_j\notin\mathbb{Z}_{2b}. \qquad (2.17)$$

Proof: Since $f_{i0}^+(x,t) \in \mathcal{H}^s\big(\overline{E}_+^{n+1},\gamma\big)$ and the space $\mathcal{H}^s\big(\overline{E}_+^{n+1},\gamma\big)$ is continuously embedded in $\mathcal{H}^{s-r,r}\big(\overline{E}_+^{n+1},\gamma\big)$ for any $r \geq 0$, it follows from Theorem IV.3 that problem (2.9), (2.10) has a unique solution $u_{j0}^+(x,t) \in \mathcal{H}^{\sigma+t_j-r,r}\big(\overline{E}_+^{n+1},\gamma\big)$, $v_{j\lambda}(x) \in H^{s+t_j-2b\lambda+b}(\mathbb{R}^n)$, $\sigma \leq -t_m+b$ and that estimate (2.11) is valid. The functions $u_{j0}^+(x,t)$ and $v_{j\lambda}(x)$ are, generally speaking, independent and may be considered as the components of elements $u_j(x,t) \in \widetilde{\mathcal{H}}_{(t_j')}^{\sigma+t_j}\big(E_+^{n+1},\gamma\big)$ with $\sigma < -t_m + b$. Thus, to prove Theorem IV.4, it suffices to show that actually

$$1. \quad u_{j0}(x,t) \in \mathcal{H}^{s+t_j}\big(E_+^{n+1},\gamma\big), \qquad \text{for all } j \text{ such that } s+t_j > b, \qquad (2.18)$$

$$D_t^{\lambda-1}u_{j0}(x,0) = v_{j\lambda}(x), \qquad \lambda = 1,\ldots,s''+t_j', \qquad (2.18')$$

$$2. \quad u_{j0}^+(x,t) \in \mathcal{H}_+^{s+t_j}\big(E^{n+1},\gamma\big), \qquad \text{for all } j \text{ such that } s+t_j < b, \qquad (2.19)$$

and verify the validity of estimate (2.17).

1. Consider the functions $u_{j0}^+(x,t)$ with j such that $s+t_j > b$. To prove (2.18) we will use Lemmas III.1, III.3, III.9, and III.10, expression (3.4) and relation (4.9) of Chapter III, which defines equivalent norms in the space $\mathcal{H}^{s+t_j}\big(E_+^{n+1}\gamma\big)$. According to relation (4.9) of Chapter III, to calculate the norm $\big\{\big\{u_{j0}(x,t), E_+^{n+1},\gamma\big\}\big\}_{s+t_j}$ we should first multiply equality (2.12) by $\Lambda^{2bs''+t_j}(i\xi,p)$, then discard the terms that are polynomial in p $\big($which corresponds to the terms with $D_t^\mu\delta(t)\big)$, and finally multiply the remaining expression by $\Lambda^{s-2bs''}(i\xi,p)$ and calculate its L_2-norm. To do this we analyze each term in the right-hand side of (2.12) and transform it as follows. If $s'' > 0$, then by (3.15) we have $\Lambda_0(D,D_t)f_{i0}^+(x,t) = (\Lambda_0 f_{i0})^+(x,t) + \Lambda_0^{(0,\sigma)}(D,D_t)f_{i\sigma}(x) \times \delta(t)$, where $\Lambda_0 = \Lambda^{2bs''}$, $f_{i\sigma}(x) = D_t^{\sigma-1}f_{i0}(x,0)$, or, in terms of Fourier and Laplace transforms,

$$\Lambda_0(i\xi,p)\widehat{f}_{i0}^+(\xi,p) = (\widehat{\Lambda_0 f_{i0}^+})(\xi,p) + \Lambda_0^{(0,\sigma)}(i\xi,p)\widetilde{f}_{i\sigma}(\xi), \qquad (2.20)$$

where repetition of σ indicates summation with respect to σ from 1 to s''. In addition, since $\Lambda^{2bs''+t_j}(i\xi,p)L_{ij}(i\xi,p)l_{iv}^{(0,\lambda)}(i\xi,p)$ are quasihomogeneous polynomials of degrees of homogeneity $2N + 2bs'' + t_\nu - 2b\lambda$, we can write

$$\Lambda^{2bs''+t_j}L_{ij}l_{iv}^{(0,\lambda)}(i\xi,p) = Q_{j\nu\lambda}(\xi,p)L(i\xi,p) + \mathcal{R}_{j\nu\lambda}(\xi,p), \qquad (2.21)$$

where $Q_{j\nu\lambda}(\xi,p)$ and $\mathcal{R}_{j\nu\lambda}(\xi,p)$ are quasihomogeneous functions of degrees of homogeneity $2bs'' + t_\nu - 2b\lambda$ and $2N + 2bs'' + t_\nu - 2b\lambda$, respectively, depending on p polynomially, with the degree of $\mathcal{R}_{j\nu\lambda}$ in p being at most $\kappa - 1$ and $Q_{j\nu\lambda} \equiv 0$ for $2bs'' + t_\nu - 2b\lambda < 0$. Next, in the expression for $\Lambda^{2bs''+t_j}u_{j0}^+$ replace $\Lambda^{2bs''}\widehat{f}_{i0}^+$ and $\Lambda^{2bs''+t_j'}L_{ij}l_{iv}^{(0,\lambda)}(i\xi,p)$ by their expressions (2.20) and (2.21) and in the terms that contain $\widehat{f}_{i\sigma}(\xi)$ isolate the terms that depend polynomially on p. This yields

$$\Lambda^{2bs''+t_j}\widehat{u}_{j0}^+ = L^{-1}\Lambda^{t_j}L_{ij}(\widehat{\Lambda^{2bs''}f_{i0}})^+ + L^{-1}\mathcal{P}_{ij\sigma}(\xi,p)\widetilde{f}_{i\sigma}(\xi)$$

$$+ L^{-1}\mathcal{R}_{j\nu\lambda}(\xi,p)\widetilde{v}_{\nu\lambda}(\xi) + Q_{j\nu\lambda}(\xi,p)\widetilde{v}_{\nu\lambda}(\xi) + \mathcal{N}_{ij\sigma}(\xi,p)\widetilde{f}_{i\sigma}(\xi), \qquad (2.22)$$

where $\mathcal{P}_{ij\sigma}(\xi,p)$ are polynomials in p (the remainders of the division of $\Lambda^{t_j} L_{ij}\Lambda_0^{0,\sigma}(\xi,p)$ by $L(i\xi,p)$ in p), and $\mathcal{N}_{ij\sigma}(\xi,p)$ are also polynomials in p.

Now discard the polynomial (in p) terms in the right-hand side of (2.22), multiply the remaining terms by $\Lambda^{s-2bs''}(i\xi,p)$ and estimate the L_2-norm of each term using Lemma III.1. Since $\big|[f_{i\sigma},\mathbb{R}^n]\big|_{s-2b\sigma+b} < C_s\{\{f_{i0},E_+^{n+1},\gamma\}\}_s$, $\widetilde{v}_{\nu\lambda}(\xi) = \widetilde{\psi}_{\nu\lambda}(\xi)$, we infer from (2.22) that

$$\{\{u_{j0}(x,t),E_+^{n+1},\gamma\}\}_{s+t_j}$$

$$< C\left(\sum_{i=1}^{m}\{\{f_{i0}^{+},E_+^{n+1},\gamma\}\}_s + \big|[\psi_{\nu\lambda}(x),\mathbb{R}^n]\big|_{s+t_j-2b\lambda+b}\right),$$

$$\forall j: s+t_j > b,\ s+t_j \notin \mathbb{Z}_{2b}. \qquad (2.23)$$

This proves the inclusion (2.18).

To prove the validity of equalities (2.18′) we proceed as follows. First note that if $s > 0$ then, by Theorem IV.3, for arbitrary sufficiently smooth functions $f_{i0}(x,t)$ and $\psi_{j\lambda}(x)$ that belong to the spaces indicated in Theorem IV.3, there exists a unique solution $\big(u_{j0}^{+}(x,t),v_{i\lambda}(x),\lambda=1,\ldots,t_j'\big)$ of problem (2.9), (2.10) which satisfies (2.18). Consequently, by Lemma IV.1,

$$u_{j0}^{+}(x,t) = \big(u_{j0}^{+}(x,t)\big)^{+}, \quad v_{j\lambda}(x) = \psi_{j\lambda}(x) = D_t^{\lambda-1}u_{j0}(x,0), \quad \lambda=1,\ldots,t_j'.$$

Now let $s < 0$. Consider the sequences of smooth functions $f_{i0}^{+(\sigma)}(x,t)$ and $\psi_{j\lambda}^{(\sigma)}(x)$, $\sigma=1,2,\ldots$, which converge to $f_{i0}^{+}(x,t)$ and $\psi_{j\lambda}(x)$ in $\mathcal{H}^s\big(E_+^{n+1},\gamma\big)$ and $H^{s+t_j-2b\lambda+b}(\mathbb{R}^n)$, respectively. Then, by (2.11), the sequence of solutions $\big(u_{j0}^{+(\sigma)}(x,t),v_{j\lambda}^{(\sigma)}(x),\lambda=1,\ldots,t_j'\big)$ converges to $\big(u_{j0}^{+}(x,t),v_{j\lambda}(x),\lambda=1,\ldots,t_j'\big)$, the solution of the problem (2.9), (2.10). Morevoer, by (2.23), $u_{j0}^{(\sigma)}(x,t) \to u_{j0}(x,t)$ in $\mathcal{H}^{s+t_j}\big(E_+^{n+1},\gamma\big)$ as $\sigma \to \infty$, for any j such that $s+t_j > b$. Then, by Theorem II.1 on traces, $D_t^{\lambda-1}u_{j0}^{(\sigma)}(x,t) \to D_t^{\lambda-1}u_{j0}(x,0)$ as $\sigma \to \infty$ for any λ such that $s+t_j-2b\lambda+b > 0$. By construction, $v_{j\lambda}^{(\sigma)}(x) \to v_{j\lambda}(x)$. On the other hand, $v_{j\lambda}^{(\sigma)}(x) = D_t^{\lambda-1}u_{j0}^{(\sigma)}(x,+0) \to D_t^{\lambda-1}u_{j0}(x,0)$, which implies equalities (2.18′).

2. Now let us establish the inclusions (2.19) and the estimate of norms $\{\{u_{j0}^{+}(x,t),E^{n+1},\gamma\}\}_{s+t_j}$ for any j such that $s+t_j < b$. Since $s < b$, the norms $\{\{u_{j0}^{+}(x,t),E^{n+1},\gamma\}\}_{s+t_j}$, $s+t_j < b$ cannot be estimated directly by using (2.12) as the L_2-norms of the functions $\Lambda^{s+t_j}(i\xi,p)\widehat{u}_{j0}^{+}(\xi,p)$, because in general the L_2-norms of $\Lambda^{s+t_j}(i\xi,p)L^{-1}(i\xi,p)L_{ij}(i\xi,p)\,l_{iv}^{(0,\lambda)}(i\xi,p)\widetilde{v}_{\nu\lambda}(\xi)$ with ν,λ satisfying $t_\nu-t_j-2b\lambda+b > 0$ are not bounded. To estimate these norms, we observe that for $s''+t_m' \geq 0$ we have $\big(|\xi|^{2b}+|p|\big)^{s''+t_m'} \sim |\xi|^{2bs''+t_m}+|p|^{s''+t_m'}$, and consequently

$$\{\{u_{j0}^{+}(x,t),E^{n+1},\gamma\}\}_{s+t_j} \sim \{\{u_{j0}^{+}(x,t),E^{n+1},\gamma\}\}_{s-2bs''+t_j-t_m,\,2bs''+t_m}$$

$$+ \{\{D_t^{s''+t_m'}u_{j0}^{+},E^{n+1},\gamma\}\}_{s-2bs''+t_j-t_m}. \qquad (2.24)$$

The finiteness of the first norm in the right-hand side of (2.24) and its estimate follow from Theorem IV.3. Hence, the estimation of the norm $\{\{u_{j0}^+, E^{n+1}, \gamma\}\}_{s+t_j}$ reduces to the estimation of the second norm in the right-hand side of (2.24), which in turn is achieved by using the system of parabolic equations in the new functions $\{D_t^{s''+t'_m} u_{j0}^+(x,t), j : s+t_j < b;\ D_t^{t'_m - t'_j}\big(D_t^{s''+t'_j} u_{j0}(x,t)\big)^+, j : s+t_j > b\}$ obtained from system (2.9) as follows.

Let $j(s)$ be the number of the functions $u_{j0}^+(x,t)$ for which $s+t_j < b$ and write $l_{ij}(D, D_t) = l_{ij\alpha}(D)D_t^\alpha$, where one sums with respect to α from 0 to t'_j. By what has been proved above, $u_{j0}^+(x,t) = \big(u_{j0}(x,t)\big)^+$ for any $j > j(s)$ and the inclusions (2.18) as well as the equalities (2.18′) hold. Thus, replacing in the system (2.9) $l_{ij}(D, D_t)u_{j0}^+(x,t)$ with $j > j(s)$ by their expressions provided by formula (2.15) we get

$$\sum_{j \leq j(s)} l_{ij}(D, D_t)u_{j0}^+(x,t) + \sum_{j > j(s)} l_{ij}^{(0,s''+t'_j)}(D, D_t)\big(D_t^{s''+t'_j} u_{j0}(x,t)\big)^+$$

$$= f_{i0}^+(x,t) + \sum_{j \leq j(s)} \sum_{\lambda=1}^{t'_j} l_{ij}^{(0,\lambda)}(D, D_t)v_{j\lambda}(x) \times \delta(t)$$

$$+ \sum_{j > j(s)} \sum_{\lambda=s''+t'_j+1}^{t'_j} l_{ij}^{(0,\lambda)}(D, D_t)v_{j\lambda}(x) \times \delta(t) \tag{2.25}$$

$$+ \sum_{j > j(s)} \sum_{\alpha=0}^{s''+t'_j-1} l_{ij\alpha}(D)\big(D_t^\alpha u_{j0}\big)^+$$

$$\equiv F_i^+(x,t), \qquad i = 1, \ldots, m.$$

Now differentiate each equality (2.25) $s'' + t'_m$ times with respect to t. Using the identities

$$l_{ij}^{(0,s''+t'_j)}(D, D_t)D_t^{s''+t'_j} = l_{ij}(D, D_t) - \sum_{\alpha=0}^{s''+t'_j-1} l_{ij\alpha}(D)D_t^\alpha$$

we obtain

$$\sum_{j \leq j(s)} l_{ij}(D, D_t)D_t^{s''+t'_m} u_{j0}^+ + \sum_{j > j(s)} l_{ij}(D, D_t)D_t^{t'_m - t'_j}\big(D_t^{s''+t'_j} u_{j0}\big)^+$$

$$= D_t^{s''+t'_m} F_i(x,t) + \sum_{j > j(s)} \sum_{\alpha=0}^{s''+t'_j-1} l_{ij\alpha}(D)D_t^\alpha\big(D_t^{s''+t'_j} u_{j0}(x,t)\big)^+ \tag{2.26}$$

$$\equiv \mathcal{F}_i^+(x,t), \qquad i = 1, \ldots, m.$$

Let us regard (2.26) as a system of parabolic equations in the whole space E^{n+1}, with the unknowns $D_t^{s''+t'_m} u_{j0}^+(x,t)$, $D_t^{t'_m-t'_j}\big(D_t^{s''+t'_j} u_{j0}\big)^+$. Theorem IV.3 yields

$$\big\{\big\{D_t^{s''+t'_m} u_j^+, E^{n+1}, \gamma\big\}\big\}_{s-2bs''+t_j-t_m} < C_s\big\{\big\{\mathcal{F}_i^+(x,t), E^{n+1}, \gamma\big\}\big\}_{s-2bs''-t_m}, \tag{2.27}$$

where C_s is a constant that does not depend on u. It remains to estimate the right-hand side of (2.27). By the boundedness of the operator of multiplication by $\theta(t)$ in $\mathcal{H}^{s-2bs''}(E^{n+1}, \gamma)$ for $s \notin \mathbb{Z}_{2b}$, inequality (2.27) yields

$$\big\{\big\{\mathcal{F}_i^+(x,t), E^{n+1}, \gamma\big\}\big\}_{s-2bs''-t_m}$$
$$< C_s\left(\sum_{i=1}^{m} \big\{\big\{F_i^+(x,t), \overline{E}_+^{n+1}, \gamma\big\}\big\}_s + \sum_{j>j(s)} \big\{\big\{u_{j0}, E_+^{n+1}, \gamma\big\}\big\}_{s+t_j}\right). \tag{2.28}$$

Since, by Lemmas III.1 and III.16,

$$\big\{\big\{l_{ij}^{(0,\lambda)}(D, D_t)v_{j\lambda}(x) \times \delta(t), E^{n+1}, \gamma\big\}\big\}_s < C_s\big[\!\big[v_{j\lambda}(x), \mathbb{R}^n\big]\!\big]_{s+t_j-2b\lambda+b}$$

for all j, λ such that $s + t_j - 2b\lambda + b < 0$, and

$$\big\{\big\{l_{ij\alpha}(D)\big(D_t^\alpha u_{j0}\big)^+, E^{n+1}, \gamma\big\}\big\}_s < C\big[\!\big[v_{j\lambda}(x), \mathbb{R}^n\big]\!\big]_{s+t_j-2b\lambda+b}$$

for all $j > j(s)$, we see that the norms $\big\{\big\{F_i^+, E^{n+1}, \gamma\big\}\big\}_s$ obey the bounds

$$\big\{\big\{F_i^+(x,t), E^{n+1}, \gamma\big\}\big\}_s < C_s\left(\sum_{i=1}^{m} \big\{\big\{f_{i0}^+(x,t), E^{n+1}, \gamma\big\}\big\}_s\right.$$
$$\left. + \sum_{j>j(s)} \big\{\big\{u_{j0}(x,t), E_+^{n+1}, \gamma\big\}\big\}_{s+t_j} + \sum_{j,\lambda} \big[\!\big[v_{j\lambda}(x), \mathbb{R}^n\big]\!\big]_{s+t_j-2b\lambda+b}\right). \tag{2.29}$$

Estimates (2.28) and (2.29) in conjunction with (2.27) yield

$$\big\{\big\{D_t^{s''+t'_m} u_j^+, E^{n+1}, \gamma\big\}\big\}_{s-2bs''+t_j-t_m}$$
$$< C\left(\sum_{i=1}^{m} \big\{\big\{f_{i0}^+(x,t), E^{n+1}, \gamma\big\}\big\}_s + \sum_{j,\lambda} \big[\!\big[v_{j\lambda}(x), \mathbb{R}^n\big]\!\big]_{s+t_j-2b\lambda+b}\right). \tag{2.30}$$

Finally, the desired estimate for $\big\{\big\{u_{j0}^+(x,t), E^{n+1}, \gamma\big\}\big\}_{s+t_j}$ follows from (2.30) and the equivalence of norms (2.24). This proves (2.19).

IV.2.3. Theorem on the solvability of the general initial value problem

Now we return to the analysis of the initial boundary value problem (2.6)–(2.8).

Theorem IV.5. *Let $\mathcal{L}(D, D_t)$ be a parabolic operator, and suppose that the operator $C(D, D_t)$ satisfies the complementarity condition with respect to $\mathcal{L}(D, D_t)$. Further, let $\psi_\alpha(x) \in H^{s-\rho_\alpha-b}\left(\mathbb{R}^n\right)$, and $\left(f_{i0}(x, t), f_{i1}(x), \ldots, f_{i,-s_i'}(x)\right)$ the components of the element $f_i(x, t) \in \widetilde{\mathcal{H}}^{s-s_i}_{(-s_i')}\left(\overline{E}^{n+1}_+, \gamma\right)$. Then for any $s \in \mathbb{R}^1$ the problem (2.6)–(2.8) has a unique solution $u(x, t) = (u_1, \ldots, u_m)$, $u_j(x, t) \in \widetilde{\mathcal{H}}^{s+t_j}_{(t_j')}\left(\overline{E}^{n+1}_+, \gamma\right)$, which obeys the estimate*

$$
\left|\{\{u_j(x, t), E^{n+1}_+, \gamma\}\}\right|_{s+t_j,(t_j')} < C_s \Big(\left|\{\{f_i(x, t), E^{n+1}_+, \gamma\}\}\right|_{s-s_i,(-s_i')}
$$
$$
+ \left|[\psi_\alpha(x), \mathbb{R}^n]\right|_{s-\rho_\alpha-b} \Big), \quad s \notin \mathbb{Z}_{2b}, \tag{2.31}
$$

where repetition of the indices i, j, α indicates summation.

Proof: The proof can be broken into two cases ($s < -t_m + b$ and $s > -t_m + b$) and follows the same pattern as the proofs of Theorems IV.3 and IV.4, with some technical differences. The main point is that, in contrast to the Cauchy problem, where $v_{j\lambda}(x)$ are given functions, in the general initial value problem (2.6)–(2.8) the functions $v_{j\lambda}(x)$ must be found from the system of linear differential equations (2.4), (2.5) in the unknowns $v_{j\lambda}(x)$ that appear in (2.6). Thus, refering the reader to the considerations made in Subsection IV.2.1 concerning the formulation of the general initial value problem (2.6)–(2.8), we give here only a brief outline of the proof.

First note that in the case of the general parabolic system (2.6), too, the functions $u^+_{j0}(x, t)$ can be uniquely expressed in terms of the functions $v_{j\lambda}(x)$, $f_{i0}(x, t)$, $f_{i\lambda}(x, t)$, and $\psi_\alpha(x)$. Moreover, formulas (2.12) for determining $\widehat{u}^+_{j0}(\xi, p)$, on which the estimation of the norms $\left\{\{u^+_{j0}(x, t), E^{n+1}, \gamma\}\right\}_{s+t_j}$ relies, remain valid. Hence, to prove the theorem for $s < -t_m + b$ it suffices to express $v_{j\lambda}(x)$ in terms of $f_{i\lambda}(x, t)$ and $\psi_\alpha(x)$ and estimate the norms $\left|[v_{j\lambda}(x), \mathbb{R}^n]\right|_{s+t_j-2b\lambda+b}$.

By the discussion in Subsection IV.2.1, all the functions $v_{j\lambda}(x)$ that figure in system (2.6) are determined uniquely from the system (2.4), (2.5) as follows. For each fixed $\rho' = -t_m', \ldots, -1$ the matrix composed of the coefficients of the unknown functions $v_{j,\rho'+t_j'+1}(x)$ is a nonsingular matrix of scalars. Hence, one can solve each system (2.4), (2.5) for $v_{j,\rho'+t_j'+1}(x)$ in terms of $v_{j\lambda}(x)$ with $\lambda < \rho'+t_j'+1$, $f_{i,\rho'-s_i'+1}(x)$, and $\psi_\alpha(x)$. But since for $\rho' = -t_m'$ the functions $v_{j1}(x)$, $j = \nu(-t_m')+1, \ldots, m$, are expressed solely through $f_{i1}(x)$ with $i = 1, \ldots, \mu(-t_m')$, and $\psi_\alpha(x)$ with α such that $\rho_\alpha' = -t_m'$, we can calculate these functions and substitute them in all the other equations. Then, letting $\rho' = -t_m' + 1$, we obtain $v_{j,\rho'+t_j'+1}(x)$, $j = \nu(-t_m')+1, \ldots, m$ and substitute them in the remaining equations. Iterating this procedure we obtain all the functions $v_{j\lambda}(x)$, expressed solely in terms of

$f_{i\lambda}(x,t)$ and $\psi_\alpha(x)$. Furthermore, thanks to the quasihomogeneity of the operators $l_{ij\lambda}(D)$ and $c_{\alpha j\lambda}(D)$, the functions $v_{j\lambda}(x)$ obey the estimates

$$\big\|[v_{j\lambda}(x),\mathbb{R}^n]\big\|_{s+t_j-2b\lambda+b}$$
$$< C\Big(\big\|[f_{iq}(x),\mathbb{R}^n]\big\|_{s+s_i-2bq+b} + \big\|[\psi_\alpha(x),\mathbb{R}^n]\big\|_{s-\rho_\alpha+b}\Big), \quad \forall s \in \mathbb{R}^1. \qquad (2.32)$$

Estimating the norms $\big\{\{u_{j0}^+(x,t), E_+^{n+1}, \gamma\}\big\}_{s+t_j}$ by the corresponding norms of $f_{i0}^+(x,t)$ and $v_{j\lambda}(x)$ with the help of (2.12) and taking into account estimates (2.32) for $v_{j\lambda}(x)$, we obtain the assertion of Theorem IV.5 for $s < -t_m + b$.

The proof of the theorem for $s > -t_m + b$ parallels the proof of Theorem IV.4.

IV.3. Model parabolic boundary value problems in $\overline{\mathbb{R}}_{++}^{n+1}$

In this section we investigate model parabolic boundary value problems in the spaces $\widetilde{\mathcal{H}}_s$. To avoid rather tedious analytical calculations that play no fundamental role, we start our analysis by posing and studying the model boundary value problem for systems parabolic in the sense of Petrovskiĭ in the spaces of distributions $\widetilde{\mathcal{H}}^s$. Then we will pose the model boundary value problem for general parabolic systems and formulate the main theorem on its solvability.

IV.3.1.   Formulation of the model boundary value problem in the spaces $\widetilde{\mathcal{H}}_s$ for a system parabolic in the sense of Petrovskiĭ

Let $\mathcal{L}(D, D_t)$ be a quasihomogeneous operator parabolic in the sense of Petrovskiĭ, with constant coefficients (ord $l_{ij} = t_j = 2bt_j'$, $\sum t_j = 2b\kappa = 2N$), and let $\mathcal{B}(D, D_t) = \big(b_{qj}(D, D_t)\big)$, $q = 1, \ldots, N$ be a quasihomogeneous matrix linear differential operator with constant coefficients (ord $b_{qj} = \sigma_q + t_j$) that satisfies the Lopatinskiĭ condition with respect to $\mathcal{L}(D, D_t)$.

The *model parabolic boundary value problem* in $\mathbb{R}_{++}^{n+1}$ in the space

$$\mathcal{H}^s = \prod_{j=1}^{m} \mathcal{H}^{s+t_j}\big(\mathbb{R}_{++}^{n+1}, \gamma\big)$$

of sufficiently smooth ($s > \sigma_0 = \max_q(0, \sigma_q+1)$) solutions is the problem of finding a vector-valued function $u_0(x,t) = (u_{10}, \ldots, u_{m0}) \in \mathcal{H}^s$ that satisfies the system

$$\mathcal{L}_i(D, D_t)u_0(x,t) \equiv l_{ij}(D, D_t)u_{j0}(x,t) = f_{i0}(x,t) \in \mathcal{H}^s\big(\mathbb{R}_{++}^{n+1}, \gamma\big), \qquad (3.1)$$
$$i = 1, \ldots, m,$$

in $\mathbb{R}_{++}^{n+1}$, together with the boundary conditions on E_+^n:

$$\mathcal{B}_q(D, D_t)u_0(x,t)\big|_{E_+^n} \equiv b_{qj}(D, D_t)u_{j0}(x,t)\big|_{E_+^n} =$$

$$= \varphi_q(x', t) \in \mathcal{H}^{s-\sigma_q-1/2}(E_+^n, \gamma), \qquad (q = 1, \ldots, N), \qquad (3.2)$$

and the initial conditions at $t = 0$:

$$D_t^{\lambda-1} u_{j0}(x, 0)\big|_{x \in \mathbb{R}_+^n} = \psi_{j\lambda}(x) \in H^{s+t_j-2b\lambda+b}(\mathbb{R}_+^n), \qquad (3.3)$$

$$\lambda = 1, \ldots, t_j', \quad j = 1, \ldots, m,$$

where the boundary and the initial conditions (3.2), (3.3) are understood as an equality between the traces of the corresponding functions on E_+^n and $\mathbb{R}_+^n$.

In the indicated classes of functions the model boundary value problem (3.1)–(3.3) has a unique solution $u_0(x, t)$, which satisfies the a priori estimate

$$\left\| u_{j0}(x, t), \mathbb{R}_{++}^{n+1}, \gamma \right\|_{s+t_j} < C\left(\sum_i \left\| f_{i0}(x, t), \mathbb{R}_{++}^{n+1}, \gamma \right\|_s \right.$$

$$\left. \left[\psi_{j\lambda}(x), \mathbb{R}_+^n \right]_{s+t_j-2b\lambda+b} + \left\langle\!\left\langle \varphi_q(x', t), E_+^n, \gamma \right\rangle\!\right\rangle_{s-\sigma_q-1/2} \right),$$

where $s > \sigma_0$ and f_{i0}, φ_q, and $\psi_{j\lambda}$ satisfy certain compatibility conditions for $t = 0$ (see [1, 94]).

If one discards the requirement that the solution of the boundary value problem (3.1)–(3.3) be sufficiently smooth up to the boundary (with the aim of considering noncompatible or nonsmooth right-hand sides), two essentially new circumstances arise. First, the boundary value problem (3.1)–(3.3) becomes meaningless, since in this case (3.2) and (3.3), being equalities between the traces of the corresponding functions, generally speaking, become meaningless. Second, we will see that, even in the case when equalities (3.2) and (3.3) are meaningful, the problem

$$\mathcal{L}_i u_0(x, t) = f_{i0}'(x, t), \quad \mathcal{B}_q u_0\big|_{E_+^n} = \varphi_q'(x', t), \qquad D_t^{\lambda-1} u_{j0}(x, 0) = \psi_{j\lambda}'(x),$$

$$(i = 1, \ldots, m), \qquad (q = 1, \ldots, N), \qquad (\lambda = 1, \ldots, t_j', \ j = 1, \ldots, m),$$

where $f_{i0}'(x, t)$, $\varphi_q'(x', t)$, and $\psi_{j\lambda}'(x)$ are distributions supported on $\overline{E}_+^n \cup \overline{\mathbb{R}}_{+}^n$, $\mathbb{R}^{n-1}$, and $\mathbb{R}^{n-1}$, respectively, has a nontrivial solution $u_0(x, t)$, infinitely differentiable in $\mathbb{R}_{++}^{n+1}$, that solves the homogeneous problem (3.1)–(3.3). Consequently, discarding the requirement that the solution $u_0(x, t)$ be sufficiently smooth up to the boundary of $\mathbb{R}_{++}^{n+1}$ results in the presence of a nontrivial kernel of the operator $\mathfrak{U}_0$ of problem (3.1)–(3.3). This points to the necessity of considering boundary value problems in "closed domains", in order to take into account the contribution of right-hand sides $f_i'(x, t)$, $\varphi_q'(x', t)$, and $\psi_{j\lambda}'(x)$ that are supported on $\overline{E}_+^n \cup \overline{\mathbb{R}}_+^n$, $\mathbb{R}^{n-1}$, and $\mathbb{R}^{n-1}$, respectively.

Here we will use the Green formulas (3.14)–(3.16) of Chapter III and the operations $\mathcal{L}_i u\big|_{\mathbb{R}_{++}^{n+1}}$ and $\mathcal{B}_q u\big|_{E_+^n}$ based on these formulas to formulate correctly the model boundary value problem in the spaces $\widetilde{\mathcal{H}}^s$. For simplicity, we will assume that in the operators $b_{qj}(D, D_t)$ the order of differentiation of $u_j(x, t)$ with respect to t is at most t'_j. For $u_j(x, t) \in C^\infty_{(0)}(\overline{\mathbb{R}}^{n+1}_{++})$ we introduce the notation

$$u_{j0}^{++}(x, t) = \theta(x_n)\theta(t)u_j(x, t),$$

$$v_{j\lambda}^+(x) = \theta(x_n)D_t^{\lambda-1}u_j(x, 0),$$

$$\omega_{j\lambda k}(x') = D_n^{k-1}D_t^{\lambda-1}u_j(x', 0, 0).$$

If $u_0(x, t) = (u_{10}, \ldots, u_{m0})$ is a smooth solution of the problem (3.1)–(3.3), then, by the Green formulas (3.16), (3.15),

$$\mathcal{L}_i u_0\big|_{\overline{\mathbb{R}}^{n+1}_{++}} \equiv l_{ij}(D, D_t)u_{j0}^{++}(x, t) - l_{ij}^{(k,0)}(D, D_t)u_{jk}^+(x', t) \times \delta(x_n)$$

$$- l_{ij}^{(0,\lambda)}(D, D_t)v_{j\lambda}^+(x) \times \delta(t) + l_{ij}^{(k,\lambda)}(D, D_t)\omega_{j\lambda k}(x') \times \delta(x_n, t)$$

$$= f_{i0}^{++}(x, t), \qquad i = 1, \ldots, m, \qquad (3.4)$$

$$\mathcal{B}_q u_0\big|_{\overline{E}^n_+} \equiv b_{qjk}(D', D_t)u_{jk}^+(x', t) - b_{qjk}^{(\lambda)}(D', D_t)\omega_{j\lambda k}(x') \times \delta(t) = \varphi_q^+(x', t)$$

$$(q = 1, \ldots, N), \qquad (3.5)$$

$$D_t^{\lambda-1}u_{j0}(x, 0)\big|_{\overline{\mathbb{R}}^n_+} \equiv v_{j\lambda}^+(x) = \psi_{j\lambda}^+(x), \qquad \lambda = 1, \ldots, t'_j, \quad j = 1, \ldots, m, \qquad (3.6)$$

where $f_{i0}^{++} = \theta(x_n)\theta(t)f_{i0}$, $\varphi_q^+(x', t) = \theta(t)\varphi_q(x', t)$, $v_{j\lambda}^+(x) = \theta(x_n)v_{j\lambda}(x)$. By analogy to the case of the initial value problem (2.6)–(2.8), we regard the system (3.4)–(3.6) as a system in the unknown functions $u_{j0}^{++}(x, t)$, $u_{jk}^+(x', t)$, $v_{j\lambda}^+(x)$, and $\omega_{j\lambda k}(x')$. If the equations (3.5) involve derivatives of $u_{jk}(x', t)$ of orders $k > t_j + 1$ (i.e., $\sigma_0 > 0$), then the functions $u_{jk}(x', t)$ with $k = t_j, \ldots, \sigma_0 + t_j = \tau_j$ can be found by differentiating the equations of the system $\mathcal{L}u_0 = f_0$ with respect to x_n at $x_n = 0$:

$$\left(D_n^{\sigma-1}\mathcal{L}_i u_0\right)_{E^n_+} = D_n^{\sigma-1}f_{i0}(x, t)\big|_{E^n_+} \equiv f_{i\sigma}(x', t), \quad \sigma = 1, \ldots, \sigma_0, \ i = 1, \ldots, m. \qquad (3.7)$$

Let $l_{ij}(D, D_t) = l_{ijk}(D', D_t)D_n^k$. Then the Green formulas for equations (3.7) take the form

$$\left(D_n^{\sigma-1}\mathcal{L}_i u_0\right)_{\overline{E}^n_+} \equiv l_{ijk}(D', D_t)u_{j\sigma+k}^+(x', t) - l_{ijk}^{(\lambda)}(D', D_t)\omega_{j\lambda\sigma+k}(x') \times \delta(t)$$

$$= f_{i\sigma}^+(x', t), \qquad \sigma = 1, \ldots, \sigma_0, \ i = 1, \ldots, m. \qquad (3.8)$$

Since the order of differentiation of u_{j0} with respect to t in the boundary operators $\mathcal{B}_q$ is at most t'_j, it follows that for any j equations (3.4), (3.5), and (3.8) contain the functions $\omega_{j\lambda k}(x')$ with $\lambda \leq t'_j$, $k \leq \tau_j$. Therefore, in the case of smooth solutions $u_{j0}(x,t)$ the functions $\omega_{j\lambda k}(x')$ can be found by differentiating the initial conditions (3.6) with respect to x_n at $x_n = 0$:

$$\omega_{j\lambda k}(x') = \psi_{j\lambda k}(x') \ \left(\equiv D_n^{k-1} v_{j\lambda}(x)\big|_{x_n=0} = D_n^{k-1}\psi_{j\lambda}(x)\big|_{x_n=0}\right) \quad (3.9)$$

$$(k,\lambda) \in P_j = \big\{(k,\lambda) : k + 2b\lambda \leq \tau_j, \ 1 \leq \lambda \leq t'_j\big\}.$$

For each fixed $j = 1,\ldots,m$ denote by P_j the set of all pairs of indices (k,λ) of the functions $\omega_{j\lambda k}(x')$ that figure in the Green formulas (3.4), (3.5), and (3.8). Also, denote by $u_j(x,t)$, $f_i(x,t)$, and $\psi_{j\lambda}(x)$ the elements of the spaces $\widetilde{\mathcal{H}}^{s+t_j}_{(t'_j,\tau_j,P_j)}(\overline{\mathbb{R}}^{n+1}_{++}\gamma)$, $\widetilde{\mathcal{H}}^s_{(0,\sigma_0,\emptyset)}(\overline{\mathbb{R}}^{n+1}_{++}\gamma)$ and $\widetilde{H}^{s+t_j-2b\lambda+b}_{(\tau_j-2b\lambda)}(\overline{\mathbb{R}}^n_+)$, respectively. Let

$$U_j(x,t) = \big(u_{j0}^{++}(x,t), u_{j1}^+(x',t),\ldots,u_{j\tau_j}^+(x',t),$$

$$v_{j1}^+(x),\ldots,v_{jt'_j}^+(x),\omega_{j\lambda k}(x'), \ (k,\lambda) \in P_j\big) = \chi u_j,$$

$$F_i(x,t) = \big(f_{i0}^{++}(x,t), f_{i1}^+(x',t),\ldots,f_{i\sigma_0}^+(x',t)\big) = \chi f_i(x,t),$$

$$V_{j\lambda}(x) = \big(v_{j\lambda}^+(x),\omega_{j\lambda 1}(x'),\ldots\big) = \chi_0 v_{j\lambda}(x)$$

be the images of u_j, f_i, and $v_{j\lambda}$ under the corresponding isometric isomorphisms; let us agree to identify u_j, f_i, and $v_{j\lambda}$ with their images.

Equalities (3.4)–(3.6), (3.8), and (3.9) can be extended by continuity from the set of sufficiently smooth functions $u_j(x,t)$ to the whole function space $\widetilde{\mathcal{H}}^s = \prod_{j=1}^m \widetilde{\mathcal{H}}^{s+t_j}_{(t'_j,\tau_j,P_j)}(\overline{\mathbb{R}}^{n+1}_{++},\gamma)$. By Lemmas III.16–III.19, they define a bounded operator $\mathfrak{U}$ that acts from all the space $\widetilde{\mathcal{H}}^s$ to the space of vector-valued functions $\mathcal{F} = \big(F_1,\ldots,F_m, \varphi_1,\ldots,\varphi_N, V_{j\lambda}(x), \lambda = 1,\ldots,t'_j, j = 1,\ldots,m\big)$, with $f_{i0}^{++}(x,t) \in \mathcal{H}^s(\overline{\mathbb{R}}^{n+1}_{++},\gamma)$, $f_{i\sigma}^+(x',t) \in \mathcal{H}^{s-\sigma+\frac{1}{2}}(\overline{E}^n_+,\gamma)$, $\varphi_q^+(x',t) \in \mathcal{H}^{s-\sigma-\frac{1}{2}}(\overline{E}^n_+,\gamma)$, $\psi_{j\lambda}(x) \in H^{s+t_j-2b\lambda+b}(\overline{\mathbb{R}}^n_+)$, $\psi_{j\lambda k}(x') \in H^{s+t_j-k-2b\lambda+b+1/2}(\mathbb{R}^{n-1})$, whose components satisfy the compatibility conditions that follow from the theorem on traces. If $s < -t_m + \frac{1}{2}$, then the right-hand sides of (3.4), (3.5), (3.8), and (3.9) are, generally speaking, arbitrary distributions, which are not compatible to one another. If $s > -t_m + \frac{1}{2}$, they satisfy the conditions $f_{i\sigma}(x',t) = D_n^{\sigma-1} f_{i0}(x,t)\big|_{x_n=0}$, $\sigma < s + \frac{1}{2}$, $\psi_{j\lambda k}(x') = D_n^{k-1}\psi_{j\lambda}(x',0)$, for all λ, k such that $k + 2b\lambda < s + b + \frac{1}{2}$. This means that for the indicated values of s, σ, λ, and k the corresponding conditions (3.8) and (3.9) are consequences of conditions (3.4) and (3.6). Hence, the number of independent additional conditions (3.8) and (3.9) depends on s. These compatibility conditions of the right-hand sides of (3.4), (3.6), (3.8), and (3.9) are evidently necessary for the inclusion $u(x,t) \in \widetilde{\mathcal{H}}^s$ to hold.

The foregoing analysis leads to the following natural formulation of the model parabolic boundary value problem in $\widetilde{\mathcal{H}}^s$.

Given vector functions $F_i(x,t) = \big(f_{i0}(x,t), f_{i1}(x',t),\ldots,f_{i\sigma_0}(x',t)\big)$, $i = 1,\ldots,m$, $\Phi(x',t) = (\varphi_1,\ldots,\varphi_N)$, and $\Psi_{j\lambda}(x) = \big(\psi_{j\lambda}^+(x),\psi_{j\lambda 1}(x'),\ldots\big)$ such that

$F_i(x,t) = \chi f_i(x,t),\ \Psi_{j\lambda}(x) = \chi_0 \psi_{j\lambda}(x)$, where

$$f_i(x,t) \in \widetilde{\mathcal{H}}^s_{(0,\sigma_0,\emptyset)}\big(\mathbb{R}^{n+1}_{++}\gamma\big), \quad \psi_{j\lambda}(x) \in \widetilde{H}^{s+t_j-2b\lambda+b}_{(\tau_j-2b\lambda)}\big(\overline{\mathbb{R}}^n_+\big),$$

find elements $u_j(x,t) \in \widetilde{\mathcal{H}}^{s+t_j}$ whose components satisfy the set of equations*)

$$\mathcal{L}_i(D,D_t)u(x,t)\big|_{\overline{\mathbb{R}}^{n+1}_{++}} = f^{++}_{i0}(x,t) \in \mathcal{H}^s\big(\overline{\mathbb{R}}^{n+1}_{++},\gamma\big), \quad i=1,\ldots,m, \quad (3.10)$$

$$\big(D^{\sigma-1}_n \mathcal{L}_i u\big)_{\overline{E}^n_+} = f^+_{i\sigma}(x',t) \in \mathcal{H}^{s-\sigma+1/2}\big(\overline{E}^n_+,\gamma\big), \quad (3.11)$$

$$\sigma = 1,\ldots,\sigma_0,\ i=1,\ldots,m,$$

$$\mathcal{B}_q(D,D_t)u\big|_{\overline{E}^n_+} = \varphi^+_q(x',t) \in \mathcal{H}^{s-\sigma_q-1/2}\big(\overline{E}^n_+,\gamma\big), \quad q=1,\ldots,N, \quad (3.12)$$

$$v_{j\lambda}(x)\big|_{\overline{\mathbb{R}}^n_+} = \psi^+_{j\lambda}(x) \in H^{s+t_j-2b\lambda+b}\big(\overline{\mathbb{R}}^n_+\big), \quad (3.13)$$

$$\lambda = 1,\ldots,t'_j,\ j=1,\ldots,m,$$

$$\omega_{j\lambda k}(x') = \psi_{j\lambda k}(x) \in H^{s+t_j-2b\lambda-k+b+1/2}\big(\mathbb{R}^{n-1}\big), \quad (3.14)$$

$$(k,\lambda) \in P_j,\ j=1,\ldots,m.$$

The problem (3.10)–(3.14) will be referred to as the *model boundary value problem in* $\overline{\mathbb{R}}^{n+1}_{++}$ for a system parabolic in the sense of Petrovskiĭ in the space $\widetilde{\mathcal{H}}^s$, and $u = (u_1,\ldots,u_m) \in \widetilde{\mathcal{H}}^s$ will be called its solution.

Let us pause for a moment to sketch how problem (3.10)–(3.14) can be studied in the spaces $\widetilde{\mathcal{H}}^s$, $s \in \mathbb{R}^1$. In the case $s \le -t_m + \frac{1}{2}$ both the components of $u_j(x,t)$ and the right-hand sides of equations (3.10)–(3.14) are, generally speaking, arbitrary distributions and satisfy no compatibility conditions. Therefore, for $s \le -t_m + \frac{1}{2}$ problem (3.10)–(3.14) may be regarded as a system of equations in u^{++}_{j0}, u^+_{jk}, $v^+_{j\lambda}$, $\omega_{j\lambda k}$, the unknown components of the solution $u(x,t) \in \widetilde{\mathcal{H}}^s$ one is looking for, provided that $\operatorname{supp} u^{++}_{j0} \subseteq \overline{\mathbb{R}}^{n+1}_{++}$. When $\operatorname{supp} u^{++}_{j0} \subseteq \overline{\mathbb{R}}^{n+1}_{++}$ the system (3.10)–(3.14) can be solved explicitly by using the integral Fourier transformation in x and the Laplace transformation in t. The formulas for the explicit representation of the solution yield an a priori estimate of the solution in $\widetilde{\mathcal{H}}^s$. Further, in the case $s > -t_m + \frac{1}{2}$, one proceeds as follows. Regard the smoother, compatible right-hand sides of (3.10)–(3.14), as vector-valued functions that belong to the same spaces, but with s replaced by $\sigma < -t_m + \frac{1}{2}$ for s. By what has been said above, for these (noncompatible) right-hand sides there exists a unique solution $u(x,t) \in \widetilde{\mathcal{H}}^\sigma$. Then, using the smoothness and compatibility of the right-hand sides of (3.10)–(3.14) and a special procedure for increasing the smoothness up to the boundary, one proves that the obtained solution $u(x,t) \in \widetilde{\mathcal{H}}^\sigma$ belongs in

---

*) See the footnote at the beginning of Subsection IV.2.2.

fact to $\widetilde{\mathcal{H}}^s$ and the corresponding a priori estimate holds in $\widetilde{\mathcal{H}}^s$. Based on these considerations, we break the analysis of the boundary value problem (3.10)–(3.14) in $\widetilde{\mathcal{H}}^s$ into two cases: 1) $s \leq -t_m + \frac{1}{2}$ and 2) $s > -t_m + \frac{1}{2}$.

IV.3.2. Reduction of the boundary value problem to a system of linear algebraic equations

In this section we show how, by using the integral Fourier transform in x and the Laplace transform in t, problem (3.10)–(3.14) can be reduced to a system of linear algebraic equations in $\widehat{u}_{jk}^{+}(\xi', p)$, the Fourier transforms in x and the Laplace transforms in t of the components $u_{jk}^{+}(x', t)$ of the sought-for solution $u(x, t) \in \widetilde{\mathcal{H}}^s$. En route we will see that the estimation of the norms $\left\| u_{j0}^{++}(x, t), \mathbb{R}^{n+1}, \gamma \right\|_{s+t_j-r,r}$, for any $s < -t_m + \frac{1}{2}$, $r \geq 0$ can be reduced to that of $\left\langle\!\left\langle u_{jk}^{+}(x', t), E^n, \gamma \right\rangle\!\right\rangle_{s+t_j-k+1/2}$. We wish to emphasize that the availability of a priori estimates precisely for the norms $\left\| u_{j0}^{++}(x, t), \mathbb{R}^{n+1}, \gamma \right\|_{s+t_j-r,r}$ of the distributions $u_{j0}^{++}(x, t)$, which have weighted smoothness index $s + t_j - r$ with respect to all the variables (x, t) and an additional smoothness in the "tangential" variables (x', t), plays a key role and is essentially used below in order to improve the smoothness of the sought-for solution $u(x, t)$ when the smoothness and the compatibility of the right-hand sides of equations (3.10)–(3.14) are increased.

Thus, let $-\infty < s < -t_m + \frac{1}{2}$ and consider the set of equations (3.10)–(3.14) in the unknown components $u_{j0}^{++}(x, t)$, $u_{jk}^{+}(x', t)$, $v_{j\lambda}^{+}(x)$, and $\omega_{j\lambda k}(x')$. Eliminate from equations (3.10)–(3.14) the distributions $v_{j\lambda}^{+}(x)$ and $\omega_{j\lambda k}(x')$ by expressing them from equations (3.13), (3.14) and substituting the results in (3.10)–(3.12). Then equations (3.10)–(3.12) become

$$l_{ij} u_{j0}^{++}(x, t) = \mathcal{F}_i^{++}(x, t) \equiv f_{i0}^{++}(x, t) + l_{ij}^{(k,0)}(D, D_t) u_{jk}^{+}(x', t) \times \delta(x_n)$$

$$+ l_{ij}^{(0,\lambda)}(D, D_t)\psi_{j\lambda}^{+}(x) \times \delta(x_n) - l_{ij}^{(k,\lambda)}(D, D_t)\psi_{j\lambda k}(x') \times \delta(x_n, t), \quad i = 1, \ldots, m, \tag{3.15}$$

$$b_{qjk}(D', D_t) u_{jk}^{+}(x', t) = \Phi_q^{+}(x', t)$$

$$\equiv \varphi_q^{+}(x', t) + b_{qjk}^{(\lambda)}(D', D_t)\psi_{j\lambda k}(x') \times \delta(t) \quad (q = 1, \ldots, N), \tag{3.16}$$

$$l_{ij\sigma k}(D', D_t) u_{jk}^{+}(x', t) = \mathcal{F}_{i\sigma}^{+}(x', t) \equiv f_{i\sigma+1}^{+}(x', t)$$

$$+ l_{ij\sigma k}^{(\lambda)}(D', D_t)\psi_{j\lambda k}(x') \times \delta(t) \quad (\sigma = 0, 1, \ldots, \sigma_0, \ i = 1, \ldots, m). \tag{3.17}$$

Consider the set of equations (3.15)–(3.17) in the remaining unknown functions $u_{j0}^{++}(x, t)$ and $u_{jk}^{+}(x', t)$. Note that if $\omega_{j\lambda k}(x') \in H^{s+t_j-2b\lambda-k+b+1/2}(\mathbb{R}^{n-1})$, $v_{j\lambda}^{+}(x) \in H^{s+t_j-2b\lambda+b}(\overline{\mathbb{R}}_{+}^n)$, $u_{jk}^{+}(x', t) \in \mathcal{H}^{s+t_j-k+1/2}(\overline{E}_{+}^n, \gamma)$, then, by the quasihomogeneity of the operators $l_{ij}^{(k,\lambda)}(D, D_t)$ and Lemma III.1, the following estimates hold for any $s < -t_m + \frac{1}{2}$ and $r \geq 0$:

$$\left\| l_{ij}^{(k,0)} u_{jk}^{+} \times \delta(x_n), \mathbb{R}^{n+1}, \gamma \right\|_{s-r,r} \leq C \left\langle\!\left\langle u_{jk}^{+}(x',t), \overline{E}_{+}^{n}, \gamma \right\rangle\!\right\rangle_{s+t_j-k+1/2}, \qquad (3.18)$$

$$\left\| l_{ij}^{(0,\lambda)} v_{j\lambda}^{+}(x) \times \delta(t), \mathbb{R}^{n+1}, \gamma \right\|_{s-r,r} \leq C \left| \left[v_{j\lambda}^{+}(x), \mathbb{R}^{n} \right] \right|_{s+t_j-2b\lambda+b}, \qquad (3.19)$$

$$\left\| l_{ij}^{(k,\lambda)} \omega_{j\lambda k}(x') \times \delta(x_n,t), \mathbb{R}^{n+1}, \gamma \right\|_{s-r,r}$$

$$\leq C \left\langle\!\left\langle \omega_{j\lambda k}(x'), \mathbb{R}^{n-1} \right\rangle\!\right\rangle'_{s+t_j-2b\lambda-k+b+1/2}, \qquad (3.20)$$

where C are constants that do not depend on u_{jk}^{+}, $v_{j\lambda}^{+}$, $\omega_{j\lambda k}$.

The estimates (3.18)–(3.20) imply that $\mathcal{F}_i^{++}(x,t) \in \mathcal{H}^{s-r,r}(\mathbb{R}^{n+1}, \gamma)$. Hence, Theorem IV.1 yields at once the estimate

$$\left\| u_{j0}^{++}(x,t), \mathbb{R}^{n+1}, \gamma \right\|_{s+t_j-r,r} < C \sum_{i=1}^{m} \left\| \mathcal{F}_i^{++}(x,t), \mathbb{R}^{n+1}, \gamma \right\|_{s-r,r}$$

$$< C \left(\left\langle\!\left\langle u_{jk}^{+}(x',t), \overline{E}_{+}^{n}, \gamma \right\rangle\!\right\rangle_{s+t_j-k+1/2} + \left| \left[v_{j\lambda}^{+}(x), \mathbb{R}^{n} \right] \right|_{s+t_j-2b\lambda+b} \right.$$

$$\left. + \left\langle\!\left\langle \omega_{j\lambda k}(x'), \mathbb{R}^{n-1} \right\rangle\!\right\rangle'_{s+t_j-2b\lambda-k+b+1/2} \right). \qquad (3.21)$$

Moreover, in order for $\operatorname{supp} u_{j0}^{++}(x,t) \subseteq \overline{\mathbb{R}}_{n+1}^{++}$ it is necessary and sufficient that conditions (1.4) be satisfied. Now in (1.4) replace $\mathcal{F}_i^{++}(x,t)$ by their expressions, move all the known functions to the right-hand sides, and denote

$$d_{jk}^{\nu l}(\xi',p) = \int_{-\infty}^{\infty} \Lambda_{++}^{s-r+t_\nu-N}(i\xi,p) L^{-1}(i\xi,p) L_{i\nu}(i\xi,p) l_{ij}^{(k,0)}(i\xi,p) \xi_n^{l-1} d\xi_n, \qquad (3.22)$$

$$\widehat{\beta}_{\nu l}^{+}(\xi',p) = -\int_{-\infty}^{\infty} \Lambda_{++}^{s-r+t_\nu-N}(i\xi,p) L^{-1}(i\xi,p) L_{i\nu}(i\xi,p)$$

$$\times \left[\widehat{f}_{i0}^{++}(\xi,p) + l_{ij}^{(0,\lambda)}(i\xi,p)\widetilde{v}_{j\lambda}^{+}(\xi) - l_{ij}^{(k,\lambda)}(i\xi,p)\widetilde{\omega}_{j\lambda k}(\xi') \right] \xi_n^{l-1} d\xi_n, \qquad (3.23)$$

where repetition of all indices except for ν indicates summation with respect to them; also, in all arguments $i = \sqrt{-1}$. Then conditions (1.4) take the form

$$d_{jk}^{\nu l}(\xi',p)\widehat{u}_{jk}^{+}(\xi',p) = \widehat{\beta}_{\nu l}^{+}(\xi',p), \qquad \nu = 1,\ldots,m, \ l = 1,\ldots,N. \qquad (3.24)$$

Further, applying the Fourier transformation in x' and the Laplace transformation in t we can write equations (3.16) and (3.17) as

$$b_{qjk}(i\xi',p)\widehat{u}_{jk}^{+}(\xi',p) = \widehat{\varphi}_q^{+}(\xi',p) + b_{qjk}^{(\lambda)}(i\xi',p)\widetilde{\omega}_{j\lambda k}(\xi') = \widehat{\Phi}_q^{+}(\xi',p),$$

$$q = 1,\ldots,N, \qquad (3.25)$$

$$l_{ij\sigma k}(i\xi',p)\widehat{u}^{+}_{jk}(\xi',p) = \widehat{f}^{+}_{i\sigma+1}(\xi',p) + l^{(\lambda)}_{ij\sigma k}(i\xi',p)\widetilde{\omega}_{j\lambda k}(\xi') \equiv \widehat{\mathcal{F}}^{+}_{i\sigma}(\xi',p)$$
$$(\sigma = 0,1,\ldots,\sigma_0, \ \ i = 1,\ldots,m). \tag{3.26}$$

Thus, the model boundary value problem (3.10)–(3.14) reduces to the system (3.24)–(3.26), which consists of $\sum \tau_j + N(m-1) = m(n+\sigma_0) + N$ linear (algebraic) equations in $2N + m\sigma_0 = \tau$ unknown functions $\widehat{u}^{+}_{jk}(\xi',p)$. Consequently, the uniqueness of the solution of the original problem (3.10)–(3.14) is equivalent to the uniqueness of the solution of the system of equations (3.24)–(3.26), and so we are led to the following condition.

Condition IV.1. *For any pair (ξ',p) with $\xi' \in \mathbb{R}^{n-1}$ and $\operatorname{Re} p \geq \gamma$, the system of linear algebraic equations (3.24)–(3.26) has a unique solution.*

IV.3.3. Analysis of the algebraic system; construction of a solution analytic in p

The system of algebraic equations (3.24)–(3.26) is overdetermined and has the following structure. By Theorem IV.2, for any pair (ξ',p) with $\xi' \in \mathbb{R}^{n-1}$ and $\operatorname{Re} p \geq \gamma$, among the mN equations (3.24) there are exactly N linearly independent equations, and the remaining ones can be expressed through these independent equations. Condition IV.1 means that these N linearly independent equations complement the system (3.24)–(3.26) to a uniquely solvable system in $\tau = 2N + m\sigma_0$ unknown functions $\widehat{u}^{+}_{jk}(\xi',p)$.

Let $\mathfrak{U}(\xi',p)$ be the matrix composed of the coefficients of the unknown functions $\widehat{u}^{+}_{jk}(\xi',p)$ of the system (3.24)–(3.26). Condition IV.1 implies that for any pair (ξ',p) as above there exists a nonsingular $\tau \times \tau$ matrix $A(\xi',p)$ composed of rows of the matrix $\mathfrak{U}(\xi',p)$, by means of which one obtains the unique solution $\widehat{u}^{+}_{jk}(\xi',p)$ of the system (3.24)–(3.26). However, there is no guarantee that we can choose *one* such matrix $A(\xi',p)$ that is nonsingular for *all pairs* (ξ',p) under consideration, as is necessary in order to estimate the norms $\langle\!\langle \widehat{u}^{+}_{jk}(x',t), E^n, \gamma \rangle\!\rangle_{s+t_j-k+1/2}$. For this reason we will indicate how to construct a left inverse $\mathfrak{U}^{-1}_{\mathrm{L}}(\xi',p)$ to the matrix $\mathfrak{U}(\xi',p)$ for all (ξ',p), and then derive formulas for the solution $\widehat{u}^{+}_{jk}(\xi',p)$ of the system (3.24)–(3.26) and establish the analyticity of $\widehat{u}^{+}_{jk}(\xi',p)$ in p for $\operatorname{Re} p > \gamma$.

For the sake of definiteness we assume that the unknown functions $\widehat{u}_{jk}(\xi',p)$ are arranged in the ascending order of $j,k : k = 1,\ldots,\tau_j, \ j = 1,\ldots,m$, and equations (3.24) are arranged in the ascending order of the indices $\nu = 1,\ldots,m$, $l = 1,\ldots,m$. Formulas (3.22) imply that $d^{\nu l}_{jk}(\xi',p)$ are quasihomogeneous functions of degrees $s - r - N + l - k + t_j$ and $b_{qjk}(\xi',p)$, $l_{ij\sigma k}(\xi',p)$ are quasihomogeneous functions of degrees $\sigma_q + t_j - k + 1$ and $\sigma + t_j - k + 1$, respectively. Hence, as it can be easily verified, the matrix of the system (3.24)–(3.26) has a structure similar to that of the matrix of a general parabolic system [94]. This means that there exist real numbers $a_\alpha(s)$ and $b_\beta(s)$ such that the elements $\eta_{\alpha\beta}(\xi',p)$ (on the α-th row and the β-th column) are quasihomogeneous functions of degrees $a_\alpha(s) + b_\beta(s) : \eta_{\alpha\beta}(\rho\xi',\rho^{2b}p) = \rho^{a_\alpha(s)+b_\beta(s)}\eta_{\alpha\beta}(\xi',p)$, $\rho > 0$. To construct the matrix $\mathfrak{U}^{-1}_{\mathrm{L}}(\xi',p)$ we will use the method proposed in [94].

Thus, let $\mathfrak{U}_q(\xi',p)$, $q=1,\ldots,Q$ be the list of nonsingular $\tau\times\tau$ matrices composed of rows of the matrix $\mathfrak{U}(\xi',p)$: $\Delta_q(\xi',p)=\det\mathfrak{U}_q(\xi',p)\neq 0$. By the foregoing analysis, each $\Delta_q(\xi',p)$ is a quasihomogeneous function: $\Delta_q(\rho\xi',\rho^{2b}p)=\rho^{k_q(s)}\Delta_q(\xi',p)$, $\rho>0$. Without loss of generality, we may assume that all the $k_q(s)$ are different from zero; otherwise, in place of the system (3.24)–(3.26) we may consider the equivalent system obtained by multiplying each equation (3.24) by $\left(|\xi'|^{2b}+p\right)^h$, $h<0$, and for this new system all $k_q(s)\neq 0$.

Let $k_q(s)\neq 0$ and let $k_{(q)}(s)$ be numbers such that $k_q(s)k_{(q)}(s)$ are constants, $q=1,\ldots,Q$. By Condition IV.1, the functions $\Delta_q^{k_{(q)}(s)}(\xi',p)$ have no common zeroes, and consequently

$$\Delta(\xi',p)=\Delta_q^{k_{(q)}(s)}(\xi',p)\overline{\Delta}_q^{k_{(q)}(s)}(\xi',p)=\left|\Delta_q(\xi',p)\right|^{2k_{(q)}(s)}>0,\qquad q=1,\ldots,Q.$$

Denote by $\mathcal{D}_q(\xi',p)$ the $\tau\times N(m-1)$-matrices composed of $N(m-1)$ zero columns and the columns of the adjoint matrix $\widehat{\mathfrak{U}}(\xi',p)$ to $\mathfrak{U}(\xi',p)$, so that $\mathcal{D}_q(\xi',p)\mathfrak{U}(\xi',p)=\Delta_q(\xi',p)E$, where E is the τ-dimensional unit matrix. Also, let

$$\Delta_{(q)}(\xi',p)=\Delta^{-1}(\xi',p)\Delta_q^{k_{(q)}(s)-1}(\xi',p)\overline{\Delta}_q^{k_{(q)}(s)}(\xi',p).$$

It is readily verified that the matrix $\mathfrak{U}_{\mathrm{L}}^{-1}(\xi',p)$ composed of $\Delta_{(q)}(\xi',p)\mathcal{D}_q(\xi',p)$, $q=1,\ldots,Q$, is a left inverse for $\mathfrak{U}$: $\mathfrak{U}_{\mathrm{L}}^{-1}(\xi',p)\mathfrak{U}(\xi',p)=E$. The matrix $\mathfrak{U}(\xi',p)$ is a block matrix, with blocks corresponding to the equations (3.24)–(3.26). For the sake of convenience we will accordingly divide the matrix $\mathfrak{U}_{\mathrm{L}}^{-1}(\xi',p)$ into blocks. Namely, if $d_{jk}^{\nu l}(\xi',p)$, $b_{qjk}(i\xi',p)$, or $l_{ij}(i\xi',p)$ is the element of the β-th column and the α-th row of the matrix $\mathfrak{U}(\xi',p)$, then the element of the matrix $\mathfrak{U}_{\mathrm{L}}^{-1}(\xi',p)$ located in the β-th row and the α-th column will be denoted by $\mathcal{D}_{l\nu}^{kj}(\xi',p)$, $b_q^{kj}(\xi',p)$, and $l_{\sigma i}^{kj}(\xi',p)$, respectively. Then $\widehat{u}_{jk}^{+}(\xi',p)$, the unique solution of the system (3.24)–(3.26), can be written as

$$\widehat{u}_{jk}^{+}(\xi',p)=\mathcal{D}_{l\nu}^{kj}(\xi',p)\widehat{\beta}_{\nu l}^{+}(\xi',p)+b_q^{kj}(\xi',p)\widehat{\Phi}_q^{+}(\xi',p)+l_{\sigma i}^{kj}(\xi',p)\widehat{\mathcal{F}}_{i\sigma}^{+}(\xi',p);\quad(3.27)$$

note that $\mathcal{D}_{l\nu}^{kj}(\xi',p)$, $b_q^{kj}(\xi',p)$ $(q=1,\ldots,N)$, $l_{\sigma i}^{kj}(\xi',p)$ are quasihomogeneous functions of degrees $-s+r+N-l+k-t_j$, $-\sigma_q-t_j+k-1$, $-t_j-\sigma+k-1$, respectively.

Now let us prove the analyticity of $\widehat{u}_{jk}^{+}(\xi',p)$ in p in the half-plane $\operatorname{Re}p>\gamma$. To do this, we make use of the arbitrariness of the choice of a left inverse $\mathfrak{U}_{\mathrm{L}}^{-1}(\xi',p)$ (which is a consequence of the assumption that the solution is unique) and argue in much the same manner as in [94]. Specifically, the unique solution $\widehat{u}_{jk}^{+}(\xi',p)$ is defined by (3.27) with an arbitrary left-inverse matrix $\mathfrak{U}_{\mathrm{L}}^{-1}(\xi',p)$. Hence, in a neighborhood of an arbitrary fixed point p_0 with $\operatorname{Re}p_0>\gamma$ we can take for a left inverse the matrix

$$\mathfrak{U}_{\mathrm{L}}^{-1}(\xi',p)=\Delta_{(q)}(\xi',p,p_0)\mathcal{D}_q(\xi',p),\quad q=1,\ldots,Q,$$

where

$$\Delta_{(q)}(\xi', p, p_0) = \Delta^{-1}(\xi', p, p_0)\Delta_q^{k(q)-1}(\xi', p)\overline{\Delta}_q^{k(q)}(\xi', p_0),$$

$$\Delta(\xi', p, p_0) = \Delta_q^{k(q)}(\xi', p)\overline{\Delta}_q^{k(q)}(\xi', p_0).$$

Note that $\Delta(\xi', p, p_0) > 0$ for any p_0 with $\operatorname{Re} p_0 \geq \gamma$. Hence, $\Delta(\xi', p, p_0) > 0$ in a neighborhood of p_0 thanks to the continuous dependence of $\Delta(\xi', p, p_0)$ on p. Since $\overline{\Delta}_q(\xi', p_0)$ does not depend on p, the elements of the matrix $\mathfrak{U}_{\mathrm{L}}^{-1}(\xi', p, p_0)$ are analytic functions of p in a neighborhood of p_0. Moreover, $\widehat{\beta}_{\nu l}^+(\xi', p)$, $\widehat{\Phi}_q^+(\xi', p)$, and $\widehat{\mathcal{F}}_{i\sigma}^+(\xi', p)$ are analytic functions of p for $\operatorname{Re} p > \gamma$. Hence, if in a neighborhood of p_0 we take the matrix $\mathfrak{U}_{\mathrm{L}}^{-1}(\xi', p, p_0)$ as a left inverse, then the functions $\widehat{u}_{jk}^+(\xi', p)$, defined with its help and formulas (3.27), will be analytic in p in a neighborhood of p_0. Since p_0 is an arbitrary point of the domain $\operatorname{Re} p > \gamma$, we get the analyticity of $\widehat{u}_{jk}^+(\xi', p)$ in p for $\operatorname{Re} p > \gamma$.

IV.3.4. Theorem on well-posedness of the model parabolic boundary value problem in the spaces $\widetilde{\mathcal{H}}^s$, $s < -t_m + \frac{1}{2}$

Let us now proceed to the analysis of the problem (3.10)–(3.14), keeping in mind the discussion in Subsection IV.3.2, the results of the investigation of the algebraic system (3.24)–(3.26), and formulas (3.27) for its solution $\widehat{u}_{jk}^+(\xi', p)$. For convenience, we denote by

$$\left\| u_j(x, t), \overline{\mathbb{R}}_{++}^{n+1}, \gamma \right\|_{s-r+t_j, \, r;(\kappa, \tau, P_j)}$$

the norm defined by equality (5.3) of Chapter III in the case $\Omega = \mathbb{R}_+^n \times \mathbb{R}_+^1$, with the norm $\left\| u_{j0}^{++}(x, t), \overline{\mathbb{R}}_{++}^{n+1}, \gamma \right\|_{s+t_j}$ replaced by $\left\| u_{j0}^{++}(x, t), \overline{\mathbb{R}}_{++}^{n+1}, \gamma \right\|_{s-r+t_j, \, r}$. Also, we denote

$$\left\| f_i(x, t), \overline{\mathbb{R}}_{++}^{n+1}, \gamma \right\|_{s-r, r;(\sigma_0)}$$

$$= \left\| f_{i0}(x, t), \overline{\mathbb{R}}_{++}^{n+1}, \gamma \right\|_{s-r, r} + \sum_{\sigma=1}^{\sigma_0} \left\langle\!\left\langle f_{i\sigma}^+(x', t), \overline{E}_+^n, \gamma \right\rangle\!\right\rangle_{s-\sigma+1/2}.$$

Theorem IV.6. *Let $\mathcal{L}$ be a parabolic operator and $\mathcal{B}$ be an operator that satisfies the Lopatinskiĭ condition. Suppose that the right-hand sides of (3.10)–(3.14) belong to the following spaces:*

$$f_{i0}^{++}(x, t) \in \mathcal{H}^s\big(\overline{\mathbb{R}}_{++}^{n+1}, \gamma\big),$$

$$f_{i\sigma}^+(x', t) \in \mathcal{H}^{s-\sigma+1/2}\big(\overline{E}_+^n, \gamma\big),$$

$$\varphi_q^+(x', t) \in \mathcal{H}^{s-\sigma_q-1/2}\big(\overline{E}_+^n, \gamma\big), \tag{3.28}$$

$$\psi_{j\lambda}^+(x) \in H^{s+t_j-2b\lambda+b}\big(\overline{\mathbb{R}}_+^n\big),$$

$$\psi_{j\lambda k}(x') \in H^{s+t_j-k-2b\lambda+b+1/2}\big(\mathbb{R}^{n-1}\big), \quad \gamma > \gamma_0 > 0.$$

Then for any $s \in \mathbb{R}^1$, $s < -t_m + \frac{1}{2}$ the problem (3.10)–(3.14) has a unique solution $u = (u_1, \ldots, u_m)$, with $u_j(x,t) \in \widetilde{\mathcal{H}}^{s+t_j}_{(t'_j,\tau,P_j)}(\overline{\mathbb{R}}^{n+1}_{++}, \gamma)$, which obeys the estimate

$$\left\| u_j(x,t), \overline{\mathbb{R}}^{n+1}_{++}, \gamma \right\|_{s-r+t_j,\, r;\,(t'_j,\tau_j,P_j)} < C \left(\sum_{i=1}^{m} \left\| f_i(x,t), \overline{\mathbb{R}}^{n+1}_{++}, \gamma \right\|_{s-r,r;(\sigma_0)} \right.$$

$$+ \left\langle\!\!\left\langle \varphi_q^+(x',t), \overline{E}^n_+, \gamma \right\rangle\!\!\right\rangle_{s-\sigma_q-1/2} + \left[\!\left[\psi_{j\lambda}^+(x), \overline{\mathbb{R}}^n_+ \right]\!\right]_{s+t_j-2b\lambda+b}$$

$$\left. + \left\langle\!\!\left\langle \psi_{j\lambda k}(x'), \mathbb{R}^{n-1} \right\rangle\!\!\right\rangle'_{s+t_j-2b\lambda-k+b+1/2} \right),$$

$$\tag{3.29}$$

where repetition of i, j, λ, q, and k indicates summation over all their corresponding values.

Proof: We prove the theorem under the assumption that Condition IV.1 holds for problem (3.10)–(3.14). We will show below that Condition IV.1 is equivalent to the Lopatinskiĭ condition.

As we have seen in Subsection IV.3.2, for given distributions $v_{j\lambda}^+(x)$ and $\omega_{j\lambda k}(x')$, the existence of a unique solution $u(x,t)$ to problem (3.10)–(3.14) in the indicated space is equivalent to the condition that the system (3.24)–(3.26) be uniquely solvable. Further, formula (3.21) shows that deriving the estimate (3.29) amounts to estimating the norms $\left\langle\!\!\left\langle u_{jk}^+(x',t), \overline{E}^n_+, \gamma \right\rangle\!\!\right\rangle_{s+t_j-k+1/2}$. Under Condition IV.1, formulas (3.27) are valid, and so, thanks to the quasihomogeneity of $d^{\nu l}_{jk}(\xi',p)$, $b_{qjk}(i\xi',p)$, and $l_{ij\sigma k}(i\xi',p)$ and Lemma III.1,

$$\left\langle\!\!\left\langle u_{jk}^+(x',t), E^n, \gamma \right\rangle\!\!\right\rangle_{s+t_j-k+1/2} < C \left(\sum_{\nu,l} \left\langle\!\!\left\langle \beta_{\nu l}^+(x',t), E^n, \gamma \right\rangle\!\!\right\rangle_{N-l+r+1/2} \right.$$

$$\tag{3.30}$$

$$\left. + \left\langle\!\!\left\langle \Phi_q^+(x',t), E^n, \gamma \right\rangle\!\!\right\rangle_{s-\sigma_q-1/2} + \sum_{i,\sigma} \left\langle\!\!\left\langle \mathcal{F}_{i\sigma}^+(x',t), E^n, \gamma \right\rangle\!\!\right\rangle_{s-\sigma-1/2} \right).$$

Let us estimate the norms in the right-hand side of (3.30). Note that if $\omega_{j\lambda k}(x') \in H^{s+t_j-2b\lambda-k+b+1/2}(\mathbb{R}^{n-1})$ and $s_{j\lambda k} = s + t_j - 2b\lambda - k + b + \frac{1}{2} < 0$, then $\omega_{j\lambda k}(x') \times \delta(t) \in \mathcal{H}^{s_{j\lambda k}-b}(E^n, \gamma)$ for any $\gamma > 0$ and

$$\left\langle\!\!\left\langle b^{(\lambda)}_{qjk}\omega_{j\lambda k} \times \delta(t), E^n, \gamma \right\rangle\!\!\right\rangle_{s-\sigma_q-1/2} < C \left\langle\!\!\left\langle \omega_{j\lambda k}(x'), \mathbb{R}^{n-1} \right\rangle\!\!\right\rangle'_{s_{j\lambda k}}, \tag{3.31}$$

$$\left\langle\!\!\left\langle l^{(\lambda)}_{ij\sigma k}\omega_{j\lambda k} \times \delta(t), E^n, \gamma \right\rangle\!\!\right\rangle_{s-\sigma-1/2} < C \left\langle\!\!\left\langle \omega_{j\lambda k}(x'), \mathbb{R}^{n-1} \right\rangle\!\!\right\rangle'_{s_{j\lambda k}}, \tag{3.32}$$

Consequently, for $s < -t_m + b + \frac{1}{2}$,

$$\left\langle\!\!\left\langle \Phi_q^+, E^n, \gamma \right\rangle\!\!\right\rangle_{s-\sigma_q-1/2} < C \left(\left\langle\!\!\left\langle \varphi_q^+, E^n, \gamma \right\rangle\!\!\right\rangle_{s-\sigma_q-1/2} \right.$$

$$\left. + \sum_{j,\lambda,k} \left\langle\!\!\left\langle \omega_{j\lambda k}(x'), \mathbb{R}^{n-1} \right\rangle\!\!\right\rangle'_{s_{j\lambda k}} \right), \tag{3.33}$$

$$\langle\!\langle \mathcal{F}^+_{i\sigma}, E^n, \gamma\rangle\!\rangle_{s-\sigma-1/2} < C\bigg(\sum_i f^+_{i\sigma}(x',t), E^n, \gamma\rangle\!\rangle_{s-\sigma-1/2}$$
$$+ \sum_{j,\lambda,k} \langle\!\langle \omega_{j\lambda k}(x'), \mathbb{R}^{n-1}\rangle\!\rangle'_{s_{j\lambda k}} \bigg). \tag{3.34}$$

From the estimates (3.19) and (3.20) it follows that the functions

$$\mathcal{E}^{++}_i(x,t) = f^{++}_{i0}(x,t) + l^{(0,\lambda)}_{ij}(D,D_t)v^+_{j\lambda}(x) \times \delta(t)$$
$$- l^{(k,\lambda)}_{ij}(D,D_t)\omega_{j\lambda k} \times \delta(x_n,t) \tag{3.35}$$

belong to $\mathcal{H}^{s-r,r}(\overline{\mathbb{R}}^{n+1}_{++},\gamma)$ for $s < -t_m + b$ and

$$\|\mathcal{E}^{++}_i(x,t), \mathbb{R}^{n+1}, \gamma\|_{s-r,r} < C\bigg(\|f^{++}_{i0}(x,t), \mathbb{R}^{n+1}, \gamma\|_{s-r,r}$$
$$+ \|[v^+_{j\lambda}(x), \mathbb{R}^n]\|_{s+t_j-2b\lambda+b} + \langle\!\langle \omega_{j\lambda k}(x'), \mathbb{R}^{n-1}\rangle\!\rangle'_{s_{j\lambda k}} \bigg), \tag{3.36}$$

where the summation with respect to (λ, k) is carried out over all $(\lambda, k) \in P_j$, $j = 1, \ldots, m$.

Next, if in relation (3.23) we use the quasihomogeneity of the functions $\Lambda^{s-r+t_\nu-N}_{++}(i\xi,p)L^{-1}(i\xi,p)L_{i\nu}(i\xi,p)$ and the Cauchy-Bunyakovskiĭ inequality, we obtain

$$|\hat{\beta}^+_{\nu l}(\xi',p)|^2 < C(|\xi'|^2 + |p|^{1/b})^{-N+l-1/2} \int\limits_{-\infty}^{\infty} (|\xi'|^2 + |p|^{1/b})^{s-r}|\hat{\mathcal{E}}^{++}_i(\xi,p)|^2\, d\xi_n.$$

Therefore,

$$\langle\!\langle \beta^+_{\nu l}(x',t), E^n, \gamma\rangle\!\rangle_{N-l+r+1/2} < C\|\mathcal{E}^{++}_i(x,t), \mathbb{R}^{n+1}, \gamma\|_{s-r,r}. \tag{3.37}$$

Consequently, for $s < -t_m + \frac{1}{2}$ inequalities (3.33), (3.34), (3.36), and (3.37) are valid and we finally obtain

$$\langle\!\langle u^+_{jk}(x',t), E^n, \gamma\rangle\!\rangle_{s+t_j-k+1/2} < K, \tag{3.38}$$

where K designates the right-hand side of (3.29). Thus, if Condition IV.1 holds for the problem (3.10)–(3.14) and the right-hand sides belong to the spaces indicated in (3.28), then estimates (3.21) and (3.28) hold for any $s \in \mathbb{R}^1$, $s < -t_m + \frac{1}{2}$. These estimates yield (3.29) for any $r \geq 0$, while the analyticity of $\hat{u}^+_{jk}(\xi',p)$ for $\mathrm{Re}\, p > \gamma$ yields the inclusion $\mathrm{supp}\, u^+_{jk}(x',t) \subseteq \overline{E}^n_+$. To complete the proof it remains to apply the fact that Condition IV.1 is equivalent to the Lopatinskiĭ condition.

IV.3.5. Analysis of the model boundary value problem in $\widetilde{\mathcal{H}}^s$ with data compatible with zero at $t = 0$

Let us turn to the investigation of problem (3.10)–(3.14) in the spaces $\widetilde{\mathcal{H}}^s$ for $s > -t_m + \frac{1}{2}$. We have proved in Subsection IV.3.4 that the validity of the parabolicity and the Lopatinskiĭ conditions guarantees the existence of a unique solution $u(x,t)$ to problem (3.10)–(3.14) for arbitrary right-hand sides that belong to the spaces indicated in (3.28) with a finite $s < -t_m + \frac{1}{2}$. In particular, if the inclusions (3.28) hold for some $s > -t_m + \frac{1}{2}$, then, generally speaking, the solution $u(x,t)$ furnished by Theorem IV.6 does not lie in $\widetilde{\mathcal{H}}^s$. As it follows from the theorems on traces of functions in $\mathcal{H}^s$ and H^s, in order for $u(x,t) \in \widetilde{\mathcal{H}}^s$ it is necessary that the right-hand sides of the problem satisfy certain compatibility conditions. Since the compatibility condition in the general case can be reduced to the more simple situation of compatibility with zero at $t = 0$, we begin the analysis of the boundary value problem (3.10)–(3.14) with the case when the right-hand sides are compatible with zero at $t = 0$.

First let us introduce some notations. Since for the problem (3.10)–(3.14) (t'_j, τ_j, P_j) are fixed, we will write $\widetilde{\mathcal{H}}^{s+t_j}(\overline{\mathbb{R}}^{n+1}_{++}, \gamma)$ instead of $\widetilde{\mathcal{H}}^{s+t_j}_{(t'_j, \tau_j, P_j)}(\overline{\mathbb{R}}^{n+1}_{++}, \gamma)$. Also, we will denote by $\widetilde{\mathcal{H}}^{s+t_j}_{+}(\overline{\mathbb{R}}^{n+1}_{++}, \gamma)$ the subspace $\widetilde{\mathcal{H}}^{s+t_j}(\mathbb{R}^{n+1}_{++}, \gamma)$ obtained by completion, with respect to the norm of $\widetilde{\mathcal{H}}^{s+t_j}(\overline{\mathbb{R}}^{n+1}_{++}, \gamma)$, of the set of all functions $u_j(x,t)$ that are infinitely differentiable in $\overline{\mathbb{R}}^{n+1}_{++}$ and vanish at $t = 0$ together with all their derivatives: $v^+_{j\lambda}(x) \equiv 0$, for all λ such that $s + t_j - 2b\lambda + b > 0$, $\omega_{j\lambda k}(x') \equiv 0$, for all (λ, k) such that $s + t_j - 2b\lambda - k + b + \frac{1}{2} > 0$. It can be easily seen that $\widetilde{\mathcal{H}}^{s+t_j}_{+}(\overline{\mathbb{R}}^{n+1}_{++}, \gamma)$ consists of the elements $u_j(x,t)$ whose components $v^+_{j\lambda}(x)$ and $\omega_{j\lambda k}(x')$ vanish identically for the indicated values of λ and k. Moreover, the components $u_{j0}(x,t)$ and $u_{jk}(x',t)$, continued by zero for $t < 0$, retain their smoothness, that is, $u^+_{j0}(x,t) \in \mathcal{H}^{s+t_j}(\overline{\mathbb{R}}^{n+1}_{+}, \gamma)$, $u^+_{jk}(x',t) \in \mathcal{H}^{s+t_j-k+1/2}(E^n, \gamma)$, $k = 1, \ldots, \tau_j$. For $s < -t_j + \frac{1}{2}$ we have $\mathcal{H}^{s+t_j}_{+}(\overline{\mathbb{R}}^{n+1}_{++}, \gamma) = \mathcal{H}^{s+t_j}(\mathbb{R}^{n+1}_{++}, \gamma)$, while for $s > -t_j + \frac{1}{2}$ we have $\mathcal{H}^{s+t_j}_{+} \subseteq \mathcal{H}^{s+t_j}(\mathbb{R}^{n+1}_{++}, \gamma)$.

Further, let $\mathfrak{U}_0$ be the operator of the problem (3.10)–(3.14). Denote $\widetilde{\mathcal{H}}^s = \prod\limits_{j=1}^{m} \widetilde{\mathcal{H}}^{s+t_j}(\overline{\mathbb{R}}^{n+1}_{++}, \gamma)$, $\widetilde{\mathcal{H}}^s_{+} = \prod\limits_{j=1}^{m} \widetilde{\mathcal{H}}^{s+t_j}_{+}$, $\widetilde{\mathcal{K}}^s = \mathfrak{U}_0\widetilde{\mathcal{H}}^s$, $\widetilde{\mathcal{K}}^s_{+} = \mathfrak{U}_0\widetilde{\mathcal{H}}^s_{+}$, and for $u(x,t) \in \widetilde{\mathcal{H}}^s$ consider the vector $\mathfrak{U}_0 u(x,t) = \mathcal{F}(x,t) \in \widetilde{\mathcal{K}}^s$. The components of $\mathcal{F}(x,t)$ obviously satisfy on $\overline{E}^n_{+}$, $\overline{\mathbb{R}}^n_{+}$, $\mathbb{R}^{n-1}$ the compatibility conditions that follow from the theorem on traces:

$$f^+_{i\sigma}(x',t) = \left(D_n^{\sigma-1} f_{i0}(x',0,t)\right)^+, \quad \forall \sigma : s - \sigma + \tfrac{1}{2} > 0,$$

$$v_{j\lambda}(x) = \left(D_t^{\lambda-1} u_{j0}(x,0)\right)^+, \quad \forall \lambda : s + t_j - 2b\lambda + b > 0,$$

$$\omega_{j\lambda k}(x') = D_n^{k-1} D_t^{\lambda-1} u_{j0}(x',0,0) = D_t^{\lambda-1} u_{jk}(x',0,0) =$$

$$= D_n^{k-1} v_{j\lambda}(x',0), \quad \forall (\lambda, k) : s + t_j - 2b\lambda - k + b + \tfrac{1}{2} > 0.$$

In particular, for $u(x,t) \in \widetilde{\mathcal{H}}_+^s$, the corresponding components $v_{j\lambda}^+(x)$ and $\omega_{j\lambda k}(x')$ of the vector $\mathfrak{U}_0 u = \mathcal{F} \in \widetilde{\mathcal{K}}_+^s$ vanish identically. This means that for the existence of a solution $u(x,t)$ of problem (3.10)–(3.14) in the space $\widetilde{\mathcal{H}}_+^s$ with $s > -t_m + \frac{1}{2}$ it is necessary that the corresponding components $v_{j\lambda}^+(x)$ and $\omega_{j\lambda k}(x')$ vanish identically.

Definition IV.1. *The right-hand sides of equations (3.10)–(3.14) are said to be compatible with zero at $t = 0$ to order s if $\mathcal{F}(x,t) \in \widetilde{\mathcal{K}}_+^s$. In other words, the right-hand sides of the boundary value problem (3.10)–(3.14) are compatible with zero at $t = 0$ to order s if the corresponding components $v_{j\lambda}^+(x)$ and $\omega_{j\lambda k}(x')$ vanish identically and*

$$f_{i0}^+(x,t) \in \mathcal{H}_+^s(\overline{\mathbb{R}}_+^{n+1}, \gamma),$$

$$f_{i\sigma}^+(x',t) \in \mathcal{H}_+^{s-\sigma+1/2}(E^n, \gamma),$$

$$\varphi_q^+(x',t) \in \mathcal{H}_+^{s-\sigma_q-1/2}(E^n, \gamma).$$

We will show below that increasing the smoothness of the data of the boundary value problem (3.10)–(3.14) and their compatibility with zero at $t = 0$ results in a corresponding increase in the smoothness of the solution in $\overline{\mathbb{R}}_{++}^{n+1}$.

The proof will be carried out in two steps, by analogy to the case of the Cauchy problem.

First let us show that the solution $u(x,t)$ furnished by Theorem IV.6 is smoother in the "tangential" variables (x',t). Then we will use this observation to prove that in fact $u(x,t)$ is smooth in all variables (x,t) in $\overline{\mathbb{R}}_{++}^{n+1}$.

Theorem IV.7. *Let $\mathcal{L}$ be a parabolic operator, $\mathcal{B}(D, D_t)$ an operator that satisfies the Lopatinskiĭ condition and let $\mathcal{F}(x,t)$, the vector of the right-hand sides of (3.10)–(3.14), belong to the space $\widetilde{\mathcal{K}}_+^s$. Then the unique solution $u(x,t)$ of problem (3.10)–(3.14), furnished by Theorem IV.6, belongs to $\widetilde{\mathcal{H}}_+^s$ for $s \notin \mathbb{Z}_{1,2b}$ and obeys the estimate*

$$\left\|\!\left\| u_j(x,t), \overline{\mathbb{R}}_{++}^{n+1}, \gamma \right\|\!\right\|_{s-s'+t_j-t_m,\, s'+t_m;\, (t_j', \tau_j, P_j)} < C\Bigg(\left\langle\!\left\langle \varphi_q, \overline{E}_+^n, \gamma \right\rangle\!\right\rangle_{s-\sigma_q-1/2}$$

$$+ \sum_{i=1}^m \left\|\!\left\| f_i, \overline{\mathbb{R}}_{++}^{n+1}, \gamma \right\|\!\right\|_{s-s'-t_m,\, s'+t_m;\, (\sigma_0)}$$

$$+ \left|\!\left[\psi_{j\lambda}^+(x), \overline{\mathbb{R}}_+^n \right]\!\right|_{s-2b\lambda+t_j+b}$$

$$+ \left\langle\!\left\langle \psi_{j\lambda k}, \mathbb{R}^{n-1} \right\rangle\!\right\rangle_{s+t_j-2b\lambda-k+b+1/2} \Bigg),$$

(3.39)

where C is a constant that depends on s and γ.

Proof: Let $\mathcal{F}(x,t) \in \widetilde{\mathcal{K}}_+^s$, $s > -t_m + \frac{1}{2}$. Then the components $f_{i0}^{++}(x,t)$ and $f_{i\sigma}^+(x',t)$ belong to the spaces indicated in (3.28) and the initial conditions (3.13) and (3.14) imply that $v_{j\lambda}^+(x) \equiv 0$ for $s + t_j - 2b\lambda + b > 0$, $\omega_{j\lambda k}(x') \equiv 0$ for all (λ, k) such that $s + t_j - 2b\lambda - k + b + \frac{1}{2} > 0$. Next, from the continuity of the embedding $\mathcal{H}_{++}^s(\mathbb{R}^{n+1}, \gamma) \subseteq \mathcal{H}_{++}^{s-r,r}(\mathbb{R}^{n+1}, \gamma)$ for any $r \geq 0$ it readily follows that if $f_{i0}^{++}(x, t) \in \mathcal{H}_{++}^s(\mathbb{R}^{n+1}, \gamma)$, then

$$f_{i0}^{++}(x, t) \in \mathcal{H}_{++}^{s-s'-t_m, s'+t_m}(\mathbb{R}^{n+1}, \gamma), \tag{3.40}$$

$$\left\| f_{i0}^{++}(x, t), \mathbb{R}^{n+1}, \gamma \right\|_{s-s'-t_m, s'+t_m} \leq \left\| f_{i0}(x, t), \overline{\mathbb{R}}_{++}^{n+1}, \gamma \right\|_s. \tag{3.41}$$

From these relations and the foregoing discussion it follows that the functions $\mathcal{E}_i^{++}(x, t)$, defined by (3.35), belong to the spaces $\mathcal{H}_{++}^{s-s'-t_m, s'+t_m}(\mathbb{R}^{n+1}, \gamma)$. Therefore, the assertion of Theorem IV.7 follows from Theorem IV.6 for $r = s' + t_m$.

We see that an increase in the smoothness of the right-hand sides $\mathcal{F}(x, t)$ of problem (3.10)–(3.14) in the tangential variables (x', t) and in the order of their compatibility with zero at $t = 0$ results in a corresponding increase in the smoothness of the solution $u(x, t)$ in (x', t). The next step in the analysis of the problem (3.10)–(3.14) is improving the smoothness of $u(x, t)$ in all the variables (x, t). To do this we need an analogue of Lemma IV.1.

Lemma IV.2. *Let $\mathcal{L}(D, D_t)$ be a parabolic operator and suppose that the distributions $u_{j0}^+(x, t) \in \mathcal{H}^{s+t_j}(\overline{\mathbb{R}}_+^{n+1}, \gamma)$, $u_{jk}(x', t) \in \mathcal{H}^{s+t_j-k+1/2}(E_n, \gamma)$, $s < -t_m + \frac{1}{2}$ form a solution of the following system in $\overline{\mathbb{R}}_+^{n+1}$:*

$$l_{ij}(D, D_t)u_{j0}^+(x, t) - l_{ij}^{(k,0)}(D, D_t)u_{jk}(x', t) \times \delta(x_n) = f_{i0}^+(x, t)$$
$$f_{i0}^+(x, t) \in \mathcal{H}_+^q(\mathbb{R}^{n+1}, \gamma), \qquad q > -\tfrac{1}{2}. \tag{3.42}$$

If $u_{j0}(x, t) \in \mathcal{H}^{q+t_j}(\mathbb{R}_+^{n+1}, \gamma)$, where $u_{j0}(x, t)$ is the restriction of $u_{j0}^+(x, t)$ to $\mathbb{R}_+^{n+1}$, then $u_{j0}^+(x, t) = \left(u_{j0}(x, t)\right)^+$ and $u_{jk}(x', t) = D_n^{k-1}u_{j0}(x', +0, t)$, $k = 1, \dots, m$.

Proof: Since $u_{j0}(x, t) \in \mathcal{H}^{q+t_j}(\mathbb{R}_+^{n+1}, \gamma)$, $q > -\frac{1}{2}$, it follows that there exist the traces $u_{jk}'(x', t) = D_n^{k-1}u_{j0}(x', +0, t)$ and $u_{jk}(x', t) \in \mathcal{H}^{q+t_j-k+1/2}(E_n, \gamma)$, $k = 1, \dots, t_j$. Thus, by formula (3.14) of Chapter III,

$$l_{ij}(D, D_t)\left(u_{j0}(x, t)\right)^+ - l_{ij}^{(k,0)}(D, D_t)u_{jk}'(x', t) \times \delta(x_n)$$
$$= \left(f_{i0}(x, t)\right)^+ \in \mathcal{H}^q(\mathbb{R}_+^{n+1}, \gamma), \qquad q > -\tfrac{1}{2}. \tag{3.43}$$

Subtracting (3.43) from (3.42) we get

$$l_{ij}(D, D_t)w_j^+(x, t) - l_{ij}^{(k,0)}(D, D_t)u_{jk}''(x', t) \times \delta(x_n) = 0, \tag{3.44}$$

where $w_j^+(x, t) = u_{j0}^+ - (u_{j0})^+$ and $u_{jk}''(x', t) = u_{jk}(x', t) - u_{jk}'(x', t)$. The inclusions $w_j^+(x, t) \in \mathcal{H}^{s+t_j}(\mathbb{R}_+^{n+1}, \gamma)$, $s < -t_m + \frac{1}{2}$ and the fact that $w_j^+(x, t) = 0$ in $\mathbb{R}_+^{n+1}$ yield

$$w_j^+(x, t) = \sum_{\sigma=0}^{t_m-t_j} \beta_{j\sigma}(x', t)D_n^{t_m-t_j-\sigma}\delta(x_n), \qquad j = 1, \dots, m.$$

Let $l_{ij}(D, D_t) = \sum_{r=0}^{t_j} l_{ijr}(D', D_t)D_n^{t_j-r}$, $l_{ij}^{(k,0)} = \sum_{r=0}^{t_j-k} l_{ijr}D_n^{t_j-r-k}$. Substitute in (3.44) the expressions for $w_j^+(x,t)$, l_{ij}, and $l_{ij}^{(k,0)}$, and then apply to these equation the Fourier and the Laplace transformations and equate the coefficients of the like terms in powers of $i\xi_n$ in each of the resulting equation. This yields a system of linear equations in the unknown functions $\widehat{\beta}_{j\sigma}(\xi',p)$, $\widehat{u}''_{jk}(\xi',p)$, from which they can be found successively, as in Lemma IV.1. Hence, repeating the argument used to prove Lemma IV.1 we get $\widehat{\beta}_{j\sigma}(\xi',p) \equiv 0$, $\widehat{u}''_{jk}(\xi',p) \equiv 0$ for any j, σ, and k. This implies the assertion of Lemma IV.2.

We are ready to prove the main theorem of this subsection.

Theorem IV.8. *Let $\mathcal{L}$ and $\mathcal{B}$ be operators that satisfy the assumptions of Theorem IV.6, $\mathcal{F}(x,t) \in \widetilde{\mathcal{K}}_+^s$, and let $s \notin \mathbb{Z}_{1,2b}$. Then the solution $u(x,t)$ of problem (3.10)–(3.14) belongs to the space $\widetilde{\mathcal{H}}_+^s$ and satisfies the estimate*

$$\left\| u_j(x,t), \overline{\mathbb{R}}_{++}^{n+1}, \gamma \right\|_{s+t_j, (t'_j, \tau_j, P_j)} < C_s \left(\sum_{i=1}^m \left\| f_i(x,t), \overline{\mathbb{R}}_{++}^{n+1}, \gamma \right\|_{s,(\sigma_0)} \right.$$

$$+ \left\langle\!\!\left\langle \varphi_q(x',t), \overline{E}_+^n, \gamma \right\rangle\!\!\right\rangle_{s-\sigma_q-1/2} + \left[\!\left[\psi_{j\lambda}(x), \overline{\mathbb{R}}_+^n \right]\!\right]_{s+t_j-2b\lambda+b} \tag{3.45}$$

$$\left. + \left\langle\!\!\left\langle \psi_{j\lambda k}(x'), \mathbb{R}^{n-1} \right\rangle\!\!\right\rangle'_{s+t_j-2b\lambda-k+b+1/2} \right).$$

Proof: Note that, since $\widetilde{\mathcal{K}}^s = \widetilde{\mathcal{K}}_+^s$ for $s < -t_m + \frac{1}{2}$, for these values of s the assertion of the theorem is a consequence of Theorem IV.6. This means that for $s < -t_m + \frac{1}{2}$ no compatibility conditions of the right-hand sides are necessary for the solvability of the problem (3.10)–(3.14). Thus it remains to prove the theorem for $s > -t_m + \frac{1}{2}$.

Let $s > -t_m + \frac{1}{2}$ and assume that the right-hand sides of (3.10)–(3.14) belong to the indicated spaces and are compatible with zero at $t = 0$ to order s. As noted in Theorem IV.7, the continuity of the embeddings $\mathcal{H}_{++}^s(\mathbb{R}^{n+1}, \gamma) \subseteq \mathcal{H}_{++}^{s-r,r}(\mathbb{R}^{n+1}, \gamma)$ for any $r \geq 0$ implies the validity of the inclusion (3.40), the estimate (3.41), the inclusions $u_{jk}(x',t) \in \mathcal{H}_+^{s+t_j-k+1/2}(E^n, \gamma)$, and the inequalities $\left\langle\!\!\left\langle u_{jk}(x',t), E^n, \gamma \right\rangle\!\!\right\rangle_{s+t_j-k+1/2} < A$, where A designates the right-hand side of (3.39). Therefore, to prove Theorem IV.8 it suffices to show that

(a)

$$u_{j0}^+(x,t) \in \mathcal{H}_+^{s+t_j}(\mathbb{R}^{n+1}, \gamma), \quad u_{j0}^{++}(x,t) = \theta(x_n)u_{j0}^+(x,t), \tag{3.46}$$

$$D_n^{k-1}u_{j0}^+(x',+0,t) = u_{jk}^+(x',t), \quad k = 1,\ldots, s'+t_j, \ \forall j: s+t_j \geq 0, \tag{3.47}$$

where $u_{j0}^+(x,t)$ is the restriction of the distribution $u_{j0}^{++}(x,t)$ to $\mathbb{R}_+^{n+1}$;

(b)

$$u_{j0}^{++}(x,t) \in \mathcal{H}_{++}^{s+t_j}(\mathbb{R}^{n+1}, \gamma), \quad \forall j: s+t_j < 0. \tag{3.48}$$

Before embarking upon the proof of relations (3.46)–(3.48) we note that the distributions $u_{j0}^{++}(x,t)$ are solutions of the system

$$l_{ij}(D,D_t)u_{j0}^{++}(x,t) = \mathcal{F}_i^{++}(x,t) \equiv l_{ij}^{(k,0)}u_{jk}^+(x',t) \times \delta(x_n)$$

$$+l_{ij}^{(0,\lambda)}v_{j\lambda}^+(x) \times \delta(t) - l_{ij}^{(k,\lambda)}\omega_{j\lambda k}(x') \times \delta(x_n,t) + f_{i0}^{++}(x,t)$$

$$\equiv l_{ij}^{(k,0)}(D,D_t)u_{jk}^+(x',t) \times \delta(x_n) + \mathcal{E}_i^{++}(x,t) + f_{i0}^{++}(x,t), \qquad i = 1,\ldots,m, \quad (3.49)$$

whose right-hand sides $\mathcal{F}_i^{++}(x,t)$ (thanks to the condition of compatibility with zero at $t = 0$ to order s) contain only the distributions $v_{j\lambda}^+(x)$ and $\omega_{j\lambda k}(x')$ with indices j,λ,k such that $s+t_j-2b\lambda+b < 0$, $s+t_j-2b\lambda-k+b+\frac{1}{2} < 0$ respectively. In particular, for $s > b+\frac{1}{2}$, $\mathcal{F}_i^{++}(x,t)$ does not contain $v_{j\lambda}^+(x)$ and $\omega_{j\lambda k}(x')$ at all.

(a) The proof of (3.46) and the estimate of the norms $\|u_{j0}^+(x,t),\mathbb{R}_+^{n+1},\gamma\|_{s+t_j}$ are based on the explicit representation formulas for the solution of (3.49) in terms of the Fourier and Laplace transforms,

$$\widehat{u}_{j0}^{++}(\xi,p) = L_{ij}(i\xi,p)L^{-1}(i\xi,p)\widehat{\mathcal{E}}_i^{++}(\xi,p) + L_{ij}(i\xi,p)L^{-1}(i\xi,p)\widehat{f}_{i0}^{++}(\xi,p)$$

$$+ L_{ij}(i\xi,p)L^{-1}(i\xi,p)l_{i\nu}^{(k,0)}(i\xi,p)\widehat{u}_{jk}^+(\xi',p), \qquad (3.50)$$

where repetition of i,ν,k indicates summation over the corresponding values. By formula (4.8) of Chapter III, (3.50) yields

$$\|u_{j0}^+(x,t),\mathbb{R}^{n+1},\gamma\|_{s+t_j} \sim \|\Lambda_{++}^{s-s'}(D,D_t)P_n^+\Lambda_{++}^{s'+t_j}(D,D_t)u_{j0}^{++}(x,t),\mathbb{R}^{n+1},\gamma\|_0$$

$$\leq\|\Lambda_{++}^{s-s'}(D,D_t)P_n^+\Lambda_{++}^{s'+t_j}(D,D_t)L^{-1}(D,D_t)L_{ij}(D,D_t)\mathcal{E}_i^{++}(x,t),\mathbb{R}^{n+1},\gamma\|_0$$

$$+\|\Lambda_{++}^{s-s'}(D,D_t)P_n^+L_{ij}(D,D_t)L^{-1}(D,D_t)\Lambda_{++}^{s'+t_j}(D,D_t)f_{i0}^{++}(x,t),\mathbb{R}^{n+1},\gamma\|_0$$

$$+\|\Lambda_{++}^{s-s'}(D,D_t)P_n^+L^{-1}(D,D_t)L_{ij}(D,D_t)\Lambda_{++}^{s'+t_j}(D,D_t)l_{i\nu}^{(k,0)}(D,D_t)u_{jk}^+(x',t)$$

$$\times \delta(x_n),\mathbb{R}^{n+1},\gamma\|_0 = N_{j1} + N_{j2} + N_{j3}.$$

$$(3.51)$$

Let us estimate separately each of the norms $N_{j\alpha}$, $\alpha = 1,2,3$. To estimate N_{j1} we use the above-mentioned fact that the terms $\mathcal{E}_i^{++}(x,t)$ contain only the distributions $v_{j\lambda}^+(x)$ and $\omega_{j\lambda k}(x')$ with those j,λ, and k for which $s+t_j-2b\lambda+b < 0$ and $s + t_j - 2b\lambda - k + b + \frac{1}{2} < 0$, respectively, and $\mathcal{E}_i^{++}(x,t) \equiv 0$ for $s > b+\frac{1}{2}$. Since for such $v_{j\lambda}^+(x)$ and $\omega_{j\lambda k}(x')$ we have

$$v_{j\lambda}^+(x) \times \delta(t) \in \mathcal{H}^{s+t_j-2b\lambda}(E^{n+1},\gamma),$$

$$\omega_{j\lambda k}^+(x') \times \delta(x_n,t) \in \mathcal{H}^{s+t_j-2b\lambda-k}(E^{n+1},\gamma),$$

for arbitrary $\gamma > 0$, it follows that $\mathcal{E}_i^{++}(x,t) \in \mathcal{H}^s(E^{n+1},\gamma)$ and $\operatorname{supp}\mathcal{E}_i^{+}(x,t) \subseteq \overline{E}^{n+1}_+$. The functions $\Lambda^{s'+t_j}_{++}(i\xi,p)L^{-1}(i\xi,p)L_{ij}(i\xi,p)$ are quasihomogeneous of degree s', satisfy the assumptions of Lemma III.1 of Chapter III and are analytic in p for $\operatorname{Re} p > 0$. By that lemma, $\Lambda^{s'+t_j}_{++}L^{-1}L_{ij}(D,D_t)$ are bounded operators from $\mathcal{H}^s_+(E^{n+1},\gamma)$ to $\mathcal{H}^{s-s'}_+(E^{n+1},\gamma)$. Hence, from the boundedness of the operator P^+_n in $\mathcal{H}^{s-s'}(E^{n+1},\gamma)$ it immediately follows that

$$N_{j1} < C\Big([v^+_{j\lambda}(x),\mathbb{R}^n]_{s+t_j-2b\lambda+b} + \langle\!\langle\omega_{j\lambda k}(x'),\mathbb{R}^{n-1}\rangle\!\rangle'_{s+t_j-2b\lambda-k+b+1/2}\Big). \quad (3.52)$$

Let us turn now to the norms N_{j2} and N_{j3}. First we transform the expressions containing $f^+_{i0}(x,t)$ and $u^+_{jk}(x,t)$ on which P^+_n acts in the expressions for N_{j2} and N_{j3}. By the commutation formula for the operator $\Lambda^{s'}_{++}(D,D_t)$ and the operator of multiplication by the function $\theta(x_n)$ and by the compatibility of $f^{++}_{i0}(x,t)$ with zero at $t=0$ to order $s > \frac{1}{2}$, we have

$$\Lambda^{s'}_{++}(D,D_t)f^{++}_{i0}(x,t) = \Lambda^{s'}_{++}(D,D_t)\big(lf^+_{i0}(x,t)\big)^+$$

$$+ \sum_{\mu=1}^{s'} J^\mu_n \Lambda^{s'}_{++}(D,D_t)f^+_{i\mu}(x',t) \times \delta(x_n), \quad (3.53)$$

where $lf^+_{i0}(x,t)$ is a smooth extension of $f^{++}_{i0}(x,t)$ from $\mathbb{R}^{n+1}_+$ to $\mathbb{R}^{n+1}$.

For $s > 0$ we replace $\Lambda^{s'}_{++}f^{++}_{i0}(x,t)$ in (3.51) by the right-hand side of (3.53). Let $\mathcal{M}_{ij\mu}(\zeta,p) = m_{ij\mu r}(\zeta',p)\zeta_n^{r-1}$ $(r = 1,\dots,N)$ be the remainders of the division (in ζ_n) of the polynomials $L_{ij}(\zeta,p)\Lambda^{t_j}_{++}(\zeta,p)J^\mu_n\Lambda^{s'}_{++}(\zeta,p)$ by $L(\zeta,p)$ and $Q_{j\nu k}(\zeta,p) = q_{j\nu k r}(\xi',p)\zeta_n^{r-1}$ $(r = 1,\dots,N)$ be the remainders of the division of the polynomials $L_{ij}(\zeta,p)\Lambda^{s'+t_j}_{++}(\zeta,p)l^{(k,0)}_{i\nu}(\zeta,p)$ by $L(\zeta,p)$. Then, by the definition of the operator P^+_n,

$$N_{j2} = \Big\|\Lambda^{s-s'}_{++}P^+_n L_{ij}L^{-1}\Lambda^{t_j}_{++}\Big[\big(\Lambda^{s'}_{++}lf^+_{i0}\big)^+ + J^\mu_n\Lambda^{s'}_{++}f^+_{i\mu}(x',t)\times\delta(x_n)\Big],\mathbb{R}^{n+1},\gamma\Big\|_0$$

$$\leq \Big\|\Lambda^{s-s'}_{++}P^+_n L_{ij}L^{-1}\Lambda^{t_j}_{++}\big(\Lambda^{s'}_{++}lf^+_{i0}\big)^+,\mathbb{R}^{n+1},\gamma\Big\|_0 +$$

$$+ \Big\|\Lambda^{s-s'}_{++}P^+_n L_{ij}L^{-1}\Lambda^{t_j}_{++}J^\mu_n\Lambda^{s'}_{++}\big(f^+_{i\mu}(x',t)\times\delta(x_n)\big),\mathbb{R}^{n+1},\gamma\Big\|_0$$

$$= \Big\|\Lambda^{s-s'}_{++}P^+_n L_{ij}L^{-1}\Lambda^{t_j}_{++}\big(\Lambda^{s'}_{++}lf^+_{i0}\big)^+,\mathbb{R}^{n+1},\gamma\Big\|_0 +$$

$$+ \Big\|\Lambda^{s-s'}_{++}P^+_n L^{-1}D_n^{r-1}m_{ij\mu r}(D',D_t)f^+_{i\mu}(x',t)\times\delta(x_n),\mathbb{R}^{n+1},\gamma\Big\|_0,$$

$$N_{j3} = \Big\|\Lambda^{s-s'}_{++}P^+_n L^{-1}D_n^{r-1}q_{j\nu k r}(D',D_t)u^+_{jk}(x',t)\times\delta(x_n),\mathbb{R}^{n+1},\gamma\Big\|_0.$$

$$(3.54)$$

Since $L_{ij}(\zeta,p)\Lambda^{t_j}_{++}(\zeta,p)J^\mu_n\Lambda^{s'}_{++}(\zeta,p)$ and $L_{ij}(\zeta,p)\Lambda^{s'+t_j}_{++}(\zeta,p)l^{(k,0)}_{i\nu}(\zeta,p)$ are quasihomogeneous functions of degree $2N+s'-\mu$ and $2N+s'+t_\nu-k$, respectively, it follows that $m_{ij\mu r}(\zeta', p)$ and $q_{j\nu kr}(\zeta', p)$ are quasihomogeneous functions of degree $2N + s' - \mu - r + 1$ and $2N + s' + t_\nu - k - r + 1$, respectively, and satisfy the assumption of Lemma III.1. Hence,

$$\langle\!\langle m_{ij}(D', D_t)f_{i\mu}^+(x', t), E^n, \gamma\rangle\!\rangle_{r-2N+s-s'-1/2}$$

$$< C\langle\!\langle f_{i\mu}^+(x', t), E^n, \gamma\rangle\!\rangle_{s-\mu+1/2}, \tag{3.55}$$

$$\langle\!\langle q_{j\nu kr}(D', D_t)u_{\nu k}^+(x', t), E^n, \gamma\rangle\!\rangle_{r-2N+s-s'-1/2}$$

$$< C\langle\!\langle u_{\nu k}^+(x', t), E^n, \gamma\rangle\!\rangle_{s-t_\nu-k+1/2}. \tag{3.56}$$

The boundedness of the operator of multiplication by $\theta(x_n)$ in $\mathcal{H}^{s-s'}(\mathbb{R}^{n+1}, \gamma)$ and the smoothness of the extension $lf_{i0}^+(x, t)$ imply that

$$\left\|(\Lambda_{++}^{s'}(D, D_t)lf_{i0}^+(x, t))^+, \mathbb{R}^{n+1}, \gamma\right\|_{s-s'} < C\left\|f_{i0}^+(x, t), \mathbb{R}^{n+1}, \gamma\right\|_s.$$

By Theorem II.1 on traces (see Chapter II),

$$\langle\!\langle f_{i\mu}^+(x', t), E^n, \gamma\rangle\!\rangle_{s-\mu+1/2} < C\left\|f_{i0}(x, t), \mathbb{R}_+^{n+1}, \gamma\right\|_s.$$

Therefore, estimating N_{j2} and N_{j3} directly with the help of (3.55)–(3.56) and taking into account that the operators figuring in N_{j2} and N_{j3} are bounded in the corresponding space, we obtain

$$N_{j2} \leq \left\|L_{ij}L^{-1}\Lambda_{++}^{t_j}(\Lambda_{++}^{s'}lf_{i0}^+)^+, \mathbb{R}^{n+1}, \gamma\right\|_{s-s'}$$

$$+ \left\|L_{ij}L^{-1}J_n^\mu\Lambda_{++}^{s'}f_{i\mu}^+(x', t) \times \delta(x_n), \mathbb{R}^{n+1}, \gamma\right\|_{s-s'}$$

$$\leq C\left\|(\Lambda_{++}^{s'}lf_{i0}^+)^+, \mathbb{R}^{n+1}, \gamma\right\|_{s-s'}$$

$$+ C\left\|L^{-1}(D, D_t)D_n^{r-1}m_{j\mu r}(D', D_t)f_{i\mu}^+(x', t) \times \delta(x_n), \mathbb{R}^{n+1}, \gamma\right\|_{s-s'}$$

$$\leq C\left\|f_{i0}^+(x, t), \mathbb{R}^{n+1}, \gamma\right\|_s$$

$$+ \left\|m_{ij\mu r}(D', D_t)f_{i\mu}^+(x', t) \times \delta(x_n), \mathbb{R}^{n+1}, \gamma\right\|_{s-s'+r-2N-1}$$

$$\leq C\left\|f_{i0}^+(x, t), \mathbb{R}_+^{n+1}, \gamma\right\|_s. \tag{3.57}$$

Estimates (3.51)–(3.57) yield the first part of (3.46). It remains to prove equalities (3.46) and (3.47). To do this, we use Lemma IV.2 and the method of approximating the solution of the boundary value problem (3.10)–(3.14) by smooth solutions, constructed for smooth compatible right-hand sides.

Thus, for any right-hand sides of (3.10)–(3.14), compatible with zero at $t = 0$ to order s, that is, for $f_{i0}^+(x, t) \in \mathcal{H}_+^s(\mathbb{R}_+^{n+1}, \gamma)$, $\varphi_q^+(x', t) \in \mathcal{H}_+^{s-\sigma_q-1/2}(E^n, \gamma)$,

$\psi_{j\lambda}^+(x) \in H^{s+t_j-2b\lambda+b}(\overline{\mathbb{R}}_+^n, \gamma)$, $\psi_{j\lambda k}(x') \in H^{s+t_j-2b\lambda-k+b+1/2}(\mathbb{R}^{n-1})$ $\left(\psi_{j\lambda}^+(x) \equiv 0,\right.$ $\psi_{j\lambda k}(x') \equiv 0$ for $s+t_j-2b\lambda+b > 0$, $s+t_j-2b\lambda-k+b+\frac{1}{2} > 0$) problem (3.10)–(3.14) has a unique solution $u(x,t) = (u_1, \ldots, u_m)$ and for any j such that $s + t_j > \frac{1}{2}$ we have $u_{j0}^+(x,t) \in \mathcal{H}^{s+t_j}(\mathbb{R}_+^{n+1}, \gamma)$, $u_{jk}^+(x',t) \in \mathcal{H}_+^{s+t_j-k+1/2}(E^n, \gamma)$, $k = 1, \ldots, t_j$. In particular, if the order of compatibility of the right-hand sides with zero at $t = 0$ is higher that $\sigma_0 = \max_q(0, \sigma_q)$, then the inclusions (3.46) hold for any $j = 1, \ldots, m$; from Lemma IV.2 it immediately follows that $u_{j0}^{++}(x,t) = \theta(x_n)u_{j0}^+(x,t)$ and that equality (3.47) holds for any $j = 1, \ldots, m$.

Let us regard the components $u_{j0}^{++}(x,t)$, $u_{j0}^+(x,t)$ of the solution $u = (u_1, \ldots, u_m) \in \prod_{j=1}^m \widetilde{\mathcal{H}}_+^{s+t_j}$ as a solution of the system (3.15)–(3.17). From the estimates (3.38) and (3.39) it follows, in particular, that the norms $\left\| u_{j0}^+(x,t), \mathbb{R}_+^{n+1}, \gamma \right\|_{s+t_j}$ and $\left\langle\!\left\langle u_{jk}^+(x',t), E^n, \gamma \right\rangle\!\right\rangle_{s+t_j-k+1/2}$ depend continuously on the right-hand sides of equations (3.10)–(3.14) in the corresponding spaces. Using this observation and the fact that the smooth functions that vanish together with all their derivatives at $t = 0$ and $x_n = 0$ are dense in the spaces $\mathcal{H}_+^{s+t_j}(\mathbb{R}_+^{n+1}, \gamma)$, $\mathcal{H}_+^{s+t_j-k+1/2}(E^n, \gamma)$, $\mathcal{H}_+^{s-\sigma_q-1/2}(E^n, \gamma)$, and $H_+^{s+t_j-2b\lambda+b}(\mathbb{R}^n)$, respectively, we construct sequences of smooth functions $u_{j0}^{(p)++}(x,t)$ and $u_{jk}^{(p)+}(x',t)$ that converge as $p \to \infty$ to $u_{j0}^{++}(x,t)$ and $u_{jk}^+(x',t)$, respectively, in the sense described below.

Specifically, let $s > -t_m + \frac{1}{2}$ and suupose that, for any j, λ, k for which $\psi_{j\lambda}^+ \not\equiv 0$, $\psi_{j\lambda k}(x') \not\equiv 0$, the functions $\psi_{j\lambda}^{(p)+}(x) \in C_{(0)}^\infty(\overline{\mathbb{R}}_+^n)$, $\varphi_q^{(p)+}(x',t) \in C_{(0)}^\infty(\overline{E}_+^n)$, and $\psi_{j\lambda k}(x') \in C_0^\infty(\mathbb{R}^{n-1})$ with supports in $\mathbb{R}_+^n$, E_+^n, and $\mathbb{R}^{n-1}$, converge to $\psi_{j\lambda}^+(x)$, $\varphi_q^+(x',t)$, and $\psi_{j\lambda k}(x')$ in the spaces $H^{s+t_j-2b\lambda+b}(\overline{\mathbb{R}}_+^n)$, $\mathcal{H}_+^{s-\sigma_q-1/2}(E^n, \gamma)$, and $H^{s+t_j-k-2b\lambda+b+1/2}(\mathbb{R}^{n-1})$, respectively. Denote by $\delta_p(x_n) \in C_0^\infty(\mathbb{R}_+^1)$ and $\delta_p(t) \in C_0^\infty(\mathbb{R}_+^1)$ δ-like sequences that converge to $\delta(x_n)$ and $\delta(t)$, respectively, as $p \to \infty$. Finally, let $f_i^{(p)++}(x,t) \in C_{(0)}^\infty(\overline{\mathbb{R}}_{++}^{n+1})$ be functions that converge to $f_i^{++}(x,t)$ in the spaces $\widetilde{\mathcal{H}}^s$.

Consider the distributions

$$F_i^{++}(x,t) = f_{i0}^{++}(x,t) + l_{ij}^{(0,\lambda)}(D, D_t)\psi_{j\lambda}^+(x) \times \delta(t)$$

$$- l_{ij}^{(k,\lambda)}(D, D_t)\psi_{j\lambda k}(x') \times \delta(x_n, t),$$

$$\Phi_q^+(x',t) = \varphi_q^+(x',t) + b_{qj\lambda}^{(k)}(D', D_t)\psi_{j\lambda k}(x') \times \delta(t),$$

$$\mathcal{F}_{i\mu}^+(x',t) = f_{i\mu+1}^+(x',t) + l_{ij\mu k}^{(\lambda)}(D', D_t)\psi_{j\lambda k}(x') \times \delta(t),$$

which figure in the right-hand sides of (3.15)–(3.17). Denote by $F_i^{(p)++}(x,t)$, $\Phi_q^{(p)+}(x',t)$, and $\mathcal{F}_{i\mu}^{(p)+}(x',t)$ the functions obtained from $F_i^{++}(x,t)$, $\Phi_q^+(x',t)$, and $\mathcal{F}_{i\mu}^+(x',t)$ upon replacing in them $f_{i0}^{++}(x,t)$, $f_{i\mu}^+(x',t)$, $\varphi_q^+(x',t)$, $\delta(x_n)$, and $\delta(t)$

by $f_{i0}^{(p)++}(x,t)$, $f_{i\mu}^{(p)+}(x',t)$, $\varphi_q^{(p)+}(x',t)$, $\delta_p(x_n)$, and $\delta_p(t)$, respectively. Obviously, $F_i^{(p)++}(x,t)$, $\Phi_q^{(p)+}(x',t)$, and $\mathcal{F}_{i\mu}^{(p)+}(x',t)$ are infinitely differentiable in $\overline{\mathbb{R}}_{++}^{n+1}$, $\overline{E}_+^n$, and $\overline{E}_+^n$. Clearly, if we replace in (3.15)–(3.17) $F_i^{++}(x,t)$, $\Phi_q^+(x',t)$, and $\mathcal{F}_{i\mu}^+(x',t)$ by $F_i^{(p)++}(x,t)$, $\Phi_q^{(p)+}(x',t)$, and $\mathcal{F}_{i\mu}^{(p)+}(x',t)$, we obtain a boundary value problem (3.10)–(3.14) whose right-hand sides are compatible with zero at $t=0$ to any order $s > \sigma_0 > 0$.

Denote by $u_{j0}^{(p)++}(x,t)$, $u_{jk}^{(p)+}(x',t)$ $(p=1,2,\ldots)$ the components of the solution $u^{(p)}(x,t)$ of problem (3.10)–(3.14) constructed for $F_i^{(p)++}(x,t)$, $\Phi_q^{(p)+}(x',t)$, and $\mathcal{F}_{i\mu}^{(p)+}(x',t)$. By the foregoing analysis and Lemma IV.2, $u_{j0}^{(p)++}(x,t)$, $u_{jk}^{(p)}(x',t)$ are smooth functions in $\overline{\mathbb{R}}_+^{n+1}$, $\overline{E}_+^n$ and

$$u_{j0}^{(p)++}(x,t) = \theta(x_n)u_{j0}^{(p)+}(x,t),$$

$$D_n^{k-1}u_{j0}^{(p)+}(x',0,t) = u_{jk}^{(p)+}(x',t), \quad k=1,\ldots,\tau_j. \tag{3.58}$$

Next, from the continuity of the embeddings

$$\mathcal{H}_+^s(\mathbb{R}^{n+1},\gamma) \subset \mathcal{H}^\sigma(\mathbb{R}_+^{n+1},\gamma),$$

$$\mathcal{H}_+^{s-\sigma_q-1/2}(E^n,\gamma) \subset \mathcal{H}_+^{\sigma-\sigma_q-1/2}(E^n,\gamma),$$

$$\mathcal{H}_+^{s-\mu-1/2}(E^n,\gamma) \subset \mathcal{H}_+^{\sigma-\mu+1/2}(E^n,\gamma), \quad \forall s > \sigma : \sigma < -t_m + \tfrac{1}{2}$$

and estimates (3.38) and (3.39) it follows that the sequences $\{u_{j0}^{(p)++}(x,t)\}_{p=1}^\infty$ converge to $u_{j0}^{++}(x,t)$ in $\mathcal{H}_+^{\sigma+t_j}(\mathbb{R}_+^{n+1},\gamma)$ and the sequences $\{u_{jk}^{(p)+}(x',t)\}_{p=1}^\infty$ converge to $u_{jk}^+(x',t)$ in $\mathcal{H}_+^{\sigma-k+1/2}(E^n,\gamma)$. Moreover, for any j such that $s+t_j \geq 0$ we have $u_{j0}^{(p)+}(x,t) \in \mathcal{H}^{s+t_j}(\mathbb{R}_+^{n+1},\gamma)$ and the following estimates hold:

$$\left\| u_{j0}^{(p)+}(x,t), \mathbb{R}_+^{n+1}, \gamma \right\|_{s+t_j} < C_s A_{(p)},$$

$$\left\langle\!\left\langle u_{jk}^{(p)+}(x',t), E^n, \gamma \right\rangle\!\right\rangle_{s+t_j-k+1/2} < C_s A_{(p)}, \tag{3.59}$$

where C_s is a constant that does not depend on $F_i^{(p)++}(x,t)$, $\Phi_q^{(p)+}(x',t)$, and $\mathcal{F}_{i\mu}^{(p)+}(x',t)$. Here

$$A_{(p)} = \sum_{i=1}^m \left\| f_i^{(p)}(x,t), \overline{\mathbb{R}}_{++}^{n+1}, \gamma \right\|_{s,(\sigma_0)} + \left\langle\!\left\langle \varphi_q^{(p)+}(x',t), E^n, \gamma \right\rangle\!\right\rangle_{s-\sigma_q-1/2}$$

$$+ \left\| [\psi_{j\lambda}^+(x), \mathbb{R}^n] \right\|_{s+t_j-2b\lambda+b} + \left\langle\!\left\langle \psi_{j\lambda k}(x'), \mathbb{R}^{n-1} \right\rangle\!\right\rangle'_{s+t_j-k-2b\lambda+b+1/2},$$

where one sums over all the values of q, j, λ, k for which the corresponding functions are not identically equal to zero. Letting $p \to \infty$ in (3.59), we obtain (3.46) and (3.47).

(b) Now let us establish the inclusions (3.48). Clearly, if $s > 0$, then $s+t_j > 0$ for any $j = 1, \ldots, m$, and the embeddings (3.46) and equalities (3.47) proved above hold for all the functions $u^+_{j0}(x,t)$. Thus, it remains to prove (3.48) and estimate the norms $\left\| u^{++}_{j0}(x,t), \mathbb{R}^{n+1}, \gamma \right\|_{s+t_j}$. In the present case, like in the analysis of the Cauchy problem, the inverse Fourier and Laplace transforms of the right-hand sides of (3.50), which are used to determine the components $u^{++}_{j0}(x,t)$, involve the terms $L^{-1}(D, D_t)L_{ij}(D, D_t)l^{(k,0)}_{i\nu}(D, D_t)u^+_{\nu k}(x',t) \times \delta(x_n)$. The norms $\left\| \cdot, \mathbb{R}^{n+1}, \gamma \right\|_{s+t_j}$ of these terms cannot be estimated directly. Indeed, for j, k such that $s+t_j-k+\frac{1}{2} > 0$ the indicated terms only belong to the spaces $\mathcal{H}^{t_j - t_\nu + k - 1/2 - \varepsilon}(\mathbb{R}^{n+1}, \gamma)$, where $\varepsilon > 0$ is an arbitrary number, but, generally speaking, do not belong to $\mathcal{H}^{s+t_j}(\mathbb{R}^{n+1}, \gamma)$. For this reason, to estimate the norms $\left\| u^{++}_{j0}, \mathbb{R}^{n+1}, \gamma \right\|_{s+t_j}$ we will rely on a method similar to that used in the analysis of the Cauchy problem.

By the equivalence $\left(|\xi|^2 + |p|^{1/b} \right)^{s'+t_m} \sim |\xi_n|^{2(s'+t_m)} + \left(|\xi'|^2 + |p|^{1/b} \right)^{s'+t_m}$, we have

$$
\begin{aligned}
\left\| u^{++}_{j0}(x,t), \mathbb{R}^{n+1}, \gamma \right\|_{s+t_j} \sim &\; \left\| u^{++}_{j0}(x,t), \mathbb{R}^{n+1}, \gamma \right\|_{s-s'+t_j-t_m, s'+t_m} \\
&+ \left\| D^{s'+t_m}_n u^{++}_{j0}(x,t), \mathbb{R}^{n+1}, \gamma \right\|_{s-s'+t_j-t_m, 0}.
\end{aligned}
\tag{3.60}
$$

Since $s - s' - t_m < -t_m + \frac{1}{2}$, the estimate of the first norm in the right-hand side of (3.60) follows from Theorem IV.6. To estimate the second norm in the right-hand side of (3.60), we transform the system (3.49) into another system, from which we can determine uniquely $D^{s'+t_m}_n u^{++}_{j0}(x,t)$ for any j with $s + t_j < 0$. First, we replace each term $l_{ij}(D, D_t)u^{++}_{j0}(x,t)$ with j such that $s + t_j \geq 0$ by its expression given by formula (3.16) of Chapter III, taking into account the compatibility conditions of the right-hand sides with zero at $t = 0$ to order s. This yields

$$
\sum_{j:s+t_j<1/2} l_{ij}(D, D_t)u^{++}_{j0}(x,t) + \sum_{j:s+t_j>1/2} l^{(s'+t_j,0)}_{ij}(D, D_t)\left(D^{s'+t_j}_n u^+_{j0}(x,t) \right)^+
$$

$$
= f^{++}_{i0}(x,t) + l^{(0,\lambda)}_{ij}(D, D_t)v^+_{j\lambda}(x) \times \delta(t) - l^{(k,\lambda)}_{ij}(D, D_t)\omega_{j\lambda k}(x') \times \delta(x_n, t)
$$

$$
+ l^{(k,0)}_{ij}(D, D_t)u^+_{jk}(x',t) \times \delta(x_n) + \sum_{j:s+t_j>1/2} \left[l_{ij}(D, D_t), \theta(x_n, t) \right]_{s+t_j} u^+_{j0}(x,t)
$$

$$
\equiv \mathcal{M}^{++}_i(x,t), \qquad i = 1, \ldots, m.
\tag{3.61}
$$

Here one sums over all values of j, λ, and k for which $s+t_j - 2b\lambda + b < 0$, $s+t_j - 2b\lambda - k + b + \frac{1}{2} < 0$, $s+t_j - k + \frac{1}{2} < 0$ respectively; $\left[l_{ij}(D, D_t), \theta(x_n, t) \right]_{s+t_j} u^+_{j0}(x,t)$ stands for the expression

$$
\sum_{\sigma=[s]+t_j-b+1}^{s'+t_j-1} l_{ij\sigma}(\cdot, D', D_t)\left(D^\sigma_n u_{j0} \right)^{++} + \sum_{\sigma=0}^{[s]+t_j-b} l^{(\sigma_{sj})}_{ij\sigma}(D', D_t)\left(D^{\sigma_{sj}}_t D^\sigma_n u_{j0} \right)^{++} +
$$

$$+ \sum_{\sigma+2b\mu\leq[s]+t_j-b} l_{ij\sigma\mu}(D')\big(D_t^\mu D_n^\sigma u_{j0}\big)^{++}, \tag{3.62}$$

which follows from formula (3.16) of Chapter III if one takes into account that $v_{j\lambda}^+(x) \equiv 0$ for any j,λ such that $s+t_j-2b\lambda+b > 0$; and $\sigma_{s_j} = \big[(s+t_j+b-\sigma)/2b\big]$. Now differentiate (3.61) $s'+t_m$ times with respect to x_n and use the identity

$$D_n^{s'+t_j} l_{ij}^{(s'+t_j,0)}(D,D_t) = l_{ij}(D,D_t) - \sum_{\sigma=1}^{s'+t_j} l_{ij\sigma}(D',D_t)D_n^{\sigma-1}.$$

This yields

$$\sum_{j:s+t_j<1/2} l_{ij}(D,D_t)D_n^{s'+t_m}u_{j0}^{++}(x,t) + \sum_{j:s+t_j>1/2} l_{ij}(D,D_t)D_n^{t_m-t_j}\big(D_n^{s'+t_j}u_{j0}^+\big)^+$$

$$= D_n^{s'+t_m}\mathcal{M}_i^{++}(x,t) + \sum_{j:s+t_j>1/2}\sum_{\sigma=1}^{s'+t_j} l_{ij\sigma}(D',D_t)D_n^{t_m-t_j+\sigma-1}\big(D_n^{s'+t_j}u_{j0}^+(x,t)\big)^+$$

$$\equiv F_i^{++}(x,t), \qquad i=1,\dots,m. \tag{3.63}$$

Thus, we see that the unknown functions $D_n^{s'+t_m}u_{j0}^{++}(x,t)$, with j such that $s+t_j < \frac{1}{2}$ and $D_n^{t_m-t_j}\big(D_n^{s'+t_j}u_{j0}^+(x,t)\big)^+$, with j such that $s+t_j > \frac{1}{2}$ satisfy the parabolic system (3.63).

From the system (3.63) we find

$$D_n^{s'+t_m}u_{j0}^{++}(x,t) = L_{ij}(D,D_t)L^{-1}(D,D_t)F_i^{++}(x,t), \quad \forall j: s+t_j < \tfrac{1}{2}, \tag{3.64}$$

where one sums with respect to i. Equalities (3.64) yield the estimates

$$\big\|D_n^{s'+t_m}u_{j0}^{++}(x,t), \mathbb{R}^{n+1}, \gamma\big\|_{s-s'+t_j-t_m} \leq C\sum_{i=1}^m \big\|F_i^{++}(x,t), \mathbb{R}^{n+1}, \gamma\big\|_{s-s'-t_m}. \tag{3.65}$$

Now it remains to estimate the norms figuring under the sum sign in the right-hand side of (3.65). To this end we use the inclusions $u_{j0}^+(x,t) \in \mathcal{H}_+^{s+t_j}(\mathbb{R}_+^{n+1},\gamma)$ established above for any j such that $s+t_j > 0$, equalities (3.62), and Lemmas III.2, III.3 on the boundedness of the operators of multiplication by $\theta(\dot{x}_n)$ and $\theta(t)$. Repeating some of the arguments used in the proof of Lemma III.16 in estimating the norms of similar terms, we obtain

$$\big\|[l_{ij}(D,D_t),\theta(x_n,t)]_{s+t_j}u_{j0}(x,t),\mathbb{R}^{n+1},\gamma\big\|_s \leq C\big\|u_{j0}^+(x,t),\mathbb{R}_+^{n+1},\gamma\big\|_{s+t_j}. \tag{3.66}$$

Therefore, directly estimating the norms $\left\|\cdot, \mathbb{R}^{n+1}, \gamma\right\|_s$ of the remaining terms of the expressions $\mathcal{M}_i^{++}(x,t)$ and taking into account estimates (3.66), we obtain

$$
\left\|\mathcal{M}_i^{++}(x,t), \mathbb{R}^{n+1}, \gamma\right\|_s \leq \Big(\left\|f_{i0}^{++}(x,t), \mathbb{R}^{n+1}, \gamma\right\|_s + \left|\left[v_{j\lambda}^+(x), \mathbb{R}^n\right]\right|_{s+t_j-2b\lambda+b}
$$
$$
+ \left\langle\!\left\langle \omega_{j\lambda k}(x'), \mathbb{R}^{n-1} \right\rangle\!\right\rangle_{s+t_j-2b\lambda-k+b+1/2}
$$
$$
+ \sum_{j:s+t_j>1/2} \left\|u_{j0}^+(x,t), \mathbb{R}_+^{n+1}, \gamma\right\|_{s+t_j}
$$
$$
+ \sum_{j,k:s+t_j-k+1/2<0} \left\langle\!\left\langle u_{jk}^+(x',t), E^n, \gamma \right\rangle\!\right\rangle_{s+t_j-k+1/2} \Big).
$$
$$(3.67)$$

Since
$$
\left\|l_{ij\sigma}(D', D_t)D_n^{t_m-t_j+\sigma-1}\big(D_n^{s'+t_j}u_{j0}^+(x,t)\big)^+, \mathbb{R}^{n+1}, \gamma\right\|_{s-s'-t_m}
$$
$$
\leq C\left\|u_{j0}^+, \mathbb{R}_+^{n+1}, \gamma\right\|_{s+t_j}, \qquad (3.68)
$$

(3.67) and (3.68) yield the estimate $\left\|F_i^{++}(x,t), \mathbb{R}^{n+1}, \gamma\right\|_{s-s'-t_m} \leq C \cdot A$, where A denotes the right-hand side of (3.39). Consequently,

$$
\left\|D_n^{s'+t_m}u_{j0}^{++}(x,t), \mathbb{R}^{n+1}, \gamma\right\|_{s-s'+t_j-t_m} < C \cdot A, \qquad \forall j : s+t_j < \tfrac{1}{2}, \qquad (3.69)
$$

Since each one of the norms $\left\|u_{j0}^+(x,t), \mathbb{R}_+^{n+1}, \gamma\right\|_{s+t_j}$, $\left\langle\!\left\langle u_{jk}^+(x',t), E^n, \gamma \right\rangle\!\right\rangle_{s+t_j-k+\frac{1}{2}}$ that figure in the right-hand side of (3.67) is estimated by the right-hand side of (3.45), we immediately obtain from (3.69) the estimate

$$
\left\|D_n^{s'+t_m}u_{j0}^{++}(x,t), \mathbb{R}^{n+1}, \gamma\right\|_{s-s'+t_j-t_m} < C \cdot K, \qquad (3.70)
$$

where K denotes the right-hand side of (3.45). Formula (3.60) and the estimates (3.39), (3.70) yield the inclusions (3.48) for any j for which $s+t_j < \tfrac{1}{2}$. Thus, if Condition IV.1 is satisfied, then the inclusions (3.46), (3.48), the equalities (3.47), and the estimate (3.45) hold.

To complete the proof of Theorem IV.8 it remains to show that Condition IV.1 is equivalent to the Lopatinskiĭ condition.

IV.3.6. Equivalence of Condition IV.1 and the Lopatinskiĭ condition

Lemma IV.3. *Condition IV.1 is equivalent to the Lopatinskiĭ condition for any* $\xi' \in \mathbb{R}^{n+1}$ *and* p *with* $\operatorname{Re} p \geq 0$, $(\xi', p) \neq 0$.

Proof: As noted in Subsection I.2.3 the Lopatinskiĭ condition is equivalent to the condition of unique solvability of the boundary value problem

$$
l_{ij}(i\xi', D_n, p)\widehat{u}_{j0}(\xi', p, x_n) = 0, \qquad x_n \in \mathbb{R}_+^1, \; i = 1, \ldots, m, \qquad (3.71)
$$
$$
b_{qj}(i\xi', D_n, p)\widehat{u}_{j0}(\xi', p, x_n)\big|_{x_n=+0} = 0, \qquad q = 1, \ldots, N, \qquad (3.72)
$$
$$
u_{j0}(\xi', p, x_n) \to 0, \; x_n \to +\infty.
$$

Therefore, the proof of Lemma IV.3 reduces to verifying the equivalence of Condition IV.1 and the condition of unique solvability of problem (3.71)–(3.72).

Let $\widehat{u}_{j0}(\xi', p, x_n)$ be a nontrivial solution of the problem (3.71)–(3.72). It is readily verified that $\widehat{u}_{jk}(\xi', p) = D_n^{k-1}\widehat{u}_{j0}(\xi', p, 0)$, $k = 1, \ldots, \tau_j$, is a nontrivial solution of the homogeneous system (3.24)–(3.26), which is possible only when Condition IV.1 is violated. Hence, violation of the Lopatinskiĭ condition implies violation of Condition IV.1.

Conversely, let us show that if for some choice of the pair $(\xi', p) \neq 0$ Condition IV.1 does not hold, then the Lopatinskiĭ condition is not satisfied for the same values of ξ' and p. Suppose Condition IV.1 is not satisfied. This means that for the pair (ξ', p) in question the homogeneous system (3.24)–(3.26) has a nontrivial solution $\widehat{u}'_{jk}(\xi', p)$, $k = 1, \ldots, \tau_j$, $j = 1, \ldots, m$.

Let us construct a nontrivial solution of the problem (3.71)–(3.72). To do this, we consider on the whole line x_n the system of equations

$$l_{ij}(i\xi', D_n, p)\widehat{u}_{j0}^{+}(\xi', p, x_n) = l_{ij}^{(k,0)}(i\xi', D_n, p)\widehat{u}'_{jk}(\xi', p) \times \delta(x_n), \qquad (3.73)$$
$$i = 1, \ldots, m.$$

Since $\widehat{u}'_{jk}(\xi', p)$ satisfy conditions (1.4), the solution $\widehat{u}_{j0}^{+}(\xi', p, x_n)$, $j = 1, \ldots, m$ is supported in the closed half-line $x_n \geq 0$. Further, arguing in exactly the same manner as in the proof of inclusions (3.46) one shows that $\widehat{u}_{j0}^{+}(\xi', p, x_n) \in H^s(\mathbb{R}^1_+)$ with respect to the variable x_n for any j and $s > 0$. It follows that the derivatives $D_n^{k-1}\widehat{u}_{j0}(\xi', p, x_n)\big|_{x_n=+0} = \widehat{u}'_{jk}(\xi', p)$ exist for any $k = 1, \ldots, \tau_j$. Hence, to prove that the solution $u_0(x, t) = (u_{10}, \ldots, u_{m0})$ is not trivial, it suffices to show that $\widehat{u}_{jk}(\xi', p) = \widehat{u}'_{jk}(\xi', p)$ for all indices j, k, indicated above. It is easy to check that the functions $\widehat{w}_j(\xi', p, x_n) = \widehat{u}_{j0}^{+}(\xi', p, x_n) - \theta(x_n)\widehat{u}_{j0}(\xi', p, x_n)$, supported at the point $x_n = 0$, are a solution of the system

$$l_{ij}(i\xi', D_n, p)\widehat{w}_j(\xi', p, x_n) = l_{ij}^{(k,0)}(i\xi', D_n, p)\widehat{u}''_{jk}(\xi', p) \times \delta(x_n), \qquad (3.74)$$
$$i = 1, \ldots, m,$$

where $\widehat{u}''_{jk}(\xi', p) = \widehat{u}_{jk}(\xi', p) - \widehat{u}'_{jk}(\xi', p)$, and that $\widehat{w}_j(\xi', p, x_n) \in H^{s+t_j}(\mathbb{R}^1_+)$ for any $s \leq -t_m$. It follows that $\widehat{w}_j(\xi', p, x_n)$ are linear combinations of the delta-function $\delta(x_n)$ and its derivatives:

$$\widehat{w}_j(\xi', p, x_n) = \sum_{q=0}^{t_m - t_j} \widehat{\beta}_{jq}(\xi', p)D_n^{t_m - t_j - q}\delta(x_n), \qquad j = 1, \ldots, m.$$

Therefore, if

$$l_{ij}(i\xi, p) = \sum_{r=0}^{t_j} l_{ijr}(i\xi', p)(i\xi_n)^{t_j - r}, \qquad l_{ij}^{(k,0)}(i\xi, p) = \sum_{r=0}^{t_j - k} l_{ijr}(i\xi', p)(i\xi_n)^{t_j - r - k},$$

then upon replacing in (3.74) $\widehat{w}_j$, l_{ij}, and $l_{ij}^{(k,0)}$ by their expression and changing the order of summation with respect to r, j, q in both sides of the equations we obtain

$$\sum_{\sigma=0}^{t_m} \left(\sum_{j:t_m-t_j\geq\sigma} \sum_{r+q=\sigma} l_{ijr}\widehat{\beta}_{jq} \right)(i\xi_n)^{t_m-\sigma}$$
$$= \sum_{j:t_j\geq 1}\sum_{\eta=1}^{t_j} \left(\sum_{r+k=\eta} l_{ijr}\widehat{u}''_{jk} \right)(i\xi_n)^{t_j-\eta}, \quad j=1,\ldots,m. \qquad (3.75)$$

As in Lemma IV.2, to determine $\widehat{\beta}_{jq}(\xi',p)$ and $\widehat{u}''_{jk}(\xi',p)$ we equate the coefficients of like powers of $i\xi_n$ in each of the equations (3.75) and use also the equations obtained by differentiating the equations of system (3.71) with respect to x_n at $x_n = 0$. The coefficients $\widehat{\beta}_{jq}$ and $\widehat{u}''_{jk}$ can be determined successively, beginning with $\widehat{\beta}_{j0}$ and $\widehat{u}''_{j1}$. By equating the coefficients of $(i\xi_n)^{t_m}$ in each of the equations (3.75) we get $\sum_{j=1}^{m} l_{ij0}(0,0)\widehat{\beta}_{j0} = 0$, $i = 1,\ldots,m$, which thanks to the parabolicity of $\mathcal{L}$ implies $\det\|l_{ij0}\| = \mathrm{const} \neq 0$ and $\widehat{\beta}_{j0} = 0$, $j = 1,\ldots,m$. Further, equating the coefficients of $(i\xi_n)^{t_m-1}$ and taking into account that $\widehat{\beta}_{j0} = 0$ we get the following system of equations for $(\widehat{\beta}_{j1}, \widehat{u}''_{j1})$:

$$\sum_{j:t_j\leq t_m-1} l_{ij0}\widehat{\beta}_{j1} + \sum_{j:t_j=t_m} l_{ij0}\widehat{u}''_{j1} = 0, \qquad i = 1,\ldots,m$$

with the same determinant $\det\|l_{ij0}\| \neq 0$. Therefore, $\widehat{\beta}_{j1} = 0$, $\widehat{u}''_{j1} = 0$. To calculate the remaining functions $\widehat{\beta}_{jq}$ and $\widehat{u}''_{jk}$ we proceed by induction. Let $\widehat{\beta}_{jq} = 0$ for any j such that $t_j \leq t_m - q$, $q = 1,\ldots,\sigma - 1$ and $\widehat{u}''_{jk} = 0$ for any j, k such that $t_j - k = t_m - q$, $q = 1,\ldots,\sigma - 1$. Now let us prove that $\widehat{\beta}_{j\sigma} = 0$ for any j such that $t_j \leq t_m - \sigma$, $\widehat{u}''_{j,t_j-t_m+\sigma}(\xi',p) = 0$ for any j such that $t_j > t_m - \sigma$, $\sigma \leq t_m$.

Indeed, since $\widehat{\beta}_{jq} = 0$ and $\widehat{u}''_{j,t_j-t_m+q} = 0$ for the indicated q, j and k, equating the coefficients of $(i\xi_n)^{t_m-\sigma}$ yields the system

$$\sum_{j:t_j\leq t_m-\sigma} l_{ij0}\widehat{\beta}_{j\sigma} - \sum_{j:t_j>t_m-\sigma} l_{ij0}\widehat{u}''_{j,t_j-t_m+\sigma} = 0, \qquad i = 1,\ldots,m,$$

from which in turn it follows that $\widehat{\beta}_{j\sigma} = 0$, $\widehat{u}''_{j,t_j-t_m+\sigma} = 0$. Thus, $\widehat{\beta}_{j\sigma} = 0$, $j = 1,\ldots,m$, $\sigma = 1,\ldots,t_m - t_j$ and $\widehat{u}''_{jk} = 0$, $k = 1,\ldots,t_j$, $j = 1,\ldots,m$. To prove that $\widehat{u}''_{jk} = 0$ for $k = t_j+1,\ldots,\tau_j$, we consider the system obtained by subtracting the equations $D_n^{\sigma-t_m-1}l_{ij}(i\xi', D_n, p)\widehat{u}_{j0}(\xi',p,x_n)\big|_{x_n=0} = 0$, $i = 1,\ldots,m$, $\sigma = t_m + 1,\ldots,\tau_j + t_m - t_j$, from the corresponding homogeneous equations (3.17) in $\widehat{u}_{jk}(\xi',p)$. Then, by what has been established above we get $l_{ij0}\widehat{u}''_{j,t_j-t_m+\sigma}(\xi',p) = 0$, $i = 1,\ldots,m$, whence $\widehat{u}''_{j,t_j-t_m+\sigma}(\xi',p) = 0$, $\sigma = t_m + 1,\ldots,\tau_j - t_m + t_j$, $j = 1,\ldots,m$. Thus, $\widehat{u}_{j0}(\xi',p,x_n)$ is a nontrivial solution of the problem (3.71)–(3.72), that is, for the above choice of the pair (ξ',p) the Lopatinskiĭ condition is violated.

IV.4. The model boundary value problem in $\overline{\mathbb{R}}_{++}^{n+1}$ for general parabolic systems

IV.4.1. Formulation of the boundary value problem

Let $\mathcal{L}(D, D_t) = \left(l_{ij}(D, D_t)\right)_{i,j=1}^{m}$ (with $\operatorname{ord} l_{ij} = s_i + t_j$, $t_j = 2bt'_j$, $s_i = 2bs'_i$, $\sum_i (s_i + t_i) = 2N = 2b_k$) be a matrix quasihomogeneous linear parabolic operator, and let $\mathcal{B}(D, D_t) = \left(b_{qj}(D, D_t)\right)$ $(\operatorname{ord} b_{qj} = \sigma_q + t_j)$, $C(D, D_t) = \left(c_{\alpha j}(D, D_t)\right)$ $(\operatorname{ord} c_{\alpha j} = \rho_\alpha + t_j)$ be matrix quasihomogeneous linear differential operators, satisfying the parabolicity condition, the Lopatinskiĭ condition, and the complementarity condition, respectively.

The following problem will be called the *model parabolic boundary value problem* in the spaces of sufficiently smooth functions $\mathcal{H}^s = \prod\limits_{j=1}^{m} \mathcal{H}^{s+t_j}(\mathbb{R}_{++}^{n+1}, \gamma)$, $(s > \sigma_0)$: Find a solution $u(x, t) = \left(u_1(x, t), \ldots, u_m(x, t)\right)$ of the system

$$\mathcal{L}_i(D, D_t)u(x, t) \equiv l_{ij}(D, D_t)u_j(x, t) = f_i(x, t), \qquad i = 1, \ldots, m, \tag{4.1}$$

satisfying the following boundary conditions at $x_n = 0$:

$$\mathcal{B}_q(D, D_t)u(x, t)\big|_{x_n=0} = b_{qj}(D, D_t)u_j(x, t)\big|_{x_n=0} = \varphi_q(x', t), \qquad q = 1, \ldots, N, \tag{4.2}$$

and the following initial conditions at $t = 0$:

$$C_\alpha(D, D_t)u(x, t)\big|_{t=0} \equiv c_{\alpha j}(D, D_t)u_j(x, t)\big|_{t=0} = \psi_\alpha(x), \qquad \alpha = 1, \ldots, \kappa. \tag{4.3}$$

Below we will use the analysis in Sections IV.2 and IV.3 to formulate the model parabolic boundary value problem in the spaces $\widetilde{\mathcal{H}}^s$ for the general parabolic system.

Let $l_{ij} = l_{ijk}(D', D_t)D_n^{k-1} = \mathcal{L}_{ijp}(D)D_t^{p-1}$; $b_{qj} = b_{qjk}(D', D_t)D_n^{k-1}$; $c_{\alpha j}(D, D_t) = c_{\alpha jr}(D)D_t^{r-1}$.

As in Section IV.3, be begin by writing the Green formulas for the equations of the system, the boundary and the initial conditions:

$$\mathcal{L}_i u\big|_{\overline{\mathbb{R}}_{++}^{n+1}} \equiv l_{ij}(D, D_t)u_{j0}^{++} - l_{ij}^{(k,0)}(D, D_t)u_{jk}^{+}(x', t) \times \delta(x_n)$$

$$- l_{ij}^{(0,\lambda)}(D, D_t)v_{j\lambda}^{+}(x) \times \delta(t) + l_{ij}^{(k,\lambda)}(D, D_t)\omega_{j\lambda k}(x') \times \delta(x_n, t)$$

$$= f_{i0}^{++}(x, t), \qquad i = 1, \ldots, m, \tag{4.4}$$

$$\mathcal{B}_q u\big|_{\overline{E}_+^n} \equiv b_{qjk}(D', D_t)u_{jk}^{+}(x', t) - b_{qjk}^{(\lambda)}(D', D_t)\omega_{j\lambda k}(x') \times \delta(t)$$

$$= \varphi_q^{+}(x', t), \qquad q = 1, \ldots, N, \tag{4.5}$$

$$C_\alpha(D, D_t)u\big|_{\overline{\mathbb{R}}_+^n} \equiv c_{\alpha j\lambda}(D)v_{j\lambda}^{+}(x) - c_{\alpha j\lambda}^{(k)}(D)\omega_{j\lambda k}(x') \times \delta(x_n)$$

$$= \psi_\alpha^{+}(x), \qquad \alpha = 1, \ldots, k, \tag{4.6}$$

where the summation with respect to j, λ and k is carried out over all their values such that $l^{(k,\lambda)}_{i,j} \neq 0$, $b^{(\lambda)}_{qjk} \neq 0$, $c^{(k)}_{\alpha j \lambda} \neq 0$. Then, we regard the Green formulas (4.4)–(4.6) as a system of equations in the components u^{++}_{j0}, u^{+}_{jk}, $v^{+}_{j\lambda}$, $\omega_{j\lambda k}$ of the solution sought $u(x,t)$ and, arguing by analogy with Section IV.3, we complement it to a uniquely solvable system in the corresponding space $\widetilde{\mathcal{H}}^s$.

Thus, let us consider the system (4.4)–(4.6). Since $\rho'_\alpha \leq -1$, $s'_i \leq 0$, it follows that for every $j = 1, \ldots, m$, equations (4.4) and (4.6) contain the components $v^{+}_{j\lambda}(x)$ with $\lambda \leq t'_j$, while equations (4.4) and (4.5) contain the components $u^{+}_{jk}(x',t)$ with $k \leq \tau_j = t_j + \sigma_0$, where $\sigma_0 = \max_q (0, \sigma_q + 1)$. By the analysis in Subsection IV.2.1, the functions $v_{j\lambda}(x)$ with $\lambda \leq t'_j$ are uniquely determined by the system of equations composed of the initial conditions (4.3) and the equations obtained by differentiating the corresponding equations of system (4.1) at $t = 0$: $(D^{\beta-1}_t \mathcal{L}_i)u\big|_{t=0} = f_{i0\beta}(x)$ $(= (D^{\beta-1}_t f_i(x,t)\big|_{t=0})$, $\beta = 1, \ldots, -s'_i$, for all i such that $s'_i \leq -1$. Therefore, the components of $v^{+}_{j\lambda}(x)$ are uniquely determined by equation (4.6) and the Green formulas corresponding to equations (4.3):

$$(D^{\beta}_t \mathcal{L}_i)u\big|_{\overline{\mathbb{R}}^n_+} \equiv \mathcal{L}_{ij\lambda}(D)v^{+}_{j\lambda+\beta}(x) - \mathcal{L}^{(k)}_{ij\lambda}(D)\omega_{j\lambda+\beta k}(x') \times \delta(x_n)$$
$$= f^{+}_{i0\beta+1}(x), \qquad \beta = 0, \ldots, -s'_i + 1, \ \forall i : s'_i \leq -1; \tag{4.7}$$

they can be expressed solely in terms of $f^{+}_{i0\beta}(x)$, $\psi^{+}_\alpha(x)$ and $\omega_{j\lambda k}(x')$.

Arguing in a manner similar to Subsection IV.3.1 we see that if $\sigma_0 > 0$, then, in order to determine the functions $u_{jk}(x',t)$ with $k > t_j$ we need to supplement the equations of system (4.1) and the boundary conditions (4.2) by the corresponding equations of system (4.1), differentiated with respect to x_n at $x_n = 0$:

$$(D^{\sigma-1}_n \mathcal{L}_i)u\big|_{E^n_+} \equiv (D^{\sigma-1}_n l_{ij})u_j\big|_{E^n_+} = f_{i\sigma 0}(x',t) \left(= D^{\sigma-1}_n f_{i0}\big|_{x_n=0}\right),$$
$$\sigma = 1, \ldots, \sigma_0 - s_i + 1. \tag{4.8}$$

Now supplement equations (4.1), (4.5) by the Green formulas for equations (4.8):

$$(D^{\sigma}_n \mathcal{L}_i)u\big|_{\overline{E}^n_+} = (D^{\sigma}_n l_{ij})u\big|_{\overline{E}^n_+} \equiv l_{ijk}(D',D_t)u^{+}_{jk+\sigma}(x',t) - l^{(\lambda)}_{ijk}\omega_{j\lambda k+\sigma}(x') \times \delta(t)$$

$$= f^{+}_{i\sigma+1}(x',t) \left(= \left(D^{\sigma}_n f_{i0}\big|_{x_n=0}\right)^{+}\right), \qquad \sigma = 0, 1, \ldots, \sigma_0 - s_i. \tag{4.9}$$

If we supplement (4.5), (4.9) by the necessary and sufficient condition (1.4) for the validity of the inclusions $\operatorname{supp} u^{++}_{j0}(x,t) \subseteq \overline{\mathbb{R}}^{n+1}_{++}$, then, as in the case of the boundary value problem (3.10)–(3.14) for a system parabolic in the sense of Petrovskiĭ, we get a system of equations from which the components $u^{+}_{jk}(x',t)$ can be uniquely expressed solely in terms of $\varphi^{+}_q(x',t)$, $f^{+}_{i\sigma 0}(x',t)$, and $\omega_{j\lambda k}(x')$.

Thus, it remains to write a system of equations in $\mathbb{R}^{n-1}$ from which one can uniquely determine the unknown components of $\omega_{j\lambda k}(x')$ and to select a corresponding space $\widetilde{\mathcal{H}}^s = \prod_{j=1}^{m} \widetilde{\mathcal{H}}^{s+t_j}_{(t'_j, \tau_j, P_j)}$, in which the solution is being sought. We will discuss this stage in more detail, starting with a description of the sets P_j of pairs of indices (k, λ) of the functions $\omega_{j\lambda k}(x')$ that figure in (4.4), (4.5) and (4.9).

Since l_{ij}, b_{qj}, and $c_{\alpha j}$ are operators of order $\sigma_i + t_j$, $\sigma_q + t_j$, and $\rho_\alpha + t_j$, respectively, it is readily verified that for each fixed j the Green formulas (4.4), (4.5), and (4.9) involve the functions $\omega_{j\lambda k}(x')$ with indices λ and k that satisfy the inequality $k + 2b\lambda \leq \tau_j$. Moreover, $1 \leq \lambda \leq \lambda_j^0 = \max(t'_j, t'_j + \rho'_0 + 1)$, where $\rho'_0 = \max \rho'_j$ and $\lambda_j = t'_j + \rho'_j + 1$ is the maximal order of differentiation of $u_{j0}(x, t)$ with respect to t in the boundary operator $\mathcal{B}$. Denote by P_j the set $\{(\lambda, k) : k + 2b\lambda \leq \tau_j, \ k \geq 1, \ 1 \leq \lambda \leq \lambda_j^0\}$.

Let $\rho' (= -t'_m, \ldots, -1)$, $\mu(\rho')$, and $\nu(\rho')$ be the numbers defined in Subsection IV.2.1. For each $\rho' = -t'_m, \ldots, -1$, consider the system (4.6), (4.7):

$$C_\alpha u\big|_{t=0} = \sum_{\lambda=1}^{s'_i + t'_j + 1} c_{\alpha j \lambda}(D) v_{j\lambda}(x) = \psi_\alpha(x), \qquad \forall \alpha : \rho'_\alpha = \rho', \qquad (4.10)$$

$$D_t^{\rho' - s'_i} \mathcal{L}_i u\big|_{t=0} = \sum_{\lambda=1}^{s'_i + t'_j + 1} \mathcal{L}_{ij\lambda}(D) D_t^{\rho' - s'_i + \lambda - 1} u_{j0}(x, 0)$$

$$= \mathcal{L}_{ij\lambda}(D) v_{j\rho' - s'i + \lambda}(x) = f_{i\rho' - s'_i + 1}(x) \ \big(\equiv D_t^{\rho' - s'_i} f_{i0}(x, 0) \big),$$

$$\forall i : i = 1, \ldots, \mu(\rho'), \qquad (4.11)$$

which involves only the functions $v_{j\lambda}(x)$ with $j = \nu(\rho') + 1, \ldots, m$. The system (4.10), (4.11) is such that the determinant of the matrix composed of the coefficients of $v_{j\rho' + t'_j + 1}(x)$ is different from zero.

Hence, if for a fixed ρ' we differentiate each of the equations of the system (4.10), (4.11) $k - 1$ times with respect to x_n at the point $x_n = 0$, then for $n \geq 2$ we obtain the system of linear differential equations

$$(D_n^{k-1} D_t^{\rho' - s'_i} \mathcal{L}_i) u\big|_{\mathbb{R}^{n-1}} = f_{i\rho' - s'_i + 1 k}(x') \big(= D_n^{k-1} D_t^{\rho' - s'_i} f_{i0}(x, t)\big|_{\mathbb{R}^{n-1}} \big),$$

$$\forall i : i = 1, \ldots, \mu(\rho'), \ \rho' - s'_i \geq 0,$$

$$(k, \lambda) \in Q_i = \{(k, \lambda) : k + 2b\lambda \leq \sigma_0 - s_i\}, \qquad (4.12)$$

$$(D_n^{k-1} C_\alpha) u\big|_{\mathbb{R}^{n-1}} = \psi_{\alpha k}(x') \big(= D_n^{k-1} \psi_\alpha(x)\big|_{\mathbb{R}^{n-1}} \big),$$

$$\forall \alpha : \rho'_\alpha = \rho', \qquad k \in N_\alpha = \{k : k = 1, \ldots, \sigma_0 - 2b(\rho'_\alpha + 1)\}, \qquad (4.13)$$

in the unknowns $\omega_{j\lambda k}(x')$ with $j = \nu(\rho') + 1, \ldots, m$, $\lambda = 1, \ldots, t'_j + \rho' + 1$. In particular, if $\rho' \geq 0$, then $\mu(\rho') = m$, and the system for determining $\omega_{jt'_j + \rho' + 1}(x')$

consists only of the equations (4.12). In the case $n = 1$ the system (4.12), (4.13) is a system of linear algebraic equations in the numbers $\omega_{j\lambda k}$.

Consider the combined system of equations (4.12), (4.13). By the foregoing discussion, this system is uniquely solvable in $\omega_{j\rho'+t_j'+1}(x')$ for any fixed ρ' and k, and $\omega_{j\rho'+t_j'+1}(x')$ are expressed uniquely in terms of $\omega_{j\lambda\sigma}(x')$, where $\sigma > k$, $\lambda < \rho'+t_j'+1$, and $f_{i,\rho'-s_i'+1}(x')$, $\psi_{\alpha k}(x')$. In particular, for $\rho' = -t_m'$ the functions $\omega_{j1k}(x')$ are expressed in terms of $f_{i,-t_m'-s_i'+1k}(x')$ with the indices i for which $s_i' = -t_m'$, and $\psi_{\alpha k}(x')$ with the indices α for which $\rho_\alpha' = -t_m'$. Therefore, for each fixed ρ', letting $k = 1,\ldots,\sigma_0 - 2b(\rho'+1)$ we express $\omega_{j\rho'+t_j'+1}(x')$ in terms of $\omega_{j\lambda\sigma}(x')$ with $\lambda < \rho' + t_j' + 1$ and $\sigma > k$. Repeating this procedure for all ρ', beginning with $\rho' = -t_m'$ and terminating with $\rho' = \rho_0'$, we uniquely express all the $\omega_{j\lambda k}(x')$ with $(k,\lambda) \in P_j$ solely in terms of the given functions $f_{i\rho'-s_i'+1k}(x')$ and $\psi_{\alpha k}(x')$.

Thus we have written the system (4.12), (4.13), from which we can uniquely determine the functions $\omega_{j\lambda k}(x')$, $(\lambda,k) \in P_j$. Let us proceed to the choice of functional spaces and to posing the model boundary value problem.

For any $s \in \mathbb{R}^1, s \notin \mathbb{Z}_{1,2b}$ we set

$$\widetilde{\mathcal{H}}^s = \prod_j \widetilde{\mathcal{H}}_{(t_j',\tau_j,P_j)}^{s+t_j}\big(\overline{\mathbb{R}}_{++}^{n+1},\gamma\big),$$

$$\mathcal{K}^s = \prod_{i=1}^m \widetilde{\mathcal{H}}_{(-s_i',\beta_i,Q_i)}^{s-s_i}\big(\overline{\mathbb{R}}_{++}^{n+1},\gamma\big) \times \prod_{q=1}^N \mathcal{H}^{s-\sigma_q-1/2}\big(\overline{E}_+^n,\gamma\big) \times \prod_{\alpha=1}^\kappa \widetilde{H}_{(N_\alpha)}^{s-\rho_\alpha-b}\big(\overline{\mathbb{R}}_+^n\big),$$

where $\beta_i = \sigma_0 - s_i$. Let $\mathcal{F}_i = \big(f_{i0},\ldots,f_{m0},f_{i0\sigma},\ldots,f_{i\beta_{i0}}(x),\ldots\big)$ be vectors, which we identify with the elements $f_i(x,t) \in \mathcal{H}_{(-s_i',\beta_i,Q_i)}^{s-s_i}\big(\overline{\mathbb{R}}_{++}^{n+1},\gamma\big)$, that is, $\mathcal{F}_i = \chi f_i$ are the isometrically isomorphic images of the elements $f_i(x,t)$.

The compatible equalities (4.4), (4.7), (4.9), (4.12) and (4.6), (4.13), regarded as the equality of the corresponding components of the elements $f_i(x,t) \in \widetilde{\mathcal{H}}_{(-s_i',\beta_i,Q_i)}^{s-s_i}$ and $\psi_\alpha(x') \in \widetilde{H}_{(N_\alpha)}^{s-\rho_\alpha-b}$, can be written in the following compact form:
$$\mathcal{L}_i u\big|_{\overline{\mathbb{R}}_{++}^{n+1}} = f_i(x,t), \ i = 1,\ldots,m, \ C_\alpha u\big|_{\overline{\mathbb{R}}_+^n} = \psi_\alpha(x), \ \alpha = 1,\ldots,\kappa.$$

By Lemmas III.16–III.19 of Chapter III, equalities (4.4)–(4.7) and (4.9), (4.12) define an operator $\mathfrak{U}$ that can be extended by continuity from C^∞ to the entire space $\widetilde{\mathcal{H}}^s$, yielding a bounded operator from $\widetilde{\mathcal{H}}^s$ to the subspace $\widetilde{\mathcal{K}}^s$ of $\mathcal{K}^s$, where $\widetilde{\mathcal{K}}^s$ consists of the vector functions $\big(f_1(x,t),\ldots,f_m(x,t),\varphi_1(x',t),\ldots,$ $\varphi_N(x',t),\psi_1(x),\ldots,\psi_\kappa(x')\big)$, satisfying the natural compatibility conditions that follow from the theorems on traces. This means, in particular, that each one of the equations (4.7)–(4.13) is a consequence of the equations of the system $\mathcal{L}_i u = f_{i0}$ or of the initial conditions $C_\alpha u\big|_{t=0} = \psi_\alpha$, provided the corresponding traces exist on E_+^n, $\mathbb{R}_+^n$, $\mathbb{R}^{n-1}$. Each one of equations (4.7)–(4.13) complements the system that consists of the previous equations, beginning with those values of s for which these equations becomes meaningless when regarded as an equality of the traces.

The *model boundary value problem for a general parabolic system in the space* $\widetilde{\mathcal{H}}^s$ is defined to be the problem of finding $u(x,t) \in \widetilde{\mathcal{H}}^s$ such that $u(x,t)$ satisfies the system

$$\mathcal{L}_i(D, D_t)u(x,t)\big|_{\overline{\mathbb{R}}^{n+1}_{++}} = f_i(x,t) \in \widetilde{\mathcal{H}}^{s-s_i}_{\cdots}\big(\overline{\mathbb{R}}^{n+1}_{++}, \gamma\big), \qquad i = 1,\ldots,m, \qquad (4.14)$$

the boundary conditions on $\overline{E}^n_+$

$$\mathcal{B}_q(D, D_t)u(x,t)\big|_{\overline{E}^n_+} = \varphi_q(x',t) \in \mathcal{H}^{s-\sigma_q-1/2}\big(\overline{E}^n_+, \gamma\big), \qquad q = 1,\ldots,N, \qquad (4.15)$$

and the initial conditions

$$C_\alpha(D, D_t)u(x,t)\big|_{\overline{\mathbb{R}}^n_+} = \psi_\alpha(x) \in \widetilde{H}^{s-\rho_\alpha-b}_{(N_\alpha)}\big(\overline{\mathbb{R}}^n_+\big), \qquad \alpha = 1,\ldots,\kappa. \qquad (4.16)$$

IV.4.2.   The model boundary value problem with data compatible with zero

For the existence of a solution $u(x,t)$ of problem (4.14)–(4.16), satisfying given smoothness requirements in $\overline{\mathbb{R}}^{n+1}_{++}$, it is necessary that a certain compatibility condition of the right-hand sides hold at $t = 0$. Let us formulate this condition.

Denote by $\widetilde{\mathcal{H}}^{s+t_j}_{+\cdots}$ ($s + t_j \notin \mathbb{Z}_{1,2b}$ for $s + t_j > 0$) the completion of the set of smooth functions that vanish at $t = 0$ up to order $s + t_j$ with respect to the norm of $\widetilde{\mathcal{H}}^{s+t_j}$. Also, let $\mathcal{F}^s$ denote the space of the vector functions $F(x,t) = (f_1,\ldots,f_m,\varphi_1,\ldots,\varphi_N,\psi_1,\ldots,\psi_\kappa)$ and $\widetilde{\mathcal{H}}^s_+ = \prod_j \widetilde{\mathcal{F}}^{s+t_j}_+$, $\widetilde{\mathcal{F}}^s_+ = \mathfrak{U}\widetilde{\mathcal{H}}^s_+$.

Definition IV.2. *The right-hand sides of (4.14)–(4.16) are said to satisfy the condition of compatibility with zero at $t = 0$ to order s if $F \in \widetilde{\mathcal{F}}^s_+$.*

This means that all the functions $\psi^+_\alpha(x)$, $\psi_{\alpha k}(x')$, $f_{i0\beta}(x)$, $f_{i\sigma\beta}(x')$ with indices α, k, σ, and β such that $s - \rho_\alpha - b > 0$, $s - \rho_\alpha - b - k - \frac{1}{2} > 0$, $s - s_i - 2b\beta + b > 0$, $s - s_i - \sigma - 2b\beta + b + \frac{1}{2} > 0$ are equal zero, and the functions $f_{i0\sigma}(x',t)$, $\varphi_q(x',t)$, extended by zero for $t < 0$, continue to belong to spaces $\mathcal{H}^{\cdots}$ with the same smoothness exponents.

Theorem IV.9. *Suppose that $\mathcal{L}(D, D_t)$ is a parabolic operator, $\mathcal{B}(D, D_t)$ satisfies the Lopatinskiĭ condition, and $C(D, D_t)$ satisfies the complementarity condition. Then for any right-hand side $F(x,t) \in \widetilde{\mathcal{F}}^s_+$, compatible at $t = 0$ with zero to order s, there exists a unique solution $u(x,t) = (u_1,\ldots,u_m) \in \widetilde{\mathcal{H}}^s_+$, which for $s + t_j \notin \mathbb{Z}_{1,2b}$, $j = 1,\ldots,m$, satisfies the estimate*

$$\big\|u_j(x,t),\overline{\mathbb{R}}^{n+1}_{++},\gamma\big\|_{s+t_j,(t'_j,\tau_j,P_j)} < C_s\Big(\big\|f_i(x,t),\overline{\mathbb{R}}^{n+1}_{++},\gamma\big\|_{s-s_i,(-s'_i,\beta_j,Q_i)}$$

$$+\big\langle\!\big\langle\varphi_q(x',t),\overline{E}^n_+,\gamma\big\rangle\!\big\rangle_{s-\sigma_q-1/2} + \big|\{\psi_\alpha(x),\overline{\mathbb{R}}^n_+,\}\big|_{s-\rho_\alpha-b,(N_\alpha)}\Big). \qquad (4.17)$$

Proof: The proof is carried out in the same way as the proof of Theorem IV.8 and can be divided into two stages. We will briefly describe them, paying attention to new moments in the proof that arise in the case of general parabolic systems.

(a) As in the case of systems parabolic in the sense of Petrovskiĭ sense, we begin the study of the general model parabolic boundary value problem (4.14)–(4.16) with the case $s < -t_m + \frac{1}{2}$. Generally speaking, for these values of s the right-hand sides of the problem satisfy no compatibility conditions, and problem (4.14)–(4.16) consists in finding the distributions $u_j^{++}(x,t)$, $u_{jk}^+(x',t)$, $v_{j\lambda}^+(x)$, and $\omega_{j\lambda k}(x')$, that satisfy the system of equations (4.4)–(4.6), (4.7), (4.9) and (4.12), (4.13). Since one knows beforehand that $\operatorname{supp} u_j^{++}(x,t) \subseteq \overline{\mathbb{R}}^{n+1}_{++}$, it follows that $u_j^{++}(x,t)$ must satisfy conditions (1.4). Thus, in order to find the sought distributions $u_j^{++}(x,t)$, $u_{jk}^+(x',t)$, $v_{j\lambda}^+(x)$, and $\omega_{j\lambda k}(x')$ we need to solve a system similar to (3.4)–(3.6), (3.8), (3.9).

This system is solved by successive elimination of the components and reduction to a system in the unknown components $u_{jk}^+(x',t)$. Note that first we use the equations of the system $\mathcal{L}_i(D, D_t)u_{++}(x,t) = \mathcal{F}_i^{++}(x,t)$ to express $u_j^{++}(x,t)$ in terms of $u_{jk}^+(x',t)$, $v_{j\lambda}^+(x)$, and $\omega_{j\lambda k}(x')$. Applying Lemma III.1 of Chapter III to the formulas obtained for $u_{j0}^{++}(x,t)$, we immediately get estimates of the norms $\left\| u_{j0}^{++}\overline{\mathbb{R}}^{n+1}_{++}, \gamma \right\|_{s+t_j}$ in terms of corresponding norms of the components $u_{jk}^+(x',t)$, $v_{j\lambda}^+(x)$, $\omega_{j\lambda k}(x')$, and $f_{i0}^{++}(x,t)$.

Next, let us eliminate the functions $v_{j\lambda}^+(x)$ and $\omega_{j\lambda k}(x')$ and estimate the norms $\left\| [v_{j\lambda}^+(x), \mathbb{R}^n] \right\|_{s+t_j-2b\lambda+b}$ and $\left\langle\!\left\langle \omega_{j\lambda k}(x'), \mathbb{R}^{n-1} \right\rangle\!\right\rangle'_{s+t_j-k+b+1/2}$. In contrast to the boundary value problem (3.10)–(3.14), in which the initial conditions had a very simple form and $v_{j\lambda}^+(x)$ and $\omega_{j\lambda k}(x')$ were directly expressed through the right-hand sides of the initial conditions on $\overline{\mathbb{R}}^n_+$ and $\mathbb{R}^{n-1}$, in order to find $v_{j\lambda}^+(x)$ and $\omega_{j\lambda k}(x')$ in the present case and to estimate their norms, we use the rather tedious procedure, described above, of successive determination of all $v_{j\lambda}^+(x)$ and $\omega_{j\lambda k}(x')$ with $(k, \lambda) \in P_j$. In this procedure, in the first step we use the initial conditions (4.12), (4.13) on $\mathbb{R}^{n-1}$ to successively express all the $\omega_{j\lambda k}(x')$ with $(k, \lambda) \in P_j$ solely in terms of $\psi_{\alpha k}(x')$ and $f_{i\sigma q}(x')$ and then to estimate the norms $\left\langle\!\left\langle \omega_{j\lambda k}(x'), \mathbb{R}^{n-1} \right\rangle\!\right\rangle'_{s+t_j-2b\lambda-k+b+1/2}$ directly, by using the obtained expressions for $\omega_{j\lambda k}(x')$. In the next step we express the distributions $v_{j\lambda}^+(x)$ in terms of $f_{i\sigma 0}^+(x)$ and $\psi_\alpha^+(x)$, $\omega_{j\lambda k}(x')$ by using the initial conditions (4.6) and (4.7). Then, using the estimates already obtained for $\left\langle\!\left\langle \omega_{j\lambda k}(x'), \mathbb{R}^{n-1} \right\rangle\!\right\rangle'_{s+t_j-2b\lambda-k+b+\frac{1}{2}}$, we estimate the norms $\left\| [v_{j\lambda}^+(x), \mathbb{R}^n] \right\|_{s+t_j-2b\lambda+b}$ solely in terms of the corresponding norms of the distributions $f_{i\sigma 0}^+(x)$, $\psi_\alpha^+(x)$, $f_{i\sigma k}(x',t)$, and $\psi_{\alpha k}(x')$.

Thus it remains to find the components $u_{jk}^+(x',t)$ and to estimate their norms $\left\langle\!\left\langle u_{jk}^+(x',t), E^n, \gamma \right\rangle\!\right\rangle_{s+t_j-k+1/2}$.

The components $u_{jk}^+(x',t)$ satisfy the boundary conditions (4.5), (4.9) and the supplementary conditions (1.4). Taking the Fourier transformation in x and the Laplace transformation in t of the conditions (4.5), (4.9) and adding conditions (1.4), we obtain for $\widehat{u}_{jk}^+(\xi',p)$ a system of linear algebraic equations similar to the system (3.24)–(3.26) studied in Subsection IV.3.3. All the constructions and arguments of Subsection IV.3.3 remain in force for this system; in particular, formulas (3.27) for determining $\widehat{u}_{jk}^+(\xi',p)$ are valid. These formulas yield estimates of the norms $\langle\!\langle u_{jk}^+(x',t), E^n, \gamma \rangle\!\rangle_{s+t_j-k+1/2}$, exactly as in Subsection IV.3.3. This completes the proof of Theorem IV.8 in the case $s < -t_m + \frac{1}{2}$.

(b) Let $s > -t_m + \frac{1}{2}$. Since the right-hand sides are compatible with zero at $t = 0$ to order s and obviously belong to the same spaces, with s replaced by an arbitrary $\sigma < -t_m + \frac{1}{2}$, it follows from the first part of Theorem IV.8 that the problem has a unique solution $u \in \widetilde{\mathcal{H}}^s$ that satisfies estimate (4.17) with s replaced by σ. Therefore, it remains to prove that the obtained solution $u(x,t)$ is in fact smoother in (x,t), i.e., belongs to the space $\widetilde{\mathcal{H}}^s$, and that estimate (4.17) holds. The proof of these statements is carried out step-by-step, as in the case of problem (3.10)–(3.14), and makes essential use of the "double anisotropy" of the spaces $H_{(x,t),(x',t)}^{s,\ r}(\mathbb{R}_+^{n+1},\gamma)$, considered in Subsection IV.3.5 in the process of increasing the smoothness of the solution $u(x,t)$ of the indicated problem.

Specifically, we note that if the data of the problem are compatible with zero at $t = 0$ to order s, then conditions (4.12) and (4.13) imply that $\omega_{j\lambda k}(x') \equiv 0$ for any j, λ, k such that $s + t_j - 2b\lambda - k + b + \frac{1}{2} > 0$ and $v_{j\lambda}^+(x) \equiv 0$ for any j, λ such that $s + t_j - 2b\lambda + b > 0$, while the remaining $\omega_{j\lambda k}(x')$ and $v_{j\lambda}^+(x)$ obey the corresponding estimates.

From this observation and the system of equations for the determination of the distributions $u_{jk}^+(x',t)$ it follows that $u_{jk}^+(x',t) \in \mathcal{H}_+^{s+t_j-k+1/2}(E^n,\gamma)$ for any j,k, and $u_{j0}^{++}(x,t) \in \mathcal{H}_{(x,t),(x',t)}^{s-\sigma+t_j,\sigma}(\mathbb{R}_+^{n+1},\gamma)$. In other words, increasing the smoothness of the data of the problem and their compatibility with zero at $t = 0$ results in a corresponding increase of the smoothness of the solution in the "tangential" variables (x',t). To complete the proof it remains to establish the inclusions (3.46) and (3.48) and to estimate the norms $\big\|u_j^+(x,t),\overline{\mathbb{R}}_+^{n+1},\gamma\big\|_{s+t_j}$.

IV.5. The model parabolic conjugation problem classes of smooth functions

IV.5.1. Formulation of the problem; the compatible covering condition

For the sake of simplicity, we denote $\Omega_1^+ = \mathbb{R}_{++}^{n+1} = \{(x,t) : x_n > 0,\ t > 0\}$, $\Omega_2^+ = \{(x,t) : x_n < 0,\ t > 0\}$. Suppose that in each domain Ω_r^+ there is given a matrix parabolic operator $\mathcal{L}_r(D, D_t)$ whose elements, $l_{(r)ij}$, are weighted homogeneous operators of orders $s_{ri} + t_{rj}$, $i,j = 1,\ldots,m_r$, $r = 1,2$, where s_{ri} and

t_{rj} are integers, $\sum(s_{ri} + t_{rj}) = 2N_r = 2b\kappa_r$. Let $\mathcal{B}_r(D, D_t)$, $C_r(D, D_t)$ be linear matrix operators whose elements $b_{(r)qj}$ and $c_{(r)\alpha j}$ are weighted homogeneous operators of orders $\sigma_q + t_{rj}$ and $\rho_{r\alpha} + t_{rj}$, respectively, where σ_q, $\rho_{r\alpha}$ are integers, $q = 1, \ldots, N_1 + N_2 = b(\kappa_1 + \kappa_2)$. As above in this book, we consider only operators $\mathcal{L}_r$ for which $s_{ri} = 2bs'_{ri}$, $t_{rj} = 2bt'_{rj}$, where $s'_{ri} \leq 0$, $t'_{rj} \geq 0$. In particular, we may assume that $s_{ri} = 0$, $i = 1, \ldots, m_r$, $r = 1, 2$. Then we obtain systems parabolic in the sense of Petrovskiĭ, and we can take the matrices that correspond to the Cauchy data at $t = 0$ as the matrices C_r.

We define the *model conjugation problem* in a class of ordinary sufficiently smooth functions as the problem of finding solutions $u_{(r)}(x, t) = (u_{r1}, \ldots, u_{rm_r})$ in the domain Ω_r^+ of the system

$$l_{(r)ij}(D, D_t)u_{(r)j}(x, t) = f_{(r)i}(x, t), \qquad i = 1, \ldots, m_r, \; r = 1, 2, \qquad (5.1)$$

which satisfy at $t = 0$ the initial conditions

$$c_{(r)\alpha j}(D, D_t)u_{(r)j}(x, t)\big|_{t=0} = \psi_{(r)\alpha}(x), \qquad \alpha = 1, \ldots, \kappa_r, \; r = 1, 2, \qquad (5.2)$$

and satisfy on the boundary separating Ω_1^+ and Ω_2^+ the conjugation conditions

$$b_{(1)qj}(D, D_t)u_{(1)j}(x, t)\big|_{x_n=+0} + b_{(2)qj}(D, D_t)u_{(2)j}(x, t)\big|_{x_n=-0} = \varphi_q(x', t),$$

$$q = 1, \ldots, N = N_1 + N_2. \qquad (5.3)$$

In (5.1)–(5.3) repetition of j (and only of j) for each fixed r indicates summation with respect to j from 0 to m_r.

Now let us formulate the conditions imposed on the operators $C_r(D, D_t)$ and $\mathcal{B}_r(D, D_t)$.

<u>Conditions on the operators C_r.</u> I. By the discussion in Subsection IV.2.1, we require that the operators $C_r(D, D_t)$ satisfy the following condition: *the rows of the matrix $D_r(0, p) = C_r(0, p)\widehat{\mathcal{L}}_r(0, p)$, $r = 1, 2$, are linearly independent modulo p^{κ_r}, where $\widehat{\mathcal{L}}_r$ denotes the adjoint matrix of $\mathcal{L}_r$.*

To formulate the conditions imposed on the operators $\mathcal{B}_r$, we first need to introduce some auxiliary objects. Set $L_r(i\xi, p) = \det \mathcal{L}_r(i\xi, p)$ and let $L_r(i\xi, p) = L_r^+(\xi, p)L_r^-(\xi, p)$ be a factorization of the polynomial $L_r(i\xi, p)$ in ξ_n, where L_r^+ [resp., L_r^-] is a polynomial in ξ_n whose zeroes satisfy $\operatorname{Im} \xi_n > 0$ [resp., $\operatorname{Im} \xi_n < 0$]. Consider the matrix

$$Q(\xi, p) = (\mathcal{B}_1\widehat{\mathcal{L}}_1, \mathcal{B}_2\widehat{\mathcal{L}}_2) = (Q_1 L_1^+ + Q_1'(\cdot, \xi_n), Q_2 L_2^- + Q_2'(\cdot, \xi_n))$$

and the matrix $Q'(\cdot, \xi_n)$ composed of the remainders $Q_1'(\cdot, \xi_n), Q_2'(\cdot, \xi_n)$ of the division of the matrices $\mathcal{B}_1\widehat{\mathcal{L}}_1$ by $L_1^+(\cdot, \xi_n)$ and $\mathcal{B}_2\widehat{\mathcal{L}}_2$ by $L_2^-(\cdot, \xi_n)$, respectively: $Q'(\cdot, \xi_n) = (Q_1'(\cdot, \xi_n), Q_2'(\cdot, \xi_n))$.

Definition IV.3. *The operators $\mathcal{B}_1(D, D_t)$ and $\mathcal{B}_2(D, D_t)$ are said to compatibly cover the operators $\mathcal{L}_1(D, D_t)$ and $\mathcal{L}_2(D, D_t)$ at $x_n = 0$ (or to satisfy the compatible covering condition at $x_n = 0$), if for any $\xi' \in \mathbb{R}^{n-1}$ and p with $\operatorname{Re} p \geq 0$, $(\xi', p) \neq 0$, the rows of the matrix $Q'(\cdot, \xi_n)$, regarded as polynomials in ξ_n, are linearly independent.*

It is readily seen that the compatible covering condition at $x_n = 0$ is equivalent to the condition that the rank of the matrix composed of the coefficients of the powers of ξ_n in the elements of the matrix $Q'(\cdot, \xi_n)$ is equal to $b(\kappa_1 + \kappa_2)$.

<u>Condition on the operators $\mathcal{B}_r$. II.</u> *The operators $\mathcal{B}_r(D, D_t)$ $(r = 1, 2)$ satisfy the compatible covering condition at $x_n = 0$.*

IV.5.2. Reduction of the model conjugation problem to an equivalent boundary value problem for a block-diagonal system

Let us show that the conjugation problem (5.1)–(5.3) is equivalent to a boundary value problem for a corresponding block-diagonal parabolic system in the half-space $x_n > 0$, with boundary conditions at $x_n = 0$ and initial conditions at $t = 0$. To this end we note that if $E(x, t, D, D_t)$ and $w(x, t)$ are a differential operator and a function, respectively, defined in Ω_2^+, then changing the variable x_n to $-x_n$ takes them into the operator $\overline{E}(x, t, D, D_t) \equiv E(x', -x_n, t, D', -D_n, D_t)$ and the function $\overline{w}(x, t) \equiv w(x', -x_n, t)$, respectively, defined for $x_n > 0$. Consider the conjugation problem (5.1)–(5.3) and pass from x_n to $-x_n$ in the operators $\mathcal{L}_2$, $\mathcal{B}_2$, C_2 and the vector-valued functions $u_2(x, t)$, $f_2(x, t)$, and $\psi_2(x)$. This yields the following boundary value problem in the half-space $x_n > 0$:

$$l_{(1)ij}(D, D_t)u_{(1)j}(x,t) = f_{(1)i}(x,t), \qquad i = 1, \ldots, m_1, \tag{5.4}$$

$$\bar{l}_{(2)ij}(D, D_t)\bar{u}_{(2)j}(x,t) = \bar{f}_{(2)i}(x,t), \qquad i = 1, \ldots, m_2, \tag{5.5}$$

$$\mathcal{B}_q u\big|_{x_n=0} = b_{(1)qj}(D, D_t)u_{(1)j}(x,t)\big|_{x_n=0}$$

$$+ \bar{b}_{(2)qj}(D, D_t)\bar{u}_{(2)j}(x,t)\big|_{x_n=0} = \varphi_q(x',t), \quad q = 1, \ldots, N \tag{5.6}$$

$$c_{(1)\alpha j}(D, D_t)u_{(1)j}(x,t)\big|_{t=0} = \psi_{(1)\alpha}(x), \qquad \alpha = 1, \ldots, \kappa_1, \tag{5.7}$$

$$\bar{c}_{(2)\alpha j}(D, D_t)\bar{u}_{(2)j}(x,t)\big|_{t=0} = \psi_{(2)\alpha}(x), \qquad \alpha = 1, \ldots, \kappa_2, \tag{5.8}$$

Obviously, to the system (5.4), (5.5) and the initial conditions (5.7), (5.8) there correspond the block-diagonal matrices

$$\mathcal{L}(D, D_t) = \begin{pmatrix} \mathcal{L}_1(D, D_t) & 0 \\ 0 & \overline{\mathcal{L}}_2(D, D_t) \end{pmatrix},$$

$$C(D, D_t) = \begin{pmatrix} C_1(D, D_t) & 0 \\ 0 & \overline{C}_2(D, D_t) \end{pmatrix},$$

and to the boundary conditions (5.6) there corresponds the block matrix $\mathcal{B}(D, D_t)$ $= \big(\mathcal{B}_1(D, D_t), \overline{\mathcal{B}}_2(D, D_t)\big)$. Hence, the boundary value problem (5.4)–(5.8) can be rewritten in the compact form

$$\mathcal{L}(D, D_t)u(x, t) = f(x, t),$$
$$\mathcal{B}(D, D_t)u(x, t)\big|_{x_n=0} = \varphi(x', t), \tag{5.9}$$
$$C(D, D_t)u\big|_{t=0} = \psi(x),$$

where $u(x, t) = \big(u_{(1)}(x, t), \bar{u}_{(2)}(x, t)\big)$, $f(x, t) = \big(f_1(x, t), \bar{f}_2(x, t)\big)$, $\psi(x) = \big(\psi_1(x),$ $\bar{\psi}_2(x)\big)$.

Lemma IV.4. *The compatible covering condition for the conjugation problem* (5.1)–(5.3) *is equivalent to the Lopatinskiĭ condition for the boundary value problem* (5.9).

Proof: Since $\mathcal{L}(D, D_t)$ is a diagonal matrix,

$$L(i\xi, p) = \det \mathcal{L}(i\xi, p) = \det \mathcal{L}_1(i\xi, p) \cdot \det \overline{\mathcal{L}}_2(i\xi, p)$$
$$= L^+(\cdot, \xi_n) \cdot L^-(\cdot, \xi_n) = L_1^+ \cdot \overline{L}_2^+ \cdot L_1^- \cdot \overline{L}_2^-,$$

and it can be easily seen that $L^+(\cdot, \xi_n)$ equals $L_1^+(\cdot, \xi_n) \cdot \overline{L}_2^-(\cdot, \xi_n)$ up to a constant factor. Since $\mathcal{L}(D, D_t)$ and $\mathcal{B}(D, D_t)$ are quasihomogeneous operators with constant coefficients, consisting only of the principal parts, the assertion that the boundary value problem (5.9) satisfies the Lopatinskiĭ condition means, by definition, that the rows of the matrix composed of the remainders of the division of the matrix $\mathcal{B}\widehat{\mathcal{L}}$ by $L^+(\cdot, \xi_n)$ are linearly independent as polynomials in ξ_n. It readily verified that

$$\mathcal{B}(i\xi, p)\widehat{\mathcal{L}}(i\xi, p) = \big(\mathcal{B}_1(i\xi, p), \overline{\mathcal{B}}_2(i\xi, p)\big) \begin{pmatrix} \widehat{\mathcal{L}}_1(i\xi, p)\overline{L}_2(\cdot, \xi_n) & 0 \\ 0 & L_1(\cdot, \xi_n)\widehat{\mathcal{L}}_2(i\xi, p) \end{pmatrix}$$

$$= \Big(\overline{L}_2(i\xi, p)\mathcal{B}_1(i\xi, p)\widehat{\mathcal{L}}_1(i\xi, p), \; L_1(i\xi, p)\overline{\mathcal{B}}_2(i\xi, p)\widehat{\overline{\mathcal{L}}}_2(i\xi, p)\Big).$$

On the one hand,

$$\mathcal{B}(i\xi, p)\widehat{\mathcal{L}}(i\xi, p) = \Big(\mathcal{P}_1 \cdot L^+(\cdot, \xi_n) + \mathcal{R}_1'(\cdot, \xi_n), \; \mathcal{P}_2 \cdot L^+(\cdot, \xi_n) + \mathcal{R}_2'(\cdot, \xi_n)\Big), \tag{5.10}$$

where $\mathcal{P}_1(\cdot, \xi_n)$, $\mathcal{P}_2(\cdot, \xi_n)$ and $\mathcal{R}_1'(\cdot, \xi_n)$, $\mathcal{R}_2'(\cdot, \xi_n)$ are the matrices composed of the integral (i.e., polynomial) parts and the remainders of the division of the matrices $\overline{L}_2(i\xi, p)\mathcal{B}_1(i\xi, p)\widehat{\mathcal{L}}_1(i\xi, p)$ and $L_1(i\xi, p)\overline{\mathcal{B}}_2(i\xi, p)\widehat{\overline{\mathcal{L}}}_2(i\xi, p)$, respectively, by $L^+(\cdot, \xi_n)$.

On the other hand, if $N_1(\cdot,\xi_n)$, $N_2(\cdot,\xi_n)$ and $\mathcal{D}'_1(\cdot,\xi_n)$, $\mathcal{D}'_2(\cdot,\xi_n)$ are the matrices composed of the integral parts and the remainders of the division of the elements of the matrices $\mathcal{B}_1(i\xi,p)\widehat{\mathcal{L}}_1(i\xi,p)$ and $\overline{\mathcal{B}}_2(i\xi,p)\cdot\widehat{\overline{\mathcal{L}}}_2(i\xi,p)$ by $L_1^+(\cdot,\xi_n)$ and respectively by $L_2^-(\cdot,\xi_n)$, then thanks to the fact that $L_2^-(\cdot,\xi_n)$ coincides up a constant with $\overline{L}_2^-(\cdot,\xi_n) = \big(\overline{L}_2(\cdot,\xi_n)\big)^+$, we obtain

$$\mathcal{B}\widehat{\mathcal{L}} = \Big(\overline{L}_2(\cdot,\xi_n)\big(N_1 L_1^+ + \mathcal{D}'_1(\cdot,\xi_n)\big),\ L_1(\cdot,\xi_n)\big(\overline{N}_2\overline{L}_2^-(\cdot,\xi_n) + \overline{\mathcal{D}}'_2(\cdot,\xi_n)\big)\Big). \quad (5.11)$$

Since $\overline{L}_2 \cdot N_1 \cdot L_1^+ = Q_1(\cdot,\xi_n)L^+(\cdot,\xi_n)$ and $L_1 \cdot \overline{N}_2 \cdot \overline{L}_2^- = Q_2(\cdot,\xi_n)L^+(\cdot,\xi_n)$, relation (5.11) can be written in the form

$$\mathcal{B}\widehat{\mathcal{L}}(i\xi,p) = \Big(Q_1 L^+ + \overline{L}_2(i\xi,p)\mathcal{D}'_1(\cdot,\xi_n),\ Q_2 L^+ + L_1(i\xi,p)\cdot\overline{\mathcal{D}}'_2(\cdot,\xi_n)\Big). \quad (5.12)$$

Now let us equate the right-hand sides of (5.11) and (5.12) and move the matrix $(Q_1\cdot M_+, Q_2\cdot M_+)$ to the left-hand side. Denoting $\mathcal{P}_1 - Q_1 = \mathcal{M}_1$, $\mathcal{P}_2 - Q_2 = \mathcal{M}_2$, we obtain

$$\big(\mathcal{M}_1 L^+ + \mathcal{R}'_1(\cdot,\xi_n),\ \mathcal{M}_2 L^+ + \mathcal{R}'_2(\cdot,\xi_n)\big)$$
$$= \big(\overline{L}_2(\cdot,i\xi_n)\mathcal{D}'_1(\cdot,\xi_n),\ L_1(\cdot,i\xi_n)\mathcal{D}'_2(\cdot,\xi_n)\big). \quad (5.13)$$

Recall that the degrees of the polynomials $\mathcal{R}'_1$ and $\mathcal{R}'_2$ are strictly smaller than that of $L^+(\cdot,\xi_n)$ and the degrees of $\mathcal{D}'_1(\cdot,\xi_n)$ and $\mathcal{D}'_2(\cdot,\xi_n)$ are strictly smaller than those of $L_1^+(\cdot,\xi_n)$ and $\overline{L}_2^-(\cdot,\xi_n)$, respectively. Let us show that the assertion of Lemma IV.4 follows from the matrix equality (5.13).

(a) Suppose the rows of the matrix $\mathcal{D}' = (\mathcal{D}'_1, \mathcal{D}'_2)$ are linearly dependent. This means that there exist constants $c_1,\ldots,c_\kappa$ such that $\sum|c_k| \neq 0$ and the corresponding linear combination of the rows of the matrix $(\mathcal{D}'_1, \mathcal{D}'_2)$ is equal to zero. Then (5.13) implies that the same linear combination of the rows of $\mathcal{R}' = (\mathcal{R}'_1, \mathcal{R}'_2)$ can be divided by $L^+(\cdot,\xi_n)$ without remainder. Hence, each root of $L^+(\cdot,\xi_n)$ is the root of the linear combination of the elements of an arbitrary column, which is impossible because the degree of the linear combination in the variable ξ_n is strictly smaller than the degree of $L^+(\cdot,\xi_n)$ in ξ_n. Consequently, the linear combination of the rows of matrix $\mathcal{R}'$ is identically equal to zero, that is, the rows of matrix $\mathcal{R}'$ are linearly dependent.

(b) Conversely, suppose now that the rows of the matrix $\mathcal{R}'$ are linearly dependent, i.e., there exists a nontrivial linear combination of rows that vanishes identically. Then from relation (5.13) it follows that each root of the polynomial $L^+(\cdot,\xi_n)$ is a root of the same nontrivial linear combination of the rows of the matrix appearing in the right-hand side of (5.13). It immediately follows that each root of $L_1^+(\cdot,\xi_n)$, $\overline{L}_2^-(\cdot,\xi_n)$ is a root of a linear combination of the rows of the matrices $\mathcal{D}'_1, \mathcal{D}'_2$, which is impossible. Hence, the linear combination in question of the rows of matrix $\mathcal{D}'$ is identically equal to zero, that is, the rows of the matrix $\mathcal{D}'$ are linearly dependent.

Remark IV.1. In the subsequent analysis we will make essential use of Lemma IV.4 in one particular case, which is encountered in the investigation of a certain class of nonlocal boundary value problems (the nonlocal conjugation problems). This is, for example, the situation when $\mathcal{L}_1$ is a block-diagonal matrix:

$$\mathcal{L}_1(D, D_t) = \begin{pmatrix} \mathcal{L}_{11}(D, D_t) & 0 \\ 0 & \mathcal{L}_{12}(D, D_t) \end{pmatrix},$$

where $\mathcal{L}_{11}$ and $\mathcal{L}_{12}$ are $m_{11} \times m_{11}$ and $m_{12} \times m_{12}$ matrices, $m_{11} + m_{12} = m_1$. In this case the system $\mathcal{L}_1 u_1 = f_1$ splits into two independent ones. Let $\mathcal{B}_1 = (\mathcal{B}_{11}, \mathcal{B}_{12})$, where $\mathcal{B}_{1r}$ are $N \times m_{1r}$ matrices $(r = 1, 2)$.

Consider the matrix $\mathcal{B}(i\xi, p)\widehat{\mathcal{L}}(i\xi, p)$ and represent it in the form

$$\begin{aligned}
\mathcal{B}\widehat{\mathcal{L}} &= (\mathcal{B}_1\widehat{\mathcal{L}}_1, \, \mathcal{B}_2\widehat{\mathcal{L}}_2) = \left(L_{12}\mathcal{B}_{11}\widehat{\mathcal{L}}_{11}, \, L_{11}\mathcal{B}_{12}\widehat{\mathcal{L}}_{12}, \, \mathcal{B}_2\widehat{\mathcal{L}}_2\right) \\
&= \left(L_1^+ Q_{11} + L_{12}\mathcal{D}'_{11}, \, L_1^+ Q_{12} + L_{11}\mathcal{D}'_{12}, \, Q_2 L_2^- + \mathcal{D}'_2\right) \quad (5.14) \\
&= \left(L_1^+ \mathcal{M}_1 + \mathcal{R}'_1, \, Q_2 L_2^- + \mathcal{R}'_2\right),
\end{aligned}$$

where $\mathcal{D}'_{11}(\cdot, \xi_n)$, $\mathcal{D}'_{12}(\cdot, \xi_n)$ and $\mathcal{R}'_1(\cdot, \xi_n)$, $\mathcal{R}'_2 \equiv \mathcal{D}'_2(\cdot, \xi_n)$ are the remainders of the division of the matrices $\mathcal{B}_{11}\widehat{\mathcal{L}}_{11}(i\xi, p)$, $\mathcal{B}_{12}\widehat{\mathcal{L}}_{12}(i\xi, p)$ and $\mathcal{B}_1\widehat{\mathcal{L}}_1(i\xi, p)$, $\mathcal{B}_2\widehat{\mathcal{L}}_2(i\xi, p)$ by the polynomials $L_{11}^+(\cdot, \xi_n)$, $L_{12}^+(\cdot, \xi_n)$ and $L_1^+(\cdot, \xi_n)$, $L_2^-(\cdot, \xi_n)$, respectively.

Equality (5.14) implies that the condition of linear independence of the rows of the matrix $(\mathcal{R}'_1, \mathcal{R}'_2)$ is equivalent to the condition of linear independence of the rows of the matrix $(\mathcal{D}'_{11}, \mathcal{D}'_{12}, \mathcal{D}'_2)$. Therefore, in the case where the block-diagonal matrix $\mathcal{L}_1$ consists of two blocks the compatible covering condition can be formulated as follows: the rows of the matrix $(\mathcal{D}'_{11}(\cdot, \xi_n), \mathcal{D}'_{12}(\cdot, \xi_n), \mathcal{D}'_2(\cdot, \xi_n))$, regarded as polynomials in ξ_n, are linearly independent.

It is easy to write down a similar condition, equivalent to the compatible covering condition in the case when $\mathcal{L}_1$ and $\mathcal{L}_2$ are block-diagonal matrices with an arbitrary number of blocks.

IV.6. Boundary value problem in $\widetilde{\mathcal{H}}_+^s(\overline{\mathbb{R}}_{++}^{n+1}, \gamma)$ for operators in which the coefficients of the highest-order derivatives are slowly varying functions

Using Theorem IV.7 on the solvability of the model boundary value problem (3.10)–(3.14) we will investigate the solvability of the boundary value problem

$$\begin{aligned}
\mathcal{L}_i u = l_{ij}(x, t, D, D_t)u_j\big|_{\overline{\mathbb{R}}_{++}^{n+1}} &= l_{ij}u_{j0}^{++}(x, t) - l_{ij}^{(k,0)}u_{jk}^+(x', t) \times \delta(x_n) \\
&\quad - l_{ij}^{(0,\lambda)}v_{j\lambda}^+(x) \times \delta(t) + l_{ij}^{(k,\lambda)}\omega_{j\lambda k}(x') \times \delta(x_n, t) \quad (6.1) \\
&= f_{i0}^{++}(x, t), \qquad i = 1, \ldots, m,
\end{aligned}$$

$$\mathcal{B}_q u\big|_{\overline{E}^n_+} = b_{qj}(x',t,D,D_t)u_j\big|_{\overline{E}^n_+} \equiv b_{qjk}(\cdot,D',D_t)u^+_{jk}(x',t) \times \delta(x_n)$$

$$-b^{(\lambda)}_{qjk}(\cdot,D',D_t)\omega_{j\lambda k}(x') \times \delta(t) = \varphi^+_q(x',t), \quad q = 1,\dots,N, \tag{6.2}$$

$$\left(D^{\sigma-1}_n \mathcal{L}_i\right)u\big|_{\overline{E}^n_+} = \left(D^{\sigma-1}_n l_{ij}(\cdot,D,D_t)\right)u_j\big|_{\overline{E}^n_+}$$

$$= f^+_{i\sigma}(x',t), \qquad \sigma = 1,\dots,\sigma_0, \ i = 1,\dots,m, \tag{6.3}$$

$$D^{\lambda-1}_t u_j\big|_{\overline{\mathbb{R}}^n_+} = v^+_{j\lambda}(x) = \psi^+_{j\lambda}(x), \qquad \lambda = 1,\dots,t'_j, \ j = 1,\dots,m, \tag{6.4}$$

$$D^{k-1}_n v_{j\lambda}(x)\big|_{\mathbb{R}^{n-1}} = \omega_{j\lambda k}(x') = \psi_{j\lambda k}(x'), \qquad (\lambda,k) \in P_j \tag{6.5}$$

with operators $\mathcal{L}$ and $\mathcal{B}$ in which the coefficients of the highest-order derivatives are slowly varying functions, under the assumption that the right-hand sides of (6.1)–(6.5) satisfy the conditions of compatibility with zero at $t = 0$. The latter are formulated in the same way as for the problem (3.10)–(3.14).

Let $l_{ij} = l_{ij\alpha\beta}(x,t)D^\alpha D^\beta_t$, $b_{qj} = b_{qj\alpha\beta}(x',t)D^\alpha D^\beta_t$, ord $l_{ij} = t_j = 2bt'_j$. As in Section IV.3, we shall assume for simplicity that the order of differentiation with respect to t in the operators b_{qj} is at most t'_j and that the coefficients of the operators l_{ij} and b_{qj} satisfy the conditions

$$l_{ij\alpha\beta}(x,t) \in C^{|s|+t_j+\sigma_0+r}\left(\overline{\mathbb{R}}^{n+1}_{++}\right), \tag{6.6}$$

$$b_{qj\alpha\beta}(x',t) \in C^{|s|+t_j-\sigma_q+r}\left(\overline{E}^n_+\right), \qquad \forall r > 0. \tag{6.7}$$

Let $\widetilde{\mathcal{H}}^s$ be the space defined in Section IV.3. By Lemmas III.16–III.19, relations (6.1)–(6.5) define a bounded operator $\mathfrak{U}(x,t,D,D_t)$, which acts from $\widetilde{\mathcal{H}}^s_+$ into $\widehat{\mathcal{K}}^s_+ = \mathfrak{U}\widetilde{\mathcal{H}}^s_+$, the subspace of the space $\mathcal{K}^s_+$ that consists of the vectors $\mathcal{F}(x,t) = \left(f^{++}_{10},\dots,f^{++}_{m0}, \varphi^+_1,\dots,\varphi^+_N, f^+_{11},\dots,f^+_{1\sigma_0},\dots,f^+_{m1},\dots,f^+_{m\sigma_0},\dots,\psi_{j\lambda k}, (\lambda,k)\in P_j\right)$, whose components satisfy the conditions indicated in Subsection III.5.2.

As in Subsection IV.4.2, we will say that the right-hand sides of the problem (6.1)–(6.5) are compatible with zero to order s at $t = 0$ if $\mathcal{F}(x,t) \in \widetilde{\mathcal{K}}^s_+$. This means that $v^+_{j\lambda} \equiv 0$ for all j,λ such that $s + t_j - 2b\lambda + b > 0$, $\omega_{j\lambda k}(x') \equiv 0$ for all j,λ,k such that $s + t_j - 2b\lambda - k + b + \frac{1}{2} > 0$, and that $f_{i0}, f_{i\sigma}$, and φ_q, continued by zero for $t < 0$, belong to the spaces $\mathcal{H}^s\left(\overline{\mathbb{R}}^{n+1}_+,\gamma\right)$, $\mathcal{H}^{s-\sigma+1/2}\left(E^n,\gamma\right)$ and $\mathcal{H}^{s-\sigma_q-1/2}\left(E^n,\gamma\right)$, respectively. To simplify notation, we denote the norms of the elements $u(x,t) \in \widetilde{\mathcal{H}}^s$ and $\mathcal{F}(x,t) \in \mathcal{K}^s$ by $\left\|\!\left\|u\right\|\!\right\|_s$ and $\left\|\!\left\|\mathcal{F}\right\|\!\right\|_s$.

Theorem IV.10. *Let $\mathcal{L}$ be an operator uniformly parabolic in the sense of Petrovskiĭ in $\overline{\mathbb{R}}^{n+1}_{++}$, let $\mathcal{B}$ be an operator that satisfies the Lopatinskiĭ condition uniformly in $(x',t) \in \overline{E}^n_+$, and let the right-hand sides of problem (6.1)–(6.5) satisfy the condition of compatibility with zero at $t = 0$ to order s. Suppose the coefficients $l_{ij\alpha\beta}(x,t)$ and $b_{qj\alpha\beta}(x',t)$ satisfy conditions (6.6), (6.7) and*

$$\sup_{(x,t)\in\mathbb{R}^{n+1}_{++}} \left|l_{ij\alpha\beta}(x,t) - l_{ij\alpha\beta}(0,0)\right| < \varepsilon, \qquad \forall\alpha,\beta : |\alpha| + 2b\beta = t_j, \tag{6.8}$$

$$\sup_{(x,t)\in\overline{E}^n_+} \big|b_{qj\alpha\beta}(x',t) - b_{qj\alpha\beta}(0,0)\big| < \varepsilon, \qquad \forall \alpha,\beta : |\alpha| + 2b\beta = \sigma_q + t_j. \quad (6.9)$$

Then there exist numbers $\varepsilon_0 > 0$ and $\gamma_0 > 0$ (which depend on s and the Hölder norms of the coefficients $l_{ij\alpha\beta}$ and $b_{qj\alpha\beta}$), such that for $\varepsilon < \varepsilon_0(s)$ and $\gamma > \gamma_0(s)$ problem (6.1)–(6.5) has a unique solution $u(x,t) \in \widetilde{\mathcal{H}}^s_+$, which for $s + t_j \notin \mathbb{Z}_{1,2b}$ satisfies the estimate

$$\big\|\!\big\|u(x,t)\big\|\!\big\|_s < C_s \big\|\!\big\|\mathcal{F}(x,t)\big\|\!\big\|_s. \quad (6.10)$$

Proof: Denote by l^0_{ij}, b^0_{qj}, $\big(D_n^{\sigma-1}l_{ij}\big)^0$ and l'_{ij}, b'_{qj}, $\big(D_n^{\sigma-1}l_{ij}\big)'$ the weighted highest-order and lowest-order terms of the operators l_{ij}, b_{qj}, $D_n^{\sigma-1}l_{ij}$, respectively. Further, denote

$$\Delta l^0_{ij} = \big(l^0_{ij}(x,t,\cdot) - l^0_{ij}(0,0,\cdot)\big), \quad \Delta b^0_{qj} = \big(b^0_{qj}(x',t,\cdot) - b^0_{qj}(0,0,\cdot)\big),$$

$$\Delta\big(D_n^{\sigma-1}l_{ij}\big)^0 = \big(D_n^{\sigma-1}l_{ij}\big)^0(x',t,\cdot) - \big(D_n^{\sigma-1}l_{ij}\big)(0,0,\cdot).$$

Finally, let $\Delta\mathcal{L}^0_i$, $\Delta\mathcal{B}^0_q$, $\Delta\big(D_n^{\sigma-1}\mathcal{L}_i\big)$, and $\mathcal{L}'_i$, $\mathcal{B}'_q$, $\big(D_n^{\sigma-1}\mathcal{L}_i\big)'$ be the rows of the matrices $\mathcal{L}$, $\mathcal{B}$, $D_n^{\sigma-1}\mathcal{L}_i$, respectively, composed of the indicated operators. In accordance with this notation we represent the operator $\mathfrak{U}(x,t,D,D_t)$ as $\mathfrak{U}(x,t,D,D_t) = \mathfrak{U}_0(0) + \Delta\mathfrak{U}_0(x,t) + \mathfrak{U}'(x,t)$, where $\mathfrak{U}_0(0) = \mathfrak{U}_0(0,0,D,D_t)$ is the operator of the model boundary value problem (3.10)–(3.14) with the right-hand sides compatible with zero at $t = 0$.

Theorem IV.7 asserts that for any $s \in \mathbb{R}^1$ the operator $\mathfrak{U}_0(0)$ has a bounded inverse $\mathfrak{U}_0^{-1}(0)$, defined on $\widetilde{\mathcal{K}}^s_+$, and for any s such that $s + t_j \notin \mathbb{Z}_{1,2b} \cap (0,M)$ there holds the estimate

$$\big\|\!\big\|u(x,t)\big\|\!\big\|_s = \big\|\!\big\|\mathfrak{U}_0^{-1}(0)\mathcal{F}\big\|\!\big\|_s < C_s \big\|\!\big\|\mathcal{F}(x,t), \overline{\mathbb{R}}^{n+1}_{++}, \gamma\big\|\!\big\|_s. \quad (6.11)$$

Consider the operators

$$\mathfrak{U}(x,t,\cdot)\mathfrak{U}_0^{-1}(0) = I_K + \Delta\mathfrak{U}_0(x,t)\mathfrak{U}_0^{-1}(0) + \mathfrak{U}'(x,t)\mathfrak{U}_0^{-1}(0)$$

$$= I_K + \Delta T + T' = I_K + T, \quad (6.12)$$

$$\mathfrak{U}_0^{-1}(0)\mathfrak{U}(x,t) = I_H + \mathfrak{U}_0^{-1}(0)\Delta\mathfrak{U}_0(x,t) + \mathfrak{U}_0^{-1}(0)\mathfrak{U}'(x,t)$$

$$= I_H + \Delta W + W' = I_H + W, \quad (6.13)$$

where I_K, I_H are the identity operators and T, W are bounded operators in the spaces $\widetilde{\mathcal{K}}^s_+$ and $\widetilde{\mathcal{H}}^s_+$, respectively. Note that if the norms $\|T\|_s$, $\|W\|_s$ of the operators T, W are strictly less than 1, then, by the discussion in Sections IV.2 and IV.3, the operators $I_K + T$ and $I_H + W$ have bounded inverses $(I_K + T)^{-1}$ and $(I_H + W)^{-1}$ in $\widetilde{\mathcal{K}}^s_+$ and $\widetilde{\mathcal{H}}^s_+$, respectively. Then, multiplying (6.12) from the right by $(I_K + T)^{-1}$ and (6.13) from the left by $(I_H + W)^{-1}$ we obtain $\mathfrak{U}(x,t)\mathfrak{U}_0^{-1}(0)(I_K + T)^{-1} = I_K,$

$(I_H + W)^{-1}\mathfrak{U}_0^{-1}(0)\mathfrak{U}(x,t) = I_H$. This establishes the invertibility of the operator $\mathfrak{U}(x,t,D,D_t)$ of the problem (6.1)–(6.5) and the assertion of Theorem IV.10.

Thus, to actually complete the proof of Theorem IV.10 it remains to prove the existence of the indicated numbers ε_0 and γ_0 such that for $\varepsilon < \varepsilon_0$ and $\gamma > \gamma_0$ the norms $\|T\|_s$ and $\|W\|_s$ of the operators T and W are strictly less than 1.

We wiil carry out in detail only the proof of the estimate $\|T\|_s < 1$; the estimate $\|W\|_s < 1$ is proved in exactly the same manner and follows in part from the arguments used in showing that $\|T\|_s < 1$ and from the boundedness of the operator $\mathfrak{U}_0^{-1}(0)$.

First let us examine the term ΔT. For the sake of brevity, we denote $u(x,t) = \mathfrak{U}_0^{-1}(0)\mathcal{F}(x,t)$ and note that the estimation of the norms

$$\left\| \Delta \mathcal{L}_i^0 u \big|_{\overline{\mathbb{R}}_{++}^{n+1}}, \ \mathbb{R}_{++}^{n+1}, \ \gamma \right\|_s, \quad \left\langle\!\left\langle \Delta \mathcal{B}_q u \big|_{\overline{E}_+^n}, \ \overline{E}_+^n, \ \gamma \right\rangle\!\right\rangle_{s-\sigma_q-1/2},$$

$$\left\langle\!\left\langle \Delta \left(D_n^{\sigma-1}\mathcal{L}_i\right)^0 u \big|_{\overline{E}_+^n}, \ \overline{E}_+^n, \ \gamma \right\rangle\!\right\rangle_{s-\sigma+1/2},$$

of the components of the element $\Delta\mathfrak{U}_0(x,t)u$ that correspond to the highest-order terms of the operators, which have slowly-varying coefficients, reduces to the estimation of the norms

$$\left\| \Delta l_{ij}^0 u_j \big|_{\overline{\mathbb{R}}_{++}^{n+1}}, \ \mathbb{R}_{++}^{n+1}, \ \gamma \right\|_s, \quad \left\langle\!\left\langle \Delta b_{qj}^0 u_j \big|_{\overline{E}_+^n}, \ \overline{E}_+^n, \ \gamma \right\rangle\!\right\rangle_{s-\sigma_q-1/2},$$
$$\left\langle\!\left\langle \Delta \left(D_n^{\sigma-1}l_{ij}\right)^0 u_j \big|_{\overline{E}_+^n}, \ \overline{E}_+^n, \ \gamma \right\rangle\!\right\rangle_{s-\sigma+1/2}, \tag{6.14}$$

because the components of the operator $\Delta\mathfrak{U}_0$ that correspond to the initial conditions are identically equal to zero.

1. We will deal in detail only with the norms $\left\| \Delta l_{ij}^0 u_j \big|_{\overline{\mathbb{R}}_{++}^{n+1}}, \ \mathbb{R}_{++}^{n+1}, \ \gamma \right\|_s$; the estimation of the remaining norms in (6.14) is similar, and actually even easier. By Lemma III.7, the form of the expression $\Delta l_{ij}^0 u_j \big|_{\overline{\mathbb{R}}_{++}^{n+1}}$ depends on s. For this reason, to estimate the indicated norms we will use in essential manner this lemma and the compatibility of $u_{j0}(x,t)$ with zero at $t = 0$ to order s . We will estimate these norms separately for $s + t_j < \frac{1}{2}$ and $s + t_j > \frac{1}{2}$.

(a) Let $s + t_j < \frac{1}{2}$. Then the expression $\Delta l_{ij}^0 u_j \big|_{\overline{\mathbb{R}}_{++}^{n+1}}$ is defined by (3.16) with l_{ij} replaced by Δl_{ij}^0, and the norm of each term can be estimated with the help of Lemma III.16. Indeed, since for $s + t_j < \frac{1}{2}$,

$$u_{j0}^{++}(x,t) \in \mathcal{H}_{++}^{s+t_j}(\mathbb{R}^{n+1}, \gamma),$$

$$u_{jk}^+(x',t) \times \delta(x_n) \in \mathcal{H}_{++}^{s+t_j-k+1/2}(\mathbb{R}^{n+1}, \gamma),$$

$$v_{j\lambda}^+(x) \times \delta(t) \in \mathcal{H}_{++}^{s+t_j-2b\lambda}(\mathbb{R}^{n+1}, \gamma), \tag{6.15}$$

$$\omega_{j\lambda k}(x') \times \delta(x_n, t) \in \mathcal{H}_{++}^{s+t_j-k-2b\lambda}(\mathbb{R}^{n+1}, \gamma),$$

and since Δl_{ij}^0 and $\Delta l_{ij}^{0(k,\lambda)}$ are quasihomogeneous operators of degrees t_j and $t_j - k - 2b\lambda$, respectively, we can apply inequality (2.12) of Chapter III to estimate the norm of each term. Taking into account that Δl_{ij}^0 contains only the derivatives $D^\alpha D_t^\beta u_{j0}^{++}$ with $|\alpha| + 2b\beta = t_j$, we get

$$\left\| \Delta l_{ij}^0 u_{j0}^{++}, \ \overline{\mathbb{R}}_{++}^{n+1}, \ \gamma \right\|_s \leq CA_s \| u_{j0}^+, \ \mathbb{R}_+^{n+1}, \ \gamma \|_{s+t_j}, \tag{6.16}$$

$$\left\| \Delta l_{ij}^{0(k,0)} u_{jk}^+ \times \delta(x_n), \ \overline{\mathbb{R}}_{++}^{n+1}, \ \gamma \right\|_s \leq CA_s \langle\!\langle u_{jk}^+, \ \overline{E}_+^n, \ \gamma \rangle\!\rangle_{s+t_j-k+1/2}, \tag{6.17}$$

$$\left\| \Delta l_{ij}^{0(0,\lambda)} v_{j\lambda}^+ \times \delta(t), \ \overline{\mathbb{R}}_{++}^{n+1}, \ \gamma \right\|_s \leq CA_s \big[v_{j\lambda}^+(x), \ \overline{\mathbb{R}}_+^n \big]_{s+t_j-2b\lambda+b}, \tag{6.18}$$

$$\left\| \Delta l_{ij}^{0(k,\lambda)} \omega_{j\lambda k}(x') \times \delta(x_n, t), \ \overline{\mathbb{R}}_{++}^{n+1}, \gamma \right\|_s$$
$$\leq CA_s \langle\!\langle \omega_{j\lambda k}(x'), \ \mathbb{R}^{n-1} \rangle\!\rangle'_{s+t_j-2b\lambda-k+b+1/2}, \tag{6.19}$$

where $A_s = \varepsilon + A\gamma^{-1/2b}$, $A = \max\limits_{i,j,\alpha,\beta} |l_{ij\alpha\beta}(x,t), \overline{\mathbb{R}}_{++}^{n+1}|_{|s|+r}$, $|\alpha| + 2b\beta = t_j$, and C is a constant that does not depend on u, ε and γ. From the inequalities (6.16)–(6.19) and estimate (2.29) it follows that

$$\left\| \Delta l_{ij}^0 u_j \big|_{\overline{\mathbb{R}}_{++}^{n+1}}, \overline{\mathbb{R}}_{++}^{n+1}, \ \gamma \right\|_s \leq C(\varepsilon + A\gamma^{-1/2b}) \| u_j, \ \overline{\mathbb{R}}_{++}^{n+1}, \ \gamma \|_{s+t_j}, \tag{6.20}$$

(b) Let $s + t_j > \frac{1}{2}$, $s < 0$. Then, by Lemma III.7 and the condition of compatibility with zero of u_{j0} at $t = 0$ to order s, the expression $\Delta l_{ij}^0 u_j \big|_{\overline{\mathbb{R}}_{++}^{n+1}}$ takes the form

$$\Delta l_{ij}^0 u_j \big|_{\overline{\mathbb{R}}_{++}^{n+1}} = \Delta l_{ij}^{0(s'+t_j,0)}(\cdot, D, D_t)\big(D_n^{s'+t_j} u_{j0}\big)^{++} - \sum_{k=s'+t_j+1}^{t_j} \Delta l_{ij}^{0(k,0)} u_{jk}^+ \times \delta(x_n)$$

$$- \sum_{\lambda=s''+t_j'+1}^{t_j'} \Delta l_{ij}^{0(0,\lambda)} v_{j\lambda}(x) \times \delta(t) + \sum_{k+2b\lambda>s'+t_j+2b+1} \Delta l_{ij}^{0(k,\lambda)} \omega_{j\lambda k}(x') \times \delta(x_n, t)$$

$$+ \sum_{\sigma=[s]+t_j-b+1}^{s'+t_j-1} \Delta l_{ij}^0 \big(D_n^\sigma u_{j0}\big)^{++} + \sum_{\sigma=0}^{[s]+t_j-b} \Delta l_{ij\sigma}^{0(\sigma_s+t_j)}(\cdot, D', D_t)\big(D_t^{\sigma_s+t_j} D_n^\sigma u_{j0}\big)^{++}$$

$$+ \sum_{\sigma+2b\mu\leq[s]+t_j-b} \Delta l_{ij\sigma\mu}^0(\cdot, D')\big(D_t^\mu D_n^\sigma u_{j0}\big)^{++}. \tag{6.21}$$

Now let us estimate the norm $\|\cdot, \mathbb{R}^{n+1}, \gamma\|_s$ of each term in the right-hand side of equality (6.21). Since for the indices k, λ figuring in (6.21) the distributions $u_{jk}^+(x', t) \times \delta(x_n)$, $v_{j\lambda}(x) \times \delta(t)$, and $\omega_{i\lambda k}(x') \times \delta(x_n, t)$ belong to the spaces indicated in (6.15), it follows that for the terms that contain these functions the estimates (6.17)–(6.19) remain valid.

All the other terms are of the same type and are the results of applying compositions of differential operators and operators of multiplication by characteristic functions to smooth functions. Therefore, using Lemma III.2 on the boundedness of the operator of multiplication by a characteristic function and Lemma III.16 we obtain

$$\left\|\Delta l_{ij}^{0(s'+t_j,0)}\left(D_n^{s'+t_j}u_{j0}\right)^{++},\overline{\mathbb{R}}_{++}^{n+1},\gamma\right\|_s \le CA_s\left\|u_{j0},\mathbb{R}_{++}^{n+1},\gamma\right\|_{s+t_j}, \tag{6.22}$$

$$\left\|\Delta_{ij\sigma}\left(D_n^{\sigma}u_{j0}\right)^{++},\overline{\mathbb{R}}_{++}^{n+1},\gamma\right\|_s \le CA_s\left\|u_{j0},\mathbb{R}_{++}^{n+1},\gamma\right\|_{s+t_j}, \tag{6.23}$$

$$\left\|\Delta l_{ij\sigma}^{0(\sigma_s+t_j)}\left(\cdot,D',D_t\right)\left(D_t^{\sigma_s+t_j}D_n^{\sigma}u_{j0}\right)^{++},\overline{\mathbb{R}}_{++}^{n+1},\gamma\right\|_s$$
$$\le CA_s\left\|u_{j0},\mathbb{R}_{++}^{n+1},\gamma\right\|_{s+t_j}, \tag{6.24}$$

$$\left\|\Delta l_{ij\sigma\mu}^{0}\left(\cdot,D'\right)\left(D_t^{\mu}D_n^{\sigma}u_{j0}\right)^{++},\overline{\mathbb{R}}_{++}^{n+1},\gamma\right\|_s \le CA_s\left\|u_{j0},\mathbb{R}_{++}^{n+1},\gamma\right\|_{s+t_j}. \tag{6.25}$$

Combining inequalities (6.16)–(6.19) and (6.22)–(6.25) for the corresponding values of k,λ,σ and μ we get estimate (6.20) for $s+t_j > \frac{1}{2}$, $s < 0$.

(c) If $s+t_j > \frac{1}{2}$, $s > 0$, estimate (6.20) is an obvious consequence of Lemma III.4. Indeed, in this case $\Delta l_{ij}^{0}u_j\big|_{\overline{\mathbb{R}}_{++}^{n+1}} = \left(\Delta l_{ij}^{0}u_{j0}\right)^{++}$ and $\mathcal{H}^s\left(\overline{\mathbb{R}}_{++}^{n+1},\gamma\right) = \mathcal{H}^s\left(\mathbb{R}_{++}^{n+1},\gamma\right)$, and so

$$\left\|\Delta l_{ij}^{0}u_{j0},\mathbb{R}_{++}^{n+1},\gamma\right\|_s \le CA_s\left\|u_{j0},\mathbb{R}_{++}^{n+1},\gamma\right\|_{s+t_j}. \tag{6.26}$$

Estimates (6.20) and (6.26) immediately yield the estimate

$$\left\|\Delta\mathcal{L}_i^{0}u\big|_{\overline{\mathbb{R}}_{++}^{n+1}},\overline{\mathbb{R}}_{++}^{n+1},\gamma\right\|_s \le CA_s\left\|\!\left\|u,\overline{\mathbb{R}}_{++}^{n+1},\gamma\right\|\!\right\|_s \le CA_s\left\|\!\left\|\mathcal{F}(x,t),\overline{\mathbb{R}}_{++}^{n+1},\gamma\right\|\!\right\|_s. \tag{6.27}$$

2. Arguing in a similar manner, we apply Lemmas III.3, III.4, and III.17, and performing similar (but simpler) calculations we get

$$\left\langle\!\left\langle\Delta\mathcal{B}_q^{0}u\big|_{\overline{E}_+^n},\overline{E}_+^n,\gamma\right\rangle\!\right\rangle_{s-\sigma_q-1/2} \le CB_{|s|}(\varepsilon,\gamma)\left\|\!\left\|\mathcal{F}(x,t),\overline{\mathbb{R}}_{++}^{n+1},\gamma\right\|\!\right\|_s, \tag{6.28}$$

$$\left\langle\!\left\langle\Delta\left(D_n^{\sigma-1}\mathcal{L}_i\right)^0 u\big|_{\overline{E}_+^n},\overline{E}_+^n,\gamma\right\rangle\!\right\rangle_{s-\sigma+1/2} \le CA_{|s|+\sigma_0}\left\|\!\left\|\mathcal{F}(x,t),\overline{\mathbb{R}}_{++}^{n+1},\gamma\right\|\!\right\|_s, \tag{6.29}$$

where $B_{|s|}$ and $A_{|s|+\sigma_0}$ denote the following quantities:

$$A_{|s|+\sigma_0} = \varepsilon + A\gamma^{-1/2b}, \quad A = \max_{i,j,\alpha,\beta,\sigma}\left|D_n^{\sigma-1}\left(l_{ij\alpha\beta}(x,t)\right),\overline{\mathbb{R}}_{++}^{n+1}\right|_{|s|+r},$$

$$|\alpha| + 2b\beta = t_j, \quad \sigma = 1,\dots,\sigma_0,$$

$$B_{|s|} = \varepsilon + B\gamma^{-1/2b}, \quad B = \max_{i,q,\alpha,\beta}\left|b_{qj\alpha\beta}(x',t),\overline{E}_+^n\right|_{|s-\sigma_q-1/2|+r},$$

$$|\alpha| + 2b\beta = \sigma_q + t_j, \quad \forall r > 0.$$

3. Now let us proceed to the estimation of the norms

$$\big\|\mathcal{L}'_i u\big|_{\overline{\mathbb{R}}^{n+1}_{++}}, \overline{\mathbb{R}}^{n+1}_{++}, \gamma\big\|_s, \quad \big\langle\!\big\langle \mathcal{B}'_q u\big|_{\overline{E}^n_+}, \overline{E}^n_+, \gamma\big\rangle\!\big\rangle_{s-\sigma_q-1/2},$$

$$\big\langle\!\big\langle \big(D_n^{\sigma-1}\mathcal{L}_i\big)' u\big|_{\overline{E}^n_+}, \overline{E}^n_+, \gamma\big\rangle\!\big\rangle_{s-\sigma+1/2},$$

of those components of the vector-valued function $\mathfrak{U}'(x,t,D,D_t)u(x,t)$ that do not vanish identically. Since l'_{ij}, b'_{qj}, $\big(D_n^{\sigma-1}l_{ij}\big)'$ are operators of weighted orders t_j-1, σ_q+t_j-1, $t_j+\sigma-1$, respectively, and since their coefficients satisfy the assumptions of Lemmas III.2–III.4, it follows by these lemmas that

$$l'_{ij}u_j\big|_{\overline{\mathbb{R}}^{n+1}_{++}} \in \mathcal{H}^{s+1}\big(\overline{\mathbb{R}}^{n+1}_{++},\gamma\big), \quad b'_{qj}u_j\big|_{\overline{E}^n_+} \in \mathcal{H}^{s+1-\sigma_q+1/2}\big(\overline{E}^n_+,\gamma\big),$$

$$\big(D_n^{\sigma-1}l_{ij}\big)' u_j\big|_{\overline{E}^n_+} \in \mathcal{H}^{s-\sigma+1/2}\big(\overline{E}^n_+,\gamma\big),$$

and the following estimates hold:

$$\big\|l'_{ij}u_j\big|_{\overline{\mathbb{R}}^{n+1}_{++}}, \overline{\mathbb{R}}^{n+1}_{++}, \gamma\big\|_{s+1} \le C\big\|u_j, \overline{\mathbb{R}}^{n+1}_{++}, \gamma\big\|_{s+t_j}, \tag{6.30}$$

$$\big\langle\!\big\langle b'_{qj}u_j\big|_{\overline{E}^n_+}, \overline{E}^n_+, \gamma\big\rangle\!\big\rangle_{s+1-\sigma_q-1/2} \le C\big\|u_j, \overline{\mathbb{R}}^{n+1}_{++}, \gamma\big\|_{s+t_j}, \tag{6.31}$$

$$\big\langle\!\big\langle \big(D_n^{\sigma-1}l_{ij}\big)' u_j\big|_{\overline{E}^n_+}, \overline{E}^n_+, \gamma\big\rangle\!\big\rangle_{s+1-\sigma+1/2} \le C\big\|u_j, \overline{\mathbb{R}}^{n+1}_{++}, \gamma\big\|_{s+t_j}. \tag{6.32}$$

Combining estimates (2.29), (2.31) of Chapter II, Lemmas II.16 and III.17, and inequalities (6.30)–(6.32), we obtain

$$\big\|\mathcal{L}'_i u\big|_{\overline{\mathbb{R}}^{n+1}_{++}}, \overline{\mathbb{R}}^{n+1}_{++}, \gamma\big\|_s \le C\gamma^{-1/2b}\big\|\mathcal{F}(x,t), \overline{\mathbb{R}}^{n+1}_{++}, \gamma\big\|_s, \tag{6.33}$$

$$\big\langle\!\big\langle \mathcal{B}'_q u\big|_{\overline{E}^n_+}, \overline{E}^n_+, \gamma\big\rangle\!\big\rangle_{s-\sigma_q-1/2} \le C\gamma^{-1/2b}\big\|\mathcal{F}(x,t), \overline{\mathbb{R}}^{n+1}_{++}, \gamma\big\|_s, \tag{6.34}$$

$$\big\langle\!\big\langle \big(D_n^{\sigma-1}\mathcal{L}_i\big)' u\big|_{\overline{E}^n_+}, \overline{E}^n_+, \gamma\big\rangle\!\big\rangle_{s-\sigma_q+1/2} \le C\gamma^{-1/2b}\big\|\mathcal{F}(x,t), \overline{\mathbb{R}}^{n+1}_{++}, \gamma\big\|_s. \tag{6.35}$$

From the estimates (6.27)–(6.29) and (6.33)–(6.35) it follows that in the spaces $\widetilde{\mathcal{H}}^s_+$ the norms of the operators $\Delta\mathfrak{U}_0(x,t)$ and $\mathfrak{U}'(x,t)$ obey the respective bounds

$$\big\|\Delta\mathfrak{U}_0(x,t,D,D_t)\big\|_s \le C\big(\varepsilon+\gamma^{-1/2b}\big), \quad \big\|\mathfrak{U}'(x,t,D,D_t)\big\|_s \le C\gamma^{-1/2b}. \tag{6.36}$$

Inequalities (6.36) in turn yield the following estimate for the norm $\|T\|_s$ of the operator T in $\widetilde{\mathcal{H}}^s_+$:

$$\|T\|_s \le \big\|\Delta\mathfrak{U}_0(x,t)\mathfrak{U}_0^{-1}(0)\big\|_s + \big\|\mathfrak{U}'(x,t)\mathfrak{U}_0^{-1}(0)\big\|_s \le C\big(\varepsilon+\gamma^{-1/2b}\big), \tag{6.37}$$

where C_s is a constant that does not depend on ε and γ.

Finally, let us estimate the norm $\|W\|_s$ of the operator W in $\widetilde{\mathcal{K}}_+^s$. Estimate (6.36) and the boundedness of the operator $\mathfrak{U}_0^{-1}(0)$ imply that

$$\|W\|_s \leq \left\|\mathfrak{U}_0^{-1}(0)\Delta\mathfrak{U}_0(x,t)\right\|_s + \left\|\mathfrak{U}_0^{-1}(0)\mathfrak{U}'(x,t)\right\|_s \leq C\left(\varepsilon + \gamma^{-1/2b}\right). \qquad (6.38)$$

Since the numbers ε and γ are not related in any way, it follows from (6.37) and (6.38) that there exist ε_0 and γ_0 such that for $\varepsilon < \varepsilon_0$ and $\gamma > \gamma_0$ the norms $\|T\|_s$ and $\|W\|_s$ are strictly less than 1. This completes the proof of Theorem IV.9.

Remark IV.2. It is important for the subsequent analysis that the assertion of Theorem IV.10 for the boundary value problem in $\mathbb{R}_{++}^{n+1}$ with initial conditions at $t = 0$ can be easily extended to the boundary value problem in $\mathbb{R}_+^n \times (t_0, \infty)$ with initial conditions at $t = t_0$. Indeed, the boundary value problem with initial conditions at $t = 0$ can be transformed by the shift $t \to \tau$, $\tau = t - t_0$ into the boundary value problem (6.1)–(6.5) with initial conditions given at $\tau = 0$, because the condition of compatibility with zero of the right-hand sides for $t = t_0$ does not depend on t_0. Consequently, if all initial data $v_{j\lambda}^+(x)$, $\omega_{j\lambda k}(x')$ are identically equal zero and the right-hand sides $f_{i0}^{++}(x,t)$, $f_{i\sigma}^+(x',t)$, $\varphi_q^+(x',t)$ of the problem (6.1)–(6.5) vanish for $t < t_0$, then the solution $u(x,t) \in \widetilde{\mathcal{H}}_+^s$ also vanishes for $t < t_0$, that is, $u_{j0}^{++}(x,t) = 0$, $u_{jk}^+(x',t) = 0$ for $t < t_0$.

IV.7. Conjugation problem for operators in which the coefficients of the highest-order derivatives are slowly varying

In this section we use the method of reducing the conjugation problem to a boundary value problem, Lemma IV.4, and Theorem IV.9 to formulate a theorem on the solvability of a parabolic conjugation problem for a system in which the coefficients of the highest-order derivatives are slowly varying functions. For the sake of simplicity, we restrict ourselves to the case of systems parabolic in the sense of Petrovskiĭ.

Thus, let Ω_r^+ be the same domains as in Section IV.5 and consider the operators

$$\mathcal{L}_{(r)} = \left(l_{(r)ij}(x,t,D,D_t)\right) \quad (i,j = 1,\ldots,m_r)$$

and

$$\mathcal{B}_{(r)} = \left(b_{(r)qj}(x',t,D,D_t)\right) \quad (j = 1,\ldots,m_r, q = 1,\ldots,N = N_1 + N_2),$$

where $l_{(r)ij}$ and $b_{(r)qj}$ are linear operators of weighted orders t_{rj} and $\sigma_q + t_{rj}$, respectively, $t_{rj} = 2bt'_{rj}$ (with t'_{rj} integers), $\sum_j t_{rj} = 2N_r = 2b\kappa_r$. Assume that the orders in t of the operators $b_{(r)qj}$ are at most t'_{rj}. Denote $\sigma_0 = \max_q(0, \sigma_q+1)$, $\tau_{rj} = t_{rj}+\sigma_0$. Finally, let $b_{(r)qj}=b_{(r)qjk}(\cdot,D',D_t)D_n^{k-1}$, $D_n^{\sigma-1}l_{(r)ij}=l_{(r)ij\sigma k}(\cdot,D',D_t)D_n^{k-1}$, where repetition of k for fixed r, q, j, and r, q, j, σ indicates summation with respect to k from 1 to $\sigma_q + t_{rj} + 1$ and from 1 to $\sigma_q + t_{rj}$, respectively.

Our analysis of the formulation of the parabolic boundary value problem in the spaces $\widetilde{\mathcal{H}}^s$ and of the reduction of the conjugation problem to a boundary value problem leads to the following formulation of the parabolic conjugation problem in $\widetilde{\mathcal{H}}^s$:

In $\overline{\Omega}_1^+$ and $\overline{\Omega}_2^+$ find solutions

$$u_{(1)}(x,t) \in \widetilde{\mathcal{H}}_1^s = \prod_{j=1}^{m_1} \widetilde{\mathcal{H}}_{(t'_{1j}, \tau_{1j}, P_{1j})}^{s+t_{1j}}(\Omega_1^+, \gamma),$$

$$u_{(2)}(x,t) \in \widetilde{\mathcal{H}}_2^s = \prod_{j=1}^{m_2} \widetilde{\mathcal{H}}_{(t'_{2j}, \tau_{2j}, P_{2j})}^{s+t_{2j}}(\Omega_2^+, \gamma),$$

of the system

$$l_{(1)ij}u_{(1)j}\big|_{\overline{\Omega}_1^+} \equiv l_{(1)ij}u_{(1)j0}^{++} - l_{(1)ij}^{(k,0)}u_{(1)jk}^{+}(x',t) \times \delta(x_n) - l_{(1)ij}^{(0,\lambda)}v_{(1)j\lambda}^{+}(x) \times \delta(t)$$

$$+ l_{(1)ij}^{(k,\lambda)}\omega_{(1)j\lambda k}(x') \times \delta(x_n, t) = f_{(1)i0}^{++}(x,t), \qquad i = 1, \ldots, m_1, \tag{7.1}$$

$$l_{(2)ij}u_{(2)j}\big|_{\Omega_2^+} \equiv l_{(2)ij}u_{(2)j0}^{-+} - l_{(2)ij}^{(k,0)}u_{(2)ik}^{+}(x',t) \times \delta(x_n) + l_{(2)ij}^{(0,\lambda)}v_{(2)i\lambda}^{-}(x) \times \delta(t)$$

$$- l_{(2)ij}^{(k,\lambda)}\omega_{(2)j\lambda k}(x') \times \delta(x_n, t) = f_{(2)i0}^{-+}(x,t), \qquad i = 1, \ldots, m_2, \tag{7.2}$$

satisfying the conjugation conditions at $x_n = 0$

$$b_{(1)qjk}u_{(1)jk}^{+}(x',t) - b_{(1)qjk}^{(\lambda)}\omega_{(1)j\lambda k}(x') \times \delta(t) + b_{(2)qjk}u_{(2)jk}^{+}(x',t)$$

$$- b_{(2)qjk}^{(\lambda)}\omega_{(2)j\lambda k}(x') \times \delta(t) = \varphi_q^{+}(x',t), \qquad q = 1, \ldots, N, \tag{7.3}$$

the additional boundary conditions at $x_n = 0$

$$D_n^{\sigma-1}l_{(1)ij}u_{(1)j}\big|_{E_+^n} \equiv l_{(1)ij\sigma k}(x', t, D', D_t)u_{(1)jk}^{+}(x',t) - l_{(1)ij\sigma k}^{(\lambda)}\omega_{(1)j\lambda k}(x') \times \delta(t)$$

$$= f_{(1)i\sigma}^{+}(x',t), \qquad \sigma = 1, \ldots, \sigma_0, \ i = 1, \ldots, m_1, \tag{7.4}$$

$$\left(D_n^{\sigma-1}l_{(2)ij}\right)u_{(2)j}\big|_{E_+^n} \equiv l_{(2)ij\sigma k}(\cdot, D', D_t)u_{(2)jk}^{+}(x',t) - l_{(2)ij\sigma k}^{(\lambda)}\omega_{(2)j\lambda k}(x') \times \delta(t)$$

$$= f_{(2)i\sigma}^{+}(x',t), \qquad \sigma = 1, \ldots, \sigma_0, \ i = 1, \ldots, m_2, \tag{7.5}$$

as well as the initial conditions at $t = 0$

$$v_{(1)j\lambda}^{+}(x) = \psi_{(1)j\lambda}^{+}(x), \qquad \lambda = 1, \ldots, t'_{1j}, \ j = 1, \ldots, m_1, \tag{7.6}$$

$$v_{(2)j\lambda}^{-}(x) = \psi_{(2)j\lambda}^{-}(x), \qquad \lambda = 1, \ldots, t'_{2j}, \ j = 1, \ldots, m_2, \tag{7.7}$$

$$\omega_{(1)j\lambda k}(x') = \psi_{(1)j\lambda k}(x'), \qquad \forall(\lambda, k) \in P_{(1)j}, \ j = 1, \ldots, m_1, \tag{7.8}$$

$$\omega_{(2)j\lambda k}(x') = \psi_{(2)j\lambda k}(x'), \qquad \forall(\lambda, k) \in P_{(2)j}, \ j = 1, \ldots, m_2. \tag{7.9}$$

where $P_{(r)j}$ denotes the set of the indices (λ, k) of the distributions $\omega_{(r)j\lambda k}(x')$ that figure in formulas (7.1)–(7.5) for given fixed r, j.

Let us emphasize that $\operatorname{supp} u_{(2)j0}^{-+}(x, t) \subseteq \overline{\mathbb{R}}_-^n \times \overline{\mathbb{R}}_+^1 = \overline{\mathbb{R}}_{-+}^{n+1}$, and in the case of smooth functions $u_{(2)j}$ we have

$$u_{(2)jk}(x', t) = D_n^{k-1} u_{(2)j}(x, t)\big|_{x_n=0},$$

$$\omega_{(2)j\lambda k}(x') = D_n^{k-1} v_{(2)j\lambda}(x)\big|_{x_n=-0} = D_n^{k-1} D_t^{\lambda-1} u_{(2)j}\big|_{x_n=-0,\ t=+0}.$$

Further, let $l_{(r)ij} = l_{(r)ij\alpha\beta}(x, t) D^\alpha D_t^\beta$, $b_{(r)qj} = b_{(r)qj\alpha\beta}(x', t) D^\alpha D_t^\beta$, where repetition of α and β indicates summation with respect to them, and where

$$l_{(r)ij\alpha\beta}(x, t) \in C^{|s|+t_{rj}+\sigma_0+\varepsilon}(\overline{\Omega}_r^+), \qquad r = 1, 2, \tag{7.10}$$

$$b_{(r)qj\alpha\beta}(x', t) \in C^{|s|+t_{rj}-\sigma_q+\varepsilon}(\overline{E}_+^n), \qquad \varepsilon > 0. \tag{7.11}$$

Let $\widetilde{\mathcal{H}}^s = \widetilde{\mathcal{H}}_1^s \oplus \widetilde{\mathcal{H}}_2^s$, $\widetilde{\mathcal{H}}_+^s = \widetilde{\mathcal{H}}_{1+}^s \oplus \widetilde{\mathcal{H}}_{2+}^s$, where $\oplus$ stands for direct sum, and let $\widetilde{\mathcal{H}}_{r+}^s$ $(r = 1, 2)$ be the subspace of $\widetilde{\mathcal{H}}_r^s$ obtained by completion of the set of all smooth vector-valued functions $u_{(r)}(x, t)$ whose components vanish at $t = 0$ together with their derivatives $D_t^{\lambda-1} u_{(r)j}(x, 0)$, for all λ, j such that $s + t_{rj} - 2b\lambda + b > 0$. For each set $P_{(r)j}$ and each fixed λ we denote by $P_{(r)j\lambda}$ the set of all k for which $(k, \lambda) \in P_{(r)j}$. To simplify notation we omit the upper indices in the distributions $f_{(1)i0}^{++}$, $f_{(2)i0}^{-+}$, $v_{(1)j\lambda}^+$, and $v_{(2)j\lambda}^-$.

Consider the vectors

$$V_{(r)j\lambda}(x) = \big(v_{(r)j\lambda}(x), \omega_{(r)j\lambda k}(x'), k \in P_{(r)j\lambda}\big),$$

$$\lambda = 1, \ldots, t'_{rj}, \ j = 1, \ldots, m_r, \ r = 1, 2,$$

$$F_{ri}(x, t) = \big(f_{(r)i0}(x, t), f_{(r)i1}^+(x', t), \ldots, f_{(r)i\sigma_0}^+(x', t)\big),$$

$$i = 1, \ldots, m_r, \ r = 1, 2,$$

$$\Phi(x, t) = \big(\varphi_1^+(x', t), \ldots, \varphi_N^+(x', t)\big),$$

belonging to the spaces $H^{s+t_{rj}-2b\lambda+b}(G_r) \times \prod\limits_{k \in P_{(r)j\lambda}} H^{s+t_{rj}-k-2b\lambda+b+1/2}(\mathbb{R}^{n-1})$,

$\mathcal{H}^s(\Omega_r^+, \gamma) \times \prod\limits_{\sigma=1}^{\sigma_0} \mathcal{H}^{s-\sigma+1/2}(\overline{E}_+^n, \gamma)$ and $\prod\limits_{q=1}^{N} \mathcal{H}^{s-\sigma-1/2}(\overline{E}_+^n, \gamma)$, respectively; here $G_1 = \overline{\mathbb{R}}_+^n$, $G_2 = \overline{\mathbb{R}}_-^n$.

By virtue of conditions (7.10) and (7.11) and Lemmas III.16–III.19, relations (7.1)–(7.9) define a bounded operator $\mathfrak{U}$ acting from the space $\widetilde{\mathcal{H}}^s$ into the subspace $\widetilde{\mathcal{K}}^s$ of $\mathcal{K}^s$ that consists of the vectors $\mathcal{F} = \big(F_1, F_2, \Phi, V_{rj\lambda}, \lambda = 1, \ldots, t'_{rj}, j = 1, \ldots, m_r, r = 1, 2\big)$, whose components satisfy on E_+^n and $\mathbb{R}^{n-1}$

the compatibility conditions that follow from the theorems on traces. This means that the vectors $F_{ri}(x,t)$ and $V_{(r)i\lambda}(x)$ are the isometrically isomorphic images of the elements $f_{(r)i}(x,t) \in \widetilde{\mathcal{H}}^s_{(0,\sigma_0,0)}(\Omega_r^+,\gamma)$ and $v_{(r)j\lambda}(x) \in \widetilde{H}^{s+t_{rj}-2b\lambda+b}_{(P_{(r)j\lambda})}(G_r)$: $\chi_0 v_{(r)j\lambda}(x) = V_{(r)j\lambda}(x)$, $\chi f_{(r)i}(x,t) = F_{(r)i}(x,t)$, where χ and χ_0 are the maps defined in Section III.5. In particular, it follows that

$$f_{(r)i\sigma}(x,t) = D_n^{\sigma-1} f_{(r)i0}(x,t)\big|_{x_n=0}, \qquad \forall\sigma : s - \sigma + \tfrac{1}{2} > 0,$$

$$\psi_{(r)j\lambda k}(x') = D_n^{k-1}\psi_{(r)j\lambda}(x)\big|_{x_n=0}, \qquad \forall\lambda, k : s + t_{rj} - k - 2b\lambda + b + \tfrac{1}{2} > 0.$$

Define in $\mathcal{K}^s$ the norm $\left\|\mathcal{F}\right\|_s$ as the sum of the corresponding norms of the components of the vector $\mathcal{F}(x,t)$ and let $\left\|u_{(r)}\right\|_s$ be a norm in $\widetilde{\mathcal{H}}^s_r$.

Definition IV.4. *We say that the right-hand sides of the conjugation problem (7.1)–(7.9) satisfy the condition of compatibility with zero at $t = 0$ to order s if $\mathcal{F}(x,t) \in \widetilde{\mathcal{K}}^s_+ = \mathfrak{U}\widetilde{\mathcal{H}}^s_+$.*

This means that $\psi^+_{(r)j\lambda}(x) \equiv 0$, $\psi_{(r)j\lambda k}(x') \equiv 0$ for j, λ, k such that $s + t_{rj} - 2b\lambda + b > 0$, $s + t_{rj} - k - 2b\lambda + b + \tfrac{1}{2} > 0$ $(r = 1,2)$, and the functions $f^{++}_{(1)i0}(x,t)$, $f^{-+}_{(2)i0}(x,t)$, $f^+_{(r)i\sigma}(x',t)$, $\varphi^+_q(x',t)$ (extended by zero for $t < 0$) belong to the spaces $\mathcal{H}$ with the same smoothness indices in $\Omega_1 = \mathbb{R}^n_+ \times \mathbb{R}^1$, $\Omega_2 = \mathbb{R}^n_- \times \mathbb{R}^1$, and E^n respectively.

Now let us formulate the appropriate condition for the operators $\mathcal{B}_1$, $\mathcal{B}_2$. Denote $L_{(r)}(x,t,i\xi,p) = \det\|l_{(r)ij}(x',t,i\xi,p)\|$ $(r = 1,2)$ and let

$$L_{(r)}(x',t,i\xi,p) = L_r^+(x',t,\xi',p,\xi_n) \cdot L_r^-(x',t,\xi',p,\xi_n)$$

be a factorization of L_r with respect to the variable ξ_n at the point $(x',t) \in \overline{E}^n_+$.

Definition IV.5. *We say that the operators $\mathcal{B}_1$, $\mathcal{B}_2$ satisfy the compatible covering condition with respect to the operators $\mathcal{L}_1$, $\mathcal{L}_2$ at $x_n = 0$, uniformly in $(x',t) \in \overline{E}^n_+$, if the rows of the matrix $Q'(x',t,\xi',p,\xi_n) = (Q'_1(\cdot,\xi_n), Q'_2(\cdot,\xi_n))$ composed of the remainders of the division of the matrices $\mathcal{B}_1(\cdot,\xi,p)$ and $\mathcal{B}_2(\cdot,\xi,p)$ by $L_1^+(\cdot,\xi_n)$ and $L_2^-(\cdot,\xi_n)$ (regarded as polynomials in ξ_n) are linearly independent uniformly in $(x',t) \in \overline{E}^n_+$.*

This means that for any $\xi' \in \mathbb{R}^{n-1}$ and any p such that $\operatorname{Re} p > 0$ and $(\xi',p) \neq 0$ the rank of the matrix that consists of the coefficients of the powers of ξ_n in the matrix Q' equals N, and for any point $(x',t) \in \overline{E}^n_+$ there exists a minor of rank N whose modulus is bounded below by a positive constant that does not depend on $(x',t) \in \overline{E}^n_+$.

Theorem IV.11. *Let $\mathcal{L}_{(r)}$ be operators uniformly parabolic in the sense of Petrovskiĭ in $\overline{\Omega}^+_r$, and let the operators $\mathcal{B}_{(r)}$ satisfy the compatible covering condition with respect to $\mathcal{L}_{(1)}$, $\mathcal{L}_{(2)}$ at $x_n = 0$ uniformly in $(x',t) \in \overline{E}^n_+$. Assume that the* *right-hand sides of the conjugation problem (7.1)–(7.9) satisfy the conditions of compatibility with zero at $t = 0$ to order s, $s \notin \mathbb{Z}_{1,2b}$. Finally, assume that the coefficients $l_{(r)ij\alpha\beta}(x,t)$ and $b_{(r)qj\alpha\beta}(x',t)$ satisfy conditions (7.10), (7.11) and*

$$\sup_{(x,t)\in\overline{\Omega}_r^+} \left| l_{(r)ij\alpha\beta}(x,t) - l_{(r)ij\alpha\beta}(0,0) \right| < \varepsilon,$$

$$\forall \alpha, \beta : |\alpha| + 2b\beta = t_{rj}, \tag{7.12}$$

$$\sup_{(x',t)\in\overline{E}_+^n} \left| b_{(r)qj\alpha\beta}(x',t) - b_{(r)qj\alpha\beta}(0,0) \right| < \varepsilon,$$

$$\forall \alpha, \beta : |\alpha| + 2b\beta = \sigma_q + t_j. \tag{7.13}$$

Then there exist numbers $\varepsilon_0 > 0$, $\gamma_0 > 0$ (depending on s and the Hölder norms of the coefficients $l_{(r)ij\alpha\beta}(x,t)$ and $b_{(r)qj\alpha\beta}(x',t)$) such that for $\varepsilon < \varepsilon_0$ and $\gamma > \gamma_0$ problem (7.1)–(7.9) has a unique solution $u(x,t) \in \widetilde{\mathcal{H}}_+^s$, which satisfies the estimate

$$\left\| u_{(1)}, \Omega_1^+, \gamma \right\|_s + \left\| u_{(2)}, \Omega_2^+, \gamma \right\|_s < C_s \left\| \mathcal{F}(x,t) \right\|_s.$$

Proof: Consider the expressions $\mathcal{L}_{(2)i} u_{(2)}\big|_{\overline{\Omega}_2^+}$, $\mathcal{B}_{(2)q} u_{(2)}\big|_{\overline{E}_+^n}$ and replace x_n by a new variable $z_n = -x_n$. Setting $u_{(2)j}(\cdot, -z_n) = \bar{u}_{(2)j}(\cdot, z_n)$, $v_{(2)j}(\cdot, -z_n) = \bar{v}_{(2)j}(\cdot, z_n)$, we obtain $D_n^{k-1} u_{(2)j}(\cdot, x_n) = (-1)^{k-1} D_{z_n}^{k-1} \bar{u}_{(2)j}(\cdot, z_n)$. Hence,

$$u_{(2)jk}(x',t) = D_n^{k-1} u_{(2)j}(\cdot, x_n)\big|_{x_n=-0}$$

$$= (-1)^{k-1} D_{z_n}^{k-1} \bar{u}_{(2)j}(\cdot, z_n) = (-1)^{k-1} \bar{u}_{(2)jk}(x',t),$$

$$\omega_{(2)j\lambda k}(x') = D_{x_n}^{k-1} v_{(2)j\lambda}(\cdot, x_n) = (-1)^{k-1} D_{z_n}^{k-1} \bar{v}_{(2)j\lambda}(\cdot, z_n)$$

$$= (-1)^{k-1} \omega_{(2)j\lambda k}(x').$$

If in these relations we replace $D_n^{k-1} u_{(2)j}(\cdot, x_n)$, $u_{(2)jk}(x',t)$, and $\omega_{(2)j\lambda k}(x')$ by their expressions and then relabel z_n by x_n, we get

$$\mathcal{L}_{(2)i} u_{(2)}\big|_{\overline{\Omega}_2^+} \equiv l_{(2)ij} u_{(2)j}\big|_{\overline{\Omega}_2^+} = l_{(2)ij}(\cdot, \overline{D}_n) \bar{u}_{(2)j}^{++}$$

$$+ l_{(2)ij}^{(k,0)}(\cdot, \overline{D}_n)(-1)^{k-1} \bar{u}_{(2)jk}^+ \times \delta(x_n) - l_{(2)ij}^{(0,\lambda)}(\cdot, \overline{D}_n) \bar{v}_{(2)j\lambda}^+ \times \delta(t)$$

$$- l_{(2)ij}^{(k,\lambda)}(\cdot, \overline{D}_n)(-1)^{k-1} \overline{\omega}_{(2)j\lambda k}(x') \times \delta(x_n,t), \qquad (\overline{D}_n = -D_n), \tag{7.14}$$

$$\mathcal{B}_{(2)q} u\big|_{\overline{E}_+^n} \equiv b_{(2)qjk} D_n^{k-1} \bar{u}_{(2)j}^+\big|_{\overline{E}_+^n} = \mathcal{B}_{(2)}(\cdot, \overline{D}_n) \bar{u}_{(2)}\big|_{\overline{E}_+^n}. \tag{7.15}$$

Since $\left(l_{(2)ij}(\cdot, \overline{D}_n) \right)^{(k,\lambda)} = (-1)^{k-1} l_{(2)ij}^{(k,\lambda)}(\cdot, \overline{D}_n)$, it follows that the right-hand side of (7.14) equals exactly $l_{(2)ij}(\cdot, \overline{D}_n) \bar{u}_{(2)j}(x,t)\big|_{\overline{\Omega}_1^+}$.

Equalities (7.14), (7.15) obviously mean that the operations $\mathcal{L}_{(2)i}u_{(2)}\big|_{\overline{\Omega}_2^+}$ and $\mathcal{B}_{(2)q}u_{(2)}\big|_{\overline{E}_+^n}$ commute with the operation of changing the variable x_n to $-x_n$, that is, the Green formulas for the transformed expressions $\mathcal{L}_{(2)i}(\cdot,\overline{D}_n)\bar{u}_{(2)}$ and $\mathcal{B}_{(2)q}(\cdot,\overline{D}_n)\bar{u}_{(2)}$ in Ω_1^+ and on $\overline{E}_+^n$ coincide with the transformed Green formulas for $\mathcal{L}_{(2)i}u_{(2)}$ and $\mathcal{B}_{(2)q}u_{(2)}$ in $\overline{\Omega}_2^+$ and on $\overline{E}_+^n$, respectively.

Thus, the change of variables $x_n \to -x_n$ in (7.2), (7.3), (7.5), (7.7), and $\mathcal{B}_{(2)q}u_{(2)}\big|_{\overline{E}_+^n}$ transforms the conjugation problem (7.1)–(7.9) into a boundary value problem in Ω_1^+ for the block-diagonal system

$$\mathcal{L}_{(1)i}(x,t,D,D_t)u_{(1)}\big|_{\overline{\Omega}_1^+} = f_{(1)i0}^{++}(x,t), \qquad i = 1,\ldots,m_1, \tag{7.16}$$

$$\mathcal{L}_{(2)i}(\cdot,-x_n,\overline{D}_n)\bar{u}_{(2)}\big|_{\overline{\Omega}_1^+} = \bar{f}_{(2)i0}^{++}(x,t), \qquad i = 1,\ldots,m_2, \tag{7.17}$$

with the boundary conditions on $\overline{E}_+^n$

$$\mathcal{B}_{(1)q}(\cdot,D_n)u_{(1)}\big|_{\overline{E}_+^n} + \mathcal{B}_{(2)q}(\cdot,\overline{D}_n)\bar{u}_{(2)}\big|_{\overline{E}_+^n} = \varphi_q^+(x',t), \quad q = 1,\ldots,N, \tag{7.18}$$

$$D_n^{\sigma-1}\mathcal{L}_{(1)i}u_{(1)}\big|_{\overline{E}_+^n} = f_{(1)i\sigma}^+(x',t), \qquad \sigma = 1,\ldots,\sigma_0, \; i = 1,\ldots,m_1, \tag{7.19}$$

$$D_n^{\sigma-1}\mathcal{L}_{(2)i}(\cdot,-x_n,\overline{D}_n)\bar{u}_{(2)}(x,t)\big|_{\overline{E}_+^n} = \bar{f}_{(2)i\sigma}^+(x',t),$$

$$\sigma = 1,\ldots,\sigma_0, \; i = 1,\ldots,m_2, \tag{7.20}$$

and the initial conditions at $t = 0$

$$v_{(1)j\lambda}^+(x) = \psi_{(1)j\lambda}^+(x), \quad \lambda = 1,\ldots,t_{1j}',$$
$$\omega_{(1)j\lambda k}(x') = \psi_{(1)j\lambda k}(x'), \quad (\lambda,k) \in P_{(1)j}, \tag{7.21}$$

$$\bar{v}_{(2)j\lambda}^+(x) = \overline{\psi}_{(2)j\lambda}^+(x), \quad \lambda = 1,\ldots,t_{2j}',$$
$$\overline{\omega}_{(2)j\lambda k}(x') = \overline{\psi}_{(2)j\lambda k}(x') = (-1)^{k-1}\psi_{(2)j\lambda k}(x'), \quad (\lambda,k) \in P_{(2)j}. \tag{7.22}$$

Since the operators $\mathcal{L}_{(r)}$ are uniformly parabolic in $\overline{\Omega}_r^+$, it is clear that the block-diagonal system (7.16), (7.17) is uniformly parabolic in $\overline{\Omega}_1^+$. Further, since $\mathcal{B}_{(1)}$ and $\mathcal{B}_{(2)}$ satisfy the compatible covering condition with respect to the operators $\mathcal{L}_{(1)}$ and $\mathcal{L}_{(2)}$ at $x_n = 0$, uniformly in $(x',t) \in \overline{E}_+^n$, Lemma IV.4 shows that the operators $\mathcal{B} = \big(\mathcal{B}_{(1)}(\cdot,D_n),\mathcal{B}_{(2)}(\cdot,\overline{D}_n)\big)$ and $\mathcal{L} = \begin{pmatrix} \mathcal{L}_{(1)}(\cdot,D_n) & 0 \\ 0 & \mathcal{L}_{(2)}(\cdot,\overline{D}_n) \end{pmatrix}$ are related by the Lopatinskiĭ condition uniformly in $(x',t) \in \overline{E}_+^n$. Moreover, the principal parts of the operators $\mathcal{B}$ and $\mathcal{L}$ are composed of the principal parts of the operators $\mathcal{B}_{(1)}$, $\mathcal{B}_{(2)}$ and $\mathcal{L}_{(1)}$, $\mathcal{L}_{(2)}$, respectively. Therefore, all the assumptions of Theorem IV.9 hold for the boundary value problem (7.16)–(7.22), and this immediately yields the assertion of Theorem IV.10.

Remark IV.3. Note that the compatibility condition for the right-hand sides of the conjugation problem (7.1)–(7.9) (in particular, the compatibility with zero at $t = 0$) enters the picture for those values of s for which the traces of the corresponding expressions on E^n_+, $\mathbb{R}^n_+$, $\mathbb{R}^n_-$, $\mathbb{R}^{n-1}$ begin to exist. For all the other values of s no compatibility conditions are necessary, and hence the problem (7.1)–(7.9) has a unique solution in $\widetilde{\mathcal{H}}^s$. This is manifestly the case for $s < -\max_j(t_{1j}, t_{2j}) + \frac{1}{2}$.

Remark IV.4. As we have seen, the formulation of the conjugation problem in the spaces $\widetilde{\mathcal{H}}^s$ is rather tedious even for the systems parabolic in the sense of Petrovskiĭ. This is due mainly to the presence of supplementary boundary and initial conditions. The supplementary conditions arise in the case of systems of higher order with respect to t and of boundary operators $b_{(r)qj}$ that involve t-derivatives and whose orders with respect to x_n are higher than $t_{(r)j}$. The conjugation problem in $\widetilde{\mathcal{H}}^s$ becomes much simpler for systems parabolic in the sense of Petrovskiĭ that are of first-order with respect to t ($t_{rj} = 2b$, $t'_{rj} = 1$). In the latter case the operators $b_{(r)qj}$ do not depend on D_t and their orders with respect to x_n are strictly less than $2b$.

In this situation the Green formulas for the operators $\mathcal{L}_{(r)}$ in $\overline{\Omega}^+_r$ and $\mathcal{B}_{(r)}$ in $\overline{E}^+_n$ involve only the components $u^+_{(r)jk}(x', t)$ with $k = 1, \ldots, 2b$, but do not involve the functions $\omega_{(r)j\lambda k}(x')$. Then, of course, the set $P_{(r)j}$ is empty, $t'_j = 1$ and $\widetilde{\mathcal{H}}^s = \bigoplus_{r=1}^{2} \prod_{j=1}^{m_r} \widetilde{\mathcal{H}}^{s+2b}_{(1,2b,\emptyset)}(\Omega^+_r, \gamma)$. In this case the conjugation problem is to find a solution $u(x, t) \in \widetilde{\mathcal{H}}^s$ whose components satisfy the equations

$$\mathcal{L}_{(1)i} u_{(1)}\big|_{\overline{\Omega}^+_1} \equiv l_{(1)ij}(\cdot, D, D_t) u^{++}_{(1)j0}(x, t) - l^{(k,0)}_{(1)ij}(x, t, D) u^+_{(1)jk}(x', t) \times \delta(x_n)$$

$$- l^{(0,1)}_{(1)ij}(x, 0) v^+_{(1)j1}(x) \times \delta(t) = f^{++}_{(1)i0}(x, t), \quad i = 1, \ldots, m_1, \tag{7.23}$$

$$\mathcal{L}_{(2)i} u_{(2)}\big|_{\overline{\Omega}^+_2} \equiv l_{(2)ij}(\cdot, D, D_t) u^{-+}_{(2)j0}(x, t) - l^{(k,0)}_{(2)ij}(\cdot, D) u^+_{(2)ik}(x', t) \times \delta(x_n)$$

$$- l^{(0,1)}_{(2)ij}(x, 0) v^-_{(1)j1}(x) \times \delta(t) = f^{++}_{(2)i0}(x, t), \quad i = 1, \ldots, m_2, \tag{7.24}$$

$$\mathcal{B}_{(1)} u_{(1)}\big|_{\overline{E}^n_+} + \mathcal{B}_{(2)} u_{(2)}\big|_{\overline{E}^n_+} = b_{(1)qjk}(x', t, D') u^+_{(1)jk}(x', t)$$

$$+ b_{(2)qjk}(x', t, D') u^+_{(2)jk}(x', t) \tag{7.25}$$

$$= \varphi^+_q(x', t), \quad q = 1, \ldots, N = (m_1 + m_2)b,$$

$$v^\pm_{rj}(x) = \psi^\pm_{rj}(x), \quad j = 1, \ldots, m_r, \ x \in \mathbb{R}^n_\pm, \ r = 1, 2. \tag{7.26}$$

The right-hand sides of (7.23)–(7.26) belong to the space

$$\mathcal{K}^s = \bigoplus_{r=1}^{2} \prod_{i=1}^{m_r} \mathcal{H}^s(\Omega^+_r, \gamma) \times \prod_{q=1}^{N} \mathcal{H}^{s-\sigma_q-1/2}(\overline{E}^n_+, \gamma) \times \bigoplus_{r=1}^{2} H^{s-b}(G_r)$$

and satisfy the corresponding compatibility conditions.

Theorem IV.11 admits the following corollary.

Theorem IV.12. *Let $\mathcal{L}_{(r)}(x, t, D, D_t)$ be operators uniformly parabolic in the sense of Petrovskiĭ in $\overline{\Omega}_r^{+}$, and let $\mathcal{B}_{(1)}$ and $\mathcal{B}_{(2)}$ satisfy the compatible covering condition with respect to $\mathcal{L}_{(1)}$ and $\mathcal{L}_{(2)}$ at $x_n = 0$, uniformly in $(x', t) \in \overline{E}_+^n$. Assume that the right-hand sides of equations (7.23)–(7.26) are compatible with zero at $t = 0$ to order s. Finally, assume that the coefficients of $\mathcal{B}_{(1)}$, $\mathcal{B}_{(2)}$, $\mathcal{L}_{(1)}$, and $\mathcal{L}_{(2)}$ satisfy conditions (7.10), (7.11), (7.12), and (7.13). Then there exist numbers $\varepsilon_0 > 0$, $\gamma_0 > 0$ (depending on s and the Hölder norms of the coefficients of $\mathcal{L}_{(r)}$, $\mathcal{B}_{(r)}$) such that for $\varepsilon < \varepsilon_0$ and $\gamma > \gamma_0$ problem (7.23)–(7.26) has a unique solution $u(x, t) \in \widetilde{\mathcal{H}}_+^s$, which satisfies for $s \notin \mathbb{Z}_{1,2b}$ the estimate*

$$\big\| u(x, t) \big\|_{\widetilde{\mathcal{H}}_+^s} < C_s \big\| \mathcal{F}(x, t) \big\|_{\widetilde{\mathcal{K}}_+^s}.$$

Chapter V
Parabolic Boundary Value Problems in Cylindrical Domains

V.1. Boundary value problems in a semi-infinite cylinder

V.1.1. Formulation of the boundary value problem in $\widetilde{\mathcal{H}}^s(\Omega_+,\gamma)$

Let $\Omega_+ = G \times [0,\infty)$ be a semi-infinite cylinder with lateral surface $S_+ = \Gamma \times [0,\infty)$; here G is a bounded domain in $\mathbb{R}^n$ (or the exterior of a bounded domain $\mathcal{G} : G = \mathbb{R} \setminus \mathcal{G}$) with a smooth (C^∞) $(n-1)$-dimensional boundary $\Gamma = \partial G$. Suppose that in $\overline{\Omega}_+$ there is given an operator $\mathcal{L}(x,t,D,D_t)$, uniformly parabolic in the sense of Petrovskiĭ, $\operatorname{ord} l_{ij} = t_j = 2bt'_j$, and that on $\overline{S}_+$ there is given a matrix boundary operator $\mathcal{B}(x,t,D,D_t)$, $\operatorname{ord} b_{qj} = \sigma_q + t_j$, which satisfies the Lopatinskiĭ condition on $\overline{S}_+$ uniformly in $(x',t) \in S_+$. For simplicity, we assume that $\operatorname{ord}_t b_{qj} < t'_j$ and the coefficients of the operators l_{ij} and b_{qj} are infinitely smooth in x and t.

Let $x \in \Gamma$ be an arbitrary point, $\{y_1,\ldots,y_n\}$ a local coordinate system with the origin at x and the y_n-axis directed along $\nu(x)$, the inner normal to Γ, while the other axes, $y' = \{y_1,\ldots,y_{n-1}\}$, lie in the tangent plane to Γ at x. We denote by D_ν the differentiation along the normal $\nu(x)$ and by D' the differentiation with respect to y'. Further, let

$$
\begin{aligned}
l_{ij}(x,t,D',D_\nu,D_t) &= l_{ij\sigma}(\cdot,D',D_t)D_\nu^{\sigma-1}, \\
b_{qj}(x,t,D,D_t) &= b_{qjk}(x',t,D',D_t)D_\nu^{k-1}
\end{aligned}
\tag{1.1}
$$

be the representations of the operators l_{ij} and b_{qj} in the coordinate system $\{y,t\}$. Using (1.1), we define the operators $l_{ij}^{(k,\lambda)}(x,t,D',D_\nu,D_t)$ and $b_{qjk}^{(\lambda)}(\cdot,D_t)$; for each $j = 1,\ldots,m$ we denote by P_j the set of the pairs (k,λ), for which not all the operators $l_{ij}^{(k,\lambda)}$, $b_{qjk}^{(\lambda)}$ are identically equal zero on S_+. Finally, let $\tau_j = t_j + \max\limits_q(0,\sigma_q+1)$.

Like in Section IV.6, for each j we define the space $\widetilde{\mathcal{H}}^{s+t_j}_{(t'_j,\tau_j,P_j)}(\Omega_+,\gamma)$, $\gamma \geq \gamma_0$. Let $\widetilde{\mathcal{H}}^s = \prod\limits_{j=1}^{m} \widetilde{\mathcal{H}}^{s+t_j}_{(t'_j,\tau_j,P_j)}(\Omega_+,\gamma)$ and let $\widetilde{\mathcal{H}}^s_+$ be the subspace of $\widetilde{\mathcal{H}}^s$ consisting of the elements $u(x,t) \in \widetilde{\mathcal{H}}^s$ that are compatible with zero at $t = 0$ to order s:

$$v_{j\lambda}(x) = 0, \quad \forall (j,\lambda) : s + t_j - 2b\lambda + b > 0,$$

$$\omega_{j\lambda k}(x') = 0, \quad \forall (j,\lambda,k) : s + t_j - k - 2b\lambda + b + \tfrac{1}{2} > 0,$$

$$u_{j0}^+(x,t) \in \mathcal{H}_+^{s+t_j}(\Omega_+,\gamma), \quad u_{jk}^+(x',t) \in \mathcal{H}_+^{s+t_j-k+1/2}(S_+,\gamma).$$

In $\overline{\Omega}_+$ we consider the following boundary value problem: find a vector-valued function (element) $u(x,t) \in \widetilde{\mathcal{H}}^s$ that satisfies in $\overline{\Omega}_+$ the system of equations

$$l_{ij}(x,t,D,D_t)u_j(x,t)\big|_{\Omega_+} = f_{i0}(x,t), \qquad i = 1,\ldots,m, \tag{1.2}$$

the boundary conditions on $\overline{S}_+$

$$D_\nu^{\sigma-1}l_{ij}(x,t,D,D_t)u_j(x,t)\big|_{S_+} = f_{i\sigma}(x,t), \quad \sigma = 1,\ldots,\sigma_0, \; i = 1,\ldots,m, \tag{1.3}$$

$$b_{qj}(x',t,D,D_t)u_j(x,t)\big|_{S_+} = \varphi_q(x',t), \quad q = 1,\ldots,N, \tag{1.4}$$

and the initial conditions in $\overline{G}$ and on Γ, respectively,

$$D_t^{\lambda-1}u_j\big|_{\overline{G}} = v_{j\lambda}(x) = \psi_{j\lambda}(x), \qquad \lambda = 1,\ldots,t_j', \; j = 1,\ldots,m, \tag{1.5}$$

$$D_\nu^{k-1}D_t^{\lambda-1}u_j\big|_\Gamma = \omega_{j\lambda k}(x') = \psi_{j\lambda k}(x'), \quad \forall (\lambda,k) \in P_j, \; j = 1,\ldots,m. \tag{1.6}$$

Lemmas III.16–III.19 imply that the operator $\mathfrak{U}$ that corresponds to the problem (1.2)–(1.6) is a bounded operator from $\widetilde{\mathcal{H}}^s$ to the subspace $\widetilde{\mathcal{K}}^s$ of the space $\mathcal{K}^s$, where

$$\mathcal{K}^s = \prod_{i=1}^m \left(\mathcal{H}^s(\overline{\Omega}_+,\gamma) \times \prod_{\sigma=1}^{\sigma_0} \mathcal{H}^{s-\sigma+1/2}(\overline{S}_+,\gamma) \right) \times \prod_{q=1}^N \mathcal{H}^{s-\sigma_q-1/2}(\overline{S}_+,\gamma)$$

$$\times \prod_{j=1}^m \prod_{\lambda=1}^{t_j'} \left(H^{s+t_j-2b\lambda+b}(\overline{G}) \times \prod_{k\in P_{j\lambda}} H^{s+t_j-k-2b\lambda+b+1/2}(\Gamma) \right)$$

and $\widetilde{\mathcal{K}}^s$ consists of the vector-valued functions whose components satisfy the compatibility conditions that follow from the theorems on traces of functions belonging to the spaces $\mathcal{H}^s$ and H^s on S_+, G, Γ (these conditions will be given below). Here, for each fixed λ, $P_{j\lambda}$ is the set of the values of the index k in the pairs $(\lambda,k) \in P_j$, arranged in increasing order.

Clearly,

$$\mathcal{F}_i(x,t) = \big(f_{i0}(x,t), f_{i1}(x',t),\ldots, f_{i\sigma_0}(x',t)\big) = \chi f_i(x,t),$$

$$f_i(x,t) \in \widetilde{\mathcal{H}}_{(0,\sigma_0,\emptyset)}^s(\Omega_+,\gamma),$$

$$V_{j\lambda}(x) = \big(v_{j\lambda}(x), \omega_{j\lambda k}(x'), k \in P_{j\lambda}\big) = \chi_0 v_{j\lambda}(x) \in \widetilde{H}_{(P_{j\lambda})}^{s+t_j-2b\lambda+b}(\overline{G}),$$

$$\lambda = 1,\ldots,t_j', \; j = 1,\ldots,m.$$

Denote by $F(x,t) \in \mathcal{K}^s$ the vector composed of the right-hand sides of (1.2)–(1.6).

Definition V.1. *The right-hand sides of the problem* (1.2)–(1.6) *are said to be compatible with zero at $t = 0$ to order s if $F(x,t) \in \widetilde{\mathcal{K}}_+^s$.*

As in the case of a half-space, $\widetilde{\mathcal{K}}_+^s$ consists of the vectors $F(x,t)$ whose components $v_{j\lambda}(x)$ and $\omega_{j\lambda k}(x')$ vanish identically, while the components $f_{i0}^+(x,t)$, $f_{i\sigma}^+(x',t)$, $\varphi_q^+(x',t)$ (continued by zero for $t < 0$) belong to the spaces $\mathcal{H}^{\cdots}$ with the same orders of smoothness.

Definition V.2. *The right-hand sides of the problem* (1.2)–(1.6) *are said to be compatible to order s if there exists a vector-valued function (element) $w \in \widetilde{\mathcal{H}}^s$ such that $F(x,t) - \mathfrak{U}w \in \widetilde{\mathcal{K}}_+^s$.*

Remark V.1. It must be emphasized that the compatibility conditions appear only in the case when the functions $u_{j0}(x,t)$ have traces on S_+ and G, that is, beginning with the values $s > -t_m + \frac{1}{2}$, or at least with the values $s > -t_m + b$. If $s < -t_m + \frac{1}{2}$, then there are no compatibility conditions; in this case $\widetilde{\mathcal{K}}_+^s$ coincides with $\widetilde{\mathcal{K}}^s$, and therefore the theorems on the solvability of problem (1.2)–(1.6) obtained below are valid for all $F(x,t) \in \mathcal{K}^s$.

V.1.2. Boundary value problem in $\overline{\Omega}_+ = G \times [0,\infty)$ with data compatible with zero at $t = 0$. Regularizer

We begin the study of the boundary value problem (1.2)–(1.6) by considering the case when the right-hand sides are compatible with zero at $t = 0$ to order s. If G is not a bounded domain, we will assume that the coefficients of the operator $\mathcal{L}$ are constants for $t > T_0$ and $|x| > R$. By using a regularizer we will reduce the problem to operator equations in the spaces $\widetilde{\mathcal{K}}_+^s$ and $\widetilde{\mathcal{H}}_+^s$, from the solvability of which will follow the solvability of problem (1.2)–(1.6) in $\widetilde{\mathcal{H}}_+^s$.

Let us show how to construct a regularizer. Let $\{G_q'\}$ and $\{G_q''\}$ be two families of subsets, each of which cover G: $\bigcup_q G_q' = \bigcup_q G_q'' = \overline{G}$, $G_q'' \subset G_q'$; we assume that the multiplicity of each cover does not exceed a fixed number n_0 and the distance between an arbitrary point $x \in G_q''$ and $G \setminus G_q'$ is bounded below by a constant $d \geq d_0 > 0$.

In the case where G is the exterior of a bounded domain $\mathcal{G}$, we will assume then the indicated covers consist of the sets G_0' and G_0'' (the exteriors of concentric balls of radii R_0 and $R_0 + h$ centered at some point $\mathcal{M}_0$), of concentric balls G_0' and G_0'' centered at points $\mathcal{M}_q$, and of sets G_0' and G_0'' abutting on Γ that are defined in a local coordinate system with the origin at $\mathcal{M}_q \in \Gamma$ by the inequalities $|y_k| < d$, $k = 1, \ldots, n-1$, $0 \leq y_n - F(y') < 2d$, $|y_k| < 2d$, $k = 1, \ldots, n-1$, $0 \leq y_n - F(y') < 4d$, where $y_n = F(y')$ is the equation of the piece of Γ in the neighborhood of $\mathcal{M}_q \in \Gamma$.

Further, suppose that the segments I_p' and I_p'' $(I_p'' \subset I_p')$, of respective lengths d and $2d$, and centered at the points $t = t_{(p)}$, together with the half-lines $I_0' = [T_0 - h, \infty)$ and $I_0'' = [T_0, \infty)$, cover $\overline{\mathbb{R}}_+^1$.

Set $\Omega'_{qp} = G'_q \times I'_p$, $\Omega''_{qp} = G''_q \times I''_p$. Obviously, the families of subsets $\{\Omega'_{qp}\}$ and $\{\Omega''_{qp}\}$ cover $\overline{\Omega}_+$. For convenience we relabel these subsets, denoting them by Ω'_r and Ω''_r; let $\mathfrak{M}$ and $\mathfrak{N}$ be the sets of the indices r of those subsets Ω'_r and Ω''_r that lie outside S_+ and abut on S_+, respectively, and let $\mathcal{M}_{(r)}(x_{(r)}, t_{(r)}) \in \Omega''_r$. Note that if one straightens Γ in a neighborhood of $\mathcal{M}_r$ by means of the change of coordinates $y' = z'$, $z_n = y_n - F(y')$, $t = t'$, $|z'| < d$, $0 \le z_n \le 2d$, then the set Ω'_r abutting on S_+ is transformed into an $(n+1)$-dimensional cube $\check{\Omega}'_r$ that lies in $\overline{\mathbb{R}}^{n+1}_{++} = \{(z,t) : z_n \ge 0, t \ge 0\}$.

For every $r \in \mathfrak{N}$ we denote by $\mathcal{L}_{(r)}(z, t, D_z, D_t)$ and $\mathcal{B}_{(r)}(z', t, D_z, D_t)$ the operators with variable coefficients in $\overline{\mathbb{R}}^{n+1}_{++}$ that coincide in $\check{\Omega}''_r = \{(z,t) : |z'| < d, 0 \le z_n < d, |t - t_{(r)}| < d\}$ with the expressions of the operators $\mathcal{L}(x, t, D, D_t)$ and $\mathcal{B}(x, t, D, D_t)$ in the coordinate system (z, t). Similarly, for $r \in \mathfrak{M}$ we denote by $\mathcal{L}_{(r)}(x, t, D, D_t)$ the operators with variable coefficients that coincide with $\mathcal{L}(x, t, D, D_t)$ in Ω''_r and become the operators with constant coefficients equal to their values at the points $\mathcal{M}_{(r)}$ outside $\check{\Omega}'_r$ and Ω'_r, respectively.

Let $\sum h_r(x, t) \equiv 1$ be a partition of unity in $\overline{\Omega}_+$, where $h_r \in C^\infty$, $h_r(x, t) \equiv 1$ in Ω''_r and $h_r(x, t) \equiv 0$ outside Ω'_r. Denote $f_{(r)i}(x, t) = h_r f_i$, $\psi_{(r)j\lambda}(x) = h_r(x, 0)\psi_{j\lambda}(x)$, $\varphi_{(r)q} = h_r \varphi_q$, $\psi_{(r)j\lambda k}(x') = h_r(x', 0, 0)\psi_{j\lambda k}(x')$, $f_{(r)i\sigma}(x, t) = h_r f_{i\sigma}$, and continue these functions by zero outside Ω'_r. Also, denote by $f_{(r)i}(z, t)$, $f_{(r)i\sigma}(z', t)$, $\psi_{(r)j\lambda}(z)$, $\psi_{(r)j\lambda k}(z')$, $\varphi_{(r)q}(z', t)$ the representations of the corresponding functions in the coordinate system $\{z, t\}$ with the origin at the point $\mathcal{M}_r \in S_+$. Assign to every subset Ω'_r, depending on its location, one of the following boundary value problems that were formulated above:

(a) For Ω'_r abutting on S_+, consider the following boundary value problems in $\overline{\mathbb{R}}^{n+1}_{++} = \Omega_+$:

$$l_{(r)ij}(z, t, D_z, D_t)u_{(r)j}(z, t)\big|_{\overline{\Omega}_+} = f_{(r)i0}(z, t), \quad i = 1, \ldots, m, \quad (1.7)$$

$$b_{(r)qj}(z', t, D_z, D_t)u_{(r)j}(z, t)\big|_{\overline{E}^n_+} = \varphi_{(r)q}(z', t), \quad q = 1, \ldots, N, \quad (1.8)$$

$$D_n^{\sigma-1}l_{(r)ij}(z', t, D_z, D_t)u_{(r)j}(z, t)\big|_{\overline{E}^n_+} = f_{(r)i\sigma}(z', t),$$

$$\sigma = 1, \ldots, \sigma_0, \quad i = 1, \ldots, m, \quad (1.9)$$

$$D_t^{\lambda-1}u_{(r)j}(z, 0)\big|_{\overline{\mathbb{R}}^n_+} = \psi_{(r)j\lambda}(z),$$

$$D_n^{k-1}D_t^{\lambda-1}u_{(r)j}(z, t)\big|_{\mathbb{R}^{n-1}} = \psi_{(r)j\lambda k}(z'),$$

$$\lambda = 1, \ldots, t'_j, \quad (\lambda, k) \in P_j, \quad j = 1, \ldots, m, \quad (1.10)$$

with null initial conditions if $\Omega'_r \cap \Gamma = \emptyset$.

(b) For Ω'_r that do not intersect with S_+, consider the following Cauchy problems:

$$l_{(r)ij}(x, t, D, D_t)u_{(r)j}(x, t)\big|_{\overline{E}^{n+1}_+} = f_{(r)i}(x, t), \quad i = 1, \ldots, m, \quad (1.11)$$

$$D_t^{\lambda-1} u_{(r)j}(x,0) = v_{(r)j\lambda}(x) = \psi_{(r)j\lambda}(x), \quad \lambda = 1,\dots,t_j', \ j = 1,\dots,m \qquad (1.12)$$

with null initial values (1.10) if $\Omega_r' \cap \mathbb{R}^n = \emptyset$. If Γ and the coefficients of the operators $\mathcal{L}$ and $\mathcal{B}$ are assumed to be sufficiently smooth, then the coefficients of the operators $l_{(r)ij}$ and $b_{(r)qj}$ will satisfy the conditions of Section IV.6. Therefore, by Theorem IV.10, there exist numbers d_0 and $\gamma_0(s)$ such that for $d < d_0$ and $\gamma > \gamma_0(s)$ each one of the boundary value problems (1.7)–(1.10) and (1.11), (1.12) can be uniquely solved and their solutions $u_{(r)}(z,t)$, $u_{(r)}(x,t)$ satisfy corresponding a priori estimates. Let $u_{(r)j}(x,t)$ be the solution of the Cauchy problem (1.11), (1.12) or the solution of the boundary value problem (1.7)–(1.10), written in the original coordinates (x,t). Let $\eta_r(x,t)$ be infinitely differentiable functions with compact support, vanishing identically outside the d-neighborhood of Ω_r', $\eta_r(x,t) \cdot h_r(x,t) \equiv h_r(x,t)$. Define the operator $\mathcal{R}$ by the rule $u(x,t) = \mathcal{R}F(x,t) = \sum_r \eta_r(x,t) u_{(r)}(x,t)$. By Theorem IV.10, $\mathcal{R}$ is a bounded operator from $\widetilde{\mathcal{K}}_+^s$ into the

space $\widetilde{\mathcal{H}}_+^s$, $s \notin \mathbb{Z}_{1,2b}$. By analogy to the theory of smooth solutions of elliptic and parabolic boundary value problems, we call this operator a *regularizer*.

Theorem V.1. *Let the operator $\mathcal{L}$ be uniformly parabolic in $\overline{\Omega}_+$, let the operator $\mathcal{B}$ satisfy on S_+ the Lopatinskiĭ condition uniformly in $(x',t) \in S_+$, and let Γ be of class $C^{s+\tau}$. Assume that the coefficients of the operators $\mathcal{L}$ and $\mathcal{B}$ belong to the following classes:*

$$l_{ij\alpha\beta}(x,t) \in C^{|s|+t_j+\sigma_0+\varepsilon}(\overline{\Omega}_+), \quad \forall \varepsilon > 0, \qquad (1.13)$$

$$b_{qj\alpha\beta}(x,t) \in C^{|s|+t_j-\sigma_q+\varepsilon}(\overline{S}_+), \quad \forall \varepsilon > 0 \qquad (1.14)$$

and do not depend on t for $t > T_0$. Finally, let $l_{ij\alpha\beta}(x,t)$ be constant for $|x| > R_0$ if G is an unbounded domain. Then there exists a number γ_0, which depends on s and the Hölder norms of the coefficients of $\mathcal{L}$, $\mathcal{B}$, such that for $\gamma > \gamma_0$, $s+t_j \notin \mathbb{Z}_{1,2b}$

$$\mathfrak{U}\mathcal{R}F = (I_\mathcal{K} + \Phi)F, \qquad (1.15)$$

$$\mathcal{R}\mathfrak{U}u = (I_\mathcal{H} + Q)u, \qquad (1.16)$$

where $I_\mathcal{K}$, $I_\mathcal{H}$ are the identity operators in $\widetilde{\mathcal{K}}_+^s$, $\widetilde{\mathcal{H}}_+^s$ and Φ, Q are operators in $\widetilde{\mathcal{K}}_+^s$, $\widetilde{\mathcal{H}}_+^s$, of norm less than 1.

Proof: We will prove relation (1.15); (1.16) is proved in a similar fashion. Consider the expression $\mathfrak{U}\mathcal{R}F = \sum_r \mathfrak{U} \cdot \eta_r(x,t) u_{(r)}(x,t)$ and transform as follows. In each set Ω_r' that abuts on S_+ pass from the coordinates (x,t) to the coordinates (z,t) with the origin at the point $\mathcal{M}_{(r)} \in S_+ \cap \Omega_r''$. Denote by $\mathfrak{U}_r(z,t,D_z,D_t)$ the operator of the boundary value problem (1.7)–(1.10) in $\overline{\mathbb{R}}_{++}^{n+1}$, and for $r \in \mathfrak{M}$ denote by $\mathfrak{U}_r(x,t,D_x,D_t)$ the operator of the Cauchy problem (1.11), (1.12) in $\overline{E}_+^{n+1}$.

Obvious manipulations yield

$$\mathfrak{U}\mathcal{R}F = F(x,t) + \sum_{r\in\mathfrak{M}} \big(\mathfrak{U}_r(x,t)\eta_r - \eta_r\mathfrak{U}_r\big)u_{(r)}(x,t)$$

$$+ \sum_{r\in\mathfrak{N}} \big(\mathfrak{U}_r(z,t)\eta_r(z,t) - \eta_r(z,t)\mathfrak{U}_r(z,t)\big)u_{(r)}(z,t) = (I_\mathcal{K} + \Phi)F(x,t),$$

and to complete the proof of (1.15) it remains to estimate the norm of the operator Φ. Note that multiplication of $u_{(r)}(x,t)$ by the function η_r is understood below as multiplication of the components of the elements $u_j(x,t)$ by $\eta_r(x,t)$. The estimation of the norm of Φ reduces to the estimation of the norm of each component of the vector ΦF. We begin by examining the components that correspond to the initial conditions on G and Γ.

(a) Since $v_{(r)j\lambda}(x) = \eta_r v_{j\lambda}(x)$, $\omega_{(r)j\lambda k}(x') = \eta_r\omega_{j\lambda k}(x')$, and the components of the operator $\mathfrak{U}_r$ that correspond to the initial conditions on G and Γ are the operators of multiplication of $v_{j\lambda}(x)$, $\omega_{j\lambda k}(x')$ by 1, it is clear that the components of Φ that correspond to the initial conditions are identically equal zero. Let us now estimate the norms of the remaining components of ΦF.

(b) The most tedious are the calculations of the norms of the components $\big(\mathcal{L}_{(r)i}\eta_r(z,t) - \eta_r(z,t)\mathcal{L}_{(r)i}\big)u_{(r)}(z,t)$ that correspond to the equations of system (1.7) in the case $-t_m+\frac{1}{2} < s < -\frac{1}{2}$. The estimation of the norms of the components that correspond to the boundary conditions on $\overline{E}^n_+$ is considerably simpler.

Thus, assume that $-t_m + \frac{1}{2} < s < -\frac{1}{2}$ and let $l_{(r)ij\sigma}$, $l_{(r)ij\sigma\mu}$ be the operators that correspond to the representation (1.1) of the operator l_{ij} in the coordinate system (z,t) with the origin at $\mathcal{M}_{(r)} \in S_+$. Consider the component $\big(\mathcal{L}_{(r)i}\eta_r - \eta_r\mathcal{L}_{(r)i}\big)u_{(r)}(z,t)\big|_{\mathbb{R}^{n+1}_{++}} = U_{(r)i}(z,t)$, which, by Lemma III.7 and the inclusion $u_{(r)j}(z,t) \in \widetilde{\mathcal{H}}^{s+t_j}_+$, takes the form

$$U_{(r)i}(z,t) = \sum_{k\geq s'+t_j+1} \Big[l^{(k,0)}_{(r)ij}\eta_r - \eta_r l^{(k,0)}_{(r)ij}\Big]u_{(r)jk}(z',t) \times \delta(z_n)$$

$$+ \sum_{\lambda\geq s''+t_j+1} \Big[l^{(0,\lambda)}_{(r)ij}\eta_r - \eta_r l^{(0,\lambda)}_{(r)ij}\Big]v_{(r)j\lambda}(z) \times \delta(t)$$

$$+ \sum_{k+2b\lambda\geq s'+t_j+1} \Big[l^{(k,\lambda)}_{(r)ij}\eta_r - \eta_r l^{(k,\lambda)}_{(r)ij}\Big]\omega_{(r)j\lambda k}(z') \times \delta(z_n,t) + A_{(r)i}u_{(r)}(z,t), \quad (1.17)$$

$$
A_{(r)i}u_{(r)} = \left[l^{(s'+t_j',0)}_{(r)ij} \left(D_{z_n}^{s'+t_j} \eta_r u_{(r)j0}(z,t) \right)_{++} \right.
$$

$$
\left. - \eta_r l^{(s'+t_j,0)}_{(r)ij} \left(D_{z_n}^{s'+t_j} u_{(r)j0}(z,t) \right)_{++} \right]
$$

$$
+ \sum_{\sigma=[s]+t_j-b+1}^{s'+t_j-1} \left[l_{(r)ij\sigma} \left(D_{z_n}^{\sigma} (\eta_r u_{(r)j0}) \right)_{++} \right.
$$

$$
\left. - \eta_r l_{(r)ij\sigma} \left(D_{z_n}^{\sigma} u_{(r)j0}(z,t) \right)_{++} \right]
$$

$$
+ \sum_{\sigma=0}^{[s]+t_j-b} \left[l^{(\sigma_{s'}+t_j)}_{(r)ij\sigma} \left(D_t^{\sigma_{s'}+t_j} D_{z_n}^{\sigma} (\eta_r u_{(r)j0}) \right)_{++} \right.
$$

$$
\left. - \eta_r l^{(\sigma_{s'}+t_j)}_{(r)ij\sigma} \left(D_t^{\sigma_{s'}+t_j} D_{z_n}^{\sigma} u_{(r)j0} \right)_{++} \right]
$$

$$
+ \sum_{\sigma+2b\mu\le[s]+t_j-b} \left[l_{(r)ij\sigma\mu} \left(D_t^{\mu} D_{z_n}^{\sigma} (\eta_r u_{(r)j0}(z,t)) \right)_{++} \right.
$$

$$
\left. - \eta_r l_{(r)ij\sigma\mu} \left(D_t^{\mu} D_{z_n}^{\sigma} u_{(r)j0}(z,t) \right)_{++} \right]
$$

$$(1.18)$$

Here repetition of j (and only j!) indicates summation with respect to j.

Consider first the terms in the right-hand side of (1.17) that contain δ-functions. Each pair of brackets in these terms encloses the commutator of a differential operator and the operator of multiplication by $\eta_r \in C_0^\infty$. By Leibniz's formula (3.17), these commutators are operators of lower order. Hence, by the smoothness of the coefficients l_{ij}, Lemma III.16 and estimate (2.29) of Chapter II, the following estimates hold:

$$
\left\| \left(l^{(k,0)}_{(r)ij} \eta_r - \eta_r l^{(k,0)}_{(r)ij} \right) u^+_{(r)jk} \times \delta(z_n), \mathbb{R}^{n+1}, \gamma \right\|_s
$$

$$
\le C\gamma^{-1/2b} \langle\!\langle u^+_{(r)jk}, E^n, \gamma \rangle\!\rangle_{s+t_j-k+1/2}. \tag{1.19}
$$

$$
\left\| \left(l^{(0,\lambda)}_{(r)ij} \eta_r - \eta_r l^{(0,\lambda)}_{(r)ij} \right) v^+_{(r)j\lambda}(z) \times \delta(t), \mathbb{R}^{n+1}, \gamma \right\|_s
$$

$$
\le C\gamma^{-1/2b} \left[v^+_{(r)j\lambda}(z), \mathbb{R}^n \right]_{s+t_j-2b\lambda+b}, \tag{1.20}
$$

$$
\left\| \left(l^{(k,\lambda)}_{(r)ij} \eta_r - \eta_r l^{(k,\lambda)}_{(r)ij} \right) \omega_{(r)j\lambda k}(z') \times \delta(z_n,t), \mathbb{R}^{n+1}, \gamma \right\|_s
$$

$$
\le C\gamma^{-1/2b} \langle\!\langle \omega_{(r)j\lambda k}(z'), \mathbb{R}^{n-1} \rangle\!\rangle'_{s+t_j-2b\lambda-k+b+1/2}. \tag{1.21}
$$

Now let us evaluate the norms of all the terms in the expression of $A_{(r)i}u_{(r)}$. To this end, note that all the expressions in square brackets in (1.18) have the same structure and are compositions of operators of differentiation, multiplication by $\eta_r(x,t) \in C_0^\infty$, and multiplication by the characteristic function $\theta(z_n)\theta(t)$.

Hence, it suffices to examine in detail only one term of (1.18), for example, a term of the last but one sum in (1.18). Applying the Leibniz formula to the expression $D_t^{\sigma_{s'}+t_j} D_{z_n}^{\sigma}(\eta_r u_{(r)j0})$ and then commuting the operators of multiplication by $\eta_r(z,t)$ and $\theta(z_n)\theta(t)$, we obtain

$$l_{(r)ij\sigma}^{(\sigma_{s'}+t_j)}\left(D_t^{\sigma_{s'}+t_j} D_{z_n}^{\sigma}(\eta_r u_{(r)j0})\right)_{++} - \eta_r l_{(r)ij\sigma}^{(\sigma_{s'}+t_j)}\left(D_t^{\sigma_{s'}+t_j} D_{z_n}^{\sigma} u_{(r)j0}\right)_{++}$$

$$= \left(l_{(r)ij\sigma}^{(\sigma_{s'}+t_j)}\eta_r - \eta_r l_{(r)ij\sigma}^{(\sigma_{s'}+t_j)}\right)\left(D_t^{\sigma_{s'}+t_j} D_{z_n}^{\sigma} u_{(r)j0}(z,t)\right)_{++}$$

$$+ \sum_{\alpha,\beta} C_{\alpha\beta} l_{(r)ij\sigma}^{(\sigma_{s'}+t_j)}\left(D_{z_n}^{\sigma} D_t^{\beta} \eta_r D_t^{\sigma_{s'}+t_j} D_{z_n}^{\sigma-\alpha} u_{(r)j0}(z,t)\right)_{++}$$

$$= J_i u_{(r)0} = J_{i1} u_{(r)0} + J_{i2} u_{(r)0}.$$

Since the commutator $l_{(r)ij\sigma}^{(\sigma_{s'}+t_j)}\eta_r - \eta_r l_{(r)ij\sigma}^{(\sigma_{s'}+t_j)}$ is an operator of lower order compared to $l_{(r)ij\sigma}^{(\sigma_{s'}+t_j)}$, and $\alpha + 2b\beta \geq 1$, we can evaluate each term $J_{i1} u_{(r)0}$ and $J_{i2} u_{(r)0}$ with the help of Lemma III.4 and estimate (2.29) of Chapter II, keeping in mind that $u_{(r)j0}(z,t) \in \mathcal{H}_+^{s+t_j}(\mathbb{R}_+^{n+1},\gamma)$. In this way we get

$$\left\|J_i u_{(r)0}, \mathbb{R}^{n+1}, \gamma\right\|_s \leq C_s \gamma^{-1/2b}\left\|u_{(r)j0}, \overline{\mathbb{R}}_{++}^{n+1}, \gamma\right\|_{s+t_j}.$$

Estimating by the same method the norms of the other terms figuring in $A_{(r)i} u_{(r)0}$, we obtain

$$\left\|A_{(r)i} u_{(r)0}, \mathbb{R}^{n+1}, \gamma\right\|_s \leq C_s \gamma^{-1/2b}\left\|u_{(r)j0}, \overline{\mathbb{R}}_{++}^{n+1}, \gamma\right\|_{s+t_j}. \tag{1.22}$$

Estimates (1.19)–(1.21) and (1.22) yield for $s < \frac{1}{2}$

$$\left\|(\mathcal{L}_{(r)i}\eta_r - \eta_r \mathcal{L}_{(r)i})u_{(r)}\big|_{\overline{\mathbb{R}}_{++}^{n+1}}, \mathbb{R}^{n+1}, \gamma\right\|_s$$

$$\leq C_s \gamma^{-1/2b}\left\|u_{(r)j}(z,t), \overline{\mathbb{R}}_{++}^{n+1}, \gamma\right\|_{s+t_j,(t_j',\tau_j,P_j)} \tag{1.23}$$

If $s > \frac{1}{2}$, then estimate (1.23) follows from the relation $l_{(r)ij}u_{(r)j}(z,t)\big|_{\overline{\mathbb{R}}_{++}^{n+1}} = \left(l_{(r)ij}u_{(r)j0}(z,t)\right)_{++}$ and the boundedness of $l_{(r)ij}$ as an operator acting from $\mathcal{H}_+^{s+t_j}(\mathbb{R}_+^{n+1},\gamma)$ into $\mathcal{H}_+^{s}(\mathbb{R}^{n+1},\gamma)$.

(c) Arguing in much the same manner, but now in a simpler situation, we obtain the estimates

$$\left\langle\!\left\langle (D_{z_n}^{\sigma-1}\mathcal{L}_{(r)i}\eta_r - \eta_r D_{z_n}^{\sigma-1}\mathcal{L}_{(r)i})u_{(r)}\big|_{\overline{E}_+^n}, \overline{E}_+^n, \gamma\right\rangle\!\right\rangle_{s-\sigma+1/2}$$

$$\leq C_s \gamma^{-1/2b}\left\|u_{(r)j}(z,t), \overline{\mathbb{R}}_{++}^{n+1}, \gamma\right\|_{s+t_j,(t_j',\tau_j,P_j)}, \tag{1.24}$$

$$\sigma = 1,\ldots,\tau_i - t_i = \sigma_0,$$

$$\langle\langle (\mathcal{B}_{(r)q}\eta_r - \eta_r\mathcal{B}_{(r)q})u_{(r)}|_{\overline{E}_+^n}, \overline{E}_+^n, \gamma\rangle\rangle_{s-\sigma_q-1/2}$$

$$\leq C_s\gamma^{-1/2b}\||u_{(r)j}(z,t),\overline{\mathbb{R}}_{++}^{n+1},\gamma\||_{s+t_j,(t_j',\tau_j,P_j)}, \quad q=1,\ldots,N, \quad (1.25)$$

where repetition of j indicates summation with respect to j from 1 to m.

(d) For $r \in \mathfrak{M}$, we have

$$\|(\mathcal{L}_{(r)i}\eta_r - \eta_r\mathcal{L}_{(r)i})u_{(r)}(x,t)|_{\overline{E}_+^{n+1}}, \overline{E}_+^{n+1}, \gamma\rangle\rangle_s$$

$$\leq C_s\gamma^{-1/2b}\||u_{(r)j}(x,t),\Omega_+,\gamma\||_{s+t_j,(t_j',\tau_j,P_j)}. \quad (1.26)$$

Note that the constants C_s in (1.24)–(1.26) do not depend on $u_{(r)}$ and γ. Thanks to this and the finiteness of the cover $\{\Omega_r'\}$, formulas (1.23)–(1.26) and the estimates for each $u_{(r)}(x,t)$ yield the estimate $\||\Phi F\||_{\widetilde{\mathcal{K}}^s} \leq C_s\gamma^{-1/2b}\||F\||_{\widetilde{\mathcal{H}}^s}$, which completes the proof of (1.15). Relation (1.16) and the estimate of the norm of the operator Q in $\widetilde{\mathcal{H}}^s$ are established in the same manner.

The following theorem is a direct corollary of Theorem V.1.

Theorem V.2. *Let $\mathcal{L}$ be a uniformly parabolic operator in $\overline{\Omega}_+$, let the operator $\mathcal{B}$ satisfy the Lopatinskiĭ condition uniformly on $\overline{S}_+$, and assume that their coefficients satisfy conditions (1.13), (1.14). Further, assume that Γ is of class C^∞ and the right-hand sides of the problem (1.2)–(1.6) are compatible with zero to order s at $t = 0$.*

Then there exists a number γ_0 (which depends on s, the Hölder norms of the coefficients of $\mathcal{L}$, $\mathcal{B}$, and the surface Γ) such that for $\gamma > \gamma_0$ problem (1.2)–(1.6) has a unique solution $u(x,t)$ and, for $s \notin \mathbb{Z}_{1,2b}$,

$$\||u(x,t)\||_{\widetilde{\mathcal{H}}_+^s} \leq C_s\||F(x,t)\||_{\widetilde{\mathcal{K}}_+^s}. \quad (1.27)$$

Proof: Under the assumptions of Theorem V.1, relations (1.15), (1.16) are valid and there exists the number γ_0, provided by Theorem V.1, such that for $\gamma > \gamma_0$ the norms $\|\Phi\|_{\widetilde{\mathcal{K}}^s}$ and $\|Q\|_{\widetilde{\mathcal{H}}^s}$ are strictly less than 1. This proves the invertibility of the operators $I_\mathcal{K} + \Phi_s$ and $I_\mathcal{H} + Q_s$, and hence of the operator $\mathfrak{U}_s$ of the problem (1.2)–(1.6), and the assertion of Theorem V.2.

V.1.3. Boundary value problem in Ω_+ in the general case

In this section we use Theorem V.2 on the solvability of the boundary value problem with right-hand sides compatible with zero at $t = 0$ to order s and the theorem on the solvability of the Cauchy problem in order to prove an analogue of Theorem V.2 on the solvability of the boundary value problem (1.2)–(1.6) in the spaces $\widetilde{\mathcal{H}}^s$ in the general case. As mentioned in Remark V.1, for $s < -t_m + \frac{1}{2}$ the space $\widetilde{\mathcal{K}}_+^s$ coincides with $\widetilde{\mathcal{K}}^s$, and no compatibility condition is imposed on the right-hand sides of problem (1.2)–(1.6). Therefore, problem (1.2)–(1.6) has a unique solution for any right-hand side belonging to $\widetilde{\mathcal{K}}^s$ with $s < -t_m + \frac{1}{2}$, and the a priori estimate (1.27) holds. If we want the solution $u(x,t)$ to be smoother up to the boundary of Ω_+, we need to impose the above-formulated compatibility condition on the right-hand sides.

Theorem V.3. *Let the operators $\mathcal{L}$, $\mathcal{B}$, their coefficients, and the surface Γ satisfy the assumptions of Theorem V.2.*

If the right-hand sides $F(x,t)$ of problem (1.2)–(1.6) are compatible with zero at $t=0$ to order s, then there exists a number γ_0 such that for $\gamma > \gamma_0(s)$ problem (1.2)–(1.6) has a unique solution $u(x,t) \in \widetilde{\mathcal{H}}^s$ and the estimate

$$\big\| u(x,t) \big\|_{\widetilde{\mathcal{H}}^s} \leq C_s \big\| F(x,t) \big\|_{\widetilde{\mathcal{K}}^s} \tag{1.28}$$

holds for $s + t_j \notin \mathbb{Z}_{1,2b}$.

Proof: Let the right-hand sides $F(x,t)$ be compatible to order $s > -t_m + b$ and assume that $w(x,t) \in \widetilde{\mathcal{H}}^s$ is such that $\big(F(x,t) - \mathfrak{U}w\big) \in \widetilde{\mathcal{K}}^s_+$. Since $w_j(x,t) \in \widetilde{\mathcal{H}}^{s+t_j}(\Omega_+, \gamma)$, it follows that for any j such that $s+t_j > 0$ the components $w_{j0}(x,t)$, $w_{jk}(x',t)$, $\tilde{v}_{j\lambda}(x)$, and $\widetilde{\omega}_{j\lambda k}(x')$ of $w_j(x,t)$ are connected to one another by the relations following from the theorems on traces, while the remaining components are, generally speaking, independent. Note that the element $w \in \widetilde{\mathcal{H}}^s$, by means of which problem (1.2)–(1.6) with the right-hand sides compatible at $t = 0$ to order s reduces to a problem with the right-hand sides compatible with zero at $t = 0$ to order s, is not uniquely determined. Indeed, if $w^*(x,t) \in \widetilde{\mathcal{H}}^s$ is an arbitrary element such that, for any j with $s + t_j > b$,

$$\widetilde{w}^*_{jk}(x',t) = \widetilde{w}_{jk}(x',t), \qquad s + t_j - k + \tfrac{1}{2} > 0,$$

$$\tilde{v}^*_{j\lambda}(x) = \tilde{v}_{j\lambda}(x), \qquad s + t_j - 2b\lambda + b > 0,$$

$$\widetilde{\omega}^*_{j\lambda k}(x') = \widetilde{\omega}_{j\lambda k}(x'), \qquad s + t_j - 2b\lambda - k + b + \tfrac{1}{2} > 0$$

and all the other components are arbitrary, then clearly $(w_j - w^*_j) \in \widetilde{\mathcal{H}}^{s+t_j}_+(\Omega_+, \gamma)$, and so $\mathfrak{U}(w - w^*) \in \widetilde{\mathcal{K}}^s_+$. Hence, $\big(F(x,t) - \mathfrak{U}w^*\big) \in \widetilde{\mathcal{K}}^s_+$, and consequently in Definition V.2 we can replace $w(x,t)$ by an arbitrary element $w^*(x,t) \in \widetilde{\mathcal{H}}^s$. We will use this arbitrariness in $w(x,t)$ in the reduction of problem (1.2)–(1.6) to a problem with the right-hand sides compatible with zero at $t = 0$ to order s.

Specifically, we proceed as follows. For any j such that $s + t_j > b$, that is, $s'' + t'_j \geq 1$, we consider functions $v_{j\lambda}(x) \in H^{s+t_j - 2b\lambda + b}(G)$ with $s+t_j - 2b\lambda + b > 0$, and using the smoothness of Γ, extend them from G to all the $\mathbb{R}^n$, preserving the smoothness so that $\big\| v_{j\lambda}(x), \mathbb{R}^n \big\|_{s+t_j - 2b\lambda+b} \leq C \big\| v_{j\lambda}(x), G \big\|_{s+t_j - 2b\lambda+b}$. Then for these values of j we consider the Cauchy problems

$$\big[D_t + (-\Delta)^b \big]^{s''+t'_j} w_{j0}(x,t) = 0, \qquad (x,t) \in E^{n+1}_+, \tag{1.29}$$

$$D_t^{\lambda-1} w_{j0}(x,0) = v_{j\lambda}(x), \qquad \lambda = 1, \ldots, s'' + t'_j, \tag{1.30}$$

where Δ is the Laplace operator. Theorem IV.3 implies that each of the problems (1.29)–(1.30) has a unique solution $w_{j0}(x,t)$, which satisfies the estimate

$$\big\| w_{j0}(x,t), E^{n+1}_+, \gamma \big\|_{s+t_j} \leq C \sum_{\lambda=1}^{s''+t'_j} \big\| v_{j\lambda}(x), G \big\|_{s+t_j - 2b\lambda+b}.$$

Now consider the vector-valued function $w = (w_1, \ldots, w_m) \in \widetilde{\mathcal{H}}^s$ with the following components $w_j(x, t) \in \widetilde{\mathcal{H}}^{s+t_j}$:

(a) $w_j(x, t) \equiv 0$ for each j such that $s + t_j \leq b$ (that is, all these components are identically equal to zero).

(b) $w_j(x, t)$ for j such that $s + t_j > b$ are defined as the elements with the components $w_{j0}(x, t) = w_{j0}(x, t)\big|_{\Omega_+}$, $w_{jk}(x', t) = D_\nu^{k-1} w_{j0}(x, t)\big|_{S_+}$, $k = 1, \ldots, s' + t_j$, $\tilde{v}_{j\lambda}(x) = D_t^{\lambda-1} w_{j0}(x, t)$, $\lambda = 1, \ldots, s'' + t_j'$, $\widetilde{\omega}_{j\lambda k}(x') = D_\nu^{k-1} D_t^{\lambda-1} w_{j0}(x, t)\big|_\Gamma$, $k + 2b\lambda \leq s' + t_j + b$, the remaining components being identically equal to zero.

We claim that if $u(x, t)$ is a solution of problem (1.2)–(1.6), then $U(x, t) = \big(u(x, t) - w(x, t)\big) \in \widetilde{\mathcal{H}}_+^s$ is a solution of the same problem with the right-hand sides compatible with zero at $t = 0$ to order s. Indeed, it is obvious that $w_{j\lambda k}(x') - \widetilde{\omega}_{j\lambda k}(x') \equiv 0$ for any j, λ, k such that $s + t_j - k - 2b\lambda + b + \frac{1}{2} > 0$ and $v_{j\lambda}(x) - \tilde{v}_{j\lambda}(x) \equiv 0$ for any j, λ such that $s + t_j - 2b\lambda + b > 0$. Let us verify that the components $\mathcal{L}_i U\big|_{\Omega_+}$, $\big(D_\nu^{\sigma-1} \mathcal{L}_i\big) U\big|_{\overline{S}_+}$, $\mathcal{B}_q U\big|_{\overline{S}_+}$ of the vector-valued function $\mathfrak{U}u(x, t)$, continued by zero for $t < 0$, belong to the appropriate spaces $\mathcal{H}^s$ with the same order of smoothness in $\Omega = G \times \mathbb{R}^1$, $S = \Gamma \times \mathbb{R}^1$.

Consider first the components $\mathcal{L}_i U\big|_{\Omega_+}$. If $s > 0$, then, by definition, $\mathcal{L}_i U\big|_{\Omega_+} = l_{ij}(u_{j0} - w_{j0})$, and since $D_t^{\lambda-1}(u_{j0} - w_{j0})\big|_{t=0} = 0$, $\lambda = 1, \ldots, s'' + t_j'$, it follows that the functions $u_{j0} - w_{j0}$, continued by zero for $t < 0$, remain in the class $\mathcal{H}_+^{s+t_j}(\Omega, \gamma)$. Hence, $\mathcal{L}_i U\big|_{\Omega_+} \in \mathcal{H}_+^s(\Omega, \gamma)$. If $s < 0$, then $\mathcal{L}_i U\big|_{\Omega_+}$ belongs to the space $\mathcal{H}^s(\overline{\Omega}_+, \gamma)$, which by definition coincides with the subspace $\mathcal{H}_+^s(\Omega, \gamma)$.

Now consider $\mathcal{B}_q U\big|_{\overline{S}_+} = b_{qj} U_j\big|_{\overline{S}_+}$. If $s - \sigma_q - \frac{1}{2} > 0$, then $\mathcal{B}_q U\big|_{\overline{S}_+}$ vanishes at $t = 0$ together with all its derivatives $D_t^{\lambda-1}\big(\mathcal{B}_q U\big|_{S_+}\big)_{t=0}$ for any λ such that $s - \sigma_q - 2b\lambda - \frac{1}{2} - b > 0$. Therefore, the continuation of $\mathcal{B}_q U\big|_{S_+}$ by zero for $t < 0$ belongs to $\mathcal{H}_+^{s-\sigma_q-1/2}(S, \gamma)$. If $s - \sigma_q - \frac{1}{2} < 0$, then $\mathcal{B}_q U\big|_{\overline{S}_+}$ belongs to the space $\mathcal{H}^{s-\sigma_q-1/2}(\overline{S}_+, \gamma)$, which coincides with $\mathcal{H}_+^{s-\sigma_q-1/2}(S, \gamma)$.

In a similar way one proves that $D_\nu^{\sigma-1} \mathcal{L}_i U\big|_{\overline{S}_+} \in \mathcal{H}_+^{s-\sigma+1/2}(S, \gamma)$. Thus, $U \in \widetilde{\mathcal{H}}_+^s$ and the assertion of the theorem immediately follows from Theorem V.2. Indeed, by Theorem V.2, the problem $\mathfrak{U}U(x, t) = F(x, t) - \mathfrak{U}w$ has a unique solution $U(x, t)$, which satisfies the estimate

$$\big\| U(x, t) \big\|_{\widetilde{\mathcal{H}}^s} \leq C_s \Big(\big\| F(x, t) \big\|_{\widetilde{\mathcal{K}}^s} + \big\| \mathfrak{U}w \big\|_{\widetilde{\mathcal{K}}^s} \Big), \qquad s + t_j \notin \mathbb{Z}_{1,2b}. \tag{1.31}$$

Then $u(x, t) = U(x, t) + w(x, t)$ is the solution of the problem (1.2)–(1.6). Since, by construction,

$$\big\| w \big\|_{\widetilde{\mathcal{H}}^s} \leq C \big\| F(x, t) \big\|_{\widetilde{\mathcal{K}}^s}, \tag{1.32}$$

estimates (1.31) and (1.32) yield (1.28).

V.2. Nonlocal boundary value problems. Conjugation problems

V.2.1. Problem setting in classes of smooth functions. Conditions on operators

Let G be a bounded domain of $\mathbb{R}^n$ with a smooth boundary $\Gamma_0 = \partial G$. Let Γ_1 be an $(n-1)$-dimensional surface that does not intersect Γ_0 ($\Gamma_0 \cap \Gamma_1 = \emptyset$) and such that Γ_1 divides G into two subdomains G_1 and G_2: $G = \overline{G}_1 \cup G_2$. Let $\Gamma_1 = \partial G_1$, $\Gamma_2 = \partial G_2 = \Gamma_1 \cup \Gamma_0$, and assume that the surfaces Γ_0 and Γ_1 are diffeomorphic via $y = \alpha(x)$, $x \in \Gamma_1$, $y \in \Gamma_0$. Let ν_0 and ν_1 be the unit inner normals to Γ_0 and Γ_1, respectively. Then, for sufficiently small ε ($|\varepsilon| < \varepsilon_0$) the mapping $y + \varepsilon\nu_0 \mapsto x + \varepsilon\nu_1$ is a diffeomorphism of $\Gamma_{0\varepsilon}$ (the ε-neighborhood of Γ_0) onto $\Gamma_{1\varepsilon}$ (the ε-neighborhood of Γ_1). Obviously, if a distribution $u(y,t)$ and a differential operator $\mathcal{A}(y,t,D_y,D_t)$ with sufficiently smooth coefficients are defined in $S_{0\varepsilon}^+ = \Gamma_{0\varepsilon} \times \mathbb{R}_+^1$, then the distribution $\breve{u}(x,t) = u(\alpha(x),t)$ and the operator $\breve{\mathcal{A}}(x,t,D_x,D_t) = \mathcal{A}(\alpha(x),t,J{\cdot}D_x,D_t)$ are defined in $S_{1\varepsilon}^+ = \Gamma_{1\varepsilon} \times \mathbb{R}_+^1$ (here J denotes the transpose of the Jacobian matrix of the transformation $\alpha^{-1}(y)$). Further, suppose that in each domain $\Omega_r = G_r \times \mathbb{R}_+^1$ ($r = 1,2$) there is given a matrix linear differential operator $\mathcal{L}_r(x,t,D,D_t) = (l_{(r)ij}(x,t,D,D_t))$ $(i,j = 1,\dots,m_r)$, uniformly parabolic in the sense of Petrovskiĭ ($\mathrm{ord}\, l_{(r)ij} = t_{(r)j}$), and that in $S_{1\varepsilon}^+ \cap \Omega_1$, $S_{1\varepsilon}^+ \cap \Omega_2$ and $S_{0\varepsilon}^+ \cap \Omega_2$ there are given matrix linear differential operators $\mathcal{B}_1(x,t,D,D_t) = (b_{(1)qj})$, $\mathcal{B}_2(x,t,D,D_t) = (b_{(2)qj})$, and $\mathcal{B}(y,t,D_y,D_t) = (b_{qj}(y,t,D_y,D_t))$, of sizes $m_1 \times 2b\kappa$, $m_2 \times 2b\kappa$, $m_2 \times 2b\kappa$, and $\kappa = \kappa_1 + 2\kappa_2$, respectively. One assumes that there exist numbers σ_q such that the weighted orders of the operators $b_{(r)qj}$, b_{qj} are equal to $\sigma_q + t_{(r)j}$, $\sigma_q + t_{(2)j}$, respectively. To avoid rather tedious calculations that arise in the analysis of general parabolic systems and arbitrary operators, we restrict our considerations to operators $\mathcal{B}_r$ and $\mathcal{B}$ for which $\sigma_q \le 0$.

We pose the following nonlocal boundary value problem. In each domain Ω_r ($r = 1,2$), find a solution $u_r(x,t) = (u_{(r)1},\dots,u_{(r)m_r})$ of the system of parabolic equations

$$l_{(r)ij}(x,t,D,D_t)u_{(r)j}(x,t) = f_{(r)i}(x,t), \qquad i = 1,\dots,m_r, \tag{2.1}$$

such that the values of $u_r(x,t)$ on S_0 and S_1 are connected by the *nonlocal boundary conditions* (called *nonlocal conjugation conditions*)

$$b_{(1)qj}u_{(1)j}(x,t)\big|_{S_1} + b_{(2)qj}u_{(2)j}(x,t)\big|_{S_1} + \breve{b}_{qj}\breve{u}_{(2)j}(x,t)\big|_{S_1} = \varphi_q(x',t), \tag{2.2}$$

$q = 1,\dots,b\kappa$, and the initial conditions at $t = 0$

$$D_t^{\lambda-1}u_{(r)j}(x,t)\big|_{t=0} \equiv v_{(r)j\lambda}(x) = \psi_{(r)j\lambda}(x),$$
$$\lambda = 1,\dots,t'_{rj}, \quad j = 1,\dots,m_r, \quad r = 1,2 \tag{2.3}$$

are satisfied. Here and below we denote $\breve{b}_{qj}(x,t,D_x,D_t) \equiv b_{qj}(\alpha(x),t,J \cdot D_x,D_t)$, $\breve{u}(x,t) = u(\alpha(x),t)$, and repetition of j indicates summation with respect to j from 1 to m_r, $r = 1,2$.

Let us formulate the conditions to be imposed on the operators of the problem. To this end we need some preliminary considerations.

Let (x, t) be a point of S_1^+ and let (z, t) be a local coordinate system on S_1^+ with the origin at (x, t), such that in a neighborhood of the origin the piece of the surface Γ is defined by the equation $z_n = 0$. Using the diffeomorphism $y = \alpha(x)$, we first write the operators $\mathcal{L}_2(y, t, D_y, D_t)$ and $\mathcal{B}(y, t, D_y, D_t)$ in the coordinates (x, t): $\check{\mathcal{L}}_{(2)}(x, t, D_x, D_t) = \mathcal{L}_{(2)}\big(\alpha(x), t, J \cdot D_x, D_t\big)$, $\check{\mathcal{B}}(x, t, J \cdot D_x, D_t) = \mathcal{B}\big(\alpha(x), t, J \cdot D_x, D_t\big)$. Then we rewrite the operators $\mathcal{L}_r$, $\mathcal{B}_r$, and $\check{\mathcal{B}}$ in the coordinates (z, t). Denote by $\mathcal{L}_r(z, t, D_z, D_t)$, $\mathcal{L}_3(z, t, D_z, D_t)$, $\mathcal{B}_r(z, t, D_z, D_t)$, and $\mathcal{B}_3(z, t, D_z, D_t)$ the representations of the operators $\mathcal{L}_r$, $\check{\mathcal{L}}_2$, $\mathcal{B}_r$, and $\check{\mathcal{B}}$, respectively, in the coordinate system $(z, t) \in S_1^+$, and by $u_r(z, t)$ $(r = 1, 2)$, $u_{(3)}(z, t)$ the representations of the functions $u_r(x, t)$ $(r = 1, 2)$ and $\check{u}_{(2)}(x, t)$ in the same coordinate system (z, t).

Then, in a sufficiently small neighborhood of the point $(x, t) \in S_1$ problem (2.1)–(2.3) reduces to a conjugation problem for two systems of parabolic equations, defined for $z_n > 0$ and $z_n < 0$, respectively, with the conjugation conditions at $z_n = 0$. Specifically, we search for solutions $\big(u_{(1)}(z, t), u_{(3)}(z, t)\big)$ and $u_{(2)}(z, t)$ of the systems of equations

$$\begin{pmatrix} \mathcal{L}_1(z,t,D_z,D_t) & 0 \\ 0 & \mathcal{L}_3(z,t,D_z,D_t) \end{pmatrix} \begin{pmatrix} u_{(1)}(z,t) \\ u_{(3)}(z,t) \end{pmatrix} = \begin{pmatrix} f_{(1)}(z,t) \\ f_{(3)}(z,t) \end{pmatrix}, \quad z_n > 0, \quad (2.4)$$

$$\mathcal{L}_2(z,t,D_z,D_t) u_{(2)}(z,t) = f_{(2)}(z,t), \quad z_n < 0, \quad (2.5)$$

that satisfy the conjugation conditions

$$b_{(1)qj}(z', t, D_z, D_t) u_{(1)j}(z, t)\big|_{z_n=+0} + b_{(3)qj}(z', t, D_z, D_t) u_{(3)j}(z, t)\big|_{z_n=+0}$$
$$+ b_{(2)qj}(z', t, D_z, D_t) u_{(2)j}(z, t)\big|_{z_n=-0} = \varphi_q(x', t), \quad q = 1, \ldots, b\kappa, \quad (2.6)$$

and the initial conditions at $t = 0$

$$D_t^{\lambda-1} u_{(r)j}(z, t)\big|_{t=0} \equiv v_{(r)j\lambda}(z) = \psi_{(r)j\lambda}(z),$$
$$\lambda = 1, \ldots, t'_{(r)j}, \ r = 1, 3; \ j = 1, \ldots, m_r, \ z_n > 0, \quad (2.7)$$

$$D_t^{\lambda-1} u_{(2)j}(z, t)\big|_{t=0} = v_{(2)j\lambda}(z) = \psi_{(2)j\lambda}(z),$$
$$\lambda = 1, \ldots, t'_{(2)j}, \ j = 1, \ldots, m_2, \ z_n < 0. \quad (2.8)$$

Thus, by using the diffeomorphism $y = \alpha(x)$, we succeeded to locally reduce the nonlocal boundary value problem (2.1)–(2.3) (the nonlocal conjugation problem) in a neighbourhood of S_1^+ to the conjugation problem for the block-diagonal system (2.4) for $z_n > 0$ and the system (2.5) for $z_n < 0$, with the conjugation conditions (2.6) at $z_n = 0$ and the initial conditions (2.7), (2.8).

The operators $\mathcal{L}_r$ and $\mathcal{B}_r$ are defined in some neighborhood of the origin. We have studied the conjugation problem of type (2.4)–(2.8) in Section IV.7. Let us formulate for it the compatible covering condition.

Let $\mathcal{L}_k^0(z,t,D_z,D_t)$, $\mathcal{B}_k^0(z',t,D_z,D_t)$ be the weighted principal parts of the operators $\mathcal{L}_k$ and $\mathcal{B}_k$. Let $L_k(z,t,i\xi,p) = L_k^+(\cdot,\xi_n)L_k^-(\cdot,\xi_n)$ be the factorization described above of the polynomials L_k in ξ_n, $k = 1,2,3$, and let $\mathcal{D}_1'(z',t,\xi',p,\xi_n)$, $\mathcal{D}_3'(z',t,\xi',p,\xi_n)$, and $\mathcal{D}_2'(z',t,\xi',p,\xi_n)$ be the remainders of the division of the matrices $\mathcal{B}_1^0\widehat{\mathcal{L}}_1^0$, $\mathcal{B}_3^0\widehat{\mathcal{L}}_3^0$, and $\mathcal{B}_2^0\widehat{\mathcal{L}}_2^0(\cdot,\xi_n)$ by the polynomials $L_1^+(\cdot,\xi_n)$, $L_3^+(\cdot,\xi_n)$, and $L_2^-(\cdot,\xi_n)$, respectively.

According to Remark IV.1, the compatible covering condition for the problem (2.4)–(2.8) at the point (z',t) means that for any $\xi' \in \mathbb{R}^{n-1}$ and $\operatorname{Re} p \geq 0$, $(\xi',p) \neq 0$, the rows of the matrix $\mathcal{D}'(z',t,\xi',p,\xi_n) = \big(\mathcal{D}_1'(z',t,\xi',p,\xi_n), \mathcal{D}_3'(z',t,\xi',p,\xi_n), \mathcal{D}_2'(z',t,\xi',p,\xi_n)\big)$ are linearly independent as polynomials in ξ_n. This is obviously equivalent to the assertion that the rank of the matrix $\mathcal{D}''(z',t,\xi',p)$, composed of the coefficients of the powers of ξ_n, equals $b\kappa = b(\kappa_1 + 2\kappa_2)$.

Denote by $\mathcal{M}(x',t,\xi',p)$ a maximal-rank minor composed of columns of the matrix $\mathcal{D}''(z',t,\xi',p)$, expressed in the coordinates (x,t). The quasihomogeneity of the matrices $\mathcal{L}_k^0(\cdot,\xi,p)$, $\mathcal{B}_k^0(\cdot,\xi,p)$ in ξ and p and their smoothness in (x,t) imply that every maximal-rank minor $\mathcal{M}(x',t,\xi',p)$ is also a function, quasihomogeneous in ξ' and p and continuous in all the variables. Consequently, $d(x',t) = \inf_{(\xi',p)} \big|\mathcal{M}(x',t,\xi',p)\big| > 0$ for $|\xi'|^2 + |p|^{1/b} = 1$.

Definition V.2. *The operators $\mathcal{B}_1$, $\mathcal{B}_2$, and $\mathcal{B}$ of the problem (2.1)–(2.3) are said to satisfy the nonlocal compatible covering condition at the point $(x,t) \in S_1$ if the compatible covering condition holds for the problem (2.4)–(2.8) at the corresponding point $z = 0, t$, that is, $d(x',t) > 0$.*

Definition V.3. *The operators $\mathcal{B}_1$, $\mathcal{B}_2$, and $\mathcal{B}$ of the problem (2.1)–(2.3) are said to satisfy uniformly the nonlocal compatible covering condition on $\overline{S}_1$ if $d(x',t) \geq d_0 > 0$ on $\overline{S}_1$.*

Throughout this section we will assume that the following conditions are satisfied:

    1) The operators $\mathcal{L}_r(x,t,D_x,D_t)$ are uniformly parabolic in $\overline{\Omega}_r$, $r = 1,2$;

    2) The operators $\mathcal{B}_1(x,t,D_x,D_t)$, $\mathcal{B}_2(x,t,D_x,D_t)$, and $\mathcal{B}(\alpha(x),t,J\cdot D_x,D_t)$ satisfy on $\overline{S}_1$ the nonlocal compatible covering condition uniformly in $(x,t) \in S_1$.

V.2.2. The nonlocal boundary value problem in the spaces $\widetilde{\mathcal{H}}^s$

Let ν_0 and ν_1 be the unit inner normals (with respect to G_0 and G_1) to Γ_0 and Γ_1, and let $\{y\}$, $\{x\}$ be local coordinate systems with the origins at $y_0 \in \Gamma_0$ and $x_0 \in \Gamma_1$ $(y = \alpha(x))$, in which $|x_n|$ and $|y_n|$ are the distances from x,y to Γ_1 and Γ_0, respectively. Also, let $\alpha : y' + \nu_0 y_n \mapsto x' + \nu_1 x_n$ be a smooth extension of the diffeomorphism $\alpha(x)$ from the pair Γ_0, Γ_1 to the pair $\Gamma_{0\varepsilon}$, $\Gamma_{1\varepsilon}$.

Further, let $b_{qj} = b_{qjk}(y,t,D_y',D_t)D_{\nu_0}^{k-1}$, $l_{(2)ij} = l_{(2)ijk}(y,t,D_y',D_t)D_{\nu_0}^{k-1}$ be the representations of the operators b_{qj}, $l_{(2)ij}$ in a neighborhood of S_0, and

$$b_{(r)qj} = b_{(r)qjk}(x,t,D_x',D_t)D_{\nu_1}^{k-1}, \quad l_{(2)ij} = l_{(2)ijk}(x,t,D_x',D_t)D_{\nu_1}^{k-1}$$

the representations of the operators $b_{(r)qj}$, $l_{(r)ij}$ $(r = 1, 2, 3)$ in a ε-neighborhood of S_1. Denote by P_{j1}, Q_{j1}, Q_{j2} the sets of the pairs (k, λ) of indices of the functions $\omega_{1j\lambda k}(x')$, $\omega_{2j\lambda k}(x')$, $\omega_{2j\lambda k}(y')$ that figure in the Green formulas for the operators $l_{(1)ij}$ and $b_{(1)qj}$, $l_{(2)ij}$ and $b_{(2)qj}$, $l_{(2)ij}$ and b_{qj}, written near S_1 and S_0, respectively. In the domain Ω_2^+ we define, for any $s \notin \mathbb{Z}_{1,2b}$, the space $\widetilde{\mathcal{H}}^{s+t_{2j}}_{t'_{2j},\tau_{2j},Q_{2j}}(\Omega_2^+, \gamma)$ as the completion of $C^\infty_{(0)}(\Omega_2^+)$ with respect to the norm

$$\left\|u_{(2)j}, \Omega_2^+, \gamma\right\|^2_{s+t_j} = \left\|u_{(2)j0}(x,t), \Omega_2^+, \gamma\right\|^2_{s+t_{2j}} + \sum_{\lambda=1}^{t'_{2j}} \left[[v_{2j}(x), G_2]\right]^2_{s+t_{2j}-2b\lambda+b}$$

$$+ \sum_{r=0}^{1} \sum_{k=1}^{t_{2j}} \left\langle\!\!\left\langle u_{(2)jk}, S_r^+, \gamma\right\rangle\!\!\right\rangle^2_{s+t_{2j}-k+1/2}$$

$$+ \sum_{r=0}^{1} \sum_{(\lambda,k)\in Q_{rj}} \left\langle\!\!\left\langle \omega_{rj\lambda k}, \Gamma_r\right\rangle\!\!\right\rangle'^2_{s+t_{2j}-2b\lambda-k+b+1/2},$$

$$(2.9)$$

where for $Q_{rj} = \emptyset$ the corresponding sum is assumed to be identically equal zero, $\tau_{2j} = t_{2j} + \sigma_0$. For $s \in \mathbb{Z}_{1,2b}$ the space $\widetilde{\mathcal{H}}^{s+t_{2j}}$ is defined by interpolation, as in Section III.5. Clearly, $\widetilde{\mathcal{H}}^{s+t_{2j}}$ is isometrically isomorphic to the corresponding subspace $\mathcal{U}^{s+t_{2j}}$, consisting of the vector-valued functions whose components satisfy on $S_2 = S_1 \cup S_0$, G_2, and $\Gamma_2 = \Gamma_1 \cup \Gamma_0$ the compatibility conditions that follow from the theorems on traces. Set $\widetilde{\mathcal{H}}^s_1 = \prod_{j=1}^{m_1} \widetilde{\mathcal{H}}^{s+t_{1j}}_{\dots}(\Omega_1^+, \gamma)$, $\widetilde{\mathcal{H}}^s_2 = \prod_{j=1}^{m_2} \widetilde{\mathcal{H}}^{s+t_{2j}}_{\dots}(\Omega_2^+, \gamma)$. Recalling the discussion in Chapter IV, we consider the following *nonlocal boundary value problem* (*nonlocal conjugation problem*).

In the domains $\overline{\Omega}_r^+$, find solutions $u_r(x,t) \in \widetilde{\mathcal{H}}^s_r$ of the systems of equations

$$l_{(r)ij}(x,t,D,D_t)u\big|_{\overline{\Omega}_r^+} = f_{(r)i}(x,t) \in \mathcal{H}^s(\overline{\Omega}_r^+, \gamma), \qquad r = 1, 2, \qquad (2.10)$$

that satisfy the nonlocal boundary conditions (nonlocal conjugation conditions) on $\overline{S}_r^+$

$$\mathcal{B}_q u\big|_{\overline{S}_1^+} \equiv b_{(1)qj}(\cdot, D_x, D_t)u_{(1)j}(x,t)\big|_{\overline{S}_1^+} + b_{(2)qj}(\cdot, D_x, D_t)u_{(2)j}(x,t)\big|_{\overline{S}_1^+}$$

$$+ b_{qj}\big(\alpha(x), t, J \cdot D_x, D_t\big)u_{(2)j}\big(\alpha(x), t\big)\big|_{\overline{S}_1^+} = \varphi_q^+(x', t), \quad q = 1, \dots, b_k, \quad (2.11)$$

the boundary conditions on S_r^+

$$D_\nu^{\sigma-1} l_{(r)ij} u_{(r)j}\big|_{\overline{S}_r^+} = f^+_{(r)i\sigma}(x', t), \qquad \sigma = 1, \dots, \sigma_0, \ i = 1, \dots, m_r, \qquad (2.12)$$

and the initial conditions on G_r and Γ_r $(r = 1, 2)$

$$D_t^{\lambda-1} u_{(r)j}(x, 0) = \psi_{(r)j\lambda}(x), \qquad \lambda = 1, \dots, t'_{rj}, \ j = 1, \dots, m_r, \qquad (2.13)$$

$$D_\nu^{k-1} D_t^{\lambda-1} u_{(r)j}(x, t)\big|_{\Gamma_r} = \psi_{(r)j\lambda k}(x'), \quad (\lambda, k) \in P_{rj}, \ j = 1, \dots, m_r, \qquad (2.14)$$

where ν is the inner normal to Γ_r (with respect to G_r); here it is understood that for $r = 2$ conditions (2.12), (2.14) split into conditions on $\overline{S}_0^+$, $\overline{S}_1^+$ and Γ_0, Γ_1 respectively. Clearly, for the corresponding values of s, conditions (2.12), (2.14) are consequences of the equations of system (2.10) and the initial conditions (2.13).

Regarding the conditions (2.11), which are understood in the sense of the definitions of Subsection V.2.1, we need to explain why $\mathcal{B}_q$ is a bounded operator acting from $\widetilde{\mathcal{H}}^s = \widetilde{\mathcal{H}}_1^s \otimes \widetilde{\mathcal{H}}_2^s$ to $\mathcal{H}^{s-\sigma_q-1/2}(\overline{S}_1^+, \gamma)$. By definition, $b_{qj}\big(\alpha(x), t, J \cdot D_x, D_t\big) u_{(2)j}\big(\alpha(x), t\big) \equiv \check{b}_{(3)qjk}\check{u}_{(3)jk}^+ - \check{b}_{(3)qjk}^{(\lambda)}\check{\omega}_{(3)j\lambda k}\delta(t)$.

Due to the specific form of the extension of the diffeomorphism $\alpha(x)$ from Γ_0, Γ_1 to their ε-neighborhoods, the vector-valued functions $\check{U}_j = \big(\check{u}_{(2)jk}(x', t), k = 1, \ldots, \tau_{2j}\big)$ on S_1^+ are expressed in one-to-one manner in terms of the vector-valued functions $U_j(y, t) = \big(u_{(2)jk}(y', t), k = 1, \ldots, \tau_{2j}\big)$ on S_0^+. This relation is given by a matrix $A_j : \check{U}_j = A_j U_j$; the elements of A are smooth functions, depending on the derivatives of the components $\alpha_i(x)$, $i = 1, \ldots, n$, of the diffeomorphism $\alpha(x)$.

The operators b_{qj} and $\check{b}_{qj}$ are of the same order in x_n, y_n, and from the relation $\check{U}_j = A_j U_j$ it follows that the norms $\big\langle\!\big\langle \check{u}_{(2)jk}, \overline{S}_1^+, \gamma \big\rangle\!\big\rangle_{s+t_{2j}-k+1/2}$ and $\big\langle\!\big\langle u_{(2)jk}, \overline{S}_0^+, \gamma \big\rangle\!\big\rangle_{s+t_{2j}-k+1/2}$ are equivalent. A similar assertion holds for the functions $\check{\omega}_{(2)j\lambda k}(x')$ and $\omega_{(2)j\lambda k}(y')$. Therefore, $\check{b}_{qj}\check{u}_{(2)j}\big|_{\overline{S}_1^+}$ are bounded operators from $\widetilde{\mathcal{H}}_2^s$ to $\mathcal{H}^{s-\sigma_q-1/2}(\overline{S}_1^+, \gamma)$.

Now let us consider the operator $\mathfrak{U}$ of the problem (2.10)–(2.14). By Lemmas III.16–III.19 and the foregoing analysis, for any $s \notin \mathbb{Z}_{1,2b}$, the operator $\mathfrak{U}$ is bounded from $\widetilde{\mathcal{H}}^s$ to the subspace $\widetilde{\mathcal{K}}^s$ of the space $\mathcal{K}^s$, where

$$\mathcal{K}^s = \bigoplus_{r=1}^{2}\left(\mathcal{H}^s(\overline{\Omega}_r^+, \gamma) \times \prod_{\sigma=1}^{\sigma_0}\mathcal{H}^{s-\sigma+1/2}(\overline{S}_r^+, \gamma)\right) \times \prod_{q=1}^{N}\mathcal{H}^{s-\sigma_q-1/2}(\overline{S}_1^+, \gamma)$$

$$\times \bigoplus_{r=1}^{2}\left(\prod_{\lambda,j}H^{s+t_{rj}-2b\lambda+b}(\overline{G}_r) \times \prod_{\lambda,j,k}H^{s+t_{rj}-2b\lambda-k+b+1/2}(\Gamma_r)\right)$$

and $\widetilde{\mathcal{K}}^s$ consists of the vector-valued functions $\mathcal{F}(x, t)$ whose components satisfy the natural compatibility conditions following from the theorems on traces. These conditions are similar to the compatibility conditions for the boundary value problem (4.14)–(4.16).

V.2.3. The nonlocal boundary value problem in $\widetilde{\mathcal{H}}^s$ with data compatible with zero at $t = 0$

In this section we use the results and constructions of Subsections V.2.1 and V.2.2 to study the question of the well-posedness of the problem (2.10)–(2.14) in the spaces $\widetilde{\mathcal{H}}_+^s$. Since the principal arguments and constructions are similar to those of Subsections IV.5 and IV.6, we present only the main ideas and omit technical details. For simplicity we will assume that Γ_0, Γ as well as the coefficients of the operators $l_{(r)ij}$, $b_{(r)qj}$, b_{qj} are constants for $t > T_1$ and infinitely differentiable, though the results remain valid when Γ_0, Γ, and the coefficients of the operators $\mathcal{L}_r$, $\mathcal{B}_r$ belong to the corresponding classes of Hölder functions (6.6), (6.7).

Denote by $\widetilde{\mathcal{H}}_+^{s+t_{rj}}(\Omega_r^+, \gamma)$, $s + t_{rj} \notin \mathbb{Z}_{1,2b}$, the completion with respect to the norm of $\widetilde{\mathcal{H}}^{s+t_{rj}}(\Omega_r^+, \gamma)$ of the set of smooth functions that vanish up to order $s + t_{rj}$ at $t = 0$; $\widetilde{\mathcal{H}}_+^s = \widetilde{\mathcal{H}}_{1+}^s \oplus \widetilde{\mathcal{H}}_{2+}^s$. Also, let $\widetilde{\mathcal{K}}_+^s = \mathfrak{U}\widetilde{\mathcal{H}}_+^s$.

As in the case of problem (6.1)–(6.5), we say that the right-hand sides of problem (2.10)–(2.14) are *compatible with zero to order s at $t = 0$* if $\mathcal{F}(x,t) \in \mathfrak{U}\widetilde{\mathcal{H}}_+^s = \widetilde{\mathcal{K}}_+^s$. This means, in particular, that the functions $v_{(r)j\lambda}(x)$ and $\omega_{(r)j\lambda k}(x')$, which belong to the spaces $H^{\cdots}$ with positive smoothness indices, vanish identically, and the functions $f_{(r)i}(x,t)$, $f_{(r)i\sigma}(x',t)$, $\varphi_q(x',t)$, continued by zero for $t < 0$, belong to the corresponding spaces $H^{\cdots}$ with the same smoothness indices.

To construct a regularizer of the problem (2.10)–(2.14) in $\widetilde{\mathcal{H}}_+^s$, we will essentially use a special partition of unity in $\overline{\Omega}_+$, subordinate to a special covering of the domain Ω_+ and compatible with the diffeomorphism $y = \alpha(x)$.

Specifically, let $J^{(1)}$ and $J^{(2)}$ be two families of subsets, each of which covers $\overline{G}_0$, consisting of sets $g_{(k)}$ and $G_{(k)}$ with the following structure.

1) the sets $G_{(k)}$ and $g_{(k)}$ abutting on Γ_1 in a local coordinate system with the origin at the point $\mathcal{P}_k \in \Gamma_1$ are concentric cubes with edge lengths d and $d/2$;

2) the sets $G_{(k)}$ and $g_{(k)}$ that do not intersect $\Gamma_0 \cup \Gamma_1$ are also concentric cubes centered at the point $\mathcal{P}_k$, with edge lenghts d and $d/2$;

3) the sets $G_{(\check{k})}$ and $g_{(\check{k})}$ abutting on Γ_0 are defined as the images under the diffeomorphism $y = \alpha(x)$ of half-cubes $G_{(k)} \cap G_1$ and $g_{(k)} \cap G_1$, abutting on Γ_1, and $G_{(\check{k})} \in \Gamma_{0,\varepsilon}$.

Based on the partitions $J^{(1)}$ and $J^{(2)}$, we construct functions $\zeta_{(k)}(x) \in C_0^\infty$, $0 \leq \zeta_{(k)}(x) \leq 1$, with the following properties: (a) $\zeta_{(k)}(x) \equiv 1$ in $g_{(k)}$, $\zeta_{(k)}(x) \equiv 0$ outside $G_{(k)}$; (b) $\zeta_{(k)}(y) = \check{\zeta}_{(k)}(x)$ for $G_{(\check{k})}$ abutting on Γ_0.

Obviously, the functions $\eta^{(k)}(x) = \zeta_{(k)}(x) / \sum_k \zeta_{(k)}^2(x)$ vanish outside $G_{(k)}$ and $\sum_k \zeta_{(k)}(x)\eta^{(k)}(x) \equiv 1$ in $\overline{G}_0$.

Let $\{I_{(1)p}\}$, $\{I_{(2)p}\}$ be the partitions of $\overline{\mathbb{R}}_+^1$ constructed in Subsection V.1.2, and let $\sum_p \theta_p(t) \equiv 1$ be a partition of unity in $\overline{\mathbb{R}}_+^1$ such that $\theta_p(t) \in C_0^\infty$, $\theta_p(t) \equiv 1$ in $I_{(1)p}$, $\theta_p(t) \equiv 0$ outside $I_{(2)p}$, $\theta_0(0) = 1$. The families of sets $\{\Omega_{kp}^1\} = \{g_{(k)} \times I_{(1)p}\}$, $\{\Omega_{kp}^2\} = \{G_{(k)} \times I_{(2)p}\}$ form finite covers of Ω^+, and the sets Ω_{kp}^r $(r = 1, 2)$ abutting on S_0^+ are diffeomorphically mapped onto the sets $\Omega_1 \cap \Omega_{kp}^r$ that abut on S_1^+ and lie in $S_{1\varepsilon}^+$. Obviously, $\sum_{k,p} \zeta_{kp}(x,t) = \sum_{k,p} \zeta_{(k)}(x)\theta_p(t) \equiv 1$ is a partition of unity in $\overline{\Omega}^+$.

We will use the sets $\{\Omega_{kp}^r\}$ $(r = 1, 2)$ and the partition of unity in $\overline{\Omega}^+$ to construct a regularizer of the problem (2.10)–(2.14) with data compatible with zero to order s at $t = 0$. We proceed as follows. For each interior set Ω_{kp} we denote by $\mathcal{L}_{(r)}^{kp}$ the parabolic operator with slowly varying coefficients that coincides with $\mathcal{L}_{(r)}$ in Ω_{kp}^r, and we consider the Cauchy problem

$$\mathcal{L}_{(r)}^{kp}(x,t,D,D_t)u_{(r)}^{kp}(x,t) = \zeta_{kp}(x,t)f_{(r)}(x,t), \tag{2.15}$$

$$D_t^{\lambda-1}u_{(r)}^{kp}(x,0) = \zeta_{kp}(x,0)\psi_{(r)\lambda}(x), \ \lambda = 1,\ldots,t'_{rj}, \ r = 1,2, \tag{2.16}$$

where the right-hand side is compatible with zero to order s at $t = 0$.

Next, in each Ω_{kp}^2 abutting on S_0^+ we rewrite the operators $\mathcal{L}_2$ and $\mathcal{B}$ in the local coordinate system (z,t) by using the diffeomorphism $y = \alpha(x)$. As a result, we locally obtain a conjugation problem with operators $\mathcal{L}_1$, $\mathcal{L}_2$, $\check{\mathcal{L}}_2$ and $\mathcal{B}_1$, $\mathcal{B}_2$, $\check{\mathcal{B}}$, defined with respect to z in some neighborhood of the point $z = 0$. Then we extend $\mathcal{L}_1$, $\mathcal{L}_2$, $\check{\mathcal{L}}_2$, $\mathcal{B}_1$, $\mathcal{B}_2$, $\check{\mathcal{B}}$ to operators defined in the corresponding half-space and on the whole hyperplane $z_n = 0$, respectively, so that they reduce to constant operators outside Ω_{kp}^2, and the conditions of uniform parabolicity and compatible covering are preserved.

Consider the following conjugation problem, similar to problem (2.10)–(2.14):

$$\mathcal{L}_{(r)}^{kp}(z,t,D_z,D_t)u_{(r)}^{kp}(z,t)\big|_{\overline{\mathbb{R}}_{++}^{n+1}} = \zeta_{kp}(z,t)f_{(r)}^{++}(z,t),$$

$$r = 1,3; \ z_n > 0, \tag{2.17}$$

$$D_n^{\sigma-1}\mathcal{L}_{(r)}^{kp}(z,t,D_z,D_t)u_{(r)}^{kp}(z,t)\big|_{\overline{E}_+^n} = \zeta_{kp}(z',0,t)f_{(r)\sigma}^{+}(z',t),$$

$$r = 1,3 \tag{2.18}$$

$$\mathcal{L}_{(2)}^{kp}(z,t,D_z,D_t)u_{(2)}^{kp}(z,t)\big|_{\overline{\mathbb{R}}_{-+}^{n+1}} = \zeta_{kp}(z,t)f_{(2)}^{-+}(z',t), \quad z_n < 0, \tag{2.19}$$

$$D_n^{\sigma-1}\mathcal{L}_{(2)}^{kp}(z,t,D_z,D_t)u_{(2)}^{kp}(z,t)\big|_{\overline{E}_+^n} = \zeta_{kp}(z',0,t)f_{(2)\sigma}^{+}(z',t), \tag{2.20}$$

$$\mathcal{B}_1^{kp}(z,t,D_z,D_t)u_{(1)}^{kp}(z,t)\big|_{\overline{E}_+^n} + \mathcal{B}_2^{kp}(z,t,D_z,D_t)u_{(2)}^{kp}(z,t)\big|_{\overline{E}_+^n}$$

$$+\check{\mathcal{B}}^{kp}(z,t,D_z,D_t)u_{(2)}^{kp}(z,t)\big|_{\overline{E}_+^n} = \zeta_{kp}(z',0,t)\varphi(z',t), \tag{2.21}$$

$$D_t^{\lambda-1}u_{(r)}^{kp}(z,0) = \zeta_{kp}(z,0)\psi_{(r)\lambda}(z),$$

$$D_n^{k-1}D_t^{\lambda-1}u_{(r)}^{kp}(z',0) = \zeta_{kp}(z',0,0)\psi_{(r)\lambda k}(z') \tag{2.22}$$

with initial conditions compatible with zero at $t = 0$. Clearly, the initial values vanish identically for the indices p such that $\theta_p(0) = 0$.

Denote by $\mathfrak{U}_{kp}$ the operators of the problems (2.15), (2.16), and (2.17)–(2.22), which coincide with $\mathfrak{U}$ on Ω_{kp}^1, and by $\mathfrak{U}_{kp}^{-1}$ the operators that give in Ω_{kp}^2 the solutions $u^{kp}(x,t)$ of the problems (2.15), (2.16), and (2.17)–(2.22), written in the original coordinate system (x,t): $u^{kp}(x,t) = \mathfrak{U}_{kp}^{-1}(\zeta_{kp}\mathcal{F})$.

Let the functions $\eta^{kp}(x,t) \in C^\infty$ vanish outside Ω_{kp}^2 and be identically equal to 1 on the supports of $\zeta_{kp}(x,t)$. As in Subsection V.1.2, we use the equality

$u(x,t) = \sum_{k,p} \eta^{kp}(x,t)u^{kp}(x,t)$ to define an operator $\mathcal{R}$ (regularizer) for $\mathfrak{U}$. By construction, $\mathcal{R}$ acts boundedly from $\widetilde{\mathcal{K}}^s_+$ into $\widetilde{\mathcal{H}}^s_+$, $s \notin \mathbb{Z}_{1,2b}$ (for a sufficiently fine partition $\{G_{(k)}\}$). The following theorem is valid.

Theorem V.4. *Let Γ_0, $\Gamma_1 \in C^\infty$, and assume that $\mathcal{L}_{(r)}$ are uniformly parabolic operators in $\overline{\Omega}_r$. Let the operators $\mathcal{B}_1$, $\mathcal{B}_2$, $\mathcal{B}$ satisfy uniformly the nonlocally compatible covering condition with respect to $\mathcal{L}_1$, $\mathcal{L}_2$ on S_0, S_1 and let the coefficients of $\mathcal{L}_{(r)}$, $\mathcal{B}_r$, $\mathcal{B}$ be infinitely differentiable with respect to x and t and constant for $t > T_1$. Then there exists a number $\gamma_0(s)$ such that for $\gamma > \gamma_0(s)$, given any right-hand side $\mathcal{F}(x,t)$ compatible with zero at $t = 0$ to order s, that is, $\mathcal{F}(x,t) \in \widetilde{\mathcal{K}}^s_+(\Omega_+, \gamma)$, the problem (2.10)–(2.14) has a unique solution $u(x,t) \in \widetilde{\mathcal{H}}^s_+$ and the estimate $\|u(x,t)\|_{\widetilde{\mathcal{H}}^s} \leq C_s \|\mathcal{F}(x,t)\|_{\widetilde{\mathcal{K}}^s}$ holds for $s \notin \mathbb{Z}_{1,2b}$, with a constant C_s that does not depend on $\mathcal{F}(x,t)$.*

Proof: The proof of this assertion is similar to that of Theorem V.1 and reduces to establishing the relations $\mathfrak{U}\mathcal{R} = I_\mathcal{K} + \mathcal{T}$, $\mathcal{R}\mathfrak{U} = I_\mathcal{H} + \mathcal{W}$, where $I_\mathcal{K}$ and $I_\mathcal{H}$ are the identity operators in $\widetilde{\mathcal{K}}^s$ and $\widetilde{\mathcal{H}}^s$, respectively, and $\mathcal{T}$ and $\mathcal{W}$ are contractions for $\gamma > \gamma_0(s)$. In the proof, the compatibility between the partition of unity in $\overline{G}_0$ and the diffeomorphism $y = \alpha(t)$ near Γ_0, Γ_1 is used in essential manner to isolate the terms $I_\mathcal{K}$ and $I_\mathcal{H}$ in $\mathfrak{U}\mathcal{R}$ and $\mathcal{R}\mathfrak{U}$.

V.2.4. The nonlocal boundary value problem in Ω_+ in the general case

Theorem V.4 has a verbatim analogue for a problem (2.10)–(2.14) whose right-hand sides satisfy the compatibility conditions to order s at $t = 0$.

Definition V.5. *The right-hand sides of problem (2.10)–(2.14) are said to satisfy the compatibility conditions to order s at $t = 0$ if there exists an element $w(x,t) \in \widetilde{\mathcal{H}}^s$ such that $F(x,t) = \mathcal{F}(x,t) - \mathfrak{U}w$ belongs to $\widetilde{\mathcal{K}}^s_+$, that is, $F(x,t)$ is compatible with zero at $t = 0$ to order s.*

Theorem V.5. *Let Γ_0, Γ_1, $\mathcal{L}_r$, $\mathcal{B}_{(r)}$, and their coefficients satisfy the assumptions of Theorem V.4. Then there exists a number $\gamma_0(s)$ such that for any right-hand sides $\mathcal{F}(x,t)$ compatible at $t = 0$ to order s, there exists a unique solution $u(x,t) \in \widetilde{\mathcal{H}}^s$ of problem (2.10)–(2.14), and the estimate $\|u(x,t)\|_{\widetilde{\mathcal{H}}^s} \leq C_s \|\mathcal{F}(x,t)\|_{\widetilde{\mathcal{K}}^s}$ holds for $s \notin \mathbb{Z}_{1,2b}$.*

V.2.5. Parabolic conjugation problems

The class of nonlocal parabolic boundary value problems (2.10)–(2.14) considered above, as well as the problems of Bitsadze and Samarskiĭ for elliptic equations [6], contain as a particular case some problems for parabolic equations and systems frequently met in applications. We will discuss in detail two of them: (a) general local conjugation parabolic problems for equations and systems (in particular, for systems of equations with piecewise smooth coefficients, when the discontinuities of the coefficients do not reach the boundary of the domain) and (b) boundary value problems with conditions of the type of two-point boundary conditions for ordinary differential equations.

(a) Parabolic conjugation problems. Consider the problem (2.10)–(2.14) and assume that the elements of the first $b(\kappa_1 + \kappa_2)$ rows of the matrix $\mathcal{B}$ and the elements of the remaining $b\kappa_2$ rows of the matrices $\mathcal{B}_1$ and $\mathcal{B}_2$ are identically equal to zero. Then the first $b(\kappa_1 + \kappa_2)$ conditions turn into ordinary conditions of local conjugation on S_1^+, while the other $b\kappa_2$ conditions (when written in coordinates (x, t) near S_0^+) turn into ordinary boundary conditions on S_0^+. To simplify the formulation of the conjugation problem we will assume, as before, that $\mathcal{L}_1$ and $\mathcal{L}_2$ are operators parabolic in the sense of Petrovskiĭ and the orders of the operators $b_{(1)qj}$ and $b_{(2)qj}$, b_{qj} are at most $t_{1j} - 1$ and $t_{2j} - 1$, respectively, that is, $\sigma_q \le -1$ for all q. This means that no additional boundary conditions are imposed on S_0^+, S_1^+ and Γ_0, Γ_1, and we are led to the following ordinary local conjugation problem. In the domains Ω_r^+, find solutions of the system of equations

$$\mathcal{L}_r(x, t, D, D_t)u_{(r)}(x, t)\big|_{\overline{\Omega}_r^+} = f_{(r)}(x, t), \qquad r = 1, 2, \tag{2.23}$$

that satisfy the conjugation conditions on $\overline{S}_1^+$

$$b_{(1)qj}u_{(1)j}\big|_{\overline{S}_1^+} + b_{(2)qj}u_{(2)j}\big|_{\overline{S}_1^+} = \varphi_q(x', t), \qquad q = 1, \ldots, b(\kappa_1 + \kappa_2), \tag{2.24}$$

the boundary conditions on $\overline{S}_0^+$

$$b_{qj}u_{(2)j}\big|_{\overline{S}_0^+} = \Phi_q(x, t), \qquad q = 1, \ldots, b\kappa_2, \tag{2.25}$$

and the initial conditions at $t = 0$

$$D_t^{\lambda-1}u_{(r)j}(x, 0)\big|_{\overline{G}_r} = \psi_{(r)j\lambda}(x),$$
$$\lambda = 1, \ldots, t'_{rj}, \; j = 1, \ldots, m_r, \; r = 1, 2, \tag{2.26}$$
$$D_n^{k-1}D_t^{\lambda-1}u_{(r)j}(x, t)\big|_{\Gamma_r} = \psi_{(r)j\lambda k}(x'),$$
$$(\lambda, k) \in P_{(r)j}, \; j = 1, \ldots, m_r, \; r = 1, 2. \tag{2.27}$$

Repetition of j in (2.23)–(2.27) indicates summation. In the present case the nonlocal compatible covering condition is equivalent to the operators $\mathcal{B}_1$ and $\mathcal{B}_2$ satisfying the compatible covering condition with respect to the operators $\mathcal{L}_1$ and $\mathcal{L}_2$ on S_1^+, and the operators $\mathcal{B}$ and $\mathcal{L}_2$ being connected on S_0^+ by the Lopatinskiĭ condition. Theorem V.5 admits the following corollary.

Theorem V.6. *Let* Γ_0, Γ_1, *and the coefficients of the operators* $\mathcal{L}_r$, $\mathcal{B}_r$, $\mathcal{B}$ *satisfy the assumptions of Theorem V.4, let the operators* $\mathcal{B}_1$ *and* $\mathcal{B}_2$ *satisfy the compatible covering condition with respect to the operators* $\mathcal{L}_1$ *and* $\mathcal{L}_2$ *uniformly in* (x', t) *on* $\overline{S}_1^+$, *and let the operators* $\mathcal{B}$ *and* $\mathcal{L}_2$ *be connected on* $\overline{S}_0^+$ *by the Lopatinskiĭ condition uniformly in* $(x', t) \in \overline{S}_0^+$. *Then for any right-hand sides* $\mathcal{F}(x, t)$

compatible at $t = 0$ to order s there exists a unique solution of the problem (2.23)–(2.27), $u(x,t) = \big(u_{(1)}(x,t), u_{(2)}(x,t)\big) \in \widetilde{\mathcal{H}}^s(\Omega_+, \gamma)$, $\gamma > \gamma_0(s)$ and the estimate $\|u(x,t)\|_{\widetilde{\mathcal{H}}^s} \le C_s \|\mathcal{F}(x,t)\|_{\widetilde{\mathcal{K}}^s}$ holds for $s \notin \mathbb{Z}_{1,2b}$, with a constant C_s that does not depend on $\mathcal{F}(x,t)$.

If G_2 is the exterior of a bounded domain G_1, then we are led to a Cauchy problem with conjugation conditions on S_1^+, for which an appropriate analogue of Theorem V.6 is valid.

(b) Boundary value problems with two-point type boundary conditions. Let in the problem (2.23)–(2.27) $\mathcal{L}_1 \equiv 0$, $\mathcal{B}_1 \equiv 0$. Assume that Γ_1 and Γ_0 are diffeomorphic, and moreover the ε-neighborhood of Γ_0 is diffeomorphic to the ε-neighborhood of Γ_1, which lies in G_2. The problem is then to find a solution $u_{(2)}(x,t)$ of the system $\mathcal{L}_2 u_{(2)} = f_{(2)}$ in $\overline{\Omega}_2^+$, satisfying the initial conditions (2.26) at $t = 0$, and the nonlocal conjugation conditions

$$\mathcal{B}_2 u_{(2)}\big|_{\overline{S}_1^+} + \check{\mathcal{B}}\check{u}_{(2)}\big|_{\overline{S}_1^+} = \varphi(x',t), \tag{2.28}$$

which connects the values of $u_{(2)}(x,t)$ and of its derivatives on S_1^+ and S_0^+. Such problems arise naturally, for example, in the theory of heat conduction in a body whose boundary consists of a number of connected $(n-1)$-dimensional components in the case when, instead of ordinary boundary conditions on each component of the boundary (for example, on the inner and the outer sides of a plate, or the left and the right ends of a rod), one poses conditions that connect the values of the desired solutions on different components of the boundary. Problems of this kind for the heat equation on a segment in classes of smooth solutions were investigated in [49, 50].

V.3. Boundary value problems in cylindrical domains of finite height

Let $\Omega = \overline{G} \times [0,T)$, $S = \Gamma \times [0,T)$, Ω_+ and S_+ be defined as in Section V.1, $0 < T < \infty$. For any $s \in \mathbb{R}^1$ denote by $\mathcal{H}^s(\Omega_+) = \mathcal{H}^s(\Omega_+, 0)$, $\mathcal{H}^s(S) = \mathcal{H}^s(S_+, 0)$ the ordinary anisotropic spaces in Ω_+ and on S_+ with the norms $\|u, \Omega_+\|_s$ and $\langle\!\langle \Phi, S \rangle\!\rangle_s$, respectively. Also, denote by $\mathcal{H}^s(\Omega)$ and $\mathcal{H}^s(S)$ the sets of the restrictions $u(x,t)$ and $\varphi(x',t)$ of the distributions $U(x,t) \in \mathcal{H}^s(\Omega_+)$ and $\Phi(x',t) \in \mathcal{H}^s(S_+)$ to Ω and S, respectively, equipped with the quotient norms $\|u(x,t), \Omega\|_s = \inf_U \|U, \Omega_+\|_s$ and $\langle\!\langle \varphi(x,t), S \rangle\!\rangle_s = \inf_\Phi \langle\!\langle \Phi, S_+ \rangle\!\rangle_s$. Given numbers κ, τ, and a set P of pairs (k, λ), define the space $\widetilde{\mathcal{H}}^s_{(\kappa,\tau,P)}(\Omega)$ exactly as it was done for the space $\widetilde{\mathcal{H}}^s_{(\kappa,\tau,P)}(\Omega_+)$, defining the norm $\|u(x,t), \Omega\|_s$ in $\widetilde{\mathcal{H}}^s_{...}(\Omega)$ by means of the the norms $\|u, \Omega\|_s$ and $\langle\!\langle u_k, S \rangle\!\rangle_{s-k+1/2}$.

Assume that in $\overline{\Omega}_+$ and on $\overline{S}_+$ there are defined the operators $\mathcal{L}$ and $\mathcal{B}$ described in Section V.1, and that they satisfy all the conditions imposed therein.

Let $\tau_j = t_j + \max_q(0, \sigma_q + 1)$. Consider in Ω the following boundary value problem: Find $u(x,t) = \big(u_1(x,t), \dots, u_m(x,t)\big) \in \widetilde{\mathcal{H}}^s(\Omega) = \prod \mathcal{H}^{s+t_j}_{(t'_j, \tau_j, P_j)}(\Omega)$ such that $u(x,t)$ satisfies in Ω the system of equations

$$l_{ij}(x,t,D,D_t)u_j(x,t) = f_{i0}(x,t) \in \mathcal{H}^s(\Omega), \qquad i = 1, \dots, m, \tag{3.1}$$

and satisfies the boundary conditions on S

$$b_{qj}(x,t,D,D_t)u_j(x,t)\big|_S = \varphi_q(x',t) \in \mathcal{H}^{s-\sigma_q-1/2}(S), \quad q = 1, \dots, b_k, \tag{3.2}$$

$$D_\nu^{\sigma-1}l_{ij}(x,t,D,D_t)u_j(x,t)\big|_S = f_{i\sigma}(x',t) \in \mathcal{H}^{s-\sigma+1/2}(S),$$

$$\sigma = 1, \dots, \sigma_0, \ j = 1, \dots, m, \tag{3.3}$$

and the initial conditions at $t = 0$

$$D_t^{\lambda-1}u_j(x,0) = \Psi_{j\lambda}(x) \in \widetilde{H}^{s+t_j-2b\lambda+b}_{(P_{j\lambda})}(G), \quad \lambda = 1, \dots, t'_j, \ j = 1, \dots, m. \tag{3.4}$$

From the discussion in Section V.1 it follows that the operator $\mathfrak{U}$ of problem (3.1)–(3.4) acts boundedly from the space $\widetilde{\mathcal{H}}^s(\Omega)$ into the subspace $\widetilde{\mathcal{K}}^s(\Omega)$ of the right-hand sides of (3.1)–(3.4) that are compatible to order s; here the condition of compatibility to order s is formulated exactly as in Section V.1. Hence, Theorem V.3 admits the following immediate consequence.

Theorem V.7. *Let $\mathcal{L}$ be a uniformly parabolic operator in $\overline{\Omega}$, let the operator $\mathcal{B}$ satisfy the Lopatinskiĭ condition on $\overline{S}$ uniformly in $(x',t) \in \overline{S}$, and let the coefficients of these operators belong in $\overline{\Omega}$ and on $\overline{S}$ to the classes indicated in Theorem V.1. Then for any right-hand side $F(x,t)$ compatible at $t = 0$ to order s there exists a unique solution $u(x,t)$ of the problem (3.1)–(3.4), which obeys the a priori estimate*

$$\big\|u(x,t)\big\|_{\widetilde{\mathcal{H}}^s(\Omega)} < C_s \big\|F(x,t)\big\|_{\widetilde{\mathcal{K}}^s(\Omega)}, \qquad s \notin \mathbb{Z}_{1,2b},$$

where the constant C_s does not depend on $u(x,t)$ and $F(x,t)$.

For the sake of simplicity, here we consider systems parabolic in the sense of Petrovskiĭ. The boundary value problem for general parabolic systems (in Solonnikov's sense) was considered in [34]. There the reader can find the rigorous formulation of this problem, the definitions of the spaces $\widetilde{\mathcal{H}}^s(\Omega)$ and $\widetilde{\mathcal{K}}^s(\Omega)$, the exact formulation of the theorem, and a scheme of its proof.

V.4. Solvability of the parabolic boundary value problem for right-hand sides with regular singularities

As one of the various applications of Theorem V.7, we will next study the solvability of a parabolic boundary value problem whose right-hand sides that have regular (power-like) singularities on some manifolds. For simplicity we consider the case of a single equation, whose right-hand side has a regular singularity on some manifold γ inside Ω.

Let $\mathcal{L}(x, t, D, D_t)$ be a parabolic operator in $\overline{\Omega}$ of order $2m = 2br$ and let $\mathcal{B}_j$ be operators of weighted orders $m_j < 2m$, connected on S by the Lopatinskiĭ condition. Assume, for simplicity, that the coefficients of $\mathcal{L}$ and $\mathcal{B}_j$ are sufficiently smooth, and let P be the set of the pairs (k, λ) such that $\mathcal{L}^{(k,\lambda)} \not\equiv 0$ or $b_{jk}^{(\lambda)} \not\equiv 0$ on γ, $j = 1, \ldots, m$. Finally, for each fixed λ let P_λ denote the set of all k such that $(k, \lambda) \in P$.

Consider the problem of finding a solution $u(x, t)$ of the equation

$$\mathcal{L}u = f(x, t), \tag{4.1}$$

satisfying the boundary conditions

$$\mathcal{B}_j u\big|_S = \varphi_j(x', t), \quad j = 1, \ldots, m, \tag{4.2}$$

and the initial conditions

$$D_t^{\lambda-1} u(x, t)\big|_{t=0} = \Psi_\lambda(x) \in \widetilde{H}_{(P_\lambda)}^{s-2b\lambda+b}(G), \quad \lambda = 1, \ldots, r, \tag{4.3}$$

where (4.3) is understood as an equality between components of elements of the space $\widetilde{H}_{(P_\lambda)}^{s-2b\lambda+b}(G)$.

V.4.1. Anisotropic regularizations of divergent integrals

We begin our analysis of problem (4.1)–(4.3) by constructing anisotropic regularizations of divergent integrals with regular singularities. To carry out this construction and define the spaces to which these regularizations belong, we make essential use of an estimate of the "weighted" remainder of the Taylor formula for a certain class of Hölder functions $\varphi(x, t)$.

Specifically, let α, β, and s be nonnegative numbers, $C_x^{s/\beta\,s/\alpha}(\Omega)$ the space of the functions $\varphi(x, t)$ that have derivatives $D_t^{k_0} D_x^k \varphi(x, t)$, where $\alpha k_0 + \beta|k| \leq [s]$, such that $D_t^{[s/\beta]} \varphi$ are Hölder functions of order $\{s/\beta\}$ with respect to t and $D_t^{k_0} D_x^k \varphi$ with $\alpha k_0 + \beta|k| \geq [s] - \beta$ are Hölder functions of order $\{(s - \alpha k_0)/\beta\}$ with respect to x; here $[s]$ and $\{s\}$ denote the integer and the fractional part of s, respectively. For $\varphi(x, t) \in C_x^{s/\beta\,s/\alpha}$ we consider a weighted segment of the Taylor series and form the difference

$$\Phi(x, t) = \varphi(x, t) - \sum_{\alpha k_0 + \beta|k| \leq [s]} D_t^{k_0} D_x^k \varphi(0, 0) \frac{x^k t^{k_0}}{k!\, k_0!}.$$

Lemma V.1. *Let* $\varphi(x, t) \in C_x^{s/\beta\,s/\alpha}(\Omega_0)$, *where* Ω_0 *is a bounded domain that contains the point* $(0, 0)$. *Then*

$$\left|\Phi(x, t)\right| < C\big(|x|^\alpha + |t|^\beta\big)^{s/\alpha\beta}, \tag{4.4}$$

where C *is a constant that depends only on the norm of* $[\varphi, \Omega_0]_s$ *in* $C_x^{s/\beta\,s/\alpha}(\Omega_0)$.

A proof of this lemma is given in [32].

Assume, for simplicity, that $\gamma \subset \Omega$ $(\gamma \cap S = \emptyset)$ is a smooth $(n - q)$-dimensional manifold lying in the hyperplane $t = t_0 > 0$. This means that at each point $\mathcal{M} \in \gamma$ we can define a local coordinate system, in which the first $n - q$ axes are directed along γ and the remaining axes lie in the normal plane to γ at $\mathcal{M}$. Denote the first $n - q$ coordinates by $z'' = z_1, \ldots, z_{n-q}$, and the remaining ones by $(y', t) = (y_1, \ldots, y_q, t)$. Finally, let $f(x, t)$ be a function defined in $\Omega \setminus \gamma$ such that f degenerates nonsmoothly on γ in the sense that $f(x, t) \equiv f(z'', y', t) = h(x, t)\big(|y'|^{2b} + |t|\big)^{-\lambda}$, where $h(x, t) \in L_2(\Omega)$. It can be easily verified that if $2b\lambda < \frac{q}{2} + b$, then, by the Cauchy-Bunyakovskiĭ inequality, $\|f, \Omega\|_0 \leq C\|h, \Omega\|_0$. Consequently, if $f(x, t)$ has a weak singularity on γ, then $f(x, t) \in \mathcal{H}^0(\Omega)$.

Let us consider now the case $2b\lambda \geq \frac{q}{2} + b$ and construct a suitable regularization of the function $f(x, t)$. Denote by Ω_d the domain defined in the local coordinates (z'', y', t) by the inequality $|y'|^{2b} + |t| < d$ and for $\varphi(z'', y', t) = \varphi(x, t) \in C_{x\,t}^{s\,s/2b}$ set

$$\Phi_{y',t}^s(z'', y', t) = \varphi(z'', y', t) - \sum_{2bk_0 + |k| \leq [s]} \varphi^{(k_0, k)}(z'', 0, 0) \frac{y'^k t^{k_0}}{k!\, k_0!},$$

where $k = (k_1, \ldots, k_q)$, $|k| = k_1 + \ldots + k_q$. Define the regularization $F(x, t)$ of the function $f(x, t)$ by the formula

$$(F, \varphi) = \int_{\Omega \setminus \Omega_d} f(x, t)\overline{\varphi}(x, t)\, dx dt + \int_{\Omega_d} f(x, t)\overline{\Phi_{y',t}^s(z'', y', t)}\, dx dt, \qquad (4.5)$$

where $dx = dz'' dy$. Let us evaluate the norm of the regularized functional $F(x, t)$ in the space $\mathcal{H}^s(\Omega)$.

Lemma V.2. Let $2b\lambda \geq \frac{q}{2} + b$. Then the regularization $F(x, t)$ of the function $f(x, t)$, given by formula (4.5) with $s = 2b\lambda - \frac{q}{2} - b + \varepsilon$, belongs to $\mathcal{H}^{-2b\lambda-\varepsilon}(\Omega)$ and obeys the estimate $\|F(x, t), \Omega\|_{-2b\lambda-\varepsilon} < C\|h, \Omega\|_0$ where the constant C does not depend on $f(x, t)$ and $\varepsilon > 0$ is an arbitrary number.

A proof of this lemma can be found in [32].

V.4.2. The main solvability theorem

Lemma V.2 and Theorem V.7 yield the following result.

Theorem V.8. Assume that (4.1)–(4.9) is a parabolic problem and that $f(x, t) = h(x, t)\big(|y'|^{2b} + |t - t_0|\big)^{-\mu}$, where $h(x, t) \in L_2(\Omega)$, $2b\mu \geq \frac{q}{2} + b$. Let $F_\mu(x, t)$ be a fixed regularization of $f(x, t)$ with $s = 2b\mu - \frac{q}{2} - b + \varepsilon$ (where $\varepsilon > 0$ is an arbitrary number). Also, assume that $\varphi_j(x', t) \in \mathcal{H}^{2m-2b\mu-m_j-1/2+\varepsilon}(S)$, $\Psi_\lambda(x) \in \widetilde{H}^{2m-2b\lambda-2b\mu+b+\varepsilon}(G)$ and that the compatibility condition holds at $t = 0$. Then *the problem $\mathcal{L}u = F_\mu(x,t)$, $\mathcal{B}_j u\big|_S = \varphi_j(x',t)$, $j = 1,\ldots,m$, $D_t^{\lambda-1}u(x,0) = \Psi_\lambda(x)$, $\lambda = 1,\ldots,r$, is solvable and $u(x,t) \in \widetilde{\mathcal{H}}_{(r,2m,P)}^{2m-2b\mu-\varepsilon}(\Omega)$.*

Note that since the regularization of the function $f(x,t)$ is not defined uniquely, the solution $u(x,t)$ determined by the regularized right-hand side $F(x,t)$ is also not unique. Obviously, any two regularizations of $f(x,t)$ coincide outside γ and, moreover, belong to $\mathcal{H}^{-2b\mu-\varepsilon}(\Omega)$. If $F_0(x,t)$ is a fixed regularization and $F(x,t)$ is some other regularization, then $F(x,t) - F_0(x,t) = \mathcal{F}(x,t)$, where $\mathcal{F}(x,t) \in \mathcal{H}^{-2b\mu-\varepsilon}(\Omega)$ and $\operatorname{supp}\mathcal{F} \subseteq \gamma$. This observation in conjunction with Theorem V.7 implies that the solution $u(x,t)$ of problem (4.1)–(4.3) has the form $u(x,t) = u_0(x,t) + u'(x,t)$, where $u_0(x,t)$ is a solution of problem (4.1)–(4.3) and $u'(x,t)$ is a solution of problem (4.1)–(4.3) with $\varphi_j \equiv \psi_\lambda \equiv 0$, $F_\mu = \mathcal{F}$. To determine a unique solution of the problem (4.1)–(4.3) we need to impose additional conditions on γ.

We have considered the case when $f(x,t)$ is degenerate on a manifold γ that lies in a plane $t = t_0$. In a similar way one can analyze the situations when $f(x,t)$ is degenerate on other types of manifolds that lie inside Ω as well as the situations when $\varphi_j(x',t)$ or $\Psi_\lambda(x)$ have regular (power-like) singularities on the corresponding manifolds.

V.5. Green formula, boundary and initial values of weak generalized solutions

The existence of boundary values for solutions of parabolic equations and systems in various functional spaces was investigated by many mathematicians; see, for example, [17, 18, 33, 45, 66, 77, 78, 120, 126] and references therein, and also Section VI.2 of the next chapter.

In this section we briefly present, with no proofs, results of studies on the existence of weak boundary values and of limit values of weak generalized solutions of parabolic equations of arbitrary order, obtained by means of Theorem V.7. For simplicity, we consider the case of a single equation; analogous results hold for systems for which the parabolic boundary value problem can be posed.

V.5.1. Preliminary considerations. Notation

Let G be a bounded domain of $\mathbb{R}^n$ with a smooth boundary $\partial G = \Gamma$, $\Omega = G \times (0,T)$, $\widehat{\Omega} = \overline{G} \times [0,T]$, $S = \Gamma \times (0,T)$, $\widehat{S} = \Gamma \times [0,T]$. For $\rho \geq 0$, $0 \leq \tau < T$, we set $G_\rho = \{x \in G, d(x,\Gamma) > \rho\}$, $\Gamma_\rho = \partial G_\rho$, $\Omega_\rho^\tau = G_\rho \times (\tau,T)$, $\widehat{\Omega}_\rho^\tau = \overline{G}_\rho \times [\tau,T)$, $S_\rho^\tau = \Gamma_\rho \times (\tau,T)$, $\widehat{S}_\rho^\tau = \Gamma_\rho \times [\tau,T)$; here $d(x,\Gamma_\rho)$ denotes the distance from x to Γ_ρ. Finally, let $\overset{\circ}{C}{}^\infty(\Omega)$ and $\overset{\circ}{C}{}^\infty(S)$ denote the sets of functions of the indicated class that vanish for $t \geq T$. To describe the smoothness of distributions in open and closed domains, we need to define appropriate spaces of distributions, for which the notations are somewhat different from those used above.

For any $s \geq 0$ we define $H_0^s(G)$, $\mathcal{H}_0^s(\Omega)$, $\mathcal{H}_0^s(S)$, $\overset{\circ}{\mathcal{H}}{}^s(\Omega)$, $\overset{\circ}{\mathcal{H}}{}^s(S)$ as the closures of the sets $C_0^\infty(G)$, $C_0^\infty(\Omega)$, $C_0^\infty(S)$, $\overset{\circ}{C}{}^s(\overline{\Omega})$, $\overset{\circ}{C}{}^s(\overline{S})$ with respect to the norms $H^s(\overline{G})$, $\mathcal{H}^s(\overline{\Omega})$, and $\mathcal{H}^s(\overline{S})$, respectively. Here $H^s(\overline{G}) = H^s(G)$, $\mathcal{H}^s(\overline{\Omega}) = \mathcal{H}^s(\Omega)$, $\mathcal{H}^s(\overline{S}) = \mathcal{H}^s(S)$ are the spaces defined and used earlier. For $-s < 0$ we define $H^{-s}(G) = \left(H_0^s(G)\right)^*$, $H^{-s}(\overline{G}) = \left(H^s(\overline{G})\right)^*$, $\mathcal{H}^{-s}(\Omega) = \left(\mathcal{H}_0^s(\Omega)\right)^*$, $\mathcal{H}^{-s}(\widehat{\Omega}) = \left(\overset{\circ}{\mathcal{H}}{}^s(\Omega)\right)^*$, $\mathcal{H}^{-s}(S) = \left(\mathcal{H}_0^s(S)\right)^*$, $\mathcal{H}^{-s}(\widehat{S}) = \left(\overset{\circ}{\mathcal{H}}{}^s(\overline{S})\right)^*$ as the dual spaces with respect to the extensions of the corresponding inner products in L_2, equipped with the usual norms of dual spaces. Note that, with such a definition, the space $\mathcal{H}^{-s}(\overline{G})$ $(s > \frac{1}{2})$ is isometrically isomorphic to the subspace $\mathcal{H}_{\overline{G}}^{-s}(\mathbb{R}^n)$ (consisting of the distributions supported in $\overline{G}$), and the space $H^{-s}(G)$ is isometrically isomorphic to the quotient space $\mathcal{H}^{-s}(\overline{G})/M_{-s}$, where M_{-s} is the subspace of $\mathcal{H}^{-s}(\overline{G})$ consisting of the elements supported on Γ. Similarly, $\mathcal{H}^{-s}(\Omega)$, [resp., $\mathcal{H}^{-s}(S)$] $(s > \frac{1}{2})$ is isometrically isomorphic to the quotient space $\mathcal{H}^{-s}(\widehat{\Omega})/\mathcal{Q}_{-s}$, [resp., $\mathcal{H}^{-s}(\widehat{S})/\mathcal{P}_{-s}$], where $\mathcal{Q}_{-s}$ [resp., $\mathcal{P}_{-s}$] is the subspace of $\mathcal{H}^{-s}(\widehat{\Omega})$ [resp., $\mathcal{H}^{-s}(\widehat{S})$] consisting of distributions supported on $\overline{G} \cup S$ [resp., Γ], and where $\mathcal{H}^{-s}(\widehat{\Omega})$ [resp., $\mathcal{H}^{-s}(\widehat{S})$] coincides with the set of the restrictions of distributions $u(x,t) \in \mathcal{H}_{\overline{\Omega}}^{-s}(\mathbb{R}^{n+1})$ [resp., $v(x',t) \in \mathcal{H}_{\overline{S}}^{-s}(\Gamma \times \mathbb{R}^1)$] to $\widehat{\Omega}$ [resp., $\widehat{S}$], the quotient spaces being equipped with the usual quotient norms.

V.5.2. The main theorem on boundary and initial values

Let $\mathcal{L}$ be a parabolic operator with sufficiently smooth coefficients defined in $\overline{\Omega}$ (ord $\mathcal{L} = 2m = 2br$). Write $\mathcal{L}$ near S as $\mathcal{L}(x,t,D,D_t) \equiv \mathcal{L}(\cdot, D_\nu, D_t)$ and let $\mathcal{L}^{(k,\lambda)}(\cdot, D_\nu, D_t)$ be the operators introduced above, and $\mathcal{L}^{(k,\lambda)*}$ be the formally adjoint operators (in the sense of Lagrange).

For any $s \in \mathbb{R}^1$ set $s' = [s + 1/2]$, $s'' = [(s + b)/2b]$, if s, $s/2b$ are not half-integers, and $s' = [s + 1/2] - 1$, $s'' = [(s + b)/2b] - 1$ for half-integer s, $s/2b$ (here $[\cdot]$ denotes the integer part). Let P be the set of the pairs (k, λ) for which $\mathcal{L}^{(k,\lambda)} \not\equiv 0$ on Γ, and $P(s)$ the set of the pairs $(k, \lambda) \in P$ such that $k + 2b\lambda < s' + 2m + b$. Then in the adopted notation the following Green formula holds for any $w(x,t) \in \overset{\circ}{C}{}^\infty(\overline{\Omega})$:

$$(u_0, \mathcal{L}^* w) - \langle u_k, \mathcal{L}^{(k,0)*} w \rangle - [v_\lambda, \mathcal{L}^{(0,\lambda)*} w] + \langle \omega_{\lambda k}, \mathcal{L}^{(k,\lambda)*} w \rangle' = (f_0, w), \qquad (5.1)$$

where $f_0(x,t) = (\mathcal{L}u)_0$, $k = 1, \ldots, 2m$, $\lambda = 1, \ldots, r$, $(k, \lambda) \in P$, and repetition of k, λ, and (k, λ) indicates summation over all their values. In particular, $(u_0, \mathcal{L}^* w) = (f_0, w)$ for any $w \in C_0^\infty(\Omega)$, that is, $u_0(x,t) = u(x,t)\big|_{\widehat{\Omega}}$ is a solution of the equation

$$\mathcal{L}u_0 = f_0(x,t) \quad \text{in } \mathcal{D}'(\Omega). \qquad (5.2)$$

The distribution $u_0(x,t) \in \mathcal{D}'(\Omega)$ will be called a *weak generalized solution* of equation (5.2) in Ω.

Equality (5.1) can be extended by continuity to the whole space $\widetilde{\mathcal{H}}_{(r,2m,P)}^{s+2m}$, and an element $u \in \widetilde{\mathcal{H}}^{s+2m}$ that satisfies (5.1) will be referred to as a *strong generalized solution* of equation (5.2) in $\widehat{\Omega}$. Generally speaking, the components $u_k, v_\lambda, \omega_{\lambda k}$ are not the limit values of the derivatives of $u_0(x,t)$ with respect to ν and t on S, G, and Γ, respectively, but only play the role of these derivatives; let us agree to call them the *weak "boundary values"* of $u_0(x,t)$ on S, G, and Γ, respectively. A natural problem arises: given a weak solution $u_0(x,t)$ in Ω and the right-hand side $f_0(x,t)$ of equation (5.2), how to "restore" in $\widehat{\Omega}$ the strong generalized solution $u(x,t) \in \widetilde{\mathcal{H}}^{s+2m}_{(r,2m,P)}(\widehat{\Omega})$? The following assertion holds true.

Theorem V.9. *Let $u^\tau_{\rho 0}(x,t) \in \mathcal{H}^{s+2m}(\widehat{\Omega}^\tau_\rho)$ be a weak generalized solution of the equation $\mathcal{L}u^\tau_{\rho 0} = f^\tau_{\rho 0}(x,t) \in \mathcal{H}^s(\widehat{\Omega}^\tau_\rho)$. Then there exists a vector $U^\tau_\rho(x,t) = \big(u_{\rho 1}, \ldots, u_{\rho 2m},\ v_{\tau 1}, \ldots, v_{\tau r},\ \omega^\tau_{\rho\lambda k},\ (k,\lambda) \in P(s),\ \tilde{u}_{\rho 0}(x',t), \ldots, \tilde{u}_{\rho,-s'-2m-1}(x',t);$ $-s'-2m \geq 1;\ \tilde{v}_{\tau 0}(x), \ldots, \tilde{v}_{\tau,-s''-r-1}(x);\ -s''-r \geq 1)$ $(\tilde{u}_{\rho q} \equiv 0$ for $-s'-2m < 1$, $\tilde{v}_{\tau\mu} \equiv 0$ for $-s''-r < 1)$ such that the equality (Green formula)*

$$\big(u^\tau_{\rho 0}, \mathcal{L}^*w\big) - \big\langle u_{\rho k}, \mathcal{L}^{(k,0)*}w\big\rangle - \big[v_{\tau\lambda}, \mathcal{L}^{(0,\lambda)*}w\big] + \big\langle \omega^\tau_{\rho\lambda k}, \mathcal{L}^{(k,\lambda)*}w\big\rangle'$$
$$+ \big\langle \tilde{u}_{\rho q}, \overline{D}^q_\nu(\mathcal{L}^*w)\big\rangle + \big[\tilde{v}_{\tau\mu}, \overline{D}^\mu_t(\mathcal{L}^*w)\big] = \big(f^\tau_{\rho 0}, w\big) \quad (5.3)$$
$$\big(\overline{D} = -D,\ \overline{D}_t = -D_t,\ k = 1, \ldots, 2m,\ \lambda = 1, \ldots, r,\ (k,\lambda) \in P(s)\big)$$

holds for all $w \in \overset{\circ}{C}{}^\infty(\Omega^\tau_\rho)$. Moreover, there exists a constant C_s such that

$$\big\langle\!\big\langle u_{\rho k}, S^\tau_\rho \big\rangle\!\big\rangle_{s+2m-k+\frac{1}{2}} + \big\|[v_{\tau\lambda}, G_\rho]\big\|_{s+2m-2b\lambda+b} + \big\langle\!\big\langle \omega^\tau_{\rho\lambda k}, \Gamma^\tau_\rho \big\rangle\!\big\rangle'_{s+2m-2b\lambda-k+b+\frac{1}{2}}$$
$$+ \big\langle\!\big\langle \tilde{u}_{\rho q}, S^\tau_\rho \big\rangle\!\big\rangle_{s+2m+q+\frac{1}{2}} + \big\|[\tilde{v}_{\tau\mu}, G_\rho]\big\|_{s+2m+2b\mu+b}$$
$$< C_s\Big(\big\|u^\tau_{\rho 0}, \Omega^\tau_\rho\big\|_{s+2m} + \big\|f^\tau_{\rho 0}, \widehat{\Omega}^\tau_\rho\big\|_s\Big), \qquad (k,\lambda) \in P(s). \qquad (5.4)$$

The vector $U^\tau_\rho(x,t)$ is uniquely defined up to a term $U^{\tau''}_\rho(x,t)$ supported on Γ^τ_ρ, whose components satisfy the homogeneous equation (5.3) $(f^\tau_{\rho 0} \equiv 0,\ u^\tau_{\rho 0} \equiv 0)$.

A proof of Theorem V.9 for systems parabolic in the sense of Petrovskiĭ is given in [33]. Note that [33] contains an error: in the left-hand side of (5.4) the norms should be taken over S^τ_ρ and G_ρ rather than $\widehat{S}^\tau_\rho$ and $\overline{G}_\rho$.

Let us rewrite the inner products that involve $\tilde{u}_{\rho q}(x',t)$ and $\tilde{v}_{\tau\mu}(x)$ as inner products $(\cdot,\cdot)$ over Ω^τ_ρ by means δ-functions, denoting $u'_\rho(x,t) = D^q_\nu\big(\tilde{u}_{\rho q} \times \delta(S_\rho)\big)$, $v'_\tau(x,t) = D^\mu_t\big(\tilde{v}_{\tau\mu} \times \delta(t)\big)$. Then (5.3) takes the form

$$\big(u^\tau_{\rho 0} + u^{\tau'}_\rho + v'_\tau, \mathcal{L}^*w\big) - \big\langle u_{\rho k}, \mathcal{L}^{(k,0)*}w\big\rangle - \big[v_{\tau\lambda}, \mathcal{L}^{(0,\lambda)*}w\big]$$
$$+ \big\langle \omega^\tau_{\rho\lambda k} \mathcal{L}^{(k,\lambda)*}w\big\rangle' = \big(f^\tau_{\rho 0}, w\big), \qquad for\ all\ w \in \overset{\circ}{C}{}^\infty(\widehat{\Omega}^\tau_\rho), \qquad (5.5)$$

where $u'_\rho(x,t), v'_\tau(x,t)$ are distributions supported on $\overline{S}^\tau_\rho$ and $\overline{G}_\rho$, respectively, and consequently $v'_\tau(x,t) \equiv 0$ for $s \geq -2m-b$, and for $s \geq -2m-\frac{1}{2}$ also $u'_\rho(x,t) \equiv 0$, too. As shown in [33], for suitable s summation with respect to $(k, \lambda) \in P(s)$ can be replaced by summation with respect to $(k, \lambda) \in P$; as a result, $u_{\rho k}$ and $v_{\tau \lambda}$ are replaced by terms supported on Γ_ρ. Further, if we add to $u_{\rho 0}^\tau(x, t)$ the distribution $u_\rho^{\tau'}(x, t) \in \mathcal{H}^{s+2m}(\widehat{\Omega}_\rho^\tau)$, supported on $\sigma_\rho^\tau = \overline{G}_\rho \cup \overline{S}_\rho^\tau$, then in (5.5) $u_\rho'(x, t)$, $v_\tau'(x, t)$ change, and $u_{\rho k}(x', t)$ and $v_{\tau \lambda}(x)$ change by terms supported on Γ_ρ^τ. Taking the infimum over all $u_\rho^{\tau'}(x, t)$ supported on σ_ρ^τ, and all the $u'_{\rho k}(x', t)$ and all $v_{\tau \lambda}(x)$ supported on Γ_ρ^τ, we obtain the following consequence of Theorem V.9.

Theorem V.10. *Let* $u_{\rho 0}^\tau(x, t) \in \mathcal{H}^{s+2m}(\Omega_\rho^\tau)$ *be a weak generalized solution of the equation* (5.2) *in* Ω_ρ^τ. *Then there exists a vector* $U_\rho^\tau(x, t) = \left(u_{\rho 1}, \ldots, u_{\rho 2m}, v_{\tau 1}, \ldots, v_{\tau r}, \omega_{\rho \lambda k}^\tau, (k, \lambda) \in P(s)\right)$, *such that the Green formula*

$$\left(u_{\rho 0}^\tau, \mathcal{L}^* w\right) - \left\langle u_{\rho k}, \mathcal{L}^{(k,0)*} w\right\rangle - \left[v_{\tau \lambda}, \mathcal{L}^{(0,\lambda)*} w\right] + \left\langle \omega_{\rho \lambda k}^\tau \mathcal{L}^{(k,\lambda)*} w\right\rangle' = \left(f_{\rho 0}^\tau, w\right) \quad (5.6)$$

holds for any $w \in \overset{\circ}{C}{}^\infty(\widehat{\Omega}_\rho^\tau)$ *for which*

$$D_\nu^q(\mathcal{L}^* w)\big|_{\widehat{S}_\rho^\tau} = 0, \quad q = 0, 1, \ldots, -s' - 2m - 1,$$

$$D_t^\mu(\mathcal{L}^* w)\big|_{\overline{G}_\rho} = 0, \quad \mu = 0, 1, \ldots, -s'' - r - 1,$$

$$D_\nu^q(\mathcal{L}^{(0,\lambda)*} w)\big|_{\Gamma_\rho^\tau} = 0, \quad q = 0, 1, \ldots,$$

$$D_t^\mu(\mathcal{L}^{(k,0)*} w)\big|_{\Gamma_\rho^\tau} = 0, \quad \mu = 0, 1, \ldots.$$

Moreover, there holds the estimate

$$\left\langle\!\left\langle u_{\rho k}, S_\rho^\tau \right\rangle\!\right\rangle_{s+2m-k+1/2} + \left|\!\left[v_{\tau \lambda}, G_\rho \right]\!\right|_{s+2m-2b\lambda+b}$$

$$+ \left\langle\!\left\langle \omega_{\rho \lambda k}^\tau, \Gamma_\rho^\tau \right\rangle\!\right\rangle'_{s+2m-2b\lambda-k+b+1/2} < C_s \left(\left\|u_{\rho 0}^\tau, \Omega_\rho^\tau\right\|_{s+2m} + \left\|f_{\rho 0}^\tau, \widehat{\Omega}_\rho^\tau\right\|_s\right),$$

where the constant C_s *does not depend on* $u_{\rho 0}^\tau$, $f_{\rho 0}^\tau$ *and* $\tau \in [0, \tau_0]$, $\rho \in [0, \rho_0]$.

V.5.3. Limit values of weak generalized solutions on the boundary of the domain

Theorem V.10 states that every weak generalized solution $u_0(x, t) \in \mathcal{H}^{s+2m}(\Omega)$ has weak boundary values in the sense that the components u_k, v_λ, and $\omega_{\lambda k}$ satisfy the Green formula (5.6). Let us consider these weak boundary values and examine under what conditions they will be the limit values of the solution $u_0(x, t)$ and its derivatives along surfaces σ_ρ^τ parallel to the boundary $\sigma_0 = \overline{G} \cup S$. The results obtained here are similar to those of [87, 89] in the case of elliptic equations.

For simplicity, we denote $u_0(x, t) = u_0^0(x, t)$ and consider in Ω the parabolic equation (5.2) with $f_0(x, t) \in \mathcal{H}^{s+\beta}(\widehat{\Omega})$, $s \geq -\frac{1}{2}$, $\beta \geq 0$. By Theorem V.7, $u_0(x, t)$ enjoys increased regularity locally inside Ω, that is, $u_\rho^\tau(x, t)$ actually belongs to the space $\mathcal{H}^{s+2m+\beta}(\Omega_\rho^\tau)$ for any $\tau \in (0, \tau_0]$, $\rho \in (0, \rho_0]$. Hence, by the theorems on traces, the traces $u_{0k}(x', t) = D_\nu^{k-1} u_0$ on S_ρ^τ, $v_{\tau \lambda}(x) = D_t^{\lambda-1} u_0(x, t)$ at $t = \tau$, $x \in G_\rho$, $\omega_{\rho \lambda k}^\tau(x') = D_\nu^{k-1} D_t^{\lambda-1} u_0$ on Γ_ρ^τ, $(k, \lambda) \in P(s)$, exist and belong to the spaces indicated above. The following theorem on limit values holds true.

Theorem V.11. *Let $u_0(x,t) \in D'(\Omega)$ be the solution of equation (5.2) with $f_0(x,t) \in \mathcal{H}^{s+\beta}(\widehat{\Omega})$, $s + \beta > -\frac{1}{2}$, $\beta \geq 0$. Then the following conditions are equivalent:*
1) $u_0(x,t) \in \mathcal{H}^{s+2m}(\Omega)$;
2) *There exist the limits*

$$\text{(a)} \quad \lim_{\rho \to 0} \left\langle\!\left\langle u_{0k}(x',t) - u_{0k}(x' + \nu\rho, t), S \right\rangle\!\right\rangle_{s+2m-k+1/2} = 0, \ k = 1, \ldots, 2m,$$

$$\text{(b)} \quad \lim_{\tau \to 0} \left[\!\left[v_{0\lambda}(x) - v_{\tau\lambda}(x), G_0 \right]\!\right]_{s+2m-2b\lambda+b} = 0, \ \lambda = 1, \ldots, r.$$

3) *The sets*

$$N_k = \left\{ \left\langle\!\left\langle u_{\rho k}(x' + \nu\rho, t), S_\rho \right\rangle\!\right\rangle_{s+2m-k+1/2}, \ \rho \in (0, \rho_0) \right\}, \quad k = 1, \ldots, m,$$

$$\mathcal{T}_k = \left\{ \left[\!\left[v_{\tau\lambda}(x), G_0 \right]\!\right]_{s+2m-2b\lambda+b}, \ \tau \in (0, \tau_0) \right\}, \quad \lambda = 1, \ldots, r$$

are bounded.

A proof of this theorem is given in [128]. The limit values exist if the function $f_0(x,t)$ of Theorem V.11 has some minimal regularity $(s > -\frac{1}{2})$. If $f_0(x,t)$ is less regular, then in some cases the existence of the limit values $u_k(x',t)$, $v_\lambda(x)$ and $\omega_{\lambda k}(x')$ can be guaranteed for certain specific values of k and λ. In particular, using regularizations similar to those constructed in Section V.4 one can prove the existence of the limit values $u_k(x' + \nu\rho, t)$ and $v_{\tau\lambda}(x)$ for $\rho \to 0$, $\tau \to 0$ in the case when the right-hand side has regular singularities on $\overline{G} \cup \overline{S}$.

Chapter VI
The Cauchy Problem and
Parabolic Boundary Value Problems
in Spaces of Smooth Functions

In this chapter we survey investigations, done over many years, of the Cauchy and parabolic boundary value problems in the traditional spaces of smooth functions, mainly for systems parabolic in the sense of Petrovskiĭ of the type

$$\mathcal{L}(x,t;D,D_t)u \equiv D_t u - A(x,t;D)u \equiv D_t u - \sum_{|\alpha|\leq 2b} a_\alpha(x,t)D^\alpha u = f(x,t). \quad (1)$$

We restrict ourselves to the systems of type (1) because for such systems we have the most complete results, which are relatively easy to formulate; also, the equations or systems of equations that are met in applications are most frequently of the form (1).

Without, as a rule, presenting the proofs of results (due to space limitations), we have tried to discuss and analyze in detail a sufficiently large number of significant illustrative examples. Obviously, the scientific interests and tastes of the authors played an important role in selecting the results given here.

Many important results of the theory presented are given in detail in the monographs [60, 108, 117] and the papers of monographic character [48, 94].

VI.1. Fundamental solutions of the Cauchy problem

VI.1.1. Introduction

In mathematical physics, probability theory and their numerous applications a central role is played by the function

$$\Gamma_0(t-\tau, x-y) = \left(2a\sqrt{\pi(t-\tau)}\right)^{-n} \exp\left\{-|x-y|^2/4a^2(t-\tau)\right\}, \quad (1.1)$$

$0 \leq \tau < t$, $x,y \in \mathbb{R}^n$. In terms of this function one can express the solution of the Cauchy problem for the heat equation $D_t u = a^2\Delta u + f(x,t)$, $u\big|_{t=+0} = \psi(x)$ as

$$u(x,t) = \int_{\mathbb{R}^n} \Gamma_0(x-y,t)\psi(y)\,dy + \int_0^t d\tau \int_{\mathbb{R}^n} \Gamma_0(x-y,t-\tau)f(y,\tau)\,dy,$$

if, say, $\psi(x)$ is a continuous bounded function in $\mathbb{R}^n$ and the function $f(x,t)$ is defined in the layer $\Pi = \mathbb{R}^n \times [0,T]$, is continuous, bounded, and satisfies the Hölder condition in the space variables $x_1, \ldots, x_n$. A *fundamental solution* $\mathcal{Z}(x,t;y,\tau)$ of the Cauchy problem for the system

$$\mathcal{L}(x,t;D,D_t)u \equiv D_t u - A(x,t;D)u = 0 \tag{1.2}$$

is defined as an m-dimensional square matrix such that the solution of the Cauchy problem

$$u\big|_{t=\tau} = \psi(x) \tag{1.3}$$

for system (1.2) is given by the formula

$$u(x,t) = \int\limits_{\mathbb{R}^n} \mathcal{Z}(x,t;y,\tau)\psi(y)\,dy. \tag{1.4}$$

The matrix $\mathcal{Z}(x,t;y,\tau)$ is a solution of system (1.2) for $t > \tau$ and satisfies the initial condition $\mathcal{Z}\big|_{t=+\tau} = \delta(x-y)I$, where $\delta(x)$ is the Dirac δ-function.

Fundamental solutions of the Cauchy problem were constructed and investigated for the systems parabolic in the sense of Petrovskiĭ of general type. Here we confine ourselves to the case of the system (1.2). We will consider only fundamental solutions of the Cauchy problem, which for brevity will be simply referred to as fundamental solutions.

The fundamental solutions of the parabolic systems defined by Petrovskiĭ copy in an amazing way (with natural modifications) many properties of the function (1.1). Our next objective is to construct these solutions and discuss their properties. Let us mention that the proofs of a number of basic results on fundamental solutions can be found in [60, 108, 117].

VI.1.2. Systems with bounded coefficients

Consider a system uniformly parabolic in the sense of Petrovskiĭ:

$$\mathcal{L}u \equiv D_t u - \sum_{|\alpha|\leq 2b} a_\alpha(x,t)D^\alpha u = 0, \tag{1.5}$$

where $a_\alpha(x,t)$ are m-dimensional square matrices and $u(x,t)$ is an m-dimensional vector-valued function.

We begin by constructing and investigating the fundamental solutions of the model system with constant coefficients:

$$\mathcal{L}_0(D,D_t)u \equiv D_t u - \sum_{|\alpha|=2b} a_\alpha D^\alpha u = 0, \tag{1.6}$$

Performing the Fourier transformation in the space variables $x_1, \ldots, x_n$ and using the fact that the Fourier image of $\delta(x)$ is 1, we obtain the equality

$$\Gamma_0(x, t) = (2\pi)^{-n} \int_{\mathbb{R}^n} \exp\left\{i(x, \xi)\right\} \exp\left\{A_0(i\xi)t\right\} d\xi. \tag{1.7}$$

Introduce the new variables $\tilde{\xi}_j = \xi_j t^{1/2b}$, $j = 1, \ldots, n$ and rewrite (1.7) as

$$\Gamma_0(x, t) = \left(2\pi t^{1/2b}\right)^{-n} \int_{\mathbb{R}^n} \exp\left\{i(xt^{-1/2b}, \tilde{\xi})\right\} \exp\left\{A_0(i\tilde{\xi})\right\} d\tilde{\xi}.$$

The matrix-valued function $\exp\left\{A_0(i\zeta)\right\}$ is an entire (analytic in $\mathbb{C}^n$) function of the complex variables $\zeta_1, \ldots, \zeta_n$ thanks to the parabolicity condition and the homogeneity of the matrix $A_0(i\zeta)$, and satisfies the estimate

$$\left\| \exp\left\{A_0(i\zeta)\right\} \right\| \le C \exp\left\{-c_1 |\operatorname{Re}\zeta|^{2b} + c_2 |\operatorname{Im}\zeta|^{2b}\right\}. \tag{1.8}$$

Here and in what follows C and c denote positive constants, labeled by various indices whenever necessary. Using the theorem of Gel'fand and Shilov [15] on the Fourier transform of entire functions that satisfy inequalities similar to (1.8), we obtain the following result.

Theorem VI.1. *The fundamental solution $\Gamma_0(z, t)$ of system (1.6), regarded as a matrix-valued function of the arguments $z_1 t^{-1/2b}, \ldots, z_n t^{-1/2b}$, is an entire function having of growth order $q = 2b/(2b - 1)$ for complex values of the arguments and the same order of decay for real values. The derivatives of $\Gamma_0(z, t)$ obey the estimates*

$$\left| D^\alpha D_t^{\alpha_0} \Gamma_0(z, t) \right| < C_{\alpha \alpha_0} t^{-(n+|\alpha|+2b\alpha_0)/2b} \exp\left\{\left(-c_3 |\operatorname{Re} z|^q + c_4 |\operatorname{Im} z|^q\right) t^{1-q}\right\}. \tag{1.9}$$

Thus, the intrinsic properties of $\Gamma_0(z, t)$ are completely analogous to those of the function (1.1), which for comparison purposes we rewrite as

$$\left(2a\sqrt{\pi t}\right)^{-n} \exp\left\{-(1/4a^2) \sum_{j=1}^n \left(z_j t^{-1/2}\right)^2\right\}.$$

Example VI.1. Consider the fundamental solution of the parabolic equation

$$D_t u + (-1)^b \Delta^b u = 0, \tag{1.10}$$

which is given by the integral

$$\Gamma_0(x, t) = (2\pi)^{-n} \int_{\mathbb{R}^n} \exp\left\{i(x, \xi) - |\xi|^{2b} t\right\} dt. \tag{1.11}$$

Passing to spherical coordinates in the integral (1.11) and using known equalities from the theory of Bessel functions, we get

$$\Gamma_0(x,t) = (2\pi)^{-n/2} t^{-n/2b} \int_0^\infty \exp\{-r^{2b}\} r^{n/2} J_{(n-2)/2}\big(r|x|t^{-1/2b}\big)\,dr,$$

where $J_{(n-2)/2}(x)$ is the Bessel function of the first kind. Recall that for these values of the index, the Bessel function is expressible in terms of elementary functions.

In the same way as (1.11), we can transform the integral

$$J_\gamma(x,t) = (2\pi)^{-n} \int_{\mathbb{R}^n} \exp\big\{i(x,\xi) - |\xi|^\gamma t\big\}\,d\xi, \tag{1.11}$$

where γ is an arbitrary real number, $\gamma \geq 1$. We obtain

$$J_\gamma(x,t) = (2\pi)^{-n/2} \int_0^\infty \exp\{-r^\gamma\} r^{n/2} J_{(n-2)/2}\big(r|x|t^{-1/\gamma}\big)\,dr. \tag{1.12}$$

The function defined by integral (1.12) possesses a number of important properties. Namely, for $\gamma = 2b$ it satisfies inequality (1.9), that is, it decays exponentially, and for $\gamma \neq 2b$ it obeys the estimate

$$\big|J_\gamma(x,t)\big| \leq C_\gamma t^{-n/\gamma}\big(1 + |x|t^{-1/\gamma}\big)^{-n-\gamma}.$$

As was shown by M.V. Fedoryuk, this estimate is sharp [103].

The integral $J_\gamma(x,t)$ can be calculated explicitly for $\gamma = 1$ or 2. For $\gamma = 2$ it is the fundamental solution of the heat equation, i.e., the normal distribution; for $\gamma = 1$, it is the Cauchy distribution in probability theory:

$$J_1(x,t) \equiv Z_1(x,t) = \frac{\pi^{n/2}}{\Gamma(1+n/2)} t(x^2 + y^2)^{-(n+1)/2}.$$

The integral $J_\gamma(x,t)$ is a fundamental solution of the Cauchy problem for the equation $D_t u = Au$, where A is the pseudodifferential operator with the symbol $-|\xi|^\gamma$.

It is shown in [16] that for $\gamma \in [1,2]$ the function $J_\gamma(x,t)$ is nonnegative and is the density of a probability distribution, whereas for $\gamma > 2$ is a function of alternating sign. As established in [51], the fundamental solutions of the Cauchy problem for arbitrary higher-order parabolic differential equations with smooth coefficients are alternating in sign.

In [103] M. V. Fedoryuk derived precise asymptotic formulas, obtained by the saddle-point method, that describe the behavior of $\Gamma_0(x,t)$ as $|x| \to \infty$ in the case of a single equation (1.6).

To construct a fundamental solution of the system (1.2) with variable coefficients one usually employs a method proposed by E.E. Levi. This method is based on a preliminary analysis of the fundamental solutions of the systems with "frozen" coefficients (taken at parametric points (y, τ) or (y, t)) and on the theory of so-called parabolic potentials. This method is described in detail in [48, 108, 117]. Applying Levi's method in the form given in [108], we obtain the following theorem.

Theorem VI.2. *Let the coefficients $a_\alpha(x,t)$ of the system (1.5), assumed to be uniformly parabolic in the layer $\overline{\Pi} = \mathbb{R}^n \times [0, T]$, satisfy the conditions:*

β_1: *$a_\alpha(x,t)$ are continuous in $\overline{\Pi}$, and for $|\alpha| = 2b$, $a_\alpha(x,t)$ are continuous in t uniformly with respect to $x \in \mathbb{R}^n$;*

β_2: *$a_\alpha(x,t)$ are matrix-valued functions bounded in $\overline{\Pi}$ by a constant M_0.*

β_3: *$a_\alpha(x,t)$ satisfy the Hölder condition of order λ with respect to the space variables $x_1, \ldots, x_n$, that is,*

$$\left| \Delta_x^{\tilde{x}} a(x,t) \right| \le L|x - \tilde{x}|^\lambda, \qquad \text{for all } (x,t), (\tilde{x}, t) \in \overline{\Pi}.$$

Then there exists a fundamental solution $\mathcal{Z}(x,t;y,\tau)$ of system (1.5), unique in the class of bounded functions, such that

$$\left| D^\alpha \mathcal{Z}(x,t;y,\tau) \right| \le C_\alpha (t - \tau)^{-(n+|\alpha|)/2b} \exp \left\{ -c\rho(x,t;y,\tau) \right\}, \quad |\alpha| \le 2b, \qquad (1.13)$$

$$\left| \Delta_x^{\tilde{x}} D^\alpha \mathcal{Z}(x,t;y,\tau) \right| \le C_\alpha |x - \tilde{x}|^\lambda (t - \tau)^{-(n+2b+\lambda)/2b}$$

$$\times \left[\exp\left\{ -c\rho(x,t;y,\tau) \right\} + \exp\left\{ -c\rho(\xi,t;y,\tau) \right\} \right], \qquad\qquad |\alpha| = 2b, \qquad (1.14)$$

where the constants C_α depend on $n, T, \delta_0, M_0, \lambda, L$, while c depends only on the parabolicity constant δ_0; here $\rho(x,t;y,\tau) = |x - y|^q (t - \tau)^{1-q}$.

Note that in the case of a strongly parabolic system—for example, for a single equation—all the assertions remain valid if we replace β_1 by the following weaker condition:

$\widetilde{\beta}_1$: *The coefficients $a_\alpha(x,t)$, $|\alpha| \le 2b$, are continuous in $\overline{\Pi}$.*

There exists another commonly used version of the theorem on existence and uniqueness of the fundamental solution $\mathcal{Z}(x,t;y,\tau)$. In this version condition β_3 is replaced by the condition

$\widetilde{\beta}_3$: *The coefficients $a_\alpha(x,t)$ satisfy the Hölder condition of order $\lambda \in (0,1)$ with respect to the parabolic distance $d\big((x,t),(\tilde{x},\tilde{t})\big) = \big(|x-\tilde{x}|^2 + |t-\tilde{t}|^{1/b}\big)^{1/2}$, i.e.,*

$$\left| \Delta_{x,t}^{\tilde{x},\tilde{t}} a_\alpha(x,t) \right| \le L d\big((x,t),(\tilde{x},\tilde{t})\big)^\lambda.$$

Then estimate (1.4) is replaced by

$$\left|\Delta_{x,t}^{\tilde{x},\tilde{t}}D^\alpha \mathcal{Z}(x,t;y,\tau)\right| \leq C_\alpha d\big((x,t),(\tilde{x},\tilde{t})\big)^\lambda (t-\tau)^{-(n+2b+\lambda)/2b}$$
$$\times \left[\exp\left\{-c\rho(x,t;y,\tau)\right\} + \exp\left\{-c\rho(\tilde{x},\tilde{t};y,\tau)\right\}\right]. \qquad (1.\widetilde{14}')$$

The sharpness of estimates (1.14), $(1.\widetilde{14}')$, in which the order λ remains the same in the Hölder property of the highest-order derivatives of fundamental solutions, plays an extremely important role in the theory of parabolic systems.

Note that the symbol $|\cdot|$ is used to denote the absolute value of a function, length of a vector, or norm of a matrix, depending on its argument. We wish to emphasize that estimates (1.9) and (1.13) (for real values of the arguments) are identical in appearance, but the constants C_α in (1.9) depend only on M_0 and the parabolicity constant δ_0. In particular, estimates (1.9) are true in the half-space $t > 0$ of $\mathbb{R}^{n+1}$. In general, the question of what the constants in estimates (1.9) and (1.13) depend on plays an important role in this theory.

The fundamental solutions $\mathcal{Z}(x,t;y,\tau)$ possess a number of simple and useful properties. We will present now the three most essential ones, under the assumption that conditions β_1–β_3 are satisfied. Various properties of fundamental solutions are considered in [108].

Property VI.1. *The following convolution formula holds for the fundamental solution $\mathcal{Z}(x,t;y,\tau)$:*

$$\mathcal{Z}(x,t;y,\tau) = \int_{\mathbb{R}^n} \mathcal{Z}(x,t;\tilde{x},\tilde{t})\mathcal{Z}(\tilde{x},\tilde{t};y,\tau)d\tilde{x}, \qquad \textit{for all } \tilde{t} \in (\tau,t). \qquad (1.15)$$

In the theory of Markov processes relation (1.15) is called the Chapman-Kolmogorov equation for the transition probability densities of a Markov stochastic process.

Note that for the function (1.1) relation (1.15) can be verified directly; this is a simple and instructive exercise.

To formulate the next important property we need an additional assumption.
β_4: *The derivatives $D^\alpha a_\alpha(x,t)$ exist and satisfy conditions β_1–β_3.*

Under this assumption system (1.2) admits an adjoint system in the sense of Lagrange, which can be written as

$$\mathcal{L}^*(y,\tau;D_y,D_\tau)w(y,\tau) \equiv -D_\tau w - \sum_{|\alpha|\leq 2b} (-1)^{|\alpha|} D_y^\alpha\big(a_\alpha^*(y,\tau)w\big) = 0, \qquad (1.5^*)$$

where a^* is the Hermitian-conjugate matrix of a (that is, if $a = (a^{ij})$, then $a^* = (\bar{a}^{ji})$, where the bar denotes complex conjugation). Note that the Cauchy problem for the adjoint system (1.5^*) is posed as follows: find a solution for $(y,t) \in \Pi_t$ such that

$$w\big|_{\tau=t} = \psi(y). \qquad (1.16)$$

Condition β_4 and Theorem VI.2 guarantee the existence of a fundamental solution $\widetilde{\mathcal{Z}}(x,t;y,\tau)$ of problem (1.5*) and (1.16), and the validity of the estimates (1.13), (1.14). Of course, in this fundamental solution (y,τ) are the main variables, while (x,t) serve as parameters.

Property VI.2. *Let conditions β_1–β_4 be satisfied. Then*

$$\widetilde{\mathcal{Z}}(x,t;y,\tau) = \mathcal{Z}^*(x,t;y,\tau).$$

Property VI.2 is called the *normality property* of the fundamental solution: $\mathcal{Z}(x,t;y,\tau)$, regarded as a function of (x,t), is a fundamental solution of the Cauchy problem for the original system (1.5) and the Hermitian-conjugate matrix of $\mathcal{Z}(x,t;y,\tau)$, regarded as a function of (y,τ), is a fundamental solution of the adjoint Cauchy problem (1.5*), (1.16). In particular, in the case of a single equation with real-valued coefficients, the function $\mathcal{Z}(x,t;y,\tau)$ is a fundamental solution of the original Cauchy problem as a function of (x,t) and is a fundamental solution of the adjoint problem as a function of (y,τ).

Properties VI.1 and VI.2 immediately imply that if conditions β_1–β_4 are satisfied, then $\mathcal{Z}(x,t;y,\tau)$ is jointly differentiable with respect to the original and parametric variables and the estimates

$$\left|D_x^{\alpha'} D_y^{\alpha''} \mathcal{Z}(x,t;y,\tau)\right| \le C_{\alpha'\alpha''}(t-\tau)^{-(n+|\alpha'|+|\alpha''|)/2b} \exp\left\{-c\rho(x,t;y,\tau)\right\},$$

hold for $|\alpha'| \le 2b$, $|\alpha''| \le 2b$.

If conditions β_1–β_3 are satisfied, then the Cauchy problem for system (1.2) has a unique fundamental solution $\mathcal{Z}(x,t;y,\tau)$. The following property of the fundamental solution asserts that from the matrix $\mathcal{Z}(x,t;y,\tau)$ one can recover all the coefficients of system (1.2).

Property VI.3 [91]. *The following relations hold:*

$$\lim_{\tau \to t-0}(t-\tau)^{-1}\left[\int_{\mathbb{R}^n} \mathcal{Z}(x,t;y,\tau)dy - I\right] = a_0(x,t),$$

$$\lim_{\tau \to t-0}(t-\tau)^{-1}\left[\int_{\mathbb{R}^n} \prod_{s=1}^{r}(y_{i_s} - x_{i_s})\mathcal{Z}(x,t;y,\tau)dy\right] = \begin{cases} r!\,a_{i_1\ldots,i_r}(x,t), & 0 < r \le 2b, \\ 0, & r > 2b, \end{cases}$$

where $a_{i_1\ldots,i_r}(x,t)$ is the matrix coefficient of the derivative $D_{x_{i_1}} D_{x_{i_2}} \ldots D_{x_{i_r}}$ in the system (1.2).

In particular, it follows from Property VI.3 that $\int_{\mathbb{R}^n} \mathcal{Z}(x,t;y,\tau)dy = I$ if and only if $a_0(x,t) \equiv 0$.

In the language of probability theory, Property VI.3 means that it is possible to recover the parameters of a process from the transition probability density in the theory of stochastic diffusion processes and also to recover the Kolmogorov equations used for the macroscopic description of these processes.

It is natural to ask what are the minimal requirements on the coefficients that guarantee the existence of classical fundamental solutions with the aforementioned or similar properties? We will discuss below to what extent we can drop the condition that the coefficients be bounded; this will turn out to be a rather interesting problem. For now we will keep conditions β_1, β_2 and attempt to weaken condition β_3.

Definition VI.1. 1) *The function $f(x,t)$ defined in the domain $\overline{\Omega} \subset \mathbb{R}^{n+1}$ is said to belong to the class $\widetilde{\mathcal{D}}^1$ (the Dini class) if its modulus of continuity in $x_1, \ldots, x_n$,*
$$\omega(h) = \sup_{\{(x,t),(\tilde{x},\tilde{t})\}\subset Q} \left|\Delta_x^{\tilde{x}} f(x,t)\right|, \text{ satisfies the condition}$$

$$\omega_1(h) = \int\limits_0^h \frac{\omega(z)}{z}\,dz < +\infty$$

for some positive h.

2) *The function $f(x,t) \in \widetilde{\mathcal{D}}^1$ is said to belong to the class $\widetilde{\mathcal{D}}^2$ if $\int\limits_0^a \frac{\omega_1(h)}{h}\,dh < +\infty$ for some positive a.*

Example VI.2. The function $\omega_\lambda(h) = h^\lambda$, $0 < \lambda \le 1$, is an eigenfunction of the Dini operator, $J_{\widetilde{\mathcal{D}}}(\omega) = \int\limits_0^h (\omega(z)/z)\,dz$. Indeed, $J_{\widetilde{\mathcal{D}}}(\omega_\lambda) = \lambda\omega_\lambda$. The function $\widetilde{\omega}_s(h) = \left\{\ln(1/h)\right\}^s$ can be acted upon by the Dini operator once for $s > 1$ and twice for $s > 2$.

Recall that a function $\omega(h)$ can serve as a modulus of continuity if it is nonnegative, nondecreasing, bounded, and semiadditive (that is, $\omega(h_1 + h_2) \le \omega(h_1) + \omega(h_2)$).

Theorem VI.3. *Let the coefficients $a_\alpha(x,t)$ of system (1.5), which is assumed to be uniformly parabolic in the layer $\overline{\Pi}$, satisfy conditions β_1, β_2 and*
β_5: *$a_\alpha(x,t)$ belong to the class $\widetilde{\mathcal{D}}^1$ in $\overline{\Pi}$.*
β_6: *$a_\alpha(x,t)$ with $|\alpha| = 2b$ belong to the class $\widetilde{\mathcal{D}}^2$ in $\overline{\overline{\Pi}}$.*

Then there exists a fundamental solution $\mathcal{Z}(x,t;y,\tau)$ of system (1.5), unique in the class of bounded functions, which satisfies the estimates (1.13), and for which the increments of the highest-order derivatives $D_x^\alpha \mathcal{Z}$, $|\alpha| = 2b$ satisfy the inequality

$$\left|\Delta_x^{\tilde{x}}\Delta^\alpha \mathcal{Z}(x,t;y,\tau)\right| \le C_\alpha \frac{\omega(|x - \tilde{x}|)}{\omega(t - \tau)}(t - \tau)^{-(n+2b)/2b}$$

$$\times \left[\exp\left\{-c\rho(x,t;y,\tau)\right\} + \exp\left\{-c\rho(\tilde{x},t;y,\tau)\right\}\right],$$

$$(1.17)$$

where the constants C_α depend on $n, t, \delta_0, M_0, \omega, \omega_1$, while the constants c depend only on δ_0.

The proof of Theorem VI.3 is presented in [64, 65]. This proof uses a method due to E. Hopf [118]. There are examples [46, 52] showing that if we drop condition β_5, then, generally speaking, equations (1.5) may have no fundamental solution of the Cauchy problem with the above-described properties. In classical potential theory [20] as well as in various problems of the theory of parabolic and elliptic equations the Dini condition is the weakest condition among those guaranteeing the existence of solutions with classical properties. Relevant examples will be given in Section VI.2.

VI.1.3. Systems with growing coefficients

Here we study conditions for the existence of fundamental solutions for parabolic systems whose coefficients are allowed to grow with respect to space variables. The very simple examples given in the next section show that in this case the uniqueness of the solution of the Cauchy problem in classes of bounded functions (and even in $L_2(\mathbb{R}^n)$) can be violated. Clearly, in such situations one cannot expect that a fundamental solution of the Cauchy problem with the above-described properties exists. However, it can be the case that a directional growth of the coefficients guarantees the existence of fundamental solutions whose properties are, in a certain sense, better than the properties of fundamental solutions for equations with bounded coefficients. We begin our discussion with this case; subsequently, we will formulate conditions on the growth of coefficients, necessary for the existence of fundamental solutions with properties similar to those indicated the preceding subsection.

First, some simple considerations. When we consider the equation

$$D_t u = \Delta u - q^2(x)u, \tag{1.18}$$

where $q(x)$ is a real-valued function, we naturally expect that a rapid growth of the positive function $q^2(x)$ as $|x|$ increases (which means that there is a strong energy absorption when $|x| \to \infty$) will result in fast decay of the fundamental solution of this equation. The problem is to define a notion of "dissipation" for system (1.5) and show that the fundamental solution indeed decays (and indicate the law of this decay) as "dissipation" increases.

Definition VI.2. *A function $f_0(x)$ is called a dissipation characteristic if it satisfies the following three conditions: 1) $f_0(x)$ is defined and continuous in $\mathbb{R}^n$; 2) $f_0(x) \geq 1$; 3) $f_0(x) \to \infty$ as $|x| \to \infty$.*

Definition VI.3 [108, 111]. *A system (1.5) parabolic in the sense of Petrovskiĭ is called a dissipative system if there exists a dissipation characteristic $f_0(x)$ such that the following conditions are satisfied:*
β_7: The matrix functions $\tilde{a}_\alpha(x,t) = a_\alpha(x,t)f_0(x)^{|\alpha|-2b}$ are bounded in $\overline{\Pi}$;

β_8: *The polynomial* $L(x,t;\xi,\xi_{n+1},p) = \det\left(pI - \sum\limits_{|\alpha|\leq 2b} \tilde{a}_\alpha(x,t)(i\xi)^\alpha \xi_{n+1}^{2b-|\alpha|}\right)$ *is pa-*
rabolic in the sense of Petrovskiĭ, that is, the p-roots of L *satisfy the inequality*
$\operatorname{Re} p(x,t;\xi,\xi_{n+1}) \leq -\delta_0\left(|\xi|^2 + \xi_{n+1}^2\right)^b$ *for any* $(x,t) \in \overline{\Pi}$, $(\xi,\xi_{n+1}) \in \mathbb{R}^{n+1}$.
Here δ_0 *is a positive constant.*

Example VI.3. Consider equation (1.18). Let the function $q(x)$ satisfy the conditions imposed on the dissipation characteristic, and set $f_0(x) \equiv q(x)$. Then, $L(\xi,\xi_{n+1},p) \equiv p + |\xi|^2 + \xi_{n+1}^2$. Thus, (1.18) is a dissipative equation with the dissipation characteristic $q(x)$.

Our next objective is to discuss existence theorems of fundamental solutions. To this end, we must first formulate the appropriate conditions on the coefficients $a_\alpha(x,t)$ and the dissipation characteristic $f_0(x)$. There are two sets of such conditions. In the first one, relatively strong constraints are imposed on the smoothness of the coefficients, but no additional conditions on the dissipation characteristic $f_0(x)$ are set. In the second set, which originated in the spectral theory of differential operators as developed in the work of B.M. Levitan and A.G. Kostyuchenko, only minimal constraints are imposed on smoothness, but the dissipation characteristic is subject to additional conditions.

The conditions of the first set are as follows:

β_9: *The coefficients* $a_\alpha(x,t)$ *have derivatives* $D^l a_\alpha(x,t)$, $|l| \leq 2b$, *which obey the estimates*
$$\left|D^l a_\alpha(x,t)\right| \leq C f_0(x)^{2b-|\alpha|+|l|(1-\varepsilon)}, \qquad |l| \leq 2b,$$
where $\varepsilon \in (0,1)$.

β_{10}: *The derivatives* $D^l a_\alpha(x,t)$ *with* $|l| \leq |\alpha|$ *satisfy a local Hölder conditions in* x, *that is, for any* $x \in \mathbb{R}^n$ *there exists a constant* L_x *such that* $\left|\Delta_x^{\tilde{x}} D^l a_\alpha(x,t)\right| \leq L_x |x - \tilde{x}|^\lambda$ *for some* $\lambda \in (0,1]$ *and any* $\tilde{x}$ *such that* $|x - \tilde{x}| \leq 1$.

The second set of conditions is as follows:

β_{11}: $\left|\Delta_x^{\tilde{x}} a_\alpha(x,t)\right| \leq C_x |\tilde{x} - x|^\lambda f_0(x)^{2b-|\alpha|}$ *for* $|\tilde{x} - x| \leq 1$, $|\alpha| \leq 2b$.

Here one assumes that the dissipation characteristic $f_0(x)$ satisfies two additional conditions:

1) *For any* $x, \tilde{x} \in \mathbb{R}^n$ *and* $|\tilde{x} - x| \leq 1$, $f_0(x) \leq C f_0(\tilde{x})$;
2) *There exists a small positive number* μ_0 *such that for* $|\tilde{x} - x| \geq 1$ *we have*
$f_0(x) \leq C \exp\left\{\mu_0 f_0(\tilde{x})|\tilde{x} - x|\right\}$.

To formulate the desired results, we use the dissipation characteristic to define a smooth function $g(x,t)$, whose derivatives satisfy the inequalities

$$\left|D_t g(x,t)\right| \leq C f_0(x)^{2b}, \quad \left|D^\alpha g(x,t)\right| \leq C f_0(x)^{|\alpha|}, \qquad |\alpha| = 1,\ldots,2b. \qquad (1.19)$$

Theorem VI.4. *Let the coefficients of the dissipative system* (1.5) *satisfy conditions* β_9, β_{10} *or* β_{11}, *with a dissipation characteristic* $f_0(x)$ *that possesses properties* 1) *and* 2). *Then there exists a fundamental solution* $\mathcal{Z}(x,t;y,\tau)$ *of system* (1.5), *which* *obeys the estimates*

$$\left|D^\alpha \mathcal{Z}(x,t;y,\tau)\right| \leq C_\alpha \sum_{k=1}^{|\alpha|}(t-\tau)^{-(n+|\alpha|-k)/2b} f_0(x)^k$$

$$\times \exp\left\{-c\rho(x,t;y,\tau) + \eta g(x,t) - \eta g(y,\tau)\right\},$$

(1.20)

$|\alpha| \leq 2b$, *where η is a sufficiently small positive constant.*

To prove Theorem VI.4 [108, 111], one uses various modifications of Levi's method, depending on the conditions imposed, followed by a replacement of the unknown function that allows us to include the dissipation characteristic in the exponential part of estimate (1.20).

We now proceed to the case of a system (1.5) with growing coefficients. Several variants of the conditions on coefficients can be suggested. For example,

β_{12}: *The coefficients $a_\alpha(x,t)$, $|\alpha| = 2b$, satisfy conditions β_1–β_3;*
β_{13}: *The coefficients $a_\alpha(x,t)$, $|\alpha| < 2b$, satisfy the conditions*

   1) $\left|a_\alpha(x,t)\right| \leq C_\alpha \left(1+|x|\right)^{\{(2b-|\alpha|)/(2b-1)\}-\varepsilon}$ *for some positive ε;*
   2)   *for $|\tilde{x} - x| < 1$,*

$$\left|\Delta_x^{\tilde{x}} a_\alpha(x,t)\right| \leq C_x |x - \tilde{x}|^\lambda |\tilde{x}|^{(2b-|\alpha|)/(2b-1)}, \quad \lambda \in (0,1].$$

Condition β_{13} admits the following modification: in 1) one takes $\varepsilon = 0$, but one additionally requires that the derivatives with respect to $x_1, \ldots, x_n$ exist up to the order $2b$ and that the inequalities

$$\left|D^{\alpha'} a_\alpha(x,t)\right| \leq C\left(1+|x|\right)^{\{(2b-|\alpha|+|\alpha'|)/(2b-1)\}-\varepsilon_1}, \quad 1 \leq |\alpha'| \leq 2b$$

hold for some constant $\varepsilon_1 \in (0,1)$. This modification of condition β_{13} will be denoted by $\widetilde{\beta}_{13}$.

Theorem VI.5. *Let the coefficients of system (1.5), assumed to be uniformly parabolic in the sense of Petrovskiĭ, satisfy conditions β_{12}, β_{13} or $\widetilde{\beta}_{13}$. Then there exists a fundamental solution $\mathcal{Z}(x,t;y,\tau)$ of system (1.5), which obeys the estimates*

$$\left|D^\alpha \mathcal{Z}(x,t;y,\tau)\right| \leq C_\alpha \sum_{k=1}^{|\alpha|}(t-\tau)^{-(n+|\alpha|-k)/2b}\left(1+|x|\right)^{k/(2b-1)}$$

$$\times \exp\left\{-c\rho(x,t;y,\tau) + \eta|x|^q - \eta|y|^q\right\}, \qquad |\alpha| \leq 2b.$$

In the case where the coefficients $a_\alpha(x,t)$ behave like $|x|^{(2b-|\alpha|)/(2b-1)}$ as $|x| \to \infty$ (or have a somewhat slower growth) fundamental solutions have been constructed in [35, 108] by using various methods. In [108, pp. 134–135] the author suggested a simple transformation that reduces an equation with growing coefficients to a dissipative equation.

Example VI.4. Let us find a fundamental solution of the Cauchy problem for the equation

$$D_t P = \frac{1}{4}\gamma^2 N_0 \frac{\partial^2 P}{\partial x^2} + \alpha \frac{\partial}{\partial x}(x \cdot P), \qquad P(x,0) = P_0(x). \tag{1.21}$$

This interesting equation arises in statistical radio engineering [101]: its solution $P(x,t)$ represents the probability density of the event that the random voltage in simple RC and RL electrical circuits takes the value x at time t, given the initial distribution $P_0(x)$, under the effect of white noise with constant energy density N_0. To simplify notation, we denote $\frac{1}{4}\gamma^2 N_0$ by a^2. Equation (1.21) is an equation with growing coefficients of the same class as that considered in Theorem VI.5.

Performing the Fourier transformation in x, we reduce problem (1.21) into the first-order partial differential equation

$$D_t \widetilde{P} + \alpha \xi D_\xi \widetilde{P} = -a^2 \xi^2 \widetilde{P}, \qquad \widetilde{P}(\xi,0) = \widetilde{P}_0(\xi). \tag{1.22}$$

Consider the equations of characteristics $\frac{dt}{1} = \frac{d\xi}{\alpha\xi} = \frac{d\widetilde{P}}{-a^2\xi^2\widetilde{P}}$, whose first integrals have the form

$$\Psi_1(\xi,\widetilde{P},t) = \xi \exp\{-\alpha t\}, \qquad \Psi_2(\xi,\widetilde{P},t) = \widetilde{P}\exp\{a^2\xi^2/2\alpha\}, \tag{1.23}$$

Let us solve the Cauchy problem (1.22) following [99, pp. 348–349]. Setting in (1.23) $t = 0$ and indicating the corresponding values with overbars, we get $\overline{\Psi}_1 = \xi$, $\overline{\Psi}_2 = \widetilde{P}\exp\{a^2\xi^2/2\alpha\}$, whence $\widetilde{P} = \overline{\Psi}_2 \exp\{-a^2\overline{\Psi}_1^2/2\alpha\}$, $\xi = \overline{\Psi}_1$. The solution of the Cauchy problem (1.22) is given by formula (32) of [101, p. 349]:

$$\Psi_2(\xi,\widetilde{P},t) \exp\{-a^2\overline{\Psi}_1^2/2\alpha\} - \widetilde{P}_0\big(\Psi_1(\xi,\widetilde{P},t)\big) = 0,$$

that is,

$$\widetilde{P}\exp\left\{a^2\xi^2(1 - e^{-2\alpha t})/2\alpha\right\} = \widetilde{P}_0(\xi e^{-\alpha t}). \tag{1.24}$$

Expressing $\widetilde{P}$ from (1.24) and taking the inverse Fourier transformation, we obtain the following representation of the solution of the Cauchy problem (1.21):

$$P(x,t) = (2\pi)^{-1} \int_{-\infty}^{\infty} \exp\left\{ix\xi - a^2\xi^2(1 - e^{-2\alpha t})/2\alpha\right\}$$

$$\times \left(\int_{-\infty}^{\infty} \exp\left\{-iy\xi e^{-\alpha t}\right\} P_0(y)dy\right) d\xi$$

$$= \int_{-\infty}^{\infty} \left[(2\pi)^{-1} \int_{-\infty}^{\infty} \exp\left\{i(x - ye^{-\alpha t})\xi - a^2\xi^2(1 - e^{-2\alpha t})/2\alpha\right\}d\xi\right] P_0(y)dy$$

$$= \int_{-\infty}^{\infty} \mathcal{Z}(x,t;y,0)P_0(y)dy.$$

The integral that defines the fundamental solution $\mathcal{Z}(x,t;y,0)$ can be calculated directly, and we get

$$
\begin{aligned}
\mathcal{Z}(x,t;y,0) = \Big[& 2\pi \big(a\alpha^{-1/2}\big)^2 \big(1 - e^{-2\alpha t}\big) \Big]^{-1/2} \\
& \times \exp\left\{ -\big(x - ye^{-\alpha t}\big)^2 \big/ 2\big(a\alpha^{-1/2}\big)^2 \big(1 - e^{-2\alpha t}\big) \right\}
\end{aligned}
\tag{1.25}
$$

It is worth mentioning that in [101, pp. 86–88] formula (1.25) was obtained by rather tedious calculations, using the Mehler expansion of the two-dimensional normal probability density in a series of orthogonal Chebyshev-Hermite polynomials.

Formula (1.25) defines the probability density of a nonstationary Markov stochastic process. When t tends to infinity, $\mathcal{Z}(x,t;y,0)$ tends to the normal distribution $\big(2\pi a\alpha^{-1/2}\big)^2 \exp\left\{ -x^2/2 \big(a\alpha^{-1/2}\big)^2 \right\}$ with zero mean value and dispersion $\big(a\alpha^{-1/2}\big)^2$.

VI.1.4. Second-order parabolic equations

We consider next the most important case of the linear second-order parabolic equation:

$$
D_t u = \sum_{i,j=1}^{n} a_{ij}(x,t) D_{x_i} D_{x_j} u + \sum_{i=1}^{n} b_i(x,t) D_{x_i} u + c(x,t)u.
\tag{1.26}
$$

If the coefficients of equation (1.26) satisfy the conditions $\widetilde{\beta}_1$, β_2, β_3, then there exists a fundamental solution $\mathcal{Z}(x,t;y,\tau)$ that obeys the estimates (1.13), (1.14) with $b = 1$. In the case of the set of conditions $\widetilde{\beta}_1$, β_2, β_5, β_6 estimates (1.15), (1.17) hold. The following assertion holds true.

Theorem VI.6. *If the conditions $\widetilde{\beta}_1$, β_2, β_3 or $\widetilde{\beta}_1$, β_2, β_5, β_6 are satisfied, then the fundamental solution of equation (1.26) obeys the two-sided estimates*

$$
\begin{aligned}
C_1(t-\tau)^{-n/2} \exp\left\{ -c_1|x-y|^2(t-\tau)^{-1} \right\} &\leq \mathcal{Z}(x,t;y,\tau) \\
&\leq C_2(t-\tau)^{-n/2} \exp\left\{ -c_2|x-y|^2(t-\tau)^{-1} \right\}.
\end{aligned}
\tag{1.27}
$$

Here the positive constants C_1 and C_2 depend only on $n, \delta_0, T, M_0, L, \lambda$ in the first case, and on $n, \delta_0, T, M_0, \omega, \omega_1$ in the second case. The constant c_2 depends only on δ_0, while c_1 depends on δ_0, M_0.

The upper bound of $\mathcal{Z}(x,t;y,\tau)$ was established in Theorems IV.2 and IV.3, respectively, and the lower bound can be easily derived from the Theorem 1.1 of N.V. Krylov and M.V. Safonov [53, p. 163] and the analysis of the structure of the fundamental solution $\mathcal{Z}(x,t;y,\tau)$ provided by the method of E.E. Levi and E. Hopf.

The deepest and most complete results have been obtained for the parabolic equations of divergence form,

$$p(x)D_t u = \sum_{i,j=1}^{n} D_{x_i}\big(a_{ij}(x,t)D_{x_j}u\big), \tag{1.28}$$

which contain a group of terms with highest-order derivatives. Equation (1.28) is studied under the following assumptions:

β_{14}: *The functions $p(x), a_{ij}(x,t)$ are measurable in $\mathbb{R}^n$ and $\mathbb{R}^{n+1}_+ = \{(x,t) : x \in \mathbb{R}^n, t \ge 0\}$ respectively.*

β_{15}: *There exists a constant $\mu \ge 1$ such that*

$$\mu^{-1} \le p(x) \le \mu, \qquad \mu^{-1}|\xi|^2 \le \sum_{i,j=1}^{n} a_{ij}(x,t)\xi_i\xi_j \le \mu|\xi|^2, \qquad for\ all\xi \in \mathbb{R}^n.$$

Note that condition β_{15} implies the inequalities $\big|a_{ij}(x,t)\big| \le \mu$, $i,j = 1,\ldots,n$.

Since no smoothness conditions are imposed on the coefficients, we must introduce the notion of a generalized solution of equation (1.28). To this end, introduce the space $L_{2,\mathrm{loc}}\big[(0,\infty); H^1_{\mathrm{loc}}(\mathbb{R}^n)\big]$ of all functions $u(x,t)$ that are defined in $\mathbb{R}^{n+1}_+$ and have the following properties:

1) for any fixed $t \in (0,\infty)$, the function $x \mapsto u(x,t)$ belongs to the space $H^1(K)$, for any compact subset K of $\mathbb{R}^n$;

2) $u(x,t)$, regarded as an element of $H^1(K)$ depending on the parameter t, belongs to the space $L_2\big([t_1,t_2]\big)$ for any t_1, t_2, $0 < t_1 < t_2$.

Further, denote by $C_0^1\big(\mathbb{R}^{n+1}_+\big)$ the set of all functions $u(x,t)$ with compact support that have continuous first-order partial derivatives, and by $L_2^0(\mathbb{R}^n)$ the set of all functions with compact support belonging to $L_2(\mathbb{R}^n)$.

Definition VI.4. *A function $u(x,t) \in L_{2,\mathrm{loc}}\big[(0,\infty), H^1_{\mathrm{loc}}(\mathbb{R}^n)\big]$ that satisfies the integral identity*

$$\iint\limits_{\mathbb{R}^{n+1}} \left[-puD_t v + \sum_{i,j=1}^{n} a_{ij} D_{x_i}uD_{x_j}v \right] dxdt = 0 \tag{1.29}$$

for any test function $v(x,t) \in C_0^1\big(\mathbb{R}^{n+1}_+\big)$ is called a weak solution of equation (1.28) in $\mathbb{R}^{n+1}_+$.

Definition VI.5. *A function $u(x,t)$ is called a weak solution of the Cauchy problem if $u(x,t)$ is a weak solution of equation (1.28) in $\mathbb{R}^{n+1}_+$ and satisfies the initial condition*

$$u\big|_{t=0} = \psi(x) \tag{1.30}$$

in the weak sense, that is, for any function $v(x,t) \in L_2^0(\mathbb{R}^n)$,

$$\lim_{t\to+0} \int_{\mathbb{R}^n} u(x,t)v(x)dx = \int_{\mathbb{R}^n} \psi(x)v(x)dx.$$

Definition VI.6. *The function $\Gamma(x,t;y,\tau)$ is called a weak fundamental solution of the Cauchy problem* (1.28), (1.30) *if it is defined in the domain $V = \{(x,t;y,\tau) : (x,t) \in \mathbb{R}_+^{n+1}, (y,\tau) \in \mathbb{R}_+^{n+1}, \tau < t\}$ and for any bounded measurable function $\Psi(x)$ with compact support the function*

$$u(x,t) = \int\limits_{\mathbb{R}^n} \Gamma(x,t;y,\tau)\psi(y)dy \tag{1.31}$$

is a weak bounded solution of the Cauchy problem

$$u\big|_{t=+\tau} = \psi(x) \tag{1.32}$$

One can show that $\Gamma(x,t;y,\tau)$ is the limit of a sequence of classical fundamental solutions of infinitely smooth Cauchy problems [73]. This enables us to prove an assertion first in the smooth case, which simplifies substantially the original definitions and proofs.

Let us discuss the properties of weak fundamental solutions $\Gamma(x,t;y,\tau)$. The main contrast with the properties considered above is that all the estimates obtained here involve constants that depend only on n and μ, and do not depend at all on the smoothness of coefficients and the width of the strip T. Such results are due to J. Nash [123], J. Moser [121, 122], and especially to D.G. Aronson [115]. A consistent exposition of the properties of weak fundamental solutions $\Gamma(x,t;y,\tau)$ is given in the survey paper [73]. The method by which these results were obtained is based on subtle results of the theory of divergence-type equations: Harnack's inequality, the fact that weak solutions a priori possess Hölder properties, and special a priori estimates of solutions of embedding-theorems type.

If the reader is not familiar with the language of generalized solutions, she may regard the fundamental solutions as classical ones, but then their estimates should in no way depend on the smoothness of the coefficients.

In what follow the letter K (with various indices) will denote, generally speaking, different positive constants that depend only on n and μ. The following result is of crucial importance.

Theorem VI.7. *Let conditions β_{14} and β_{15} be satisfied. Then the weak fundamental solution $\Gamma(x,t;y,\tau)$ of the problem* (1.28), (1.30) *obeys the two-sided estimate*

$$K_1(t-\tau)^{-n/2}\exp\left\{-k_1|x-y|^2(t-\tau)^{-1}\right\} \le \Gamma(x,t;y,\tau)$$
$$\le K_2(t-\tau)^{-n/2}\exp\left\{-k_2|x-y|^2(t-\tau)^{-1}\right\}. \tag{1.33}$$

Property VI.4. *The function $\Gamma(x,t;y,\tau)/p(y)$ satisfies the uniform Hölder condition with respect to the set of variables $(x,t;y,\tau)$. In particular, for any $t > \tau$ and $x,\tilde{x},y$, the estimates $\left|\Delta_x^{\tilde{x}}\Gamma(x,t;y,\tau)\right| \le K|x-\tilde{x}|^\lambda(t-\tau)^{-(n+\lambda)/2}$ hold, where λ is a positive constant that dependens only on n and μ.*

Property VI.5. *The function $\widetilde{\Gamma}(x,t;y,\tau) = \big(p(x)/p(y)\big)\Gamma(x,t;y,\tau)$ is the unique fundamental solution of the Cauchy problem (in the class of bounded functions) for the equation*

$$-p(y)D_\tau w = \sum_{i,j=1}^{n} D_{y_i}\big(a_{ij}(y,\tau)D_{y_j}w\big), \qquad 0 \le \tau < t.$$

Property VI.6. $\Gamma(x,t;y,\tau) = \int\limits_{\mathbb{R}^n} \Gamma(x,t;\tilde{x},\tilde{t})\Gamma(\tilde{x},\tilde{t};y,\tau)\,d\tilde{x}$ *for all $\tilde{t} \in (\tau,t)$.*

Property VI.7. $\int\limits_{\mathbb{R}^n} \Gamma(x,t;y,\tau)\,dy = 1.$

The following property is the integral version of the estimate of the decay of the generalized gradients of the weak fundamental solution $\Gamma(x,t;y,\tau)$ as the difference $|x - y|$ grows.

Property VI.8. *For any $A \ge 0$ and any t_1, t_2 such that $\tau < t_1 < t_2$, $t_2 - t_1 \le \min\big(1, (t_2 - \tau)/2\big)$,*

$$\int\limits_{t_1}^{t_2} d\tau \int\limits_{|x-y|\ge\sqrt{A(t-\tau)}} \sum_{j=1}^{n}\big|D_{x_j}\Gamma(x,t;y,\tau)\big|^2 dx \le K(t_2 - \tau)^{-n/2}\exp\{-kA\}.$$

Additional interesting information was obtained in [73, 75] for weak fundamental solutions of stationary equations, that is, equations whose coefficients are independent of time:

$$p(x)D_t u = \sum_{i,j=1}^{n} D_{x_i}\big(a_{ij}(x)D_{x_i}u\big). \tag{1.34}$$

Theorem VI.8. *A weak fundamental solution $\Gamma(x,t,y)$ of the Cauchy problem for equation (1.34) has classical t-derivatives of any order and the estimates*

$$\big|D_t^\nu\Gamma(x,t;y)\big| \le K(\nu,\mu,n)t^{-n/2-\nu}\exp\big\{-k(\mu)|x - y|^2 t^{-1}\big\} \tag{1.35}$$

hold for any $t > 0$.

One is naturally led to asking whether similar results hold for second-order equations with lower-order terms without any assumptions on their smoothness. Results in this direction are found in the important article of D.G. Aronson [116] (see also [80]).

Consider now equations of the type

$$p(x)D_t u = \sum_{i,j=1}^{n} D_{x_j}\big(a_{ij}(x,t)D_{x_j}u + a_i(x,t)u\big) + \sum_{i=1}^{n} b_i(x,t)D_{x_i}u + c(x,t)u. \tag{1.36}$$

Assume that the coefficients of (1.36) are measurable functions in the layer $\overline{\Pi}$, conditions β_{14}, β_{15} are satisfied and that $\theta_0 \in (0,1)$, R_0, M_0 are numbers such that

β_{16}: *For some* $p \geq 2(1 - \theta_0)$ *and* $\theta_1 = \frac{1}{2} - \frac{n}{2p} - \frac{1}{q} \geq \frac{\theta_0}{2}$,

$$a_i(x,t),\ b_i(x,t) \in L_q\Big([0,T]; L_p(\{x : |x| < R\})\Big).$$

β_{17}: *For* $T_0 = \min(1, \sqrt{T})$, *for some* $p > (1/(1 - \theta_0))$ *and* $\theta_2 = 1 - \frac{n}{2p} - \frac{1}{q} \geq 0$

$$c(x,t) \in L_q\Big([0,T]; L_p(\{x : |x - y| \leq T_0\})\Big)$$

   uniformly in y.

β_{18}: *For almost all* $x, |x| > R_0,\ t \in [0,T]$,

$$\big|a_i(x,t)\big|,\ \big|b_i(x,t)\big|,\ c(x,t) \leq M_0.$$

We have listed here these important, but not sufficiently transparent conditions in order to stress once again that the conditions on the coefficients are closely connected with the structure of the equation.

We will denote by C various positive constants, depending only on the structure of equation (1.36), that is, defined only by the quantities n, μ, θ_0, R_0, T, and the norms $\|a_i\|$, $\|b_i\|$, $\|c\|$ in the corresponding spaces $L_{p,q}$.

Theorem VI.9. *Let the coefficients of equation (1.36) satisfy conditions β_{14}–β_{18}. Then the Cauchy problem for (1.36) equation has a weak fundamental solution $\Gamma(x,t;y,\tau)$, which obeys the two-sided estimates*

$$C_1(t - \tau)^{-n/2} \exp\big\{-c_1|x - y|^2(t - \tau)^{-1}\big\} \leq \Gamma(x,t;y,\tau)$$
$$\leq C_2(t - \tau)^{-n/2} \exp\big\{-c_2|x - y|^2(t - \tau)^{-1}\big\}, \qquad (1.37)$$

where C_1, C_2, c_1, c_2 *are positive constants that depend on the structure of equation* (1.36).

In spite of the formal resemblance of estimates (1.33) and (1.37), they are essentially different. In particular, estimates (1.33) hold in $\mathbb{R}_+^{n+1}$, whereas estimates (1.37) are valid in a finite layer $\overline{\Pi}$. It is interesting to formulate conditions on $a_i(x,t)$, $b_i(x,t)$, $c(x,t)$ under which an estimate of type (1.37) will hold in $\mathbb{R}_+^{n+1}$. This question is dealt with in [74, 80]. In the case of a single space variable complete results for fundamental solutions of parabolic equations were obtained by simpler methods (see, for example, [79]).

VI.1.5. Estimates for fundamental solutions of parabolic systems in $\mathbb{R}_+^{n+1}$ and elliptic systems generated by parabolic systems

As already noted, estimates (1.9) for the fundamental solution $\Gamma_0(x,t)$ of the parabolic system with constant coefficients

$$D_t u = \sum_{|\alpha|=2b} a_\alpha D^\alpha u, \tag{1.38}$$

and estimates (1.33) for the weak fundamental solution $\Gamma(x,t;y,\tau)$ of equation (1.28) hold in $\mathbb{R}_+^{n+1}$. The monograph [108] describes also other situations when the estimates of that kind are valid. Based on these estimates one can derive simple and important formulas that enable us to construct fundamental solutions of elliptic systems generated by parabolic systems of the type (1.5) in terms of fundamental solutions of parabolic systems.

Recall that the system

$$A(x,D)u \equiv \sum_{|\alpha|\leq 2b} a_\alpha(x)D^\alpha u = 0 \tag{1.38}$$

is called *elliptic* if $\det\left(\sum_{|\alpha|=2b} a_\alpha(x)\xi^\alpha\right) \neq 0$ for all $x,\xi \in \mathbb{R}^n$, $\xi \neq 0$. A fundamental solution of such system, $\mathcal{E}(x,y)$, is defined as a square $m\times m$ matrix-valued function in terms of which the solution of the elliptic system

$$-A(x,D)u = f(x) \tag{1.40}$$

is expressible as

$$u(x) = \int_{\mathbb{R}^n} \mathcal{E}(x,y)f(y)\,dy \tag{1.41}$$

for any smooth function $f(x)$ with compact support.

Theorem VI.10. *The fundamental solution $\mathcal{E}_0(x,y)$ of the elliptic system*

$$\sum_{|\alpha|=2b} a_\alpha D^\alpha u = 0, \tag{1.42}$$

generated by the parabolic system (1.38), is given by the formula

$$\mathcal{E}_0(x-y) = \int_0^\infty \left\{ \Gamma_0(x-y,t) - \sum_{s=0}^{2b-n} \frac{1}{s!}\left(\sum_{\nu=1}^{n}(x_\nu - y_\nu - a_\nu)D_{x_\nu}\right)^s \right.$$

$$\left. \times \Gamma_0(x-y,t)\big|_{x-y=a} \right\} dt, \tag{1.43}$$

where $a = (a_1,\ldots,a_n) \neq 0$.

Thus, if $2b < n$, then $\mathcal{E}_0(x-y) = \int_0^\infty \Gamma_0(x-y,t)\,dt$. If $2b \geq n$, then the integral $\int_0^\infty \Gamma_0(x-y,t)\,dt$ diverges and should be regularized by the rule (1.43).

For $x \neq y$ the fundamental solution $\mathcal{E}_0(x-y)$ admits derivatives of any order, which obey the following estimates:

$$\text{for } |x-y| \leq 1, \ n+|\alpha| < 2b: \quad \left|D^\alpha \mathcal{E}_0(x-y)\right| \leq C;$$

$$\text{for } |x-y| \leq 1, \ n+|\alpha| = 2b: \quad \left|D^\alpha \mathcal{E}_0(x-y)\right| \leq C \ln\left(1/|x-y|\right) + C_1; \quad (1.44)$$

$$\text{for } |x-y| > 2b: \quad \left|D^\alpha \mathcal{E}_0(x-y)\right| \leq C|x-y|^{-n-|\alpha|+2b}.$$

Estimates (1.44) can be derived from (1.9) by using standard arguments [108, pp. 148–154].

Example VI.5. Let us find the fundamental solutions of the Laplace equation $\Delta u = 0$ in $\mathbb{R}^3$ and $\mathbb{R}^2$, using formulas (1.1) and (1.43):

$$\mathcal{E}_0(x_1-y_1, x_2-y_2, x_3-y_3) = \left(2\sqrt{\pi}\right)^{-3} \int_0^\infty \exp\left\{-|x-y|^2/4t\right\} t^{-3/2}\,dt$$

$$= \frac{1}{4}\pi^{-3/2} \int_0^\infty \exp\left\{-|x-y|^2\tilde{\tau}/4\right\} d\tilde{\tau}$$

$$= \frac{1}{4\pi}|x-y|^{-1};$$

$$\mathcal{E}_0(x_1-y_1, x_2-y_2) = \frac{1}{4\pi} \int_0^\infty \left[\exp\left\{-|x-y|^2/4t\right\} - \exp\left\{-1/4t\right\}\right] \frac{dt}{t}$$

(for $a = \frac{1}{\sqrt{2}}(1,1)$). The last integral can be easily calculated by differentiating with respect to the parameter $|x-y|^2$, which yields

$$\mathcal{E}_0(x_1-y_1, x_2-y_2) = \frac{1}{2\pi}\ln\left(1/|x-y|\right).$$

Example VI.6. Let us construct the fundamental matrix of solutions $\mathcal{E}(x)$ of the static equations of elasticity in displacements,

$$\Delta u_i + (k+1)D_i\theta = 0, \quad i = 1,2,3; \quad \theta = D_1 u_1 + D_2 u_2 + D_3 u_3. \quad (1.45)$$

The matrix $\mathcal{E}(x)$ is known as *Somigliana's displacement tensor*. The system

$$D_t u_i = \Delta u_i + (k+1)D_i\theta, \quad i = 1,2,3, \quad (1.46)$$

is strongly parabolic (this can be verified directly).

Performing the Fourier transformation in the space variables x_1, x_2, x_3, we obtain a system of ordinary differential equations for $(\tilde{u}_1, \tilde{u}_2, \tilde{u}_3)$:

$$\frac{d\tilde{u}_1}{dt} = -\big[\xi^2 + (k+1)\xi_1^2\big]\tilde{u}_1 - (k+1)\xi_1\xi_2\tilde{u}_2 - (k+1)\xi_1\xi_3\tilde{u}_3,$$

$$\frac{d\tilde{u}_2}{dt} = -(k+1)\xi_1\xi_2\tilde{u}_1 - \big[\xi^2 + (k+1)\xi_2^2\big]\tilde{u}_2 - (k+1)\xi_2\xi_3\tilde{u}_3, \qquad (1.47)$$

$$\frac{d\tilde{u}_3}{dt} = -(k+1)\xi_1\xi_3\tilde{u}_1 - (k+1)\xi_2\xi_3\tilde{u}_2 - \big[\xi^2 + (k+1)\xi_3^2\big]\tilde{u}_3.$$

The normal fundamental system of solutions of system (1.47) is given by

$$V(\xi, t) = \exp\{-\xi^2 t\}I + F(\xi, t)(\delta_{ij}\xi_i\xi_j)_{i,j=1}^3,$$

where $F(\xi, t) = \big[\exp\{-\xi^2(k+2)t\} - \exp\{-\xi^2 t\}\big]\xi^{-2}$.

Thanks to the structure of the symmetric matrix $V(\xi, t)$, in order to find the fundamental solution $\mathcal{Z}(x, t)$ of system (1.46) it suffices to calculate its elements $\mathcal{Z}_{11}(x, t)$ and $\mathcal{Z}_{12}(x, t)$:

$$\mathcal{Z}_{11}(x, t) = (2\pi)^{-3}\int_{\mathbb{R}^3} \exp\{-i(x, \xi)\}\big[\exp\{-|\xi|^2 t + \xi_1^2\}F(\xi, t)\big]\, d\xi,$$

$$\mathcal{Z}_{12}(x, t) = (2\pi)^{-3}\int_{\mathbb{R}^3} \exp\{-i(x, \xi)\}\xi_1\xi_2 F(\xi, t)\, d\xi.$$

Simple calculations yield

$$\mathcal{Z}_{11}(x, t) = (1/8\pi^{3/2})\bigg[t^{-3/2}\exp\{-x^2/4t\}$$
$$-\frac{1}{2}\int_0^{t(k+2)} \tau^{-5/2}\exp\{-x^2/4\tau\}\big[1 - (x_1^2/2\tau)\big]\, d\tau\bigg],$$

$$\mathcal{Z}_{12}(x, t) = (x_1 x_2)(32\pi^{3/2})^{-1}\int_0^{t(k+2)} \tau^{-7/2}\exp\{-x^2/4\tau\}\, d\tau.$$

Consider now the elements $\mathcal{E}_{11}(x)$ and $\mathcal{E}_{12}(x)$ of the fundamental matrix of solutions of system (1.45). In our case $n = 3$, $2b = 2$, $n > 2b$, and so

$$\mathcal{E}_{11}(x) = \int_0^\infty \mathcal{Z}_{11}(x, t)\, dt = (1/8\pi^{3/2})\int_0^\infty t^{-3/2}\exp\{-x^2/4t\}\, dt$$

$$- (1/16\pi^{3/2})\int_0^\infty dt \int_0^{t(k+2)} \tau^{-5/2}\exp\{-x^2/4\tau\}\big[1 - (x_1^2/2\tau)\big]\, d\tau,$$

The first integral in this formula has been calculated in the previous example. To calculate the second integral we regard it as a two-dimensional integral and change the order of integration to obtain

$$\mathcal{E}_{11}(x) = (1/8\pi)\left[\frac{k+3}{k+2}\left(1/|x|\right) + \frac{k+1}{k+2}\left(x_1^2/|x|^3\right)\right].$$

Similarly,

$$\mathcal{E}_{12}(x) = (1/8\pi)\frac{k+1}{k+2}\left(x_1 x_2/|x|^3\right).$$

Thus, the fundamental matrix of solutions of the elasticity system (1.5) (*the Lamé system of equations*) is given by the formula

$$\mathcal{E}(x) = \left(8\pi(k+2)\right)^{-1}\left[\{(k+3)/|x|\}I + \{(k+1)/|x|^3\}(x_i x_j)_{i,j=1}^3\right]. \tag{1.48}$$

Sometimes it is important to select among all the fundamental matrices of solutions of elliptic systems those that behave "nicely" when $|x| \to \infty$ (if any such exist).

Definition VI.7. *The fundamental solution $\mathcal{E}(x,y)$ is called a principal fundamental solution if $\lim\limits_{|x-y|\to\infty} \mathcal{E}(x,y) = 0$ and the convergence is uniform in all directions.*

Estimates (1.44) imply that the fundamental solution $\mathcal{E}_0(x-y)$ of system (1.42) is the principal fundamental solution if the number of the space variables, n, is larger than the order of the system, $2b$. In particular, this is true for Somigliana's tensor (1.48).

Including lower-order derivatives in the system allows one, under some special conditions, to construct the principal fundamental solution for any n and $2b$. Such cases were considered in [108]. We present below one of the results in this direction.

Theorem VI.11. *Consider the system with constant coefficients parabolic in the sense of Petrovskiĭ*

$$D_t u = \sum_{|\alpha|\leq 2b} a_\alpha D^\alpha u. \tag{1.49}$$

Assume that the real parts of all p-roots of the polynomial $\det\left(pI - \sum\limits_{|\alpha|\leq 2b} a_\alpha(i\xi)^\alpha\right)$ are different from zero for any $\xi \in \mathbb{R}^n$. Let $\Gamma(x,t)$ be a fundamental solution of the Cauchy problem for system (1.49). Then $\mathcal{E}(x-y) = \int\limits_0^\infty \Gamma(x-y,t)\,dt$ is the principal fundamental solution of the elliptic system $\sum\limits_{|\alpha|\leq 2b} a_\alpha D^\alpha u = 0$, and the estimate $|\mathcal{E}(x-y)| \leq C\exp\{-c|x-y|\}$ holds for $|x-y| > 1$.

Let us formulate some corollaries that can be derived by investigating the weak fundamental solutions of equation (1.28).

In $\mathbb{R}^n$ consider the uniformly elliptic equation

$$-\sum_{i,j=1}^{n} D_{x_i}\big(a_{ij}(x)D_{x_j}\big)u + \lambda p(x)u = 0, \tag{1.50}$$

assuming that conditions β_{14} and β_{15} are satisfied. For brevity, we omit the definitions of the notions of weak solution, weak fundamental solution, and principal weak fundamental solution, since they are similar to those for parabolic systems (λ is a nonnegative real parameter).

Theorem VI.12. *In the case $n \geq 3$, and $n = 2$, $\lambda > 0$, the function*

$$\mathcal{E}(x,y;\lambda) = \int_0^\infty e^{-\lambda t}\Gamma(x,y,t)\,dt$$

is the unique principal weak fundamental solution of equation (1.50) that satisfies the two-sided estimate $\tilde{g}_1(x-y;\lambda) \leq \mathcal{E}(x,y;\lambda) \leq \tilde{g}_2(x-y;\lambda)$. Here the function

$$\tilde{g}(x,\lambda) = k\int_0^\infty t^{-n/2}\exp\big\{-\lambda t - k|x|^2 t^{-1}\big\}dt = k\big(\sqrt{\lambda}/|x|\big)^{n/2-1}K_{n/2-1}\big(k\sqrt{\lambda}|x|\big)$$

is the fundamental solution of the equation $-\Delta u + \lambda k u = 0$, $K_\nu(z)$ is the Macdonald function [67], and $\tilde{g}_1$ and $\tilde{g}_2$ are functions with different constants k_1 and k_2, depending on n and μ.

Using the asymptotics of the Macdonald functions or estimating directly, one can show that
1)   for $n \geq 3$, $\lambda \geq 0$, and also for $n = 2$, $\sqrt{\lambda}|x-y| \geq 1$,

$$\mathcal{E}(x,y;\lambda) \sim k|x-y|^{1-n}\exp\big\{-k\sqrt{\lambda}|x-y|\big\};$$

2)   for $n = 2$, $0 < \sqrt{\lambda}|x-y| \leq 1$,

$$\mathcal{E}(x,y;\lambda) \sim k + k\ln\big(1/\sqrt{\lambda}|x-y|\big).$$

Here the symbol "$\sim$" means that the second function bounds the first one both from above and from below, with different values of k.

Our study of the relationships between fundamental solutions of parabolic and elliptic systems will be continued in Chapter VII, where we will analyze the stabilization of solutions to parabolic boundary value problems. Using the Poisson matrices for such problems to construct special asymptotic expansions of solutions (for $t \to \infty$), one is naturally led to Poisson kernels for elliptic boundary value problems generated by parabolic problems. Note that Chapter VII we give the proofs of the results discussed.

VI.2. The Cauchy problem

VI.2.1. Introduction

The fundamental solutions, their analytical description and the estimates obtained in Section VI.1 enable us to obtain rather complete results on existence and uniqueness of solutions of the Cauchy problem and their continuous dependence on the initial values and the right-hand sides. It turns out that the most efficient way to deduce sharp results is to combine the method of fundamental solutions and a priori estimates in various functional spaces. It is also important that usually the solution of the Cauchy problem can be represented in terms of fundamental solutions in integral form, convenient for subsequent analysis and applications.

To establishing uniqueness of the solutions of Cauchy problems one uses the methods based on fundamental solutions and a priori estimates, as well as a number of other powerful methods based on the theory of generalized functions, theorems on quasianalytic classes of functions, and the theory of positive solutions of partial differential equations. Note that the precise uniqueness classes of solutions of Cauchy problems are often wider that the well-posedness classes; uniqueness may hold also in situations when Cauchy problems whose initial functions have a an appropriate growth have no solutions or the existence question is open.

In the last part of this section we will obtain integral representations of solutions $u(x, t)$ of parabolic systems defined in the semi-open layer $\mathbb{R}^n \times (0, T]$ in terms of the limit values of $u(x, t)$ as $t \to 0$.

VI.2.2. Well-posedness

Here we study the well-posedness of the Cauchy problem for systems parabolic in the sense of Petrovskiĭ:

$$\mathcal{L}(x, t; D, D_t)u \equiv \left(D_t - \sum_{|\alpha| \leq 2b} a_\alpha(x, t)D^\alpha \right)u = f(x, t), \qquad u\big|_{t=0} = \psi(x). \quad (2.1)$$

The estimates (1.13) for the fundamental solution $\mathcal{Z}(x, t; y, \tau)$ immediately imply that the functions $f(x, t)$, $\psi(x)$ can be allowed to grow like $\exp\left\{ a \sum_{i=1}^{n} |x_i|^q \right\}$ when $|x| \to \infty$, and then the class of solutions considered should also be subjected to this condition. Note that already in the class of functions that grow like $\exp\left\{ a \sum_{i=1}^{n} |x_i|^{q+\varepsilon} \right\}$ for any $\varepsilon > 0$, there is no uniqueness of the solutions of problem (2.1) defined in the layer Π. Simple considerations show that in this class even the existence of a solution defined in Π for arbitrarily small T is not guaranteed.

Here is an extremely simple example.

Example VI.7. Let us find the solution of the Cauchy problem for the heat equation

$$D_t u = D_x^2 u, \qquad u\big|_{t=0} = \exp\{ax^2\}. \quad (2.2)$$

The desired solution is calculated immediately:

$$u(x,t) = \frac{1}{2\sqrt{\pi t}} \int\limits_{-\infty}^{\infty} \exp\left\{-\left[|x-y|^2/4t + ay^2\right]\right\} dy$$

$$= (1-4at)^{-1/2} \exp\left\{ax^2/(1-4at)\right\}.$$

(2.3)

Formula (2.3) implies that the solution of problem (2.2) exists in the layer $\Pi_{(0,T]}$, $T = (1/4a) - \varepsilon$, with arbitrary $\varepsilon > 0$, that $u(x,t) \to \infty$ as $t \to (1/4a)$, that $u(x,t)$ has second-order growth (for $|x| \to \infty$), and that the growth type is $k_1(t,a) = a/(1-4at)$. The function $k_1(t,a)$, which plays an important role below, possesses an interesting semigroup property, $k_1\left(t - \tau, k_1(\tau,a)\right) = k_1(t,a)$, which can be verified directly.

In the case of system (2.1) the role of the function $k_1(t,a)$ is played by the function $k_b(t,a) = ca[c^{2b-1} - a^{2b-1}t]^{1-q} \equiv k(t)$, where c is a positive constant, depending on the parabolicity constant δ_0. The function $k_b(t,a)$ also possesses the semigroup property.

Lack of space prevents us from discussing various well-posedness theorems for problem (2.1) [108, 43, 65]. We restrict our considerations to the formulation and brief discussion one such result which in our opinion contains rather precise information.

Consider the Banach space $C^{2b,1}_{k(t)}(\overline{\Pi})$ of the vector-valued functions $u(x,t)$ that have in $\overline{\Pi}$ continuous derivatives with respect to $x_1, \ldots, x_n$ up to order $2b$, a continuous derivative with respect to t, and a finite norm

$$\left\|u; \Pi\right\|_{2b,1}^{k(t)} = \sup_{(x,t)\in\overline{\Pi}} \left\{\left(\sum_{|\alpha|\leq 2b} |D^\alpha u| + |D_t u|\right) E(x,t)\right\},$$

$$E(x,t) = \exp\left\{-k(t) \sum_{i=1}^{n} x_i^q\right\}.$$

Let $\omega(h)$, $F(h)$ be nonnegative nondecreasing semiadditive bounded functions (functions of modulus-of-continuity type). Let $\mathcal{D}^{2b,1}_{\omega,F}(\Pi)$ denote the subspace of the Banach space $C^{2b,1}_{k(t)}(\Pi)$ consisting of the functions $u(x,t) \in C^{2b,1}_{k(t)}(\overline{\Pi})$ for which the norm

$$\left\|u; \Pi\right\|_{2b,1}^{\omega,F} = \left\|u; \Pi\right\|_{2b,1}^{k(t)} + \left\|u; \Pi\right\|_{2b,F}^{k(t)} + \left\|u(x,0); \mathbb{R}^n\right\|_{2b,\omega}^a + \left\|\mathcal{L}u; \Pi\right\|_{0,\omega}^{k(t)}, \quad (2.4)$$

is finite; here

$$\left\|u; \Pi\right\|_{2b,F}^{k(t)} = \sum_{|\alpha|=2b} \sup_{\{(x,t),(\tilde{x},t)\}\subset\Pi} \left\{\frac{\left|\Delta_x^{\tilde{x}} D^\alpha u(x,t)\right|}{F\left(|x-\tilde{x}|\right)} \cdot \left[E(x,t) + E(\tilde{x},t)\right]\right\},$$

and the other norms in (2.4) are defined similarly. Also, denote by $D_F(\Pi)$ the subspace of the space

$$C_{k(t)}(\Pi) = \left\{ u(x,t) : \|u; \Pi\|^{k(t)} = \sup_{(x,t)\in\Pi} |u(x,t)| \cdot E(x,t) \right\}$$

consisting of the functions $u(x,t)$ with for which the quasinorm

$$\|u; \Pi\|_F^{k(t)} = \sup_{\{(x,t),(\tilde{x},t)\}\subset\Pi} \left\{ \frac{|\Delta_x^{\tilde{x}} u(x,t)|}{F(|x-\tilde{x}|)} \cdot [E(x,t) + E(\tilde{x},t)] \right\}$$

is finite. Finally, let $\mathcal{D}_{\omega,a}^{2b}(\mathbb{R}^n)$ the subspace of $C_a^{2b}(\mathbb{R}^n)$ consisting of the functions $u(x,0)$ for which the quasinorm

$$\|u(x,0); \mathbb{R}^n\|_{2b,F}^a = \sum_{|\alpha|=2b} \sup_{\{x,\tilde{x}\}\subset\mathbb{R}^n} \left\{ \frac{|\Delta_x^{\tilde{x}} D^\alpha u(x,0)|}{\omega(|x-\tilde{x}|)} \cdot [E(x,0) + E(\tilde{x},0)] \right\}$$

is finite.

To study the Cauchy problem we will use the function spaces described above in the particular case when $\omega(h)$ satisfies the Dini condition and $F(h) = \int_0^h (\omega(z)/z)\,dz \equiv \omega_1(h)$.

Define the Banach spaces $B_1 = \mathcal{D}_{\omega,\omega_1}^{2b,1}(\Pi)$, $B_2 = \mathcal{D}_{\omega_1}(\Pi) \times \mathcal{D}_{\omega,a}^{2b}(\mathbb{R}^n)$ (the same letters will also denote the corresponding spaces of vector-valued functions), the operator of the Cauchy problem (2.1), $\mathfrak{A}u = (\mathcal{L}u, Iu|_{t=+0})$, and the vector $\chi = (f(x,t), \psi(x))$.

Theorem VI.13. 1) *Let the coefficients of system* (2.1) *satisfy conditions* β_1, β_2 *and* β_5. *Then the operator* $\mathfrak{A}$ *defines a one-to-one continuous mapping between the spaces* B_1 *and* B_2.

2) *If, in addition, condition* β_6 *is satisfied, then the unique solution of* (2.1) *in* $\mathcal{D}_{\omega,\omega_1}^{2b,1}(\Pi)$ *can expressed by means of the fundamental solution* $\mathcal{Z}(x,t;y,\tau)$ *by the formula*

$$u(x,t) = \int_{\mathbb{R}^n} \mathcal{Z}(x,t;y,0)\psi(y)\,dy + \int_0^\tau d\tau \int_{\mathbb{R}^n} \mathcal{Z}(x,t;y,\tau)f(y,\tau)\,dy. \qquad (2.5)$$

Let us make a few comments on Theorem VI.13. One is that this theorem has the following consequence: Suppose that the coefficients of system (2.1) are continuous, bounded, and satisfy the Dini condition in x. Then, in the class of functions that grow like $\exp\left\{ k(t) \sum_{i=1}^n x_i^q \right\}$, there exists a unique solution of the Cauchy problem (2.1) corresponding to an initial function $\psi(x)$, of the same smoothness as the solution, and to a right-hand side $f(x,t)$ that satisfies the Dini condition in x; moreover, this solution satisfies the two-sided estimate

$$C_1\|\chi\|_{B_2} \leq \|u;\Pi\|_{2b}^{\omega,\omega_1} \leq C_2\|\chi\|_{B_2}, \tag{2.6}$$

$$\|\chi\|_{B_2} = \|\psi;\mathbb{R}^n\|_{2b}^{a} + \|\psi;\mathbb{R}^n\|_{2b}^{a,\omega} + \|f;\Pi\|_{0}^{k(t)} + \|f;\Pi\|_{\omega}^{k(t)}.$$

The proof of Theorem VI.13 is presented in [65]. It is based on a priori estimates of solutions of the Cauchy problem in the above-defined Dini spaces. As usual, the proof uses properties of various parabolic potentials, sometimes quite subtle.

Let us continue our discussion of the contents of Theorem VI.13.

Definition VI.8. *Two functions $F(h)$ and $\omega(h)$ of modulus-of-continuity type are said to be equivalent if there exist positive constants C_1 and C_2 such that $C_1\omega(h) \leq F(h) \leq C_2\omega(h)$.*

For equivalent F and ω, the spaces $\mathcal{D}_{\omega,F}^{2b,1}$ and $\mathcal{D}_{\omega,\omega}^{2b,1}$ coincide. In the case considered in Theorem VI.13, an esential role is played by the question whether the functions $\omega(h)$ and $\omega_1(h) = \int\limits_0^h (\omega(z)/z)\,dz$ are equivalent (of course, under the assumption that this integral exists).

Lemma VI.1. *The functions $\omega(h)$ and $\omega_1(h)$ are equivalent if and only one of·the following two conditions is satisfied:*
1) $\lim\limits_{z\to 0} (\omega(2z)/\omega(z)) = 1;$
2) $(\varphi(2z)/\varphi(z)) \leq C$ *for $z \in (0,A)$, where $\varphi(z)$ is the inverse function to $\omega(z)$.*

Example VI.8. Here are some functions $\omega(h)$ equivalent to $\omega_1(h)$: 1) $\omega(h) = h^\lambda$, $\lambda \in (0,1]$; 2) $\omega(h) = h^\lambda\big(\ln\frac{1}{h}\big)^\mu$, $\lambda \in (0,1)$. The validity of condition 1) of Lemma VI.1 is verified directly.

Thus, among the particular cases of Theorem VI.13 there are, for example, sharp well-posedness theorems in classes of Hölder functions and classes of functions with the modulus of continuity $\omega(h) = h^\lambda\big(\ln\frac{1}{h}\big)^\mu$.

One can also consider some other variants of well-posedness theorems for the Cauchy problem:
1) The smoothness of the initial function is not required (we can assume, for example, that it is continuous or possesses some minimal smoothness properties). Then, the description of the classes in which the Cauchy problem is well posed will involve assumptions on the behavior of the derivatives of solutions and their moduli of continuity for $t \to 0$.
2) In the hypotheses one can replace the moduli of continuity with respect to the space variables by the moduli of continuity with respect to the parabolic distance $d\big((x,t),(\tilde{x},\tilde{t})\big) = \big(|x-\tilde{x}|^2 + |t-\tilde{t}|^{1/b}\big)^{1/2}$.

3)   One can construct well-posedness classes for the Cauchy problem in the whole
     range of smooth increasing functions with the higher-order derivatives belong-
     ing to the Dini or Hölder classes.

     All the variants listed were considered in [41, 60, 64, 65, 108].

Let us give some examples that demonstrate the sharpness of the results described
above.

Example VI.9. Consider the volume heat potential in $\mathbb{R}^3$

$$u(x,t) = \int_0^t d\tau \int_{\mathbb{R}^3} \left(2\sqrt{\pi(t-\tau)}\right)^{-3} \exp\left\{-|x-y|^2/4(t-\tau)\right\} f(y)\,dy,$$

where $f(x)$ is a continuous function with bounded support that does not vanish in
the ball of radius $R_0 < 1$ centered at the origin. Change the order of integration and
in the integral with respect to τ introduce the new variable $\tilde{\tau} = |x-y|^2(t-\tau)^{-1}$.
Then $u(x,t)$ takes the form

$$\begin{aligned}
u(x,t) = {}&(1/4\pi) \int_{\mathbb{R}^3} |x-y|^{-1} f(y)\,dy \\
&- \left(2\sqrt{\pi}\right)^{-3} \int_{\mathbb{R}^3} |x-y|^{-1} f(y) \int_0^{|x-y|^2/t} \exp\left\{-\tilde{\tau}/4\right\} \tilde{\tau}^{-1/2}\,d\tilde{\tau}\,dy.
\end{aligned}$$

(2.7)

It is easy to verify that for any continuous $f(x)$, the second term has two
continuous derivatives with respect to x_1, x_2, x_3 and one derivative with respect to
t, all of which satisfy the Hölder condition of order $\frac{1}{2}$ with respect to x. Consider
the first term in (2.7):

$$J(x) = (1/4\pi) \int_{\mathbb{R}^3} |x-y|^{-1} f(y)\,dy. \tag{2.8}$$

This volume potential for the Laplace equation is studied in detail in the classical
monograph of N.M. Günter [20]. Following this monograph, let us take $f(x) = \left\{\left(3x_1^2/|x|\right) - 1\right\}\chi(|x|)$, $\chi(0) = 0$, where $\chi(r)$ is a nondecreasing continuous positive
function. It is shown in [20] that if $\chi_1^{(\varepsilon)}(h) = \int_\varepsilon^h (\chi(r)/r)\,dr$ does not tend to a finite
limit as $\varepsilon \to 0$, then $J(x)$ has no second derivatives at the origin. Therefore, in the
present situation Dini's condition is necessary in order to ensure that the potential
$J(x)$ will have two derivatives. Further analysis shows that if the integral $\chi_1(h) = \int_0^h (\chi(r)/r)\,dr$ exists but the functions $\chi(h)$ and $\chi_1(h)$ are not equivalent, then
the modulus of continuity of the second derivatives is exactly $\chi_1(h)$, in complete
agreement with Theorem VI.13.

Example VI.10 [46]. Consider the equation

$$D_t u = a(x,t)D_x^2 u, \qquad a(x,t) = 1 + b(x,t), \tag{2.9}$$

where

$$b(x,t) = -\big(\ln(1/t)\big)^{\gamma_1} \quad \text{for} \quad 0 < (x^2/t) \leq 2 - \varepsilon_1\big(\ln(1/t)\big)^{-\varepsilon\gamma},$$

$$b(x,t) = 0 \qquad\qquad \text{for} \quad 2 - 2\varepsilon_1\big(\ln(1/t)\big)^{-\varepsilon\gamma_1} \leq 1,$$

$$b(x,t) = \big(\ln(1/t)\big)^{\gamma_1} \quad \text{for} \quad (x^2/t) \geq 2 + \varepsilon.$$

For the remaining values of (x,t), $b(x,t)$ is continued as an even function of x so that it will be nondecreasing for $x > 0$, smooth, and bounded $\big(|b(x,t)| \leq \frac{1}{2}\big)$; here $0 < \gamma_1 < 2\gamma < 1$ and the numbers $\varepsilon_1, \varepsilon_2$ are selected in a special way.

As established in [46], the modulus of continuity in x of the function $a(x,t)$ is equal to $\omega(h) = \big(\ln(1/h)\big)^{\gamma_1}$. For $\gamma_1 \in (0,1]$, $a(x,t)$ does not satisfy the Dini condition. For $\gamma_1 \in (0,1)$ one can construct a smooth positive function $v(x,t)$ such that the solution $u_\tau(x,t)$ of the Cauchy problem $u_\tau\big|_{t=\tau} = v(x,\tau)$ for equation (2.9) will have the property that

$$\lim_{\tau \to 0} \left(u_\tau(x,t) \Big/ \int_{-\infty}^{\infty} v(y,\tau)dy \right) = \infty. \tag{2.10}$$

If equation (2.9) would have a fundamental solution with the usual properties (see Theorem VI.3), then

$$u_\tau(x,t) = \int_{-\infty}^{\infty} \mathcal{Z}(x,t;y,\tau)v(y,\tau)dy. \tag{2.11}$$

Since the functions $v(x,t)$ and $u_\tau(x,t)$ are positive, the estimates for the fundamental solution $\mathcal{Z}(x,t;y,\tau)$ and (2.11) would then imply the inequality

$$u_\tau(x,t) \leq (t-\tau)^{-1/2} \int_{-\infty}^{\infty} v(y,\tau)dy,$$

which contradicts (2.10). Thus we have shown that if we drop the Dini condition on the coefficients of the equation, then, generally speaking, there is no fundamental solution with the natural properties.

Example VI.11 [52]. Let $\omega(h)$ be a modulus of continuity that does not satisfy the Dini condition:

$$\int_0^2 \big(\omega(h)/h\big)dh = \infty. \tag{2.12}$$

Using $\omega(h)$, construct a convex modulus of continuity $\lambda(h)$ that has continuous derivatives of order $N \geq 2b$ on any interval $[\delta, 2] \subset (0, 2]$ and possesses the following properties: 1) (2.12) holds; 2) $\lambda(h) \leq \omega(h)$; 3) $\left|\lambda^{(k)}(h)\right| \leq Ch^{-k}\lambda(h)$, $k = 1, \ldots, 2b$; 4) $\left(\lambda(h)/h\right)' < 0$.

Then the function

$$v(x_1, x_2, x_3, t) = \varepsilon x_1^b x_2^b \int\limits_0^{\sqrt{r^2+t^2}} \left(\lambda(h)/h\right) dh + \left(x_1^{2b}/(2b)!\right) + (-1)^{b-1}t, \quad r^2 = |x|^2,$$

defined in the cylinder $\Omega = \left\{(x, t) : r^2 < \frac{1}{2}, -\frac{1}{2} \leq t \leq 0\right\}$ satisfies, for sufficiently small ε and $r^2 + t^2 \neq 0$, the uniformly parabolic equation

$$D_t u(x_1, \ldots, x_n, t) = (-1)^{b-1} B(x_1, x_2, x_3) \left[D_{x_1}^{2b} + D_{x_2}^{2b} + \sum_{i=1}^{n} D_{x_i}^{2b}\right] u(x_1, \ldots, x_n, t),$$

where B is a positive function whose modulus of continuity in Ω is no larger than $\omega(h)$. It is clear that $\lim\limits_{r^2+t^2 \to 0} D_{x_1}^b, D_{x_2}^b, v = -\infty$ and that and on the lateral surface and the lower base of the cylinder Ω the function $v(x, t)$ can be considered to be arbitrarily smooth. Thus, for parabolic equations of arbitrary order droping the Dini conditions on the coefficients results in the absence of classical solutions (in the case under consideration, those of the Dirichlet problem).

General results of the Schauder theory of parabolic boundary value problems in classes of smooth functions, presented in the next section, yield sharp well-posedness theorems of the Cauchy problem for completely general systems parabolic in the sense of Petrovskiĭ and of the initial value problem for systems parabolic in the sense of Solonnikov in classes of bounded functions. Note that Theorem VI.13 has weaker assumptions on smoothness of the coefficients in comparison with the Schauder theory, and that the Cauchy data are allowed to grow exponentially. Both in Theorem VI.13 and the Schauder theory one establishes theorems on the isomorphisms of the Banach spaces to which the Cauchy data and the corresponding solutions of the Cauchy problem belong and two-sided estimates of the type of (2.6).

VI.2.3. Existence of a solution for systems with growing coefficients

Theorems VI.4 and VI.5 yield almost at once theorems on the existence of solutions of the Cauchy problem for a system (2.1) with growing coefficients.

Consider first a system with dissipation (Definition VI.3). The dissipation characteristic $f_0(x)$ is used to define the function $g(x, t)$, which satisfies inequalities (1.19). Next, using $g(x, t)$, let us formulate the following conditions on the initial function $\psi(x)$ and the right-hand side of the system $f(x, t)$ in (2.1):

β_{19}: $\psi(x)$ and $f(x, t)$ are defined and continuous in $\mathbb{R}^n$ and $\overline{\Pi}$, respectively.

β_{20}: *The function $\psi(x)$ has finite norm*

$$\|\psi\|^{g(x,0)} = \sup_x \big\{ |\psi(x)| \exp\big(- g(x,0)\big)\big\}.$$

β_{21}: $f(x,t)$ *has finite norm*

$$\|f(x,t)\|_\lambda^{g(x,t)} = \sup_{(x,t)\in\Pi} \big\{ |f(x,t)| \exp\big(-g(x,t)\big)\big\}$$

$$+ \sup_{\{(x,t),(\tilde{x},t)\}\subset\Pi} \left\{ \frac{\Delta_x^{\tilde{x}} f(x,t)}{|x-\tilde{x}|^\lambda} \cdot \Big(\exp\big\{-g(x,t)\big\} + \exp\big\{-g(\tilde{x},t)\big\}\Big)\right\}.$$

Theorem VI.14. *Suppose that (2.1) is a dissipative system and let conditions β_9, β_{10}, or β_{11} (with the dissipation characteristic $f_0(x)$ possessing the properties 1), 2), 3) of the latter condition) be satisfied. Further, suppose that $\psi(x)$ satisfies conditions β_{19}, β_{20}, and $f(x,t)$ satisfies conditions β_{19}, β_{21}. Then there exists a solution $u(x,t)$ of problem (2.1) that can be represented by means of the fundamental solution $\mathcal{Z}(x,t;y,\tau)$ (see Theorem VI.4) via formula (2.5). Moreover, the derivatives of $u(x,t)$ satisfy the estimates*

$$\|D^\alpha u\|^{g(x,t)} \le C\left(\sum_{j=0}^{|\alpha|} t^{-(|\alpha|-j)/2b} f_0^j(x) \|\psi\|^{g(x,0)} + f_0(x)^{|\alpha|} \|f(x,t)\|_\lambda^{g(x,t)}\right),$$

$$|\alpha| \le 2b. \qquad (2.13)$$

Using a simple method suggested in [108] on can derive from Theorem VI.14 an existence theorem for problem (2.1) for a system with growing coefficients.

Theorem VI.15. *Suppose that the coefficients of the system (2.1) satisfy conditions β_{12}, β_{13} or β_{12}, $\widetilde{\beta}_{13}$ and that the functions $\psi(x)$ and $f(x,t)$ satisfy the assumptions of the previous theorem, where the role of $g(x,t)$ is played by a smooth function, equal to $a\big[1 - (Aa)^{2b-1}t\big]^{-1/(2b-1)} \sum_{j=1}^r x_j^q$ for $|x| \ge 1$; here a is an arbitrary positive number and A is a positive constant, $0 \le t \le T_1 = \frac{1}{2}(Aa)^{1-2b}$, and $f_0(x) = \big(1+|x|\big)^{1-q}$). Then the Cauchy problem (2.1) in Π_{T_1} has a solution $u(x,t)$, which can be represented by formula (2.5) in terms of the fundamental solution $\mathcal{Z}(x,t;y,\tau)$ (see Theorem VI.5). Morevoer, the derivatives of $u_2(x,t)$ satisfy estimates (2.13) with the functions $g(x,t)$ and $f_0(x)$ described above.*

Thus the class in which the solutions of the Cauchy problem exist remains the same when we pass from a system with bounded coefficients to one whose coefficients grow as $|x| \to \infty$, provided that $a_\alpha(x,t)$ grows like $|x|^{(2b-|\alpha|)/(2b-1)}$. For dissipative systems, the classes in which the solutions exist can be substantially enlarged by selecting a rapidly growing dissipation characteristic $f_0(x)$. Of course, taking this approach one can establish uniqueness and well-posedness theorems for the Cauchy problem in the indicated classes of growing functions.

VI.2.4. Uniqueness

For the case of the very simple parabolic equation

$$D_t u = (-1)^{b-1} D_x^{2b} u \tag{2.14}$$

S. Täcklind has shown [125] that for uniqueness of solutions to hold for the Cauchy problem in the class of functions

$$|u(x,t)| \le C \exp\left\{|x|h(|x|)\right\}, \tag{2.15}$$

it is necessary and sufficient that the integral

$$\int\limits_1^\infty h(r)^{1-2b} dr \tag{2.16}$$

diverges.

Täcklind's proof of this assertion uses the fundamental solution of equation (2.14), the Green formulas, and a special construction, known as "Täcklind's ladder", which connects the motion along the time and spatial coordinates. The construction of the solutions that demonstrate non-uniqueness is based on the Carleman-Ostrowski theorem on quasianalytic classes of functions.

Let us state a typical result of Täcklind type.

Theorem VI.16 [108]. *1) Let (2.1) be a system parabolic in the sense of Petrovskiĭ whose coefficients satisfy conditions β_1–β_4, and let $u(x,t)$ be a classical solution in Π of the Cauchy problem with null initial values such that $u(x,t)$ belongs to the class (2.15) with a function $h(r)$ for which the integral (2.16) diverges. Then $u(x,t)$ vanishes identically.*

2) In the class of functions (2.15) with a function $h(r)$ for which the integral (2.16) converges there exists a nontrivial solution $v(x,t)$ of the constant-coefficients parabolic system $D_t u = \sum\limits_{|\alpha|=2b} a_\alpha D^\alpha u$ with null initial values.

Presently theorems of this kind are known for quite diverse problems and systems of equations. Note, for example, the work of N.N. Chaus [107], in which Täcklind-type theorems have been proved for the Cauchy problem for arbitrary systems with constant coefficients of the form $D_t u = A(D)u$.

In our opinion, there is another aspect worth investigating. Usually one argues as follows: let $u_1(x,t)$, $u_2(x,t)$ be two solutions of a given system of equations, constructed for the same initial values. Then $u(x,t) = u_2(x,t) - u_1(x,t)$ is a solution of the homogeneous system, constructed for null initial conditions. Therefore, a class of uniqueness is defined by the fact that the problem with null initial conditions has only the trivial solution in that class. The solution corresponding to null initial conditions always exists in any class. The question is then what determines the uniqueness? Are there simply no solutions in a wider class of functions for such rapidly growing initial data, or do solutions exist, and their coincidence at the initial time implies coincidence in the whole domain considered? Interestingly, using Täcklind's arguments (see [125]) in [76], it has been proved that if the initial function $\psi(x)$ satisfies the inequality $\psi(x) \geq \exp\{|x|h(|x|)\}$, where h is such that $\lim_{r\to\infty}(h(r)/r) = \infty$, then there is no strip Π in which the Cauchy problem for the heat equation admits a solution. A natural domain in which the solution of the Cauchy problem does exist is defined in this case in a special way, depending on the function $\psi(x)$ and, generally speaking, is not a strip.

There is another important class of uniqueness theorems. The first theorem of this class is the following classical result of Widder [127]: *any solution, nonnegative in Π, of the heat equation with null initial data vanishes identically.* The assumption on positivity may be the only information available on the solution under consideration, for example, in problems of probabilistic origin. The theory of positive solutions of partial differential equations [51] enables us, in particular, to deduce various generalizations and refinements of this theorem. For the sake of simplicity, let us consider the case of a single equation parabolic in the sense of Petrovskiĭ:

$$D_t u = \sum_{|\alpha|\leq 2b} a_\alpha(x,t)D^\alpha u \tag{2.17}$$

We are interested here in *weak real-valued solutions.*

Theorem VI.17 [51]. *Let the coefficients of (2.17) satisfy the conditions $\widetilde{\beta}_1$ and β_2–β_4, as well as the condition β_{22}: $a_\alpha(x,t)$ are real-valued functions in Π. Then the following assertions are valid:*

1) If the negative part $u^-(x,t)$ of the function $u(x,t)$ is such that

$$\int_{\mathbb{R}^n} u^-(x,t)\exp\{-|x|h(|x|)\}dx < \infty, \tag{2.18}$$

where $h(r)$ is a function for which the integral (2.16) diverges, then $u(x,+0) = 0$ implies $u(x,t) \equiv 0$ in Π.

2) Let $b = 2s+1$ and let $u_1(x,t)$, $u_2(x,t)$ be two weak solutions of equation (2.17) such that their negative parts satisfy condition (2.18) with a function $h(r)$ for which the integral (2.16) diverges and $u_1(x,+0) = u_2(x,+0)$. Then $u_1(x,t) \equiv u_2(x,t)$ in Π.

Note that we have no a priori information on the difference of two positive functions, and so to prove the second assertion of Theorem VI.16 (and it is this assertion that to us seems more natura) we need first a special proposition on the possible growth of the positive part of a solution. It is also interesting that in this form of uniqueness theorems, the assumption that b is an odd number becomes essential.

The uniqueness of solutions of the Cauchy problem for systems with growing coefficients is a subject that requires a separate study.

Note, for example, that for the equation $D_t u = D_x^2 u + q(x)u$, where $q(x)$ is a smooth function that coincides with $|x|^{2+\varepsilon}$ for $|x| > 1$ and ε is an arbitrary positive number, uniqueness is lost even in the space $L_2(\mathbb{R}^n)$. On the other hand, if $q(x)$ coincides with $|x|^{2-\varepsilon}$ for $|x| > 1$, then the usual uniqueness theorems are valid. Finally, for the case when $q(x) \to -\infty$ as $|x| \to \infty$, the classes of uniqueness may be considerably enlarged for rapidly growing $|q(x)|$.

VI.2.5. Initial values for solutions of parabolic systems. Integral representation of solutions

Here we consider solutions of the system parabolic in the sense of Petrovskiĭ

$$D_t u = \sum_{|\alpha| \leq 2b} a_\alpha(x,t) D^\alpha u, \tag{2.19}$$

whose coefficients satisfy conditions β_1–β_4 in the layer $\overline{\Pi}$. By Theorem VI.2, these conditions guarantee the existence of a fundamental solution $\mathcal{Z}(x,t;y,\tau)$ that satisfies estimates (1.13), (1.14); condition β_4 implies also that the fundamental solution $\mathcal{Z}(x,t;y,\tau)$ is normal.

Here we discuss necessary and sufficient conditions under which the solutions defined in a semi-open layer $\mathbb{R}^n \times (0,T]$ can be represented by a Poisson integral with some functions or generalized measures. These functions or measures form the sets of initial values (data) of the solutions under investigation. The presentation below follows to a large extent the work of S.D. Ivasishen [45], in which such results were obtained for a wider class of $2\vec{b}$-parabolic systems, defined by one of the authors in [109] (see also [108, 41]).

Let us introduce the necessary notations and definitions. We denote by c_0 a positive constant, which is smaller than the constant c of inequality (1.13). Let a be a positive constant and let $T_0 < \min(T, c_0/a)$; put

$$\Pi_0 = \mathbb{R}^n \times (0,T_0], \qquad k(a,t) = c_0 a \big(c_0^{2b-1} - a^{2b-1}\big)^{1-q},$$

$$E(x,t,a) = \exp\left\{-k(a,t) \sum_{i=1}^{n} x_i^q\right\}.$$

For $t \in [0,T_0]$ and $1 \leq p \leq \infty$ define the norms

$$\big\|u(x,t)\big\|_p^{k(a,t)} = \big\|E(x,t,a)u(x,t); \mathbb{R}^n\big\|_{L_p}.$$

Further, denote by $L_p^{k(a,0)}$ the space of all measurable functions $\psi(x)$, for which the norm $\big\|\psi(x)\big\|_p^{k(a,0)}$ is finite.

Let $\mathcal{B}_n$ be the σ-algebra of the Borel sets of $\mathbb{R}^n$, and let M^n be the set of all countably additive functions $\nu : \mathcal{B} \to \mathbb{C}^m$ (generalized Borel measures ν) that have a finite total variation $|\nu|(\mathbb{R}^n)$. If we introduce a norm for measures ν by the rule

$\|\nu\| = |\nu|(\mathbb{R}^n)$, then M^n becomes a Banach space. This space can be identified with the dual of the space C_0^n of all continuous functions $\varphi : \mathbb{R}^n \to \mathbb{C}^m$ satisfying $|\varphi(x)| \to 0$ as $|x| \to \infty$, equipped with the uniform norm. By $M^{k(a,0)}$ we designate the set of all generalized Borel measures $\mu : \mathcal{B}_n \to \mathbb{C}^m$ with the property that the set function

$$\nu(A) = \int_A E(x,0,a)d\mu(x), \qquad A \in \mathcal{B}_n,$$

belongs to the space M^n. Then

$$\|\mu\|^{k(a,0)} = \int_{\mathbb{R}^n} E(x,0,a)d|\mu|(x) < \infty$$

for any $\mu \in M^{k(a,0)}$.

Theorem VI.18. 1. Let $\psi(\dot{x}) \in L_p^{k(a,0)}$, $1 \leq p \leq \infty$. Then the formula

$$u(x,t) = \int_{\mathbb{R}^n} \mathcal{Z}(x,t;y,0)\psi(y)dy, \qquad (x,t) \in \Pi_0, \tag{2.20}$$

gives the unique solution of system (2.19) in Π_0 that possesses the following properties:

1.1. There exists a positive constant C, which does not depend on $\psi(x)$, such that

$$\|u(x,t)\|_p^{k(a,t)} \leq C\|\psi(x)\|_p^{k(a,0)} \qquad for \ any \ t \in (0,T_0].$$

1.2. For $1 \leq p < \infty$,

$$\lim_{t \to 0} \|u(x,t) - \psi(x); \mathbb{R}^n\|_p^{k(a,t)} = 0,$$

and for $p = \infty$

$$\lim_{t \to +0} \int_{\mathbb{R}^n} (u(x,t), \varphi(x))\, dx = \int_{\mathbb{R}^n} (\psi(x), \varphi(x))\, dx \qquad for \ all \ \varphi \in L_1^{-k(a,T_0)},$$

where $L_1^{-k(a,T_0)}$ denotes the set of all measurable functions $\varphi : \mathbb{R}^n \to \mathbb{C}^m$ for which the norm

$$\int_{\mathbb{R}^1} \exp\left\{-k(a,T_0)\sum_{i=1}^n x_i^q\right\} |\varphi(x)|\, dx$$

is finite.

2. Let $\mu \in M^{k(a,0)}$. Then the formula

$$u(x,t) = \int_{\mathbb{R}^n} \mathcal{Z}(x,t;y,0)d\mu(y), \qquad (x,t) \in \Pi_0, \qquad (2.21)$$

gives the unique solution of system (2.19) in Π_0 that possesses the following properties:

2.1. There exists a positive constant C, which does not dependent on μ, such that

$$\left\| u(x,t) \right\|_1^{k(a,t)} \leq C \|\mu\|^{k(a,0)} \qquad for\ all\ t \in (0,T_0].$$

2.2. For any $\varphi \in C_0^{-k(a,T_0)}$,

$$\lim_{t \to +0} \int_{\mathbb{R}^n} \big(u(x,t), \varphi(x)\big)dx = \int_{\mathbb{R}^n} \big(\varphi(x), d\mu(x)\big),$$

where $C_0^{-k(a,T_0)}$ denotes the set of all continuous functions $\varphi : \mathbb{R}^n \to \mathbb{C}^m$ such that

$$\exp\left\{ k(a,T_0) \sum_{q=1}^n x_i^q \right\} |\varphi(x)| \to 0 \qquad as\ |x| \to \infty.$$

Theorem VI.18 is another version of well-posedness theorem for the Cauchy problem for systems (2.19) whose coefficients satisfying Hölder conditions with respect to the space variables and with initial data belonging to natural Banach spaces of functions or Borel measures. The initial conditions are satisfied with respect to the topology of these spaces. The proofs use a standard technique presented, for example, in [108, 41].

Theorem VI.18 is of special interest because it admits the following converse:

Theorem VI.19. *Let $u(x,t)$ be a solution of system (2.19) in Π_0 such that its p-traces are bounded uniformly in t, that is, there exists a positive constant C such that, for any $t \in (0,T_0]$,*

$$\left\| u(x,t) \right\|_p^{k(a,t)} \leq C. \qquad (2.22)$$

Then for $1 \leq p < \infty$ [respectively, for $p = 1$] there exists a unique function $\psi \in L_p^{k(a,0)}$ [respectively, a unique generalized measure $\mu \in M^{k(a,0)}$] such that the solution $u(x,t)$ is represented by formula (2.20) [respectively, (2.21)].

Theorems VI.18 and VI.19 admit the following immediate consequence.

Theorem VI.20. $L_p^{k(a,0)}$, $1 \leq p < \infty$, *and $M^{k(a,0)}$ coincide with the sets of initial values of classical solutions of the parabolic system (2.19) with coefficients satisfying conditions β_1–β_4 if and only if the solutions $u(x,t)$ have uniformly bounded p-traces, that is, inequalities (2.22) hold for $1 \leq p < \infty$ and $p = 1$ respectively.*

Note, finally, that positive solutions of system (1.19), $b = 2s + 1$, have totally bounded L_1-traces, that is, these solutions satisfy inequality (2.22) with $p = 1$.

Recall that the positivity of a solution is understood in the sense that the values of the vector-valued function $u(x,t)$ belong to a convex cone of the space $\mathbb{C}^m$ [51]. Thus, representation (2.21) holds for positive solutions $u(x,t)$.

VI.3. Schauder theory of parabolic boundary value problems

VI.3.1. Introduction

Here we present the basic facts of the theory of parabolic boundary value problems in Hölder spaces. They form the basis of the Schauder theory of parabolic boundary value problems. The theory is named after J.P. Schauder, who was the first to obtain, in 1934, the basic results in the case of the Dirichlet problem for second-order elliptic equations.

We will first discuss this theory for parabolic problems of the type

$$\mathcal{L}(x,t;D,D_t)u \equiv D_t u - \sum_{|\alpha|\leq 2b} a_\alpha(x,t)D^\alpha u = f(x,t), \tag{3.1}$$

$$u\big|_{t=0} = \psi(x), \tag{3.2}$$

$$\mathcal{B}_q(x,t;D,D_t)u\big|_S = \sum_{|\alpha|+2b\alpha_0\leq r_q} b^q_{\alpha\alpha_0}(x,t)D^\alpha D_t^{\alpha_0}u\big|_S = \varphi_q, \quad q=1,\ldots,bm, \tag{3.3}$$

in the cylinder $\Omega = G\times(0,T]$, where $S = \partial G\times(0,T]$. Here the coefficients a_α's are $m\times m$ square matrices and $b^q_{\alpha\alpha_0}$ are m-dimensional row-matrices. Subsequently we will formulate the main result for general systems.

VI.3.2. The well-posedness theorem

We use the notations and the information on spaces of Hölder functions presented in Subsections II.3.1 and II.3.2. Denote by $C_0^l(\overline{\Omega})$ and $C_0^l(S)$ the sets of functions u that belong to the spaces $C_0^l(\overline{\Omega})$ and $C_0^l(\overline{S})$, respectively, and satisfy the initial conditions

$$D_t^{\alpha_0-1}u\big|_{t=+0} = 0, \qquad \alpha_0 = 1,\ldots,[l/2b]+1. \tag{3.4}$$

The Schauder theory establishes the precise dependence of the differentiability properties of the solutions to the above boundary value problem on the properties of the initial and boundary conditions and on the right-hand sides of system (3.1), this dependence being expressed in terms of memebership in the corresponding Hölder spaces. The restrictions on the smoothness of the coefficients, the boundary of the domain, and the functions f, ψ, and φ_q, $q=1,\ldots,bm$, should be precisely correlated with the function space C^l in which the solutions are sought.

We use the following conditions:

β_{23}: *The elements of the matrix* $a_\alpha(x,t)$ *belong to* $C^l(\overline{\Omega})$.
β_{24}: *The elements of the matrix* $b^q_{\alpha\alpha_0}(x,t)$ *belong to* $C^{l+2b-r_q}(\overline{S})$.
β_{25}: $\partial G \in C^{l+2b}$.

Next, let us indicate what conditions should be imposed on the functions f, ψ, and φ_q, $q = 1, \ldots, bm$, to ensure that problem (3.1)–(3.3) will have a solution in $C^{l+2b}(\overline{\Omega})$. The necessary conditions are straightforward. By the definition of the Hölder spaces,

$$f \in C^l(\overline{\Omega}), \qquad \psi \in C^{l+2b}(\overline{G}), \qquad \varphi_q \in C^{l+2b-r_q}(\overline{S}), \quad q = 1, \ldots, bm,$$

$$l > l_0 = \max_q(0, r_q - 2b). \tag{3.5}$$

(Relations (3.5) are understood in the sense that each component of the indicated vectors belongs to the corresponding space, or that the vectors belong to the Cartesian product of the corresponding spaces.) Also, there exists a constant C independent of u such that

$$|f, \Omega|_l + |\psi, G|_{l+2b} + \sum_{q=1}^{bm} |\varphi_q, S|_{l+2b-r_q} \le C|u, \Omega|_{l+2b}. \tag{3.6}$$

Further, f, ψ, and φ_q must satisfy the compatibility conditions at $x \in \partial G$, $t = 0$. To explain how these conditions arise, let us examine a simple parabolic problem for the heat equation.

Example 6.12. Take

$$D_t u = a^2 \Delta u + f, \qquad u\big|_{t=+0} = \psi, \qquad u\big|_{x_n=+0} = \varphi, \qquad \Omega = \mathbb{R}^n_+ \times (0, T].$$

If the solution $u(x, t)$ belongs to $C^{l+2}(\overline{\Omega})$, then the derivatives $D_t^{\alpha_0-1} u\big|_{t=0}$, $\alpha_0 = 1, \ldots, \big[(l+2)/2\big] + 1$ can be determined from the heat equation and the initial conditions. For $x_n = 0$ they coincide with the derivatives $D_t^{\alpha_0-1} \varphi\big|_{t=0}$, $\alpha_0 = 1, \ldots, \big[(l+2)/2\big] + 1$. Therefore, one has the relations

$$\psi\big|_{x_n=0} = \varphi\big|_{t=0}, \qquad (a^2 \Delta \psi + f\big|_{t=0})\big|_{x_n=0} = D_t \varphi\big|_{t=0},$$

$$\left\{ a^2 \Delta (a^2 \Delta \psi + f\big|_{t=0}) + D_t f\big|_{t=0} \right\} = D_t^2 \varphi\big|_{t=0},$$

and so on.

In the case of problem (3.1)–(3.3) the compatibility conditions amount to the requirement that the derivatives $D_t^{\alpha_0-1} u\big|_{t=0}$, $\alpha_0 = 1, \ldots, \big[(l+2b)/2b\big] + 1$, calculated from the system (3.1) and the initial condition (3.2), satisfy the relation derived from equality (3.3) by setting $t = 0$ in the q-th equality and the equalities obtained by differentiating it with respect to t up to the order $\big[(l+2b-r_q)/2b\big]$. Such conditions are called *compatibility conditions of order* $\big[(l+2b)/2b\big]$. In particular, if $\psi \equiv 0$ and $D_t^{\alpha_0-1} f\big|_{t=0} = 0$, $\alpha_0 = 1, \ldots, [l/2b] + 1$, then $D_t^{\alpha_0-1} u\big|_{t=0} = 0$, $\alpha_0 = 1, \ldots, \big[(l+2b)/2b\big] + 1$, and so the compatibility conditions of order $\big[(l+2b)/2b\big]$ reduce to the equalities $D_t^{\alpha_0} \varphi_q\big|_{t=0} = 0$, $\alpha_0 = 0, \ldots, \big[(l+2b-r_q)/2b\big]$.

Thus, if $u \in C^{l+2b}(\overline{\Omega})$ is a solution of (3.1)–(3.3), then the functions f, ψ, and φ_q satisfy conditions (3.6) and the compatibility conditions of order $[(l+2b)/2b]$. The central result of the Schauder theory is that these necessary conditions are also sufficient for the existence of a solution in $C^{l+2b}(\overline{\Omega})$. For general parabolic boundary value problems this result was established by V.A. Solonnikov [94].

Theorem VI.21. *Consider the parabolic boundary value problem* (3.1)–(3.3) *in* $\overline{\Omega}$, *satisfying the conditions* β_{23}–β_{25}. *If the right-hand sides of the problem* (3.1)–(3.3) *satisfy conditions* (3.5) *and the compatibility conditions of order* $[(l+2b)/2b]$, *then for any noninteger* $l > l_0 = \max(0, r_1 - 2b, \ldots, r_{2m} - 2b)$ *the problem* (3.1)–(3.3) *has a unique solution* $u(x,t)$ *in the space* $C^{l+2b}(\overline{\Omega})$ *and the following estimate holds:*

$$|u, \Omega|_{l+2b} \leq C\left(|f, \Omega|_l + |\psi, G|_{l+2b} + \sum_{q=1}^{bm} |\varphi_q, S|_{l+2b-r_q}\right). \tag{3.7}$$

Here C *is a positive constant that does not depend on* u, f, ψ, *and* φ_q.

VI.3.3. On the proof of the well-posedness theorem

The most difficult step in the proof is to show that problem (3.1)–(3.3) is well posed in a cylinder Ω_h of small height $h \leq T$ in the case when $\psi = 0$, $f \in C_0^l(\overline{\Omega}_h)$, $\varphi_q \in C_0^{l+2b-r_q}(\overline{S}_h)$, $q = 1, \ldots, bm$. This problem can be regarded as the problem of finding the solution of the operator equation

$$Au = F, \tag{3.8}$$

where A is the linear operator from the linear space $B_1 = \left(C_0^{l+2b}(\overline{\Omega}_h)\right)^m$ to $B_2 = \left(C_0^l(\overline{\Omega}_h)\right)^m \times \prod_{q=0}^{bm} C_0^{l+2b-r_q}(\overline{S}_h)$ that assigns to each vector-valued function $u(x,t) \in B_1$ the vector-valued function $F = (f, \varphi_1, \ldots, \varphi_{bm})$. Here B_1 and B_2 are Banach spaces with the norms

$$|u|_{B_1} = |u, \Omega_h|_{l+2b}, \qquad |F|_{B_2} = |f, \Omega_h|_l + \sum_{q=1}^{bm} |\varphi_q, S_h|_{l+2b-r_q}.$$

Well-posedness of equation (3.8) means that the operator A has a bounded inverse operator A^{-1} (acting from the whole space B_2 onto B_1). The operator A^{-1} can be obtained as follows. First one constructs what is known as a *regularizer*. For our problem a regularizer of the operator A is defined as an operator $R : B_2 \to B_1$ such that

$$AR = I + V; \qquad RA = I + W, \tag{3.9}$$

where I is the identity operator in B_2 in the first expression and in B_1 in the second one, and where V, W are bounded operators of norm less than 1 in B_2 and

B_1, respectively. By the contraction principle, the last property implies, like in Section V.1, that the operators $I+V$, $I+W$ have bounded inverses $(I+V)^{-1}$ and $(I+W)^{-1}$. Hence, by relations (3.9), $AR(I+V)^{-1} = I$ and $(I+W)^{-1}RA = I$. This means that the operator A has both a right bounded inverse, $R(I+V)^{-1}$, and a left bounded inverse, $(I+W)^{-1}R$. As is known, these inverses necessarily coincide, and so, $A^{-1} = R(I+V)^{-1} = (I+W)^{-1}R$.

Thus, in the special case under consideration, the proof of Theorem VI.21 reduces to the construction of a regularizer. This is achieved by pasting together, with the help of a suitable partition of unity, operators that solve model problems of two types. The first is the Cauchy problem for the parabolic system $\mathcal{L}_0(x^0,0;D,D_t)u = f$, the coefficients of which are "frozen" at a point $(x^0,0)$, $x^0 \in G$. This problem was discussed in Sections VI.1 and VI.2, the results presented therein being directly applicable to the present situation. The model problems of the second type are boundary value problems in the domains $\mathbb{R}^n_+ \times (0,T]$. Usually, such problems are obtained by passing to a local coordinate system with the origin at a point $x^0 \in \partial G$ in the differential expressions figuring in the system of equations and in the initial conditions, $\mathcal{L}_0(x^0,0;D,D_t)$ and $\mathcal{B}^0_q(x^0,0;D,D_t)$, $q = 1,\ldots,bm$. To show that the operator R has the desired properties on resorts to sharp estimates of solutions of the model problems. Such estimates are in turn obtained with the help of formulas representing the solutions of these problems in a form convenient for analysis. These formulas are presented in the next subsection.

A detailed exposition of the construction and applications of regularizers in the L_2-theory of boundary value problems is made in Chapter IV.

VI.3.4. Solution of the model parabolic boundary value problem

The formulas derived below lie at the foundation of both the Schauder theory and the analogous theory for the Sobolev spaces $W^{l\ l/2b}_{p\,x\,t}$.

In $\mathbb{R}^{n+1}_{++} = \mathbb{R}^n_+ \times (0,\infty)$ consider the model parabolic problem

$$
\begin{cases}
\mathcal{L}_0\big(D_{x'},D_{x_n},D_t\big)u = f, \\[1mm]
u\big|_{t=+0} = \psi, \\[1mm]
\mathcal{B}^0_q\big(D_{x'},D_{x_n},D_t\big)u\big|_{x_n=+0} = \varphi_q, \quad q = 1,\ldots,bm.
\end{cases}
\tag{3.10}
$$

Consider first the problem

$$
\begin{cases}
\mathcal{L}_0\big(D_{x'},D_{x_n},D_t\big)u = 0, \\[1mm]
u\big|_{t=+0} = 0, \\[1mm]
\mathcal{B}^0_q\big(D_{x'},D_{x_n},D_t\big)u\big|_{x_n=+0} = \varphi_q, \quad q = 1,\ldots,bm.
\end{cases}
\tag{3.11}
$$

We seek the solution of problem (3.11) by taking smooth functions $\varphi_q(x',t)$ with compact support and by using the Laplace transformation in x and Fourier transformation in $x_1, \ldots, x_{n-1}$:

$$\widehat{u}(x_n, \xi', p) = \int_0^\infty e^{-pt}\, dt \int_{\mathbb{R}^{n-1}} e^{-i(x', \xi')} u(\xi', x_n, p)\, d\xi'.$$

This transforms problem (3.11) into a problem for a system of ordinary differential equations:

$$\mathcal{L}_0\big(i\xi', \tfrac{d}{dx_n}, p\big)\widehat{u} = 0,$$

$$\mathcal{B}_q\big(i\xi', \tfrac{d}{dx_n}, p\big)\widehat{u}\big|_{x_n=+0} = \widehat{\varphi}_q(\xi', p), \quad q = 1, \ldots, bm, \tag{3.12}$$

$$\big|\widehat{u}(x_n, \xi', p)\big| \to 0 \text{ as } x_n \to \infty.$$

Problem (3.12) can be solved by the method of residues, which is outlined in Subsection I.2.3. Then the solution is written in the form

$$\widehat{u}_j(x_n, \xi', p) = \sum_{q=1}^{bm} \widehat{G}_{jq}(x_n; \xi', p)\widehat{\varphi}_q(\xi', p), \quad j = 1, \ldots, m, \tag{3.13}$$

where the functions $\widehat{G}_{jq}(x_n; \xi', p)$, called the *elements of the Poisson basis*, can be represented as sufficiently transparent contour integrals.

Using the inversion formulas of the theory of Laplace and Fourier integral transformations, in particular, the fact that a product is transformed into a convolution, we obtain the following important formulas:

$$u_j(x,t) = \sum_{q=1}^{bm} \int_0^t d\tau \int_{\mathbb{R}^{n-1}} G_{jq}(x-y', t-\tau)\varphi_q(y', \tau)\, dy', \tag{3.14}$$

$$G_{jq}(x,t) = -i(2\pi)^{-n} \int_{\mathbb{R}^{n-1}} \exp\{i(x', \xi')\}\, d\xi' \int_{a-i\infty}^{a+i\infty} \exp\{pt\}\widehat{G}_{jq}(\xi'; x_n, p)\, dp, \tag{3.15}$$

where a is an arbitrary positive constant.

The functions $G_{jq}(x, t)$ are called the *Poisson kernels* of problem (3.11).

To obtain substantial information on Poisson kernels, one needs to carefully study in detail the elements of the Poisson basis for complex values of ξ and p, by means of a special transformation of the integration contour in the complex p-plane. The technique of this analysis is presented in Section VII.2.

Theorem VI.22 [108, 94]. *The Poisson kernels $G_{jq}(x, t)$ are defined and infinitely differentiable in $\mathbb{R}^{n+1}_{++}$; their derivatives satisfy the following inequalities (in which $q = 2b/(2b-1)$):*

$$\big|D^\alpha D_t^{\alpha_0} G_{jq}(x,t)\big| \leq C_{\alpha\alpha_0} t^{-(n+2b-r_q+|\alpha|+2b\alpha_0)/2b} \exp\big\{-c|x|^q t^{1-q}\big\}. \tag{3.16}$$

Note that for large t estimate (3.16) can be considerably sharpened in some cases. This is discussed in detail in Chapter VII.

To solve (3.10) completely, it remains to consider the problem

$$\mathcal{L}_0\big(D_{x'}, D_{x_n}, D_t\big)u = f, \quad u\big|_{t=+0} = \psi, \quad \mathcal{B}_0\big(D_{x'}, D_{x_n}, D_t\big)u\big|_{x_n=+0} = 0, \quad (3.17)$$

where f and ψ are smooth functions with compact support.

To obtain the solution of problem (3.17) in a transparent form convenient for applications we introduce the so-called homogeneous (matrix-valued) Green function.

Definition VI.9. *A homogeneous Green function $G_0(x, y, t)$ of the problem (3.10) is an $m \times m$ matrix-valued function such that the function*

$$u(x,t) = \int\limits_0^t d\tau \int\limits_{\mathbb{R}^n} G_0(x, y, t - \tau) f(y, \tau)\, dy$$

is a solution of problem (3.10) with the null initial and boundary conditions. Here it is assumed that $f(y, t)$ is a smooth function with compact support.

A homogeneous Green function is sought in the form

$$G_0(x, y, t) = \Gamma_0(x - y, t) - V(x, y, t), \tag{3.18}$$

where $\Gamma_0(x, t)$ is a fundamental solution of the Cauchy problem for the equation $\mathcal{L}_0(D, D_t)u = 0$. Then $V(x, y, t)$ must be a solution of the boundary value problem

$$\mathcal{L}_0(D, D_t)u = 0, \quad u\big|_{t=+0} = 0, \quad \mathcal{B}_0(D, D_t)u\big|_{x_n=+0} = \mathcal{B}_0(D, D_t)\Gamma_0\big|_{x_n=+0}. \tag{3.19}$$

Problem (3.19) is solved by formulas (3.14) and (3.15). Using the estimates (1.9) for the fundamental solution $\Gamma_0(x, t)$ and estimate (3.17) for the Poisson kernels, we obtain the following assertion concerning $G_0(x, y, t)$.

Theorem VI.23 [110, 44]. *There exists a homogeneous Green function $G_0(x, y, t)$ of problem (3.10), which is given by formula (3.18), is infinitely differentiable with respect to all its arguments for $\{x, y\} \subset \mathbb{R}^n_+$, $t > 0$, and satisfies the following estimates (in which $q = 2b/(2b - 1)$):*

$$\big|D_y^l D_x^\alpha D_t^{\alpha_0} G(x, y, t)\big| \le C_{l\alpha\alpha_0} t^{-(n+|l|+|\alpha|+2b\alpha_0)/2b} \exp\big\{-c|x - y|^q t^{1-q}\big\}, \quad (3.20)$$

$$\big|D_y^l D_x^\alpha D_t^{\alpha_0} V(x, y, t)\big| \le C_{l\alpha\alpha_0} t^{-(n+|l|+|\alpha|+2b\alpha_0)/2b} \exp\big\{-c\big(|x - y|^q + y_n^q\big)t^{1-q}\big\}. \tag{3.21}$$

The solution $u(x,t)$ of problem (3.10), constructed for smooth functions f, ψ, φ_q, $q = 1,\ldots,bm$ with compact support, is given by the formula

$$u(x,t) = \int_{\mathbb{R}^n} G_0(x,y,t)\psi(y)\,dy + \sum_{q=1}^{bm} \int_0^t d\tau \int_{\mathbb{R}^{n-1}} G_q(x-y',t-\tau)\varphi_q(y',\tau)\,dy'$$

$$+ \int_0^t d\tau \int_{\mathbb{R}^n} G_0(x,y,t-\tau)f(y,\tau)\,dy. \qquad (3.22)$$

Note, finally, that in order to obtain *sharp* estimates of higher-order derivatives and their Hölder constants, which are necessary in order to construct the Schauder theory, one uses a theorem on the boundedness of singular integral operators with zero-mean kernels. Let us state a proposition on averaging of kernels in the theory of model parabolic boundary value problems.

Lemma VI.2 [112]. *Let the function $\Gamma(x,t) = t^{-1-(n-1)/2b}\Omega(xt^{-1/2b})$, $(x,t) \in \mathbb{R}^{n+1}_{++}$ be a solution of the parabolic system $\mathcal{L}_0(D,D_t)u = 0$ and assume that the function $\Omega(z)$ satisfies the inequalities $\left|D_z^\alpha \Omega(z)\right| \le C_\alpha \exp\left\{-c|z|^q\right\}$. Then*

$$\int_{\mathbb{R}^{n-1}} \Omega(z',0)\,dz' = 0.$$

Example VI.13. Consider the model boundary value problems for the heat equation

$$D_t u = a^2 \Delta u, \quad B_0(D,D_t)u\big|_{x_n=+0} = \varphi(x',t), \quad u\big|_{t=+0} = 0, \qquad (3.23)$$

presented in Example I.3.

The Poisson basis of problem (3.23) consists of only one function, $\widehat{G}(x_n,\xi',p) = B_0(i\xi',-\xi_n^+,p)^{-1}\exp\{-\xi_n^+ x_n\}$, where $\xi_n^+ = \sqrt{p/a^2 + \xi'^2}$. For the Dirichlet problem $B_0 \equiv 1$, for the Neumann problem $B_0 \equiv -\xi_n^+$, and for the oblique derivative problem $B_0 = i\sum_{k=1}^{n-1} b_k\xi_k - b_n\xi_n^+$. Using formula (3.15) and the Laplace and Fourier transformations [8, 23, 58], one can verify that the Poisson kernels of the Dirichlet, Neumann, and the oblique derivative problems are given by the formulas

$$G^{(1)}(x,t) = \left(2a\sqrt{\pi}\right)^{-n} x_n t^{-(n+2)/2} \exp\left\{-x^2/4a^2 t\right\},$$

$$G^{(2)}(x,t) = -2a^2\left(2a\sqrt{\pi}\right)^{-n} \exp\left\{-x^2/4a^2 t\right\},$$

$$G^{(3)}(x,t) = -\left[2a^2/\left(2a\sqrt{\pi t}\right)^n |b|^2\right]\left(b_n \exp\left\{-|x|^2/4a^2 t\right\}\right.$$

$$\left. + \frac{|b|^2 x_n - b_n(b,x)}{a|b|\sqrt{t}}\overset{\infty}{\underset{\frac{(b,x)}{2a|b|\sqrt{t}}}{\exp\left\{-\frac{|b|^2 x^2 - (b,x)}{4a^2|b|^2 t}\right\}\int}} \exp\{-z^2\}\,dz,$$

respectively, where $b = (b_1,\ldots,b_n)$.

To construct the homogeneous Green function $G_0(x, y, t) = \Gamma_0(x - y, t) - V(x, y, t)$, we must find V, which is expressed in terms of the Poisson kernel by the formula

$$V(x, y, t) = \int\limits_0^t d\tau \int\limits_{\mathbb{R}^{n-1}} G(x - \tilde{y}', t - \tau) B_0 \Gamma_0(\tau, \tilde{y}' - y) \, d\tilde{y}'. \tag{3.24}$$

Substituting the expressions for the Poisson kernels and the function Γ_0 into (3.24), we conclude, for example, that the homogeneous Green function of the Dirichlet problem has the form

$$G_0^{(1)}(x, y, t) = \left(2a\sqrt{\pi t}\right)^{-n} \exp\left\{-|x' - y'|^2/4a^2 t\right\}$$
$$\times \left[\exp\left\{-|x_n - y_n|^2/4a^2 t\right\} - \exp\left\{-|x_n + y_n|^2/4a^2 t\right\}\right],$$

while the homogeneous Green function of the Neumann problem is given by

$$G_0^{(2)}(x, y, t) = \left(2a\sqrt{\pi t}\right)^{-n} \exp\left\{-|x' - y'|^2/4a^2 t\right\}$$
$$\times \left[\exp\left\{-|x_n - y_n|^2/4a^2 t\right\} + \exp\left\{-|x_n + y_n|^2/4a^2 t\right\}\right].$$

Of course, the last two expressions can be alternatively derived by simple arguments, known as the method of images, but we decided to present here a general technique for deriving such formulas.

VI.3.5. Necessity of the parabolicity condition

According to Theorem VI.21, from the parabolicity of a boundary value problem it follows that the Schauder theory is valid for such problems. Here we will prove the converse assertion, following [60, 2].

Consider the model parabolic boundary value problem

$$\mathcal{L}_0(D, D_t)u = f, \quad (x, t) \in \mathbb{R}_+^n \times (0, T], \tag{3.25}$$

$$u\big|_{t=+0} = \psi, \quad x \in \mathbb{R}_+^n, \tag{3.26}$$

$$\mathcal{B}_q^0(D, D_t)u\big|_{x_n=+0} = \varphi_q, \quad q = 1, \ldots, bm, \quad (x', t) \in S = \mathbb{R}^{n-1} \times (0, T]. \tag{3.27}$$

Theorem VI.24. *For the problem (3.25)–(3.27) to be parabolic it is necessary that for some noninteger number $l > l_0$ there exists a constant $C > 0$ such that the inequality*

$$|u, \Omega|_{2b+l} \leq C\left(|\mathcal{L}_0, u, \Omega|_l + |u|_{t=0}, \mathbb{R}_+^n|_{2b+l} + \sum_{q=1}^{bm} |\mathcal{B}_q^0 u, S|_{2b+l-r_q}\right) \tag{3.28}$$

holds for any function $u(x, t) \in C^{l+2}(\overline{\Omega})$.

Proof: We claim that the validity of inequality (3.28) for any function $u \in C^{l+2b}(\overline{\Omega})$ implies that: 1) the system (3.25) is parabolic in the sense of Petrovskiĭ; 2) the boundary operator (3.27), $\mathcal{B}_0 = (\mathcal{B}_{qj}^0)$, $q = 1, \ldots, bm$, $j = 1, \ldots, m$, satisfies the complementarity condition with respect to the operator $\mathcal{L}_0$. To prove this assertion, we proceed by reductio ad absurdum: we assume the contrary and then exhibit an example of solution of problem (3.25)–(3.27), $u_\lambda(x, t)$, dependent on a parameter λ, such that estimate (3.28) cannot be satisfied for large λ.

1) Suppose the system $\mathcal{L}_0(D, D_t)u = 0$ is not parabolic. This means that for some $\xi^0 \in \mathbb{R}^n$, $\xi^0 \neq 0$, the function $\det \mathcal{L}_0(\xi^0, p)$ has a p-zero p^0 with $\operatorname{Re} p^0 \geq 0$. Let $v_\lambda(x, t)$ be the m-dimensional column-matrix with all its elements equal to $\exp\{\lambda^{2b}p^0 t + i(x, \xi^0)\}$, and let $\zeta(x, t)$ be an infinitely differentiable function in with compact support in Π^+. Define the function $u_\lambda(x, t)$ by the rule $u_\lambda(x, t) = \left[\widehat{\mathcal{L}}_0(D, D_t)v_\lambda(x, t)\right]\zeta(x, t)$, where $\widehat{\mathcal{L}}_0$ is the adjoint matrix to $\mathcal{L}_0$. Since $u_\lambda\big|_{t=0} = 0$, $\mathcal{B}u_\lambda\big|_{x_n=0} = 0$, it follows that in our case estimate (3.28) takes the form

$$\left|u_\lambda, \Pi^+\right|_{l+2b} \leq C\left|\mathcal{L}_0 u_\lambda, \Pi^+\right|_l. \tag{3.29}$$

Let us examine how the norms of the functions in the two sides of (3.29) grow as $\lambda \to \infty$. The absolute value of the leading-order term in the left-hand side has the form $C\lambda^{2bm+l}\exp\{\lambda^{2b}\operatorname{Re}p^0 t\}$. Since $\mathcal{L}_0\widehat{\mathcal{L}}_0 = \det\mathcal{L}_0 I$, it follows by the choice of ξ^0 and p^0 that the absolute value of the leading-order term in the right-hand side is $C\lambda^{2bm+l-1}\exp\{\lambda^{2b}\operatorname{Re}p^0 t\}$. Therefore, inequality (3.29) is violated for large λ. Thus, system (3.25) is parabolic in the sense of Petrovskiĭ.

2) Now let us show that the boundary operator in (3.27) satisfies the complementarity conditions with respect to the operator $\mathcal{L}_0$. Assume that this is not the case. Then there exist $\xi^{0\prime} \in \mathbb{R}^{n-1}$, $p^0 \in \mathbb{C}$, $|p^0| + |\xi^{0\prime}|^{2b} > 0$, $\operatorname{Re} p^0 \geq 0$, such that the problem $\mathcal{L}_0(i\xi^{0\prime}, D_{x_n}, p^0)v = 0$, $\mathcal{B}_q^0(i\xi^{0\prime}, D_{x_n}, p^0)v\big|_{x_n=0} = 0$, $q = 1, \ldots, bm$, $v(x_n) \to 0$ as $x_n \to \infty$, has a nontrivial solution $v_0(x_n)$.

Consider the function

$$u_\lambda(x, t) = \lambda^{-l-2b}v_0(\lambda x_n)\exp\left\{i\lambda(x', \xi^{0\prime}) + \lambda^{2b}p^0(t - T)\right\}, \quad (x, t) \in \Pi^+.$$

It follows from the definition of the function $v_0(x_n)$ and the generalized homogeneity of the operators $\mathcal{L}_0$ and $\mathcal{B}_q^0$ that $u_\lambda(x, t)$ is a solution of the system $\mathcal{L}_0 u = 0$, satisfies the null boundary conditions, and $u_\lambda = \psi_\lambda$ for $t = 0$, where $\psi_\lambda(x) = \lambda^{-l-2b}\exp\left\{-\lambda^{2b}p^0 T + i\lambda(x', \xi^{0\prime})\right\}$. Consider first the case $\operatorname{Re} p^0 > 0$. Then for sufficiently large $\lambda > 0$,

$$\left|\psi_\lambda, \mathbb{R}_+^n\right|_{l+2b} \leq C \exp\left\{-\lambda^{2b}\operatorname{Re}p^0 T\right\}. \tag{3.30}$$

Let us estimate the norm $\left|u_\lambda, \Pi^+\right|_{l+2b}$ from below (recalling its definition):

$$\left|u_\lambda, \Pi^+\right|_{l+2b} \geq \left|u_\lambda(x, T), \mathbb{R}_+^n\right|_{l+2b} \geq H\left(u_\lambda(x, T)\right) = H\left(u_1(x, T)\right), \tag{3.31}$$

where

$$H(w) = \sum_{|\alpha|=2b+[l]} \sup_x \left\{\frac{\left|\Delta_x^{\tilde{x}}D^\alpha w(x)\right|}{|x - \tilde{x}|^{l-[l]}}\right\}.$$

The estimates (3.28), (3.30), and (3.31) yield the inequality $H\big(u_1(x,T)\big) \leq C \exp\big\{-\lambda^{2b}\operatorname{Re}p^0 T\big\}$, which does not hold for large λ.

Finally, in the case $\operatorname{Re}p^0 = 0$ we must consider the function $w_\lambda(x,t) = \zeta(3t/2T)u_\lambda(x,t)$, where $\zeta(\gamma)$ is an infinitely differentiable function that equals zero for $\gamma \leq 0$ and equals 1 for $\gamma \geq 1$, $0 \leq \zeta(\gamma) \leq 1$, and show by similar arguments that inequality (3.28) fails.

Since the definition of the parabolicity of a boundary value problem was introduced by considering model problems, it follows from Theorem VI.23 that if the parabolicity condition is violated at least at one point of the cylinder Ω (for an interior point of Ω this means that the system is not parabolic, whereas for the points of S this amounts to the violation of the complementarity condition), then for such a problem the Schauder theory described above does not apply. Of course, this does not necessarily mean that such a problem cannot enjoy nice properties in other functional spaces.

VI.3.6. General boundary value problems. Well-posedness theorem

We proceed now to the formulation of the main result of the Schauder theory for systems parabolic in the sense of Solonnikov (see Subsection I.2.3):

$$\sum_{j=1}^{m} l_{kj}u_j = f_k, \quad k = 1,\ldots,m; \quad \sum_{j=1}^{m} c_{\alpha j}u_j\big|_{t=0} = \psi_\alpha(x), \quad \alpha = 1,\ldots,r;$$

$$\sum_{j=1}^{m} b_{qj}u_j\big|_S = \varphi_q, \qquad q = 1,\ldots,br. \tag{3.32}$$

We will work under the following conditions on the coefficients of the operators $\mathcal{L}$, C, $\mathcal{B}$ and the boundary S:

$\widetilde{\beta}_{23}$: *The coefficients of the operator l_{kj} belong to $C^{l-s_k}(\overline{\Omega})$.*

$\widetilde{\beta}_{24}$: *The coefficients of the operator b_{qj} belong to $C^{l-\sigma_q}(\overline{S})$.*

$\widetilde{\beta}_{25}$: *The surface S is of class $C^{l+t_{\max}}$, where $t_{\max} = \max_j t_j$.*

$\widetilde{\beta}_{26}$: *The coefficients of the operator $c_{\alpha j}$ belong to $C^{l-\rho_\alpha}(\overline{G})$.*

Theorem VI.25 [60, 94]. *Let $l > l_0 = \max(0,\sigma_1,\ldots,\sigma_{br})$ and assume that conditions $\widetilde{\beta}_1$–$\widetilde{\beta}_4$ are satisfied. Then the problem (3.32) is parabolic if an only if, given arbitrary functions $f_k \in C^{l-s_k}(\Omega)$, $\psi_\alpha \in C^{l-\rho_\alpha}(G)$ and $\varphi_q \in C^{l-\sigma_q}(S)$ that satisfy the natural compatibility conditions, problem (3.32) has a unique solution $u = (u_1,\ldots,u_m)$, $u_j(x,t) \in C^{l+t_j}(\Omega)$, and the following two-sided estimate holds:*

$$C_1|F|_l \leq \sum_{j=1}^{m} |u_j,\Omega|_{l+t_j} \leq C_2|F|_l, \tag{3.33}$$

where $|F|_l = \sum_{k=1}^{m}|f_k,\Omega|_{l-s_k} + \sum_{\alpha=1}^{r}|\psi,G|_{l-\rho_\alpha} + \sum_{\alpha=1}^{br}|\varphi_q,S|_{l-\sigma_q}$.

Similar results are valid for the general parabolic conjugation problems and the nonlocal problems considered in Subsection I.2.6.

The conditions on the arrangement of the domains G_0, G_1, G_2 and on the diffeomorphism of ∂G_0 and ∂G_1 were stated in Subsection I.2.6. We search for a vector-valued function

$$u(x,t) = \begin{cases} u_{(1)}(x,t), & (x,t) \in \Omega_1, \\ u_{(2)}(x,t), & (x,t) \in \Omega_2, \end{cases}$$

which solves the following problem:

$$\sum_{j=1}^{m_\mu} l_{(\mu)kj} u_{(\mu)j} = f_{(\mu)k}, \qquad k = 1,\dots,m_\mu, \ (x,t) \in \Omega_\mu; \tag{3.34}$$

$$\sum_{j=1}^{m_\mu} c_{(\mu)\alpha j} u_{(\mu)j}\big|_{t=0} = \psi_{(\mu)\alpha}, \qquad \alpha = 1,\dots,r_\mu, \ \mu = 1,2; \tag{3.35}$$

$$\left(\sum_{j=1}^{m_1} b_{(1)qj} u_{(1)j} + \sum_{j=1}^{m_2} b_{(2)qj} u_{(2)j} + \sum_{j=1}^{m_2} b_{qj} u_{(2)j} \right)\Bigg|_{S_1} = \varphi_q, \tag{3.36}$$

$$q = 1,\dots,b(r_1 + 2r_2).$$

Here $\left(l_{(\mu)kj}\right)$ are operators parabolic in the sense of Solonnikov with weights $s_{(\mu)k}$, $t_{(\mu)j}$, $\sum_{i=1}^{m_\mu} \left(s_{(\mu)i} + t_{(\mu)i}\right) = 2br_\mu$; $\left(c_{(\mu)\alpha j}\right)$ are initial-condition operators with weights $\rho_{(\mu)\alpha}$, $\alpha = 1,\dots,r$; and $\left(b_{(\mu)qj}\right)$ are boundary operators with weights σ_q, $q = 1,\dots,b(r_1 + 2r_2)$. Denote $l_0 = \max_q(0, \sigma_q)$.

Note that in the general situation under consideration the nonlocal compatible covering condition is completely analogous to the condition formulated in Subsection I.2.6.

Theorem VI.26 [25]. *Let $l > l_0$ and assume that the coefficients of the operators $l_{(\mu)kj}$, $b_{(\mu)qj}$, b_{qj}, and $c_{\alpha j}$ satisfy conditions $\widetilde{\beta}_1$, $\widetilde{\beta}_2$, and $\widetilde{\beta}_4$ respectively, and $S_1 \in C^{l+t_{\max}}$, where $t_{\max} = \max\limits_{\mu,j} t_{(\mu)j}$. Then the problem (3.34)–(3.36) is parabolic if and only if, given arbitrary functions $f_{(\mu)k} \in C^{l-s_{(\mu)k}}(\Omega_\mu)$, $\psi_{(\mu)\alpha} \in C^{l-\rho_{(\mu)\alpha}}(G_\mu)$, $\varphi_q \in C^{l-\sigma_q}(S_1)$, $q = 1,\dots,b(r_1 + 2r_2)$ that satisfy the natural nonlocal compatibility conditions, there exists a unique solution $u(x,t)$ with $u_{(\mu)j} \in C^{l+t_{(\mu)j}}(\Omega_\mu)$, $j = 1,\dots,m_\mu$, $\mu = 1,2$, and the following two-sided estimate holds:*

$$C_1 |F|_l \leq \sum_{\mu=1}^{2} \sum_{j=1}^{m_\mu} |u_{(\mu)j}, \Omega_\mu|_{l+t_{(\mu)j}} \leq C_2 |F|_l,$$

$$|F|_l = \sum_{\mu=1}^{2} \sum_{k=1}^{m_\mu} |f_{(\mu)k}, \Omega_\mu|_{l-s_{(\mu)k}} + \sum_{\mu=1}^{2} \sum_{\alpha=1}^{r_\mu} |\psi_{(\mu)\alpha}, G_\mu|_{l-\rho_{(\mu)\alpha}} + \sum_{q=1}^{b(r_1+2r_2)} |\varphi_q, S|_{l-\sigma_q}.$$

The main result of the Schauder theory for parabolic conjugation problems is formulated in a similar manner.

Several remarks are in order.

Remark VI.1. In order to infer from Theorems VI.25 and VI.26 results for general systems parabolic in the sense of Petrovskiĭ we must take $t_i = n_i$, $s_k = 0$, take the order of the boundary operator b_{qj} equal to $\sigma_q + n_j$, and in place of the initial value problem pose the Cauchy problem

$$D_t^{\alpha_0 - 1} u_j \big|_{t=0} = \psi_j^{(\alpha_0)}, \qquad \alpha_0 = 1, \ldots, n_j, \; j = 1, \ldots, m.$$

Note that in Theorems VI.25 and VI.26 it is not required that the domain G be bounded, but its boundary ∂G must satisfy the conditions indicated above.

In particular, Theorem VI.25 implies a theorem on well-posedness of the initial value problem for systems parabolic in the sense of Solonnikov and of the Cauchy problem for systems parabolic in the sense of Petrovskiĭ in the spaces $C^l(\Pi)$.

Remark VI.2. Theorems VI.25 and VI.26 yield theorems on the increase of regularity of the solutions up to the boundary when the regularity of the data of the problem is increased.

Remark VI.3. The results discussed above remain valid for noncylindrical domains that allow local rectification by cylinders with generatrices parallel to the t-axis.

VI.4. Green functions

VI.4.1. Introduction

Theorems VI.25, VI.26 and the well-posedness theorems given in Chapter IV establish the well-posedness of parabolic boundary value problems in spaces of smooth functions and distributions. They imply that the operator of such a boundary value problem has a bounded inverse defined on the whole range of this operator. General theorems of functional analysis [5] guarantee that in this situation the inverse operator is an integral operator, the kernel of which is, generally speaking, a distribution of finite order. This kernel is called the *Green function (matrix) of the parabolic boundary value problem*. A number of questions arise that require the development of special methods for studying Green functions. Namely, it is interesting to know when the elements of a Green function are ordinary functions, and when they are distributions, to obtain precise estimates for its derivatives with respect to both the main variables and the parameters, and to construct various asymptotic representations. Investigations of the Green functions of boundary value problems for systems parabolic in the sense of Petrovskiĭ were carried out, among others, in the papers [110, 38, 96] and the monograph [44]. Therein three different approaches to the analysis of Green functions of parabolic boundary value problems were proposed. In the first approach one looks at the structure of the inverse operator, constructed by means of the method of regularizers (see Subsection VI.3.3); in the second one the Green function is obtained as the limit of solutions of parabolic boundary value problems for a Dirac δ-like sequence of data of the problem; finally, the third approach develops the method of integral operators with nonsummable singularities proposed by Yu.P. Krasovskiĭ in his investigations of Green functions of elliptic boundary value problems. It was precisely this third approach that enabled S.D. Ivasishen to obtain definitive results. However, the first two approaches, whenever applicable, gave sharper results as far as regularity conditions are concerned. The three methods indicated above will be referred to as the *method of regularizers*, the *method of δ-like sequences*, and the *method of integral operators*, respectively. In our opinion, the three methods appear to be equally powerful and will undobtedly find further interesting applications. Unfortunately, the definitive results of the theory of Green functions are very complicated to formulate and grasp. For this reason here we will merely describe them and explain the main results on a simple transparent example.

VI.4.2.   Green functions. Homogeneous Green functions

For the sake of clarity we consider a parabolic boundary value parabolic problem for an equation parabolic in the sense of Petrovskiĭ:

$$L(x,t; D, D_t)u \equiv \sum_{|\alpha|+2b\alpha_0 \leq 2br} a_{\alpha\alpha_0}(x,t)D^\alpha D_t^{\alpha_0}u(x,t) = f(x,t), \quad (x,t) \in \Omega, \quad (4.1)$$

$$D_t^{\alpha_0-1}u\big|_{t=0} = \psi_{\alpha_0}(x), \quad \alpha_0 = 1,\ldots,r, \quad x \in G, \tag{4.2}$$

$$B_q(x,t; D, D_t)u\big|_S = \sum_{|\alpha|+2b\alpha_0 \leq r_q} b^q_{\alpha\alpha_0}(x,t)D_x^\alpha D_t^{\alpha_0}u(x,t)\big|_S$$

$$= \varphi_q(x',t), \quad q = 1,\ldots,br. \tag{4.3}$$

The Green function of the problem (4.1)–(4.3) is defined to be a matrix-valued function $G = (G_0, G_1, G_2)$ by means of which the solution of this problem (in a suitable functional space) can be represented as

$$u(x,t) = \int_0^\tau d\tau \int_\Omega G_0(x,t;y,\tau)f(y,\tau)\,dy$$

$$+ \int_0^\tau d\tau \int_\Omega G_1(x,t;y',\tau)\varphi_q(y',\tau)\,dy' + \int_G G_2(x,t;y,0)\psi(y)\,dy, \tag{4.4}$$

where $\varphi = (\varphi_1, \ldots, \varphi_{br})$, $\psi = (\psi_1, \ldots, \psi_r)$. In the case under consideration, G_0 is a scalar function and $G_1 = (G_{11}, \ldots, G_{1\,br})$, $G_2 = (G_{21}, \ldots, G_{2r})$ are vector-valued functions.

If the initial and boundary data are zero, then problem (4.1)–(4.3) is called the *homogeneous boundary value problem*, and the corresponding Green function G_0 is referred to as a *homogeneous Green function*. In the case of parabolic equations and systems of equations of first order with respect to t, the homogeneous Green function (for the model problem) was the subject of Theorem VI.23. Estimates (3.20) and (3.21) enable us to investigate the properties of the homogeneous Green function and its derivatives. However, the assumption that $f(x,t)$ and $\psi(x)$ have compact supports, made in Subsection VI.3.4, prevents us from analyzing to the end all possible situations. Let us assume first that the orders of all boundary operators B_q are smaller than the order of the equation and, passing to the limit in (3.22), let us drop the assumption that the boundary values have compact supports (retaining, of course, the compatibility conditions). Now let us give an example in which the boundary conditions involve a time derivative.

Example VI.14 [38, 44]. Consider the problem

$$D_t u = D^2 u + f(x,t), \qquad u\big|_{t=0} = \psi(x), \quad D_t u\big|_{x=0} = \varphi(t). \qquad (4.5)$$

In our discussion of Example I.5 we have shown that (4.5) is a parabolic problem. It is equivalent to the problem

$$D_t u = D^2 u + f, \qquad u\big|_{t=0} = \psi(x), \quad u\big|_{x=0} = \psi(0) + \int\limits_0^t \varphi(\tau)\,d\tau. \qquad (4.6)$$

We can represent the solution of problem (4.5), (4.6) by the formula (3.22):

$$\begin{aligned}
u(x,t) = & \int\limits_0^t d\tau \int\limits_0^\infty G_0^{(1)}(x-y, t-\tau) f(y,\tau)\,dy \\
& + \int\limits_0^t G^{(1)}(x, t-\tau)\left\{ \psi(0) + \int\limits_0^\tau \varphi(\tilde{t})d\tilde{t} \right\} d\tau \\
& + \int\limits_0^\infty G_0^{(1)}(x, y, t)\psi(y)\,dy,
\end{aligned} \qquad (4.7)$$

where $G_0^{(1)}$, $G^{(1)}$ are the Green function and the Poisson kernel of the Dirichlet problem, respectively. Next, let us rewrite (4.7) in the form (4.4):

$$
u(x,t) = \int_0^t d\tau \int_0^\infty G_0^{(1)}(x - y, t - \tau) f(y, \tau)\, dy
$$

$$
+ \int_0^t \widetilde{G}(x, t - \tau) \varphi(\tau)\, d\tau
$$

$$
+ \int_0^\infty \left\{ G_0^{(1)}(x, y, t) + 2 \int_0^t G^{(1)}(x, t - \tau) d\tau \delta(y) \right\} \psi(y)\, dy,
$$

where $\widetilde{G}(x,t) = (2/\sqrt{\pi}) \int_{x/2\sqrt{t}}^\infty \exp\{-z^2\} dz$.

Thus, the Green function of problem (4.5) is given by the formula

$$
\left(G_0^{(1)}(x, y, t - \tau), \widetilde{G}(x, t - \tau), G_0^{(1)}(x, y, t) + 2 \int_0^t G^{(1)}(x, t - \tau) d\tau \delta(y) \right). \qquad (4.8)
$$

We see that the third component of the Green function of problem (4.5) is a distribution.

Now let us return to problem (4.1)–(4.3). Rewrite the boundary operator B_q in local coordinates with the origin at the point $(\tilde{x}, \tilde{t}) \in S$. Denote by $n_q(\tilde{x}, \tilde{t})$ the maximal order of the derivative at this point in the direction of the normal, set $n_q = \max_{(\tilde{x}, \tilde{t}) \in S} n_q(\tilde{x}, \tilde{t})$, $\rho_q = n_q - 2br$, $\rho_0 = \max_q \rho_q$, and let m_q be the order of the highest t-derivative appearing in the operator B_q.

Lemma VI.3 [44]. 1) *If $n_q < 2br$, $m_q \leq r$, $q = 1, \ldots, b$, then G_0 is an ordinary function. 2) If $m_q = 0$, $2b(n - 1) < -\rho_0$, then G_{2q}, $q = 1, \ldots, r$, are ordinary functions.*

In all other cases G_0 and G_{2q} involve, generally speaking, terms that are finite linear combinations of derivatives of δ-functions concentrated on S or on the hyperplane $t = 0$, $x \in G$. Then the integrals in (4.4), whose kernels are the distributions indicated above, designate the result of applying these distributions to the corresponding test functions.

Consider now the situation which is most frequently met in applications, namely, when equation (4.1) is of first order with respect to t and the orders of the boundary operators are less than the order of the equation: $r_q < 2b$. Then all the components of the Green function are ordinary functions.

In the general case, the problem adjoint to a parabolic differential problem is not necessarily a differential problem, and for a complete investigation of adjoint problems we must extend the framework of the theory of differential boundary value problems to include equations with pseudodifferential operators. Nevertheless, under specific conditions—the so-called *normality* conditions of a boundary value problem—the adjoint problem turns out to be a differential and parabolic problem (provided that we reverse the direction of time), and the (complete) Green function is determined by the homogeneous Green function G_0 [110]. Let us pause briefly to discuss this point.

For normal boundary conditions the Green formula can be rewritten as [110]

$$(Lu, v) + \sum_{\beta=1}^{b} \langle B_\beta u, C_\beta^* v \rangle + \big[u(x,0), v(x,0)\big]$$
$$= (u, L^* v) + \sum_{\beta=1}^{b} \langle C_\beta u, B_\beta^* v \rangle + \big[u(x,T), v(x,T)\big], \tag{4.9}$$

where B_β^*, C_β^*, B_β, C_β are differential operators defined on S.

Thus, we are led to the following adjoint parabolic differential boundary value problem:

$$L^*(y, \tau, D_y, D_\tau)v = F, \qquad v\big|_{t=T} = \Psi, \; B_q^*(y, \tau, D_y)v\big|_S = \Phi_q. \tag{4.10}$$

Formula (4.9) immediately implies that the homogeneous Green function $G_0(x, t; y, \tau)$ of the problem (4.1)–(4.3), regarded as a function of (y, τ), is the homogeneous Green function of the problem (4.10), and any solution $u(x, t)$ of the problem (4.1)–(4.3) can be expressed in terms of the data of the problem as

$$u(x,t) = \int_0^t d\tau \int_G G_0(x, t; y, \tau) f(y, \tau)\, dy + \int_G G_0(x, t; y, 0)\psi(y)\, dy$$
$$+ \sum_{q=1}^{b} \int_0^t d\tau \int_{\partial G} \big\{ C_q^*(y', \tau, D_y) G_0(x, t; y', \tau) \big\} \varphi_q(y', \tau)\, dy'. \tag{4.11}$$

Comparing formulas (4.4) and (4.11) one can express all the components of the function $G(x, t; y, \tau)$ in terms the homogeneous Green function $G_0(x, t; y, \tau)$.

Note, finally, that the classical problems of mathematical physics for parabolic equations usually enjoy the normality property, and consequently their solutions can be expressed via (4.11) in terms of the homogeneous Green function.

VI.4.3. The Green function for conjugation problems

Let us indicate that definitive results on Green functions for parabolic conjugation problems of the type (2.49), Subsection I.2.5, can be found in [22]. For such problems the analogue of formula (4.4) is

$$u_{(\mu)}(x,t) = \sum_{\lambda=1}^{2} \int_0^t d\tau \int_{G_\lambda} G_0^{\mu\lambda}(x,t;y,\tau) f_{(\lambda)}(y,\tau)\, dy$$

$$+ \int_0^t d\tau \int_{\partial G_1} G_1^{\mu 1}(x,t;y',\tau)\varphi_1(y',\tau)\, dy'$$

$$+ \int_0^t d\tau \int_{\partial G_0} G_1^{\mu 2}(x,t;y',\tau)\varphi_2(y',\tau)\, dy'$$

$$+ \sum_{\lambda=1}^{2} \int_{G_\lambda} G_0^{\mu\lambda}(x,t;y,0)\psi_\lambda(y)\, dy.$$

Thus, in the present case the Green function $G = (G_0, G_1, G_2)$ is the matrix-valued function

$$G = \begin{pmatrix} G_0^{11} G_0^{12} & G_1^{11} G_1^{12} & G_2^{11} G_2^{12} \\ G_0^{21} G_0^{22} & G_1^{21} G_1^{22} & G_2^{21} G_2^{22} \end{pmatrix}.$$

Chapter VII
Behaviour of Solutions of Parabolic Boundary Value Problems for Large Values of Time

VII.1. Asymptotic representations and stabilization of solutions of model problems

VII.1.1. Formulation of the problem

In $\mathbb{R}^{n+1}_{++}$ we consider the parabolic boundary value problem:

$$\sum_{j=1}^{m} l_{lkj}(D, D_t)u_j(x, t) = 0, \qquad (x, t) \in \mathbb{R}^{n+1}_{++}, \quad k = 1, \ldots, m; \qquad (1.1)$$

$$\sum_{j=1}^{m} b_{qj}(D, D_t)u_j(x, t)|_{x_n=+0} = \varphi_q(x', t), \quad (x't) \in \mathbb{R}^n_+, \quad q = 1, \ldots, br, \quad (1.2)$$

in spaces of smooth bounded Hölder functions that vanish at $t = 0$ together with all their derivatives that appear in (1.1) and (1.2). Here $l_{kj}(D, D_t)$ and $b_{qj}(D, D_t)$ are quasihomogeneous operators with constant coefficients of orders $s_k + t_j$ and $\sigma_q + t_j$, respectively, $\sum_{k=1}^{m}(s_k + t_k) = 2br$.

The Schauder theory of parabolic boundary value problems (see Section VI.3) guarantees the well-posed solvability of the problem (1.1), (1.2) in the spaces $C^l(\mathbb{R}^n_+ \times [0, T])$, $l > \max_q(0, \sigma_q)$ for functions $\varphi_q(x', t) \in C^{l-\sigma_q}(\mathbb{R}^{n+1} \times [0, T])$ that are compatible with zero at $t = 0$. The unique solution of this problem can be represented through the Poisson kernels $G_{jq}(x, t)$ of the problem (1.1), (1.2) by the formula:

$$u_j(x, t) = \sum_{q=1}^{br} \int_0^t d\tau \int_{\mathbb{R}^{n-1}} G_{jq}(x - y', t - \tau)\varphi_q(y', \tau)dy', \quad j = 1, \ldots, m. \qquad (1.3)$$

The problem we are interested in here is to find conditions on the boundary functions $\varphi_q(x', t)$ under which the solution $u_j(x, t)$ admits an asymptotic representation (as $t \to \infty$) which, in particular, yields necessary and sufficient conditions for stabilization (i.e., the existence of a limit). Performing the Laplace transformation with respect to the time variable t we obtain from (1.3) a formula for the image (Laplace transform) $\widehat{u}_j(x,p)$ of the function $u_j(x,t)$:

$$\widehat{u}_j(x,p) = \sum_{q=1}^{br} \int_{\mathbb{R}^{n-1}} \widehat{G}_{jq}(x-y',p)\widehat{\varphi}_q(y',p)\,dy'. \tag{1.4}$$

The functions $\widehat{u}_j(x,p)$ are bounded solutions of the following elliptic boundary value problem:

$$\sum_{j=1}^{m} l_{kj}(D,p)\widehat{u}_j(x,p) = 0, \quad x \in \mathbb{R}^n_+, \ k = 1,\ldots,m, \tag{1.5}$$

$$\sum_{j=1}^{m} b_{qj}(D,p)\widehat{u}_j(x,p)|_{x_n=+0} = \widehat{\varphi}_q(x',p), \quad x' \in \mathbb{R}^{n-1}, \ q = 1,\ldots,br. \tag{1.6}$$

From operational calculus it is known that the behaviour of the original as $t \to \infty$ depends essentially on the behaviour of the image as $p \to 0$. Therefore, a significant role in the ensuing analysis will be played by the bounded solutions $\widehat{u}_j(x,0) \equiv v_j(x)$, $j = 1,\ldots,m$, of the problem (1.5), (1.6) with $p = 0$. We denote this problem by (1.5_0), (1.6_0). In order to find asymptotic representations of the bounded solutions of problem (1.5), (1.6), we need to proceed in a rather standard manner, using convenient formulas for the Poisson kernels of certain model parabolic and elliptic boundary value problems. We hope that the calculations given below will be of use in studies of various specific problems of mathematical physics.

VII.1.2. Poisson kernels of an elliptic boundary value problem with a parameter

Here we embark upon the study of the functions $\widehat{G}_{jq}(x,p)$. We begin by describing the formulas that define these functions. These formulas are well-known in the theory of elliptic and parabolic boundary value problems (the reader can herself verify that they indeed yield the needed result). Thus, consider the matrix $\mathcal{L}(i\xi,p)$ of system (1.5), then the matrix $\widehat{\mathcal{L}}(i\xi,p)$ adjoint to $\mathcal{L}(i\xi,p)$, and denote $\det \mathcal{L}(i\xi,p) = L(i\xi,p)$. Now factor the ξ_n-polynomial $L(i\xi,p) \equiv L(i\xi',i\xi_n,p)$ as

$$L(i\xi',i\xi_n,p) = L_+(\xi',\xi_n,p)L_-(\xi',\xi_n,p),$$

and from the ξ_n-polynomial $L_+(\xi',\xi_n,p) = \sum_{s=0}^{br} m_s(\xi',p)\xi_n^{br-s}$ construct the ξ_n-polynomials $L_\nu^+(\xi',\xi_n,p) = \sum_{s=0}^{\nu} m_s(\xi',p)\xi_n^{\nu-s}$, $\nu = 0,\ldots,br$.

Denote by $L_{kj}(i\xi,p)$ the elements of the adjoint matrix $\widehat{\mathcal{L}}(i\xi,p)$. Consider the $(br \times m)$ matrix $\mathcal{R}(\xi,p) = \mathcal{B}(i\xi,p)\widehat{\mathcal{L}}(i\xi,p)$. Further, let $\mathcal{R}'(\xi',p)$ be the matrix composed of the remainders of the division of the elements of $\mathcal{R}(\xi, p)$ (regarded as polynomials in ξ_n) by the ξ_n-polynomial $L_+(\xi', \xi_n, p)$. Now consider the rectangular matrix $\mathcal{D}(\xi', p)$ composed of the coefficients of the ξ_n-polynomials that form $\mathcal{R}'(\xi, p)$. In [94, p. 43] it is shown that $\mathcal{D}(\xi', p)$ has a right inverse matrix. Denote by $d_\nu^{(k,q)}(\xi', p)$ the elements of the latter.

Now introduce the functions

$$N_{kq}(\xi', \xi_n, p) = \sum_{\nu=1}^{br} d_\nu^{(k,q)}(\xi', p) L_{br-\nu}^+(\xi', \xi_n, p),$$

$$\widehat{g}_{jq}(\xi', \xi_n, p) = (2\pi i)^{-1} \sum_{k=1}^{m} L_{kj}(i\xi', i\xi_n, p) N_{kq}(\xi', \xi_n, p)/L_+(\xi', \xi_n, p). \quad (1.7)$$

The functions $\widehat{G}_{jq}(x, p)$ are defined by the formulas

$$\widehat{G}_{jq}(x, p) = \int_{\mathbb{R}^{n-1}} d\xi' \int_{\gamma_+} \widehat{g}_{jq}(\xi', \xi_n, p)) \exp\{ix \cdot \xi\} \, d\xi_n, \quad (1.8)$$

where γ_+ is a simple closed Jordan contour in the complex ξ_n-plane that encircles all the ξ_n- zeroes of the polynomial $L(i\xi', i\xi_n, p)$ with positive imaginary parts, and $x \cdot \xi \equiv (x, \xi) = \sum_{j=1}^{n} x_j \xi_j$.

To study the functions $\widehat{G}_{jq}(x, p)$ we need the following information on the functions $\widehat{g}_{jq}(\xi', \xi_n, p)$.

Lemma VII.1. 1) *The functions $\widehat{g}_{jq}(\xi', \xi_n, p)$ are homogeneous in a generalized sense, of order $-\sigma_q - t_j - 1$, that is, $\widehat{g}_{jq}(\lambda \xi', \lambda \xi_n, \lambda^{2b} p) = \lambda^{-\sigma_q - t_j - 1} \widehat{g}_{jq}(\xi', \xi_n, p)$;*
2) *$\widehat{g}_{jq}(\zeta', \zeta_n, p)$ can be analytically continued to the domain*

$$A_{\kappa, \varepsilon} = \left\{ (\zeta', \zeta_n, p) : \operatorname{Re} p > -\kappa |\operatorname{Im} p|, \ \zeta' = \xi' + i\eta', |\eta'| \le \varepsilon (|\xi'| + |p|^{1/2b}), \zeta_n \in \gamma_+ \right\}$$

of the space $\mathbb{C}^{n+1}$ for some positive κ, ε.

Proof: Assertion 1) is proved by successively verifying the generalized homogeneity of the functions in terms of which $\widehat{g}_{jq}(\xi', \xi_n, p)$ are defined.

Assertion 2) is based on an important theorem stating that if the zeroes of a polynomial whose coefficients depend polynomially on parameters form a cycle, then any symmetric function of these zeroes is analytic in the parameters in question. This shows that the functions $m_s(\xi', p)$ are analytic. After this, we successively establish the analyticity of all functions that enter formula (1.7). First, all this is carried out for real ξ' and for p with $\operatorname{Re} p > 0$.

We then observe that the entire argument remains valid in the case when the variables change in the domain $A_{\kappa, \varepsilon}$ for sufficiently small positive values of κ and ε, for which the function $L_+(\zeta', \zeta_n, p)$ is different from zero.

Using Lemma VII.1, one establishes the following statement, which will play an important role below.

Lemma VII.2. *There exists a positive constant κ_0 such that the functions $\widehat{G}_{jq}(x,p)$ are analytic in p in the domain $A_{\kappa_0} = \big\{p : \operatorname{Re} p > -\kappa_0 |\operatorname{Im} p|\big\}$ for each $x \in \mathbb{R}_+^n$. Morevoer, for each $(x,p) \in A_{\kappa_0} \times \mathbb{R}_+^n$ the following estimates hold:*

$$|\widehat{G}_{jq}(x,p)| \leq \begin{cases} C|p|^{-(\sigma_q+t_j-n+1)/2b} \exp\big\{-c|x||p|^{1/2b}\big\}, & \sigma_q + t_j - n + 1 > 0, \\[2mm] C(1 + |\ln|x\cdot p||)\exp\big\{-c|x||p|^{1/2b}\big\}, & \sigma_q + t_j - n + 1 = 0, \\[2mm] C|x|^{\sigma_q+t_j-n+1}\exp\big\{-c|x||p|^{1/2b}\big\}, & \sigma_q + t_j - n + 1 < 0, \end{cases}$$
$$(1.9)$$

where C and c are positive constants.

Proof: The analyticity of the function $\widehat{G}_{jq}(x,p)$ in $p \in A_{\kappa_0}$ follows from the properties of the function $\widehat{g}_{jq}(\xi',\xi_n,p)$ and the uniform convergence in the complex variable p of the improper integral that defines $\widehat{G}_{jq}(x,p)$ via (1.8).

Now let us establish the estimates (1.9). This is done by a standard argument which will be given in detail. For the sake of definiteness we shall assume that $|x_{n-1}| = \max\big(|x_1|,\dots,|x_{n-1}|\big)$. In the integral defining $\widehat{G}_{jq}(x,p)$ (formula (1.8)), we pass from integration over the real axis ξ_{n-1} to integration over the line $\zeta_{n-1} = \xi_{n-1} + i\eta_{n-1}$ in the complex ζ_{n-1}-plane, $\eta_{n-1} = \delta_0 \cdot \operatorname{sgn} x_{n-1} \cdot \big(|\xi'| + |p|^{1/2b}\big)$, where $\delta_0 > 0$ is a sufficiently small constant. That such a transition is allowed follows from the analyticity of $\widehat{g}_{jq}(\zeta',\zeta_n,p)$ in the variable ξ_{n-1}, Cauchy's theorem and the fact that the modulus of the integrand in (1.8) tends to zero as $|\xi_{n-1}| \to \infty$. Furthermore, exploiting the arbitrariness of the contour γ_+ in (1.8), we take for γ_+ the contour formed by the chord $\operatorname{Im} \zeta_n = c_1 \lambda$ and the arc of the circumference $|\zeta_n| = c_2 \lambda$, $c_2 \gg c_1$, $\lambda = |\zeta'| + |p|^{1/2b}$. All this enables us to represent $\widehat{G}_{jq}(x,p)$ as

$$\widehat{G}_{jq}(x,p) = \int_{\mathbb{R}^{n-2}} d\xi'' \int_{-\infty}^{\infty} \exp\big\{ix'\cdot\xi' - \delta_0\big(|\xi'| + |p|^{1/2b}\big)|x_{n-1}|\big\}\, d\xi_{n-1}$$

$$\times \int_{\gamma_+} \widehat{g}_{jq}(\xi'',\zeta_{n-1},p)\exp\{i\zeta_n x_n\}\, d\zeta_n, \quad \xi'' = (\xi_1,\dots,\xi_{n-2}).$$

Now use the generalized homogeneity of the functions $\widehat{g}_{jq}$ to recast the last formula as

$$\widehat{G}_{jq}(x,p) = \int_{\mathbb{R}^{n-2}} d\xi'' \int_{-\infty}^{\infty} \exp\big\{ix'\cdot\xi' - \delta_0\big(|\xi'| + |p|^{1/2b}\big)|x_{n-1}|\big\}\, d\xi_{n-1}$$

$$\times \int_{\gamma_+} \lambda^{-\sigma_q-t_j-1}\widehat{g}_{jq}(\lambda^{-1}\xi'',\lambda^{-1}\zeta_{n-1},\lambda^{-1}\zeta_n,\lambda^{-2b}p)\exp\{i\zeta_n x_n\}\, d\zeta_n.$$

For the values of ξ'', ζ_{n-1} and p under consideration, the variables $\tilde{\zeta}' = \lambda^{-1}\zeta'$, $\tilde{p} = \lambda^{-2b}p$ satisfy the relation $|\tilde{\xi}'| + |\tilde{p}|^{1/2b} = 1$ and the variable $\lambda^{-1}\zeta_n$ runs through a fixed closed contour γ that encircles all the ζ_n-zeroes of the function $L_+(\tilde{\zeta}',\zeta_n,\tilde{p})$ for $|\tilde{\zeta}'| + |\tilde{p}|^{1/2b} = 1$. Since the functions $\widehat{g}_{jq}$ are continuous on a compact set, they are bounded, and so

$$|\widehat{G}_{jq}(x,p)| \leq C\int_{\mathbb{R}^{n-1}} \lambda^{-\sigma_q-t_j} \exp\Big\{\big(-\delta_0|x_{n-1}| - c_1 x_n\big)\big(|\xi'| + |p|^{1/2b}\big)\Big\}\, d\xi'.$$

Now observe that $\lambda = |\zeta'| + |p|^{1/2b} \geq |\xi'| + |p|^{1/2b}$; also, since we assumed that $|x_{n-1}| = \max\left(|x_1|, \ldots, |x_{n-1}|\right)$, we have $|x_{n-1}| \geq |x'|/\sqrt{n-1}$, $\delta_0|x_{n-1}| + c_1 x_n \geq \delta_0/\sqrt{n-1}|x'| + c_1 x_n \geq \min\left(\delta_0/\sqrt{n-1}, c\right)|x| = c_3|x|$. Taking this into account, we arrive at the inequality

$$|\widehat{G}_{jq}(x,p)| \leq C \exp\left\{-c_3|p|^{1/2b}|x|\right\} \int_{\mathbb{R}^{n-1}} \exp\left\{-c_3|x||\xi'|\right\}\left(|\xi'| + |p|^{1/2b}\right)^{-\sigma_q - t_j} d\xi'.$$

$$(1.10)$$

In the last integral we introduce the new integration variables $\beta' = \xi'|p|^{-1/2b}$, and then in the integral with respect to $\beta_1, \ldots, \beta_1$ we pass to spherical coordinates. This yields

$$\int_{\mathbb{R}^{n-1}} \exp\left\{-c_3|x||\xi'|\right\}\left(|\xi'| + |p|^{1/2b}\right)^{-\sigma_q - t_j} d\xi' \leq C|p|^{-(\sigma_q + t_j - n + 1)}$$

$$\times \int_0^\infty \rho^{n-2}(1+\rho)^{-\sigma_q - t_j} \exp\{-c_3 A\rho\} d\rho, \quad A = |x|\,|p|^{1/2b}. \qquad (1.11)$$

Now consider the integral

$$\mathcal{I}(A) = \int_0^\infty \rho^{n-2}(1+\rho)^{-\sigma_q - t_j} \exp\left\{-c_3 A\rho\right\} d\rho,$$

where $n \geq 2$ (the case $n = 1$ is examined separately). An elementary analysis yields the estimates

$$\mathcal{I} \leq \begin{cases} C & \text{for } \sigma_q + t_j - n + 1 > 0, \\ C_1\left(1 + |\ln|x \cdot p||\right) & \text{for } \sigma_q + t_j - n + 1 = 0, \\ C\left(|x||p|^{1/2b}\right)^{\sigma_q + t_j - n + 1} & \text{for } \sigma_q + t_j - n + 1 < 0. \end{cases} \qquad (1.12)$$

Estimates (1.10)–(1.12) imply (1.9).

VII.1.3. Asymptotic representation of Poisson kernels of an elliptic boundary value problem with a parameter

Let us study the Poisson kernels $\widehat{G}_{jq}(x,p)$ in the domain $\mathfrak{A} = \mathbb{R}_+^n \times A_{\kappa_0}$. Recall that these kernels are analytic functions of p for any fixed $x \in \mathbb{R}_+^n$. The following theorem plays a very important role in the sequel.

Theorem VII.1. *In the domain $\mathfrak{A}$ the functions $\widehat{G}_{jq}(x,p)$ can be represented in the form*

$$\widehat{G}_{jq}(x,p) = \sum_{s=0}^{\sigma_q + t_j - n} \mathcal{P}_{jq}^s(x)p^{-\alpha_{jqs}} - \left(\frac{1}{2b}\ln p + c_{jq}\right)F_{jq}(x)$$

$$+\mathcal{K}_{jq}(x) + P_{jq}(x) + \mathcal{R}_{jq}(x,p), \qquad (1.13)$$

where $\mathcal{P}_{jq}^s(x)$ are homogeneous polynomials of degree s,

$$\alpha_{jqs} = (\sigma_q + t_j - n + 1 - s)/2b,$$

$$c_{jq} = (\sigma_q + t_j - n + 1) \int_0^1 (1 - \beta)^{\sigma_q + t_j - n} \ln \beta \, d\beta + C,$$

C is the Euler constant, $F_{jq}(x)$, $P_{jq}(x)$ are homogeneous polynomials of degree $\sigma_q + t_j - n + 1$, and $\mathcal{K}_{jq}(x)$ are the Poisson kernels of the elliptic boundary value problem (1.5_0), (1.6_0); $p^{-\alpha_{jqs}}$ and $\ln p$ are understood as the principal branches of these multi-valued functions of the complex variable p. Moreover, the following estimates hold:

$$|\mathcal{R}_{jq}(x, p)| \le C_{jq}(x)|p|^{1/2b}. \tag{1.14}$$

If $\sigma_q + t_j - n + 1 = 0$, then in (1.13) one has $\mathcal{P}_{jq}^s(x) \equiv 0$; if $\sigma_q + t_j - n + 1 < 0$, then $\mathcal{P}_{jq}^s(x) \equiv F_{jq}(x) \equiv P_{jq}(x) \equiv 0$.

We note that in the course of the proof we will present formulas for the computation of the polynomials that appear in (1.13). Formula (1.13) provides, in particular, information about the behaviour of the functions $\widehat{G}_{jq}(x, p)$ as the complex variable p tends to zero under the assumption that it does so along the paths lying inside the domain A_{κ_0} of the complex p-plane.

We now turn to the proof of Theorem VII.1. To make the argument more accessible, we break it into a number of steps.

1) Let $\sigma_q + t_j + 1 \ge n$. Making use of the properties of the functions $\widehat{g}_{jq}(\xi', \xi_n, p)$, we transform the integral (1.8) as follows.

First we write the Maclaurin formula for the exponential function,

$$e^z = \sum_{s=0}^{\sigma_q + t_j - n} \frac{z^s}{s!} + \frac{z^{\sigma_q + t_j - n + 1}}{(\sigma_q + t_j - n + 1)!} \int_0^1 (1 - \beta)^{\sigma_q + t_j - n} e^{\beta z} \, d\beta$$

and use it to recast (1.8) as

$$\widehat{G}_{jq}(x, p) = \sum_{s=0}^{\sigma_q + t_j - n} \frac{1}{s!} \int_{\mathbb{R}^{n-1}} d\xi' \int_{\gamma_+} (ix \cdot \xi)^s \widehat{g}_{jq}(\xi', \xi_n, p) \, d\xi_n$$

$$+ \frac{1}{(\sigma_q + t_j - n + 1)!} \int_0^1 (1 - \beta)^{\sigma_q + t_j - n} \, d\beta$$

$$\times \int_{\mathbb{R}^{n-1}} d\xi' \int_{\gamma_+} (ix \cdot \xi)^{\sigma_q + t_j - n + 1} \widehat{g}_{jq}(\xi', \xi_n, p) e^{i\beta x \cdot \xi} d\xi_n. \tag{1.15}$$

Our next objective is to analyze successively the terms of the formula (1.15).

2) We begin by transforming the integrals

$$\mathcal{I}_s(x,p) = \frac{1}{s!} \int_{\mathbb{R}^{n-1}} d\xi' \int_{\gamma_+} (ix \cdot \xi)^s \widehat{g}_{jq}(\xi', \xi_n, p)\, d\xi_n$$

which appear under the sum symbol in (1.15).

Let us assume (for the moment) that the variable p takes positive values. Introducing a new integration variable $\zeta = p^{-1/2b}\xi$ and making use of the generalized homogeneity of the functions $\widehat{g}_{jq}(\xi', \xi_n, p)$, we rewrite $\mathcal{I}_s(x,p)$ in the form

$$\mathcal{I}_s(x,p) = \frac{1}{s!} p^{-\alpha_{jqs}} \int_{\mathbb{R}^{n-1}} d\zeta' \int_{\gamma_+} (ix \cdot \zeta)^s \widehat{g}_{jq}(\zeta', \zeta_n, 1)\, d\zeta_n, \qquad (1.16)$$

where γ_+ is any simple Jordan contour in the half-plane $\operatorname{Im} \zeta_n > 0$ of the complex ζ_n-plane which encircles the ζ_n-zeroes of the polynomial $L_+(\zeta', \zeta_n, 1)$.

It follows from (1.8) and the assumption $\sigma_q + t_j - n + 1 > 0$ that $\widehat{G}_{jq}(x,1)$ has derivatives with respect to $x_1, \ldots, x_n$ of order $|k|$, $|k| \le \sigma_q + t_j - n$, $k = (k_1, \ldots, k_n)$, whose values at $x = 0$ are given by the formulas

$$\widehat{G}_{jq}^{(k)}(0,1) = \int_{\mathbb{R}^{n-1}} d\xi' \int_{\gamma_+} (i\xi)^k \widehat{g}_{jq}(\xi', \xi_n, 1)\, d\xi_n. \qquad (1.17)$$

Now introduce the following homogeneous polynomials in $x_1, \ldots, x_n$:

$$\mathcal{P}_{jq}^s(x) = \sum_{|k|=s} \widehat{G}_{jq}^{(k)}(0,1) \frac{x^k}{k!} = \frac{1}{s!} \int_{\mathbb{R}^{n-1}} d\xi' \int_{\gamma_+} (ix \cdot \xi)^s \widehat{g}_{jq}(\xi', \xi_n, 1)\, d\xi_n. \qquad (1.18)$$

Comparing formulas (1.16), (1.18), one concludes that

$$\mathcal{I}_s(x,p) = p^{-\alpha_{jqs}} \mathcal{P}_{jq}^s(x). \qquad (1.19)$$

Recall that so far formula (1.19) is established only for positive p.

In the case of an arbitrary $p \in A_{\kappa_0}$ the formula follows from the uniqueness of the analytic continuation of the function $\mathcal{I}_s(x,p)$ defined by (1.19), from the positive half-line of the complex p-plane to the whole domain A_{κ_0}; here it is essential to fix the principal branch of the multi-valued function $p^{-\alpha_{jqs}}$.

Thus, (1.15) can be rewritten in the form

$$\widehat{G}_{jq}(x,p) = \sum_{s=0}^{\sigma_q + t_j - n} p^{-\alpha_{jqs}} \mathcal{P}_{jq}^s(x) + \hat{\mathcal{G}}_{jq}(x,p), \qquad (1.20)$$

where

$$\hat{\mathcal{G}}_{jq}(x,p) = \left[(\sigma_q + t_j - n)!\right]^{-1} \int_0^1 (1 - \beta)^{\sigma_q + t_j - n} \Phi_{jq}(\beta, x, p)\, d\beta, \qquad (1.21)$$

$$\Phi_{jq}(\beta, x, p) = \int_{\mathbb{R}^{n-1}} d\xi' \int_{\gamma_+} Q_{jq}(x, p, \xi) e^{i\beta x \cdot \xi}\, d\xi_n, \qquad (1.22)$$

$$Q_{jq}(x, \xi, p) = (ix \cdot \xi)^{\sigma_q + t_j - n + 1} \widehat{g}_{jq}(\xi', \xi_n, p). \qquad (1.23)$$

3) Now we turn to the study of the functions $\Phi_{jq}(\beta, x, p)$. We introduce a new integration variable $\tilde{\xi} = \xi |p|^{-1/2b}$ in formula (1.22). Denote $p_0 = p/|p|$. Since $Q_{jq}(x, \xi, p)$ is a generalized homogeneous function of order zero,

$$\Phi_{jq}(\beta, x, p) = \int_{\mathbb{R}^{n-1}} d\tilde{\xi}' \int_{\gamma_+} Q_{jq}(x, \tilde{\xi}, p_0) e^{i\beta x \cdot \tilde{\xi}|p|^{1/2b}} d\tilde{\xi}_n \qquad (1.24)$$

where the contour γ_+ can be chosen to be independent of p_0.

Now let us represent $\Phi_{jq}(\beta, x, p)$ in the form

$$\Phi_{jq}(\beta, x, p) = \int_{|\tilde{\xi}'| \le 1} d\tilde{\xi}' \ldots + \int_{|\tilde{\xi}'| > 1} d\tilde{\xi}' \ldots = \Phi_{jq}^{(1)}(\beta, x, p) + \Phi_{jq}^{(2)}(\beta, x, p). \qquad (1.25)$$

Since $\Phi_{jq}^{(1)}$ for any $\beta \in [0, 1]$, p_0 and x can be written as an improper integral of a smooth function, $\Phi_{jq}^{(1)}(\beta, x, p)$ has a finite limit as $|p| \to 0$ for fixed p_0 (so that $p \in A_{\kappa_0}$), β and x, and the following estimate holds:

$$|\Phi_{jq}^{(1)}(\beta, x, p)| \le C|x|^{\sigma_q + t_j - n + 1}. \qquad (1.26)$$

Consider now the function $\Phi_{jq}^{(2)}(\beta, x, p)$. Again, we write it as

$$\Phi_{jq}^{(2)}(\beta, x, p) = \int_{|\tilde{\xi}'| > 1} d\tilde{\xi}' \int_{\gamma_+} (ix \cdot \tilde{\xi})^{\sigma_q + t_j - n + 1} \hat{g}_{jq}(\tilde{\xi}', \tilde{\xi}_n, p_0) e^{i\beta x \cdot \tilde{\xi}|p|^{1/2b}} d\tilde{\xi}_n. \qquad (1.27)$$

Let us introduce the spherical coordinates ω', ρ, where $|\omega'| = 1$, $\rho \in [1, \infty)$ in the integral with respect to $\tilde{\xi}'$ and the new variable $\omega_n = \tilde{\xi}_n/\rho$ in the integral with respect to $\tilde{\xi}_n$. As the contour γ_+ we can take the closed curve formed by the chord $\mathrm{Im}\, \tilde{\xi}_n = c_2(1 + \rho)$, $|\tilde{\xi}_n| = c_1(1 + \rho)$, $c_1 \ll c_2$. This enables us to integrate with respect to ω_n over a standard (independent of ρ) contour γ_0 that encircles the ω_n-zeroes of the polynomial $L_+(\omega', \omega_n, p)$ for $|p| \le 1$. Denote $\Omega = \Omega' \times \gamma_0$, $\Omega' = \{\omega' : |\omega'| = 1\}$, $d\omega = d\omega' d\omega_n$. First we integrate with respect to ρ, then with respect to ω. Then formula (1.27) can be recast as

$$\Phi_{jq}^{(2)}(\beta, x, p) = \int_{\Omega} (ix \cdot \omega)^{\sigma_q + t_j - n + 1} d\omega \int_1^\infty \hat{g}_{jq}(\omega', \omega_n, p_0 \cdot \rho^{-2b}) e^{i\beta x \cdot \omega |p|^{1/2b}} \frac{d\rho}{\rho}. \qquad (1.28)$$

It follows from the properties of $\hat{g}_{jq}$ described in Lemma VII.1 that the limit $\lim_{\rho \to \infty} \hat{g}_{jq}(\omega', \omega_n, p_0 \cdot \rho^{-2b}) = \hat{g}(\omega', \omega_n, 0)$ exists for any fixed direction p_0 in A_{κ_0}.

Next, denote $\alpha = \beta x \cdot \omega |p|^{1/2b}$ and rewrite $\Phi_{jq}^{(2)}$ as

$$\Phi_{jq}^{(2)}(\beta, x, p) = \int_{\Omega} (ix \cdot \omega)^{\sigma_q + t_j - n + 1} \left(\int_1^\infty e^{i\alpha\rho} \frac{d\rho}{\rho} \right) \hat{g}_{jq}(\omega', \omega_n, 0)\, d\omega$$

$$+ \int_{\Omega} (ix \cdot \omega)^{\sigma_q + t_j - n + 1} \int_1^\infty \left[\hat{g}_{jq}(\omega', \omega_n, p_0 \rho^{-2b}) - \hat{g}_{jq}(\omega', \omega_n, 0) \right] e^{i\alpha\rho} \frac{d\rho}{\rho}\, d\omega \qquad (1.29)$$

Note that the last integral with respect to ρ converges absolutely, uniformly in $p_0 \in A_{\kappa_0}$ and $\beta \in [0, 1]$.

Now recall the formula ([21, 23]):

$$\int_0^\infty e^{-\alpha\rho} \ln \rho \, d\rho = -\frac{1}{\alpha}(\ln \alpha + C),$$

valid for $\operatorname{Re}\alpha > 0$, where C is the Euler constant, and perform the following transformation:

$$\int_1^\infty e^{i\alpha\rho} \frac{d\rho}{\rho} = -i\alpha \int_1^\infty e^{i\alpha\rho} \ln \rho \, d\rho$$

$$= -i\alpha \int_0^\infty e^{i\alpha\rho} \ln \rho \, d\rho + i\alpha \int_0^1 e^{i\alpha\rho} \ln \rho \, d\rho \qquad (1.30)$$

$$= -\ln \frac{\alpha}{i} - C + i\alpha \int_0^1 e^{i\alpha\rho} \ln \rho \, d\rho.$$

It follows from (1.24)–(1.30) that

$$\Phi_{jq}(\beta, x, p) = -\int_\Omega Q_{jq}(x, \omega, 0)\left(\ln\left(\frac{\alpha}{i}\right) + C\right) d\omega$$

$$+ \int_{|\zeta'|\leq 1} d\zeta' \int_{\gamma_+} Q_{jq}(x, \zeta, p) e^{i\alpha\zeta} \, d\zeta_n$$

$$+ \int_\Omega (ix \cdot \omega)^{\sigma_q + t_j - n + 1} \int_1^\infty \left[\widehat{g}_{jq}(\omega', \omega_n, p_0\rho^{-2b}) - \widehat{g}_{jq}(\omega', \omega_n, 0)\right] e^{i\alpha\rho}\left(\frac{d\rho}{\rho}\right) d\omega$$

$$+ \int_\Omega i\alpha Q_{jq}(x, \omega, 0)\left(\int_0^1 e^{i\alpha\rho} \ln \rho \, d\rho\right) d\omega.$$

4) Now substitute $\Phi_{jq}(\beta, x, p)$ in (1.21). In the second and third terms we first integrate with respect to β and then with respect to ω and ρ; then in the integral with respect to β we perform an additional integration by parts. This yields the following representation for $\widehat{\mathcal{G}}_{jq}(x, p)$:

$$\widehat{\mathcal{G}}_{jq}(x, p) = -1[(\sigma_q + t_j - n + 1)!]^{-1} \int_\Omega Q_{jq}(x, \omega, 0)\left(\ln \frac{x \cdot \omega}{i} + \frac{1}{2b} \ln p + c_{2jq}\right) d\omega$$

$$+ \mathcal{P}_{jq}(x, p_0) + F_{jq}(x)\left(\frac{1}{2b} \ln p + c_{2jq}\right) + \mathcal{K}_{jq}(x) + \mathcal{R}_{jq}(x, p), \quad (1.31)$$

where

$$\mathcal{P}_{jq}(x, p_0) = \left[(\sigma_q + t_j - n + 1)!\right]^{-1} \left\{ \frac{1}{2b} \ln p_0 \int_\Omega Q_{jq}(x, \omega, 0)\, d\omega \right.$$

$$+ \int_{|\zeta'|\leq 1} d\zeta' \int_{\gamma_+} Q_{jq}(x, \zeta, p_0)\, d\zeta_n \tag{1.32}$$

$$\left. + \int_\Omega (ix \cdot \omega)^{\sigma_q + t_j - n + 1} \left\{ \int_1^\infty [\widehat{g}_{jq}(\omega', \omega_n, p_0\rho^{-2b}) - \widehat{g}_{jq}(\omega', \omega_n, 0)] \frac{d\rho}{\rho} \right\} d\omega \right\},$$

$$\mathcal{R}_{jq}(x, p) = \left[(\sigma_q + t_j - n + 1)!\right]^{-1} |p|^{1/2b} \left\{ \int_{|\zeta'|\leq 1} d\zeta' \int_\gamma (ix \cdot \zeta)^{\sigma_q + t_j - n + 2} \right.$$

$$\times \int_0^1 (1 - \beta)^{\sigma_q + t_j - n + 1} e^{i\beta x \cdot \zeta |p|^{1/2b}}\, d\beta\, d\zeta_n$$

$$+ \int_\Omega (ix \cdot \omega)^{\sigma_q + t_j - n + 2} \int_1^\infty [\widehat{g}_{jq}(\omega', \omega_n, p_0 \cdot \rho^{-2b}) - \widehat{g}_{jq}(\omega', \omega_n, 0)]\, d\rho \tag{1.33}$$

$$\times \int_0^1 (1 - \beta)^{\sigma_q + t_j - n + 1} e^{i\beta x \cdot \omega |p|^{1/2b}}\, d\beta\, d\omega + \int_0^1 (1 - \beta)^{\sigma_q + t_j - n + 1} \beta d\beta$$

$$\left. \times \int_\Omega (ix \cdot \omega)^{\sigma_q + t_j - n + 2} \widehat{g}_{jq}(\omega', \omega_n, 0)\, d\omega \int_0^1 \ln \rho\, e^{i\beta x \cdot \omega |p|^{1/2b}}\, d\rho \right\}$$

It immediately follows from (1.33) that for any $(x, p) \in \mathfrak{A}$

$$|\mathcal{R}_{jq}(x, p)| \leq C|p|^{1/2b}|x|^{\sigma_q + t_j - n + 2}. \tag{1.34}$$

Further, relations (1.32) and (1.33) imply that along any ray lying in A_{κ_0},

$$\lim_{p \to 0} (\mathcal{P}_{jq}(x, p) + \mathcal{R}_{jq}(x, p)) = \mathcal{P}_{jq}(x, p_0). \tag{1.35}$$

Let us examine representation (1.31). The function $\widehat{\mathcal{G}}_{jq}(x, p)$ and the first term in the right-hand side sum are analytic functions of the complex variable p in the domain A_{κ_0}. Therefore, the function $(\mathcal{P}_{jq}(x, p_0) + \mathcal{R}_{jq}(x, p))$ is also analytic in $p \in A_{\kappa_0}$.

By making use of the structure of the domain A_{κ_0} and applying Fatou's theorem [63, p.238], we conclude that the limit $\lim_{p \to 0}(\mathcal{P}_{jq}(x, p_0) + \mathcal{R}_{jq}(x, p))$

exists and is independent of p_0. Since, by (1.34), $\lim_{p \to 0} \mathcal{R}_{jq} = 0$, it follows that $\mathcal{P}_{jq}(x, p_0) \equiv \mathcal{P}_{jq}(x, 1) \equiv P_{jq}(x)$. Note, finally, that the functions

$$- \left[(\sigma_q + t_j - n + 1)! \right]^{-1} \int_\Omega Q_{jq}(x, \omega, 0) \ln \frac{x \cdot \omega}{i} \, d\omega \equiv \mathcal{K}_{jq}(x) \qquad (1.36)$$

appearing in (1.31) are the Poisson kernels of the elliptic boundary value problem (1.5_0), (1.6_0) (see [93, I], [114, II]).

Formulas (1.20), (1.31)–(1.36) directly yield the assertion of Theorem VII.1 for $\sigma_q + t_j - n + 1 \geq 0$ if we set

$$F_{jq}(x) = \left[(\sigma_q + t_j - n + 1) \right]^{-1} \int_\Omega (ix \cdot \omega)^{\sigma_q + t_j - n + 1} \widehat{g}_{jq}(\omega', \omega_n, 0) \, d\omega.$$

5) Now let $\sigma_q + t_j - n + 1 = 0$. In this case the calculations are carried out in the same way; the first term is absent, i.e., $\mathcal{P}^s_{jq}(x) \equiv 0$ in the final formula.

6) There remains only the case $\sigma_q + t_j - n + 1 < 0$. Write again (1.8):

$$\widehat{G}_{jq}(x, p) = \int_{\mathbb{R}^{n-1}} d\xi' \int_{\gamma_+} \widehat{g}_{jq}(\xi', \xi_n, p)) e^{ix \cdot \xi} \, d\xi$$

and express $\widehat{G}_{jq}(x, p)$ as

$$\widehat{G}_{jq}(x, p) = \widehat{G}_{jq}(x, 0) + \mathcal{G}_{jq}(x, p), \qquad (1.37)$$

where

$$\mathcal{G}_{jq}(x, p) = \widehat{G}_{jq}(x, p) - \widehat{G}_{jq}(x, 0)$$

$$= \int_0^p d\eta \int_{\mathbb{R}^{n-1}} e^{ix' \cdot \xi'} d\xi' \int_{\gamma_+} \frac{\partial}{\partial \eta} \widehat{g}_{jq}(\xi', \xi_n, \eta) e^{ix_n \xi_n} \, d\xi_n. \qquad (1.38)$$

The function $\partial \widehat{g}_{jq}(\xi', \xi_n, \eta)/\partial \eta$ is quasihomogeneous of order $-\sigma_q - t_j - 2b - 1$, hence an appropriately modified form of Lemma VII.1 is holds for it. We conclude that that a modified version of Lemma VII.2 holds for the function

$$\mathcal{G}'_{jq}(x, \eta) \equiv \frac{\partial}{\partial \eta} \widehat{G}_{jq}(x, \eta)$$

$$\equiv \int_{\mathbb{R}^{n-1}} \exp\{ix' \cdot \xi'\} d\xi' \int_{\gamma_+} \frac{\partial}{\partial \eta} \widehat{g}_{jq}(\xi', \xi_n, \eta) \exp\{ix_n \xi_n\} d\xi_n$$

In particular, the following estimates hold:

$$
|\mathcal{G}'_{jq}(x,\eta)| \leq
\begin{cases}
C|\eta|^{\frac{\sigma_q+t_j-n+1+2b}{/2b}} \exp\left\{-c|x||\eta|^{1/2b}\right\}, & \sigma_q+t_j-n+1+2b > 0, \\
C(1+|\ln|x|\,|\eta|) \exp\left\{-c|x||\eta|^{1/2b}\right\}, & \sigma_q+t_j-n+1+2b = 0, \\
C|x|^{\sigma_q+t_j-n+1+2b} \exp\left\{-c|x||\eta|^{1/2b}\right\}, & \sigma_q+t_j-n+1+2b < 0.
\end{cases}
\tag{1.39}
$$

As the contour of integration with respect to η in (1.38) we take the ray in the complex η-plane connecting $\eta_0 = 0$ and $\eta = p_0$. Using the estimates (1.39), we obtain the following inequalities for the functions $\mathcal{G}_{jq}(x,p)$:

1) if $\sigma_q+t_j-n+1+2b > 0$, then, taking into account that $\sigma_q+t_j-n+1 < 0$, we have

$$
|\mathcal{G}_{jq}(x,p)| \leq C \int\limits_0^{|p|} \rho^{-(\sigma_q+t_j-n+1+2b)/2b} d\rho = C_1 |p|^{-(\sigma_q+t_j-n+1)/2b};
\tag{1.40_1}
$$

2) if $\sigma_q+t_j-n+1+2b = 0$, then

$$
|\mathcal{G}_{jq}(x,p)| \leq C \int\limits_0^{|p|} (1+|\ln p|+|\ln|x||)d\rho \leq C_2|p|(1+|\ln|x||+|\ln|p||);
\tag{1.40_2}
$$

3) if $\sigma_q+t_j-n+1+2b < 0$, then

$$
|\mathcal{G}_{jq}(x,p)| \leq C|x|^{\sigma_q+t_j-n+1+2b}|p|.
\tag{1.40_3}
$$

Thus $\lim\limits_{p\to 0} \mathcal{G}_{jq}(x,p) = 0$. By (1.37), $\widehat{G}_{jq}(x,0) \equiv \mathcal{K}_{jq}(x)$ are the Poisson kernels of the elliptic problem (1.5_0), (1.6_0). Thus (1.37) is identical to (1.13) for $\sigma_q+t_j-n+1 < 0$.

VII.1.4. Definition of the class of the boundary functions used here

Let us describe the class of the boundary data $\varphi_q(x',t)$ for which we will derive an asymptotic representation (for $t \to \infty$) of the solutions of the problem (1.1), (1.2) given in the form (1.3).

Definition VII.1. *A function $\varphi(x',t)$ defined in $E_+^n = \mathbb{R}^{n-1} \times (0,\infty)$ is said to belong to the class $L_{\alpha,m}(E_+^n)$, where $\alpha = (m-n+1)/2b$, m is a non-negative integer, and $\alpha \geq 0$, if the following conditions are satisfied:*

β_1. *For any fixed $x \in \mathbb{R}^{n-1}$, $\varphi(x',t)$ has a Laplace transform $\hat{\varphi}(x',p)$ that is analytic in the half-plane $\operatorname{Re} p > 0$ of the complex p-plane.*

β_2. *For any $\varepsilon > 0$ and any $x' \in \mathbb{R}^{n-1}$ there exists a positive constant C_ε such that*

$$
|\hat{\varphi}(x',p)| \leq C_\varepsilon \exp\left\{\varepsilon|p|^{1/2b}\right\}.
\tag{1.41}
$$

β_3. *The following asymptotic representation holds for $p \to 0$:*

$$\hat{\varphi}(x',p) = \sum_{s=0}^{M} a_s(x')p^{\gamma_s} + \Phi(x',p), \quad -1 = \gamma_0 < \gamma_1 < \ldots < -1 + \alpha; \quad (1.42)$$

here $a_s(x')$ and $\Phi(x',p)$ are such that

$$\int_{\mathbb{R}^{n-1}} |a_s(x')|\,|x'|^{m-n+2}\,dx' < \infty,$$

$$\int_{\mathbb{R}^{n-1}} |x'|^{m-n+2}|\Phi(x',p)|\,dx' < C_\delta|p|^{-1+\alpha+\delta}. \qquad (1.43_1)$$

In the case of one space variable (1.43_1) takes the form:

$$|\Phi(p)| \le C_\delta|p|^{-1+\alpha+\delta} \qquad (1.43_2)$$

for all $\delta > 0$.

The problem of obtaining a direct complete description of the class $L_{\alpha,m}(E_+^n)$ is still open. Here we restrict ourselves to the study of some simple examples showing that this class may contain functions that behave differently as $t \to \infty$.

Example VII.1. Consider the functions $\varphi_k(x't) = \chi(x')\varphi_k(t)$, where $\chi(x')$ is a smooth function for which the integral

$$\int_{\mathbb{R}^{n-1}} |\chi(x')|\,|x'|^{m-n+2}dx' \qquad (1.44)$$

converges for some $m \ge n-1$; we will choose $\varphi_k(t)$ using operational calculus tables [21, 23, 58] so that $\varphi_k(x',t)$ will belong to $L_{\alpha,m}(E_+^n)$. Thus, for instance, if $\varphi_1(t) = \mathrm{erf}(\sqrt{t})$, $\varphi_2(t) = e^t\mathrm{erfc}(a/2\sqrt{t})$, $\varphi_3(t) = \exp\{-a^2t^2\}$ and $\varphi_4(t) = J_n(at)/t$, $n > 0$, where $J_n(at)$ is the n-th order Bessel function of the first kind, $a > 0$, then the functions $\varphi_k(x',t)$, $k = 1,2,3,4$ belong to the class $L_{\alpha,m}(E_+^n)$ for any $m \ge n-1$, $\alpha = (m - n + 1)/2b$. Here the exponents of p in representation (1.42) assume the following values: for $\hat{\varphi}_1(x',p)$, $\gamma_0 = 0$, $\gamma_1 = -\frac{1}{2}$, $a_1(x') = \chi(x')$, $\gamma_s = 0$ for any $s > 1$; for $\hat{\varphi}_2(x',p)$, $\gamma_0 = -1$, $\gamma_1 = -\frac{1}{2}$, $a_0(x') = \chi(x')$, $a_1(x') = -a\chi(x')$, $\gamma_s = 0$ for any $s > 1$; also, for the function $\hat{\varphi}_4(x',p)$ the sum in (1.42) is absent.

Example VII.2. Now let $\varphi_5(x',t) = \sum_{s=0}^{M} \chi_s(x')t^{-1-\gamma_s}$ where $\chi_s(x')$ are functions such that the integral (1.44) converges, and γ_s is taken from the condition β_3 (with $\alpha = (m - n + 1)/2b$). Then $\varphi_5(x',t) \in L_{\alpha,m}(E_+^n)$, and in (1.42) $\Phi(x',p) \equiv 0$, all the terms are present in the sum, and $a_s(x') = \Gamma(-1 - \gamma_s)\chi_s(x')$.

Example VII.3. Start with the bounded function $\varphi_6(t) = \sin t^2$. Its Laplace transform is $\hat{\varphi}_6(p) = \frac{1}{2}\left(\frac{\pi}{2} - \int_0^p \cos\frac{p^2-z^2}{4}dz\right)$ (cf. [23, p. 198]). In the present case $\hat{\varphi}_6(p)$

is an entire function of order two in the complex variable p. It is immediately seen that $\hat{\varphi}_6(p)$ is bounded on the imaginary axis. But then, by the Phragmén-Lindelöf theorem, $\hat{\varphi}_6(p)$ is bounded also on the half-plane $\operatorname{Re} p \geq 0$. Therefore, $\hat{\varphi}_6(p) \in L_{\alpha,m}(E_+^1)$ for any $m \geq 0$ and $\alpha = m/2b$.

The function $\varphi_{6,k} = t^k \sin t^2$ has the Laplace transform $(-1)^k \hat{\varphi}_6^{(k)}(p)$, which in the half-plane $\operatorname{Re} p \geq 0$ grows no faster than $|p|^k$. Therefore, $t^k \sin t^2$, a function that obviously does not stabilize, also belongs to the class $L_{\alpha,m}(E_+^1)$.

The following result obtained A.G. Ramm [82] is of interest for the topics discussed here.

Lemma VII.3. *Let $\hat{\varphi}(p)$ be the Laplace transform of a function $\varphi(t)$ with the following properties: 1) $\hat{\varphi}(p)$ is an analytic function of p for $\operatorname{Re} p > 0$; 2) there exists a positive constant γ such that $\overline{\lim}_{\operatorname{Re} p \geq 0} \sup_{\operatorname{Im} p} \left\{ |p|^\gamma |\hat{\varphi}(p)| \right\} < \infty$; 3) there exist a positive number α and a complex number a such that $\lim_{p \to 0} p^\alpha \hat{\varphi}(p) = a$. Then*

$$\lim_{T \to \infty} \left\{ T^{-\alpha} \int\limits_0^T \varphi(t)\,dt \right\} = \frac{a}{\Gamma(1+\alpha)}.$$

VII.1.5. Asymptotic representation of solutions

Here we will establish the central result of the present section. In order to formulate it we need a number of conditions and notations. First, the boundary functions $\varphi_q(x',t)$ will be taken from the classes $L_{\alpha_q,m_q}(E_+^n)$, with $\alpha_q = (m_q - n + 1)/2b$, $m_q = \max_j(\sigma_q + t_j)$, $\alpha_{jqs} = (\sigma_q + t_j - n + 1 - s)/2b$. We denote the coefficients of the expansion (1.42) by $a_{qs}(x')$, the powers in the expansion by $\gamma_{q0}, \gamma_{q1}, \ldots$, M by M_q, $\Phi(x',p)$ by $\Phi_q(x',p)$. Also, $\Gamma(a)$ is the Euler function. Further, let $\tilde{c}_{jq} = c_{jq} - C/2b$, where c_{jq} are taken from (1.13). The convolution with respect to x' of two functions, $f(x)$ and $g(x')$, will be denoted by $f(x) * g(x')$; in particular, we put $M_{jq}^{sl}(x) = P_{jq}^s(x) * a_{ql}(x')$, $M_{jq}(x) = F_{jq}(x) * a_{q0}(x')$, $v_{jq}(x) = \mathcal{K}_{jq}(x) * a_{q0}(x')$. Finally, we introduce the function

$$\tilde{Q}_{jq}(x) = P_{jq}(x) * a_{q0}(x') + \sum_{\gamma_{ql} - \alpha_{jqs} = -1} M_{jq}^{sl}(x),$$

where the last sum is extended to all combinations of s and l (for fixed j and q).

Theorem VII.2. *Let $n \geq 2$, $\varphi_q(x',t) \in L_{\kappa_q,m_q}(E_+^n)$, $\kappa_q = \max(0,\alpha_q)$. Then the functions*

$$u_{jq}(x,t) = \int\limits_0^t d\tau \int\limits_{\mathbb{R}^{n-1}} G_{jq}(x - y', t - \tau)\varphi_q(y',\tau)\,dy'$$

admit the following representation (for any $x \in \mathbb{R}_+^n$):

$$u_{jq}(x,t) = \sum_{\gamma_{ql} - \alpha_{jqs} < -1} M_{jq}^{sl}(x)t^{\alpha_{jqs} - \gamma_{ql} - 1} + M_{jq}(x)\left(\frac{1}{2b}\ln t + \tilde{c}_{jq}\right)$$

$$+ v_{jq}(x) + \tilde{Q}_{jq}(x) + o_{jq}(x,t),$$

(1.45)

where $o_{jq}(x,t) \to 0$ as $t \to \infty$; if $\alpha_{jqs} = 0$, then $M_{jq}^{sl}(x) \equiv 0$, and if $\alpha_q - n + 1 < 0$, then $M_{jq}^{sl}(x) \equiv M_{jq}(x) \equiv Q_{jq}(x) \equiv 0$.

Proof: 1) Consider those $u_{jq}(x,t)$ for which $\sigma_q + t_j - n + 1 \geq 0$. Using the inversion formula of operational calculus, we write:

$$u_{jq}(x,t) = (2\pi i)^{-1} \int\limits_{a-i\infty}^{a+i\infty} e^{pt} \widehat{u}_{jq}(x,p)\,dp, \qquad a > 0. \tag{1.46}$$

The estimates (1.34) and (1.41) yield the following estimate for $\widehat{u}_{jq}(x,p)$:

$$|\widehat{u}_{jq}(x,p)| = |\widehat{G}_{jq}(x,p) * \hat{\varphi}_q(x',p)| = \left| \int\limits_{\mathbb{R}^{n-1}} \widehat{G}_{jq}(x-y',p)\hat{\varphi}_q(y',p)\,dy' \right|$$

$$\leq \int\limits_{\mathbb{R}^{n-1}} |\widehat{G}_{jq}(x-y',p)||\hat{\varphi}_q(y',p)|dy' \leq C_\varepsilon e^{\varepsilon|p|^{1/2b}} C|p|^{-(\sigma_q+t_j-n+1)/2b}$$

$$\times \int\limits_{\mathbb{R}^{n-1}} c^{-c|x'-y'||p|^{1/2b}}\,dy' \cdot c^{-cx_n|p|^{1/2b}} = C_\varepsilon|p|^{-(\sigma_q+t_j)/2b}e^{(\varepsilon-cx_n)|p|^{1/2b}}.$$

$$\tag{1.47}$$

Using the Cauchy integral theorem and (1.47), in the integral figuring in (1.46) one can replace integration over the line $\operatorname{Re} p = a$ in the complex p-plane by the integration over the contour consisting of the half-lines $(-i\infty, -i)$, $(i, i\infty)$ of the imaginary axis and of the semi-circle $\Gamma = \{p : |p| = 1, \operatorname{Re} p \geq 0\}$:

$$u_{jq}(x,t) = \frac{1}{2\pi i} \int\limits_{\Gamma} e^{pt} \widehat{u}_{jq}(x,p)\,dp + \frac{1}{2\pi} \int\limits_{|\omega|>1} e^{i\omega t} \widehat{u}_{jq}(x,i\omega)\,d\omega \tag{1.48}$$

$$= u'_{jq}(x,t) + u''_{jq}(x,t).$$

Consider the function $u''_{jq}(x,t)$ for $x_n > 0$. The estimate (1.47) allows us to apply Lebesgue's lemma, according to which

$$\lim_{t\to\infty} u''_{jq}(x,t) = 0. \tag{1.49}$$

Now let us examine the term $u'_{jq}(x,t)$. First let us write the decomposition (1.13) of $\widehat{G}_{jq}(x-y',p)$,

$$\widehat{G}_{jq}(x-y',p) = \sum_{s=0}^{\sigma_q+t_j-n} \mathcal{P}_{jq}^s(x-y')p^{-\alpha_{jqs}} - \left(\frac{1}{2b}\ln p + c_{jq}\right)F_{jq}(x-y') \tag{1.50}$$

$$+ \mathcal{K}_{jq}(x-y') + P_{jq}(x-y') + \mathcal{R}_{jq}(x-y',p)$$

and the representation (1.42) of $\hat{\varphi}_q(y', p)$,

$$\hat{\varphi}_q(y', p) = \sum_{s=0}^{M_q} a_{qs}(y') p^{\gamma_{qs}} + \Phi_q(y', p), \tag{1.51}$$

where $-1 = \gamma_{q0} < \gamma_{q1} < \ldots < -1 + \alpha_q$, $\alpha_q = (m_q - n + 1)/2b$ and $m_q = \max_j(\sigma_q + t_j)$. Multiplying $\widehat{G}_{jq}(x - y', p)$ by $\hat{\varphi}_q(y', p)$ and integrating with respect to y', we get

$$\hat{u}_{jq}(x, p) = \int_{\mathbb{R}^{n-1}} \widehat{G}_{jq}(x - y', p) \hat{\varphi}_q(y', p) \, dy'.$$

Grouping the terms with like negative powers of p, we obtain

$$\begin{aligned}
\hat{u}_{jq}(x, p) = &\sum_{\gamma_{ql} - \alpha_{jqs} < -1} M_{jq}^{sl}(x) p^{\gamma_{ql} - \alpha_{jqs}} + M_{jq}(x) \left(\frac{1}{2b} \ln p + c_{jq} \right) \\
&+ \frac{1}{p} \left(v_{jq}(x) + \widetilde{Q}_{jq}(x) + T_{jq}(x, p) \right),
\end{aligned} \tag{1.52}$$

where $T_{jq}(x, p)$ is a sum of terms which, thanks to the assumptions on $\varphi_q(x', p)$, admits the bound:

$$|T_{jq}(x, p)| \le C(1 + |x|)^{m_j - n + 2} |p|^b, \quad b > -1. \tag{1.53}$$

Next, substitute the function $\hat{u}_{jq}(x, p)$ defined by (1.52) in the formula

$$u'(x, t) = \frac{1}{2\pi i} \int_\Gamma e^{pt} \hat{u}_{jq}(x, p) \, dp.$$

This yields

$$\begin{aligned}
u'(x, t) = &\sum_{\gamma_{ql} - \alpha_{jqs} < -1} M_{jq}^{sl}(x) \frac{1}{2\pi i} \int_\Gamma p^{\gamma_{ql} - \alpha_{jqs}} e^{pt} \, dp \\
&+ (v_{jq}(x) + \widetilde{Q}_{jq}(x)) \frac{1}{2\pi i} \int_\Gamma \frac{e^{pt}}{p} \, dp + M_{jq}(x) \frac{1}{2\pi i} \int_\Gamma e^{pt} \left(\frac{1}{2b} \ln p + c_{jq} \right) \frac{dp}{p} \\
&+ \frac{1}{2\pi i} \int_\Gamma e^{pt} T_{jq}(x, p) \, dp.
\end{aligned} \tag{1.54}$$

Next, using the fact that the functions $(\ln p/p)$ and $p^{-\nu}$ $(\operatorname{Re} p > 0)$ have the inverse transforms $-\ln t - C$ and $(t^{\nu-1}/\Gamma(\nu))$, respectively [23, formulas 147 and 150], we easily obtain the asymptotics

$$\frac{1}{2\pi i} \int_{\Gamma} \frac{e^{pt}}{p} \left(\frac{1}{2b} \ln p + c_{jq} \right) dp = -\frac{1}{2b} \ln t + \widetilde{c}_{jq} + o(t), \tag{1.55}$$

$$\frac{1}{2\pi i} \int_{\Gamma} e^{pt} p^{-\nu} \, dp = \frac{t^{\nu-1}}{\Gamma(\nu)} + o(t), \tag{1.56}$$

where $o(t) \to 0$ as $t \to \infty$.

Finally, consider the integral $\frac{1}{2\pi i} \int_{\Gamma} e^{pt} T_{jq}(x,p) \, dp$. Using the analyticity of $T_{jq}(x,p)$ for $\operatorname{Re} p > 0$, the estimate (1.53), the Cauchy integral theorem, and then Lebesgue's theorem, we conclude that

$$\frac{1}{2\pi i} \int_{\Gamma} e^{pt} T_{jq}(x,p) \, dp = (2\pi)^{-1} \int_{-1}^{1} e^{i\omega t} T_{jq}(x, i\omega) \, d\omega \to 0 \tag{1.57}$$

as $t \to \infty$.

Replacing each of the integrals in (1.54) by its value and using (1.55) and (1.56), we get the representation (1.45) for the case $\sigma_q + t_j - n + 1 > 0$.

2) Now let us consider those $u_{jq}(x,t)$, for which $\sigma_q + t_j - n + 1 < 0$.

Using (1.40_1)–(1.40_3) and repeating the calculations that led to (1.47), we obtain the estimate:

$$|\hat{u}_{jq}(x,p)| \leq C(1 + |\ln x_n|)|p|^{-(\sigma_q + t_j)/2b} \exp\left\{ -cx_n|p|^{1/2b} \right\}; \tag{1.58}$$

if $\sigma_q + t_j > 0$, the logarithm in (1.58) vanishes.

The remaining calculations, arguments and estimates parallel those given in part 1).

VII.1.6. Necessary and sufficient conditions of stabilization

Here we continue our analysis of the functions

$$u_{jq}(x,t) = \int_{0}^{t} d\tau \int_{\mathbb{R}^{n-1}} G_{jq}(x - y', t - \tau)\varphi_q(y', \tau) \, dy',$$

in term of which the bounded solution of the problem (1.1), (1.3) is defined by means of formula (1.3).

The representation (1.45) immediately implies the following assertion.

Theorem VII.3. *Let $n \geq 2$ and $\varphi_q(x', t) \in L_{\kappa_q, m_q}(E^n_+)$, $\kappa_q = \max(0, \alpha_q)$, $\alpha_q = (m_q - n + 1)/2b$, $m_q = \max_j(\sigma_q + t_j)$. For $u_{jq}(x, t)$ to stabilize it is necessary and sufficient that, for any $x \in \mathbb{R}^n_+$,*

$$M_{jq}(x) \equiv 0, \quad Q^\mu_{jq}(x) \equiv \sum_{\gamma_{ql} - \alpha_{jqs} = \mu} M^{sl}_{jq}(x) \equiv 0, \tag{1.59}$$

where μ is any positive number for which there exist γ_{ql} and α_{jqs} such that $\gamma_{ql} - \alpha_{jqs} = \mu$.

If these conditions are satisfied, then

$$\lim_{t \to \infty} u_{jq}(x, t) = v_{jq}(x) + \widetilde{Q}_{jq}(x), \tag{1.60}$$

where $v_{jq}(x) = \int_{\mathbb{R}^{n-1}} \mathcal{K}_{jq}(x - y') a_{q0}(y') dy'$, $\mathcal{K}_{jq}(x)$ is an element of the Poisson kernel (Poisson matrix) of the elliptic boundary value problem (1.50_0), (1.60_0), and $\widetilde{Q}_{jq}(x)$ are polynomials. The functions $u_{jq}(x, t)$ stabilize uniformly on any compact subset of the half-space $\{x : x_n \geq \delta\}$, where δ is an arbitrary positive number.

Proof: Let us examine the conditions (1.59). The assertion that the polynomials $M_{jq}(x)$ and $Q^\mu_{jq}(x)$ vanish identically means that all their coefficients are equal to zero. In order to understand what conditions follow from this fact, let us write these polynomials in more detail:

$$Q^\mu_{jq} = \sum_{\gamma_{ql} - \alpha_{jqs} = \mu} M^{sl}_{jq}(x) = \sum_{\gamma_{ql} - \alpha_{jqs} = \mu} \sum_{|k| = s} \frac{1}{k!} \widehat{G}^{(k)}_{jq}(0, 1) \int_{\mathbb{R}^{n-1}} (x - y')^k a_{ql}(y') \, dy'.$$

The polynomials $M_{jq}(x)$ have a similar form. These formulas imply that the necessary and sufficient conditions of stabilization take the form of systems of linear algebraic equations in the moments

$$a^{\nu'}_{qs} = \frac{(-1)^{\nu'}}{\nu'!} \int_{\mathbb{R}^{n-1}} (y')^{\nu'} a_{qs}(y') \, dy'$$

of the functions $a_{qs}(x')$ appearing in (1.42).

In the general case these conditions are, of course, difficult to verify. Therefore, the particular cases in which one can obtain transparent criteria of stabilization of solutions of model parabolic boundary value problems are quite important.

VII.1.7. The case of a single equation and boundary data whose mean values have a limit

To make the results more accessible, let us examine the case of a single equation:

$$L(D, D_t)u \equiv \Big(D_t^r - \sum_{\substack{|\alpha|+2b\alpha_0=2br \\ \alpha_0 < r}} a_{\alpha\alpha_0} D^\alpha D_t^{\alpha_0} \Big) u = 0 \tag{1.61}$$

$$B_q(D, D_t)u|_{x_n=0} = \sum_{|\alpha|+2b\alpha_0=r_q} b^q_{\alpha\alpha_0} D^\alpha D_t^{\alpha_0} u|_{x_n=0}$$

$$= \varphi_q(x', t), \quad q = 1, \ldots, br, \tag{1.62}$$

$$D_t^{\lambda-1} u|_{t=0}, \quad \lambda = 1, \ldots, r. \tag{1.63}$$

Definition VII.2. *A function $\varphi_q(x', t) \in L_{\kappa_q, r_q}(E_+^n)$, where $\kappa_q = \max(0, (r_q - n + 1)/2b)$, is said to belong to the set S_q if for any $x' \in \mathbb{R}^{n-1}$ there exists the limit*

$$\lim_{T\to\infty} \Big(T^{-1} \int_0^T \varphi_q(x', t)dt \Big),\ \text{and for any } l \text{ with } l = 1, \ldots [\alpha_q], \text{ the integrals}$$

$$m_l[\varphi_q] - \frac{(-1)^{l-1}}{(l-1)!} \int_0^\infty t^{l-1}\big[\varphi_q(x', t)\quad \varphi_q(x')\big]dt \tag{1.64}$$

converge.

For such functions the parameters in (1.42) equal $\gamma_{ql} = l-1$, $a_{ql}(x') = m_l[\varphi_q]$. It follows that

$$\gamma_{ql} - \alpha_{qs} = \big[(r_q - n + 1 - s - 2bl)/2b\big] - 1 = -1 + \tilde{\alpha}_{q\mu},$$

where we denoted $\mu = s + 2bl$. Then in the asymptotic representation of $u_q(x, t)$ (formula (1.45)) the double sum with respect to s and t which contains the powers of t takes the form

$$\sum_{\mu=1}^{r_q-n} \sum_{s+2bl=\mu} \mathcal{P}_q^s(x) * m_l[\varphi_q] \frac{t^{\tilde{\alpha}_{q\mu}}}{\Gamma(1 + \tilde{\alpha}_{q\mu})} = \sum_{\mu=1}^{r_q-n} M_q^\mu(x) t^{\tilde{\alpha}_{q\mu}}. \tag{1.65}$$

Now let us describe in more concrete terms the convolution $\mathcal{P}_q^s * m_l[\varphi_q]$. Recall that the polynomial $\mathcal{P}_q^s$ is given by the formula

$$\mathcal{P}_q^s(x) \equiv \sum_{|k|=s} \widehat{G}_q^{(k)}(0, 1) \frac{x^k}{k!} \equiv \mathcal{P}_s(\widehat{G}_q, x).$$

The identity

$$\mathcal{P}_q^{(s)}(x - y') = \sum_{|k|=0}^{s} \mathcal{P}_{s-|k|}(\widehat{G}_q^{(k)}, -y') \frac{x^k}{k!} \tag{1.66}$$

is readily verified.

Therefore, replacing $\mathcal{P}_q^{(s)}(x-y')$ by its expression from (1.66) and introducing the notation

$$m_{l\nu'}[\varphi_q] = \int (x')^{\nu'} m_l[\varphi_q(y',t)]\,dy', \tag{1.67}$$

we find that

$$M_q^\mu(x) = \sum_{0\leq|k|\leq\mu}\left(\sum_{s+2bl=\mu}\sum_{\nu'=s-|k|}\widehat{G}_q^{k+\nu'}(0,1)m_{l\nu'}[\varphi_q]\right)\frac{x^k}{k!}. \tag{1.68}$$

Similarly,

$$M_q(x) = \sum_{0\leq|k|\leq r_q-n+1}\sum_{|\nu'|=r_q-n+1-|k|} m_{k+\nu'}[\widehat{g}_q]\,m_{0\nu'}[\varphi_q]\frac{x^k}{k!}, \tag{1.69}$$

where

$$m_{k+\nu'}[\widehat{g}_q] = \frac{1}{(2\pi)^n i}\int_\Omega (i\omega)^{k+\nu'}\widehat{g}_q(\omega,0)\,d\omega. \tag{1.70}$$

Formulas (1.67)–(1.70) allow us render Theorem VII.3 concrete as follows.

Theorem VII.4. *Let $n \geq 2$ and $\varphi_q \in S_q$, $q = 1,\ldots,br$. Then in order that the bounded solutions of the problem (1.61)–(1.63) stabilize it is necessary and sufficient that the moments $m_{l\nu'}[\varphi_q]$ satisfy the following system of algebraic equations:*

$$\sum_{|\nu'|=r_q-n+1-|k|} m_{k+\nu'}[\widehat{g}_q]\,m_{0\nu'}[\varphi_q] = 0, \quad |k| = 0,1,\ldots,r_q-n+1, \tag{1.71_1}$$

$$\sum_{s+2bl=\mu}\sum_{\nu'=s-|k|}\widehat{G}_q^{k+\nu'}(0,1)m_{l\nu'}[\varphi_q] = 0, \quad |k| = 0,1,\ldots,r_q-n. \tag{1.71_2}$$

Remark VII.1. Conditions (1.71_1) and (1.71_2) can be recast in yet another form. Namely, substituting in these formulas the expressions for the moments $m_{l\nu'}$ and switching the order of integration and summation, we obtain

$$\int_{\mathbb{R}^{n-1}}\left(\sum_{|\nu'|=\mu-|k|}\widehat{G}_q^{(k+\nu')}(0,1)\frac{(x')^{\nu'}}{\nu'!}\right)\varphi_q(x')\,dx' = 0, \tag{1.72_1}$$

$$\int_{\mathbb{R}^{n-1}}\left(\sum_{|\nu'|=r_q-n+1-|k|} m_{k+\nu'}[\widehat{g}_q]\frac{(x')^{\nu'}}{\nu'!}\right)\varphi_q(x')\,dx' = 0. \tag{1.72_2}$$

Conditions (1.72_1) and (1.72_2) express the orthogonality of the limiting mean values of the boundary data and special polynomials whose coefficients are determined by the Poisson kernels and basis of the corresponding elliptic boundary value problem.

Remark VII.2. There exist examples of boundary value problems for which the derivatives $\widehat{G}_q^{(k)}(0,1) = 0$ for $|k| \leq \tau_q$; in this case equations (1.71) are identically satisfied for such $|k|$. The numbers τ_q play an essential role in the study of the behavior of the Poisson kernels for large values of time [27]. Note, in particular, that for the Dirichlet problem all the coefficients of the algebraic system (1.71) equal zero, and so in the case of a single equation the solution of the Dirichlet problem constructed for boundary data $\varphi_q(x',t) \in S_q$ stabilizes to the solution of the Dirichlet problem defined by the formula

$$u(x) = \sum_{q=1}^{br} \int_{\mathbb{R}^{n-1}} \mathcal{K}_q(x-y')\varphi_q(y')\,dy',$$

where $\varphi_q(x')$ are the limiting mean values of the functions $\varphi_q(x',t)$ with respect to t and $\mathcal{K}_q(x)$ are the Poisson kernels of the Dirichlet problem for the elliptic equation $\sum_{|\alpha|=2br} a_{\alpha 0}D^\alpha u = 0$.

VII.1.8. The case of a single space variable

Consider the problem (1.61)–(1.63) in the case of a single space variable x. In this case in formula (1.8) for the Poisson kernels $\widehat{G}_q(x,p)$ of the elliptic problem with a parameter the integral over $\mathbb{R}^{n-1}$ is absent. Arguing as in Subsections VII.1.1–VII.1.3 (with substantial simplifications), we obtain the following version of the fundamental representation (1.13):

$$\widehat{G}_q(x,p) = \sum_{s=0}^{r_j} \widehat{G}_q^{(s)}(0,1)\frac{x^s}{s!}\,p^{-(r_q-s)/2b} + \widehat{\mathcal{G}}_q(x,p), \tag{1.73}$$

where the function $\mathcal{G}_q(x,p)$ is analytic in p in the right-hand half-plane of the complex p-plane and $\lim_{p\to 0}\widehat{\mathcal{G}}_q(x,p) = 0$.

Definitions VII.1 and VII.2 remain valid for $n = 1$.

The representation (1.73) and the properties of the functions $\varphi_q(t)$ yield the following result.

Theorem VII.5. *Let* $n = 1$, $\varphi_q(t) \in S_q$. *Then for any* $x > 0$, $u_q(x,t)$ *can be represented (as* $t \to \infty$*) by the asymptotic formula*

$$u_q(x,t) = \sum_{\alpha_{qs}-\gamma_{qj}} \widehat{G}_q^{(s)}(0,1)m_j[\varphi_q]\,\frac{t^{\alpha_{qs}-\gamma_{qj}-1}}{\Gamma(\alpha_{qs}-\gamma_{qj})} + o(x,t), \tag{1.74}$$

where $\alpha_{qs} = (r_q - s)/2b$ *and* $o(x,t) \to 0$ *as* $t \to \infty$.

The proof is an essentially simplified version of the previous arguments.

Theorem VII.5 directly implies that for the component $u_q(x,t)$ of the solution to stabilize it is necessary and sufficient that $\hat{G}_q^s(0,1) = 0$, $s = 0,1,\ldots,r_q - 1$.

The following assertion holds true.

Theorem VII.6. *In the case $n = 1$, in order that the solutions $u_q(x,t)$ of the problem (1.61)–(1.63) stabilize for any $\varphi_q \in S_q$ it is necessary and sufficient that $r_q = q - 1$, i.e., that the problem (1.61)–(1.63) be equivalent to the Dirichlet problem.*

Proof: Denote by $\mathcal{B}'_q(\xi,p) = \sum\limits_{s=1}^{br} b_{qs}(p)\xi^{s-1}$ the remainders of the division of the polynomials $\mathcal{B}_q(i\xi,p)$ by the polynomial $L_+(\xi,p)$, and let $\mathcal{B} = \|b_{qs}(p)\|^{br}_{q,s=1}$ and $\mathcal{B}^{-1}(p) = \|b^{sq}(p)\|^{br}_{s,q=1}$.

Sufficiency. Let $m_q = q - 1$. Then the matrix $\mathcal{B}(p)$ is lower triangular and $b_{ss}(p) \neq 0$, $s = 1, \ldots, br$. Consequently, $\mathcal{B}^{-1}(p)$ is an lower triangular matrix and $b^{ss}(p) \neq 0$, $s = 1, \ldots, br$. Since $\hat{G}_q^{(s)}(0,1) = i^s b^{s+1,q}(1)$, $s = 0, 1, \ldots, br - 1$, it follows that $\hat{G}_q^{(s)}(0,1) = 0$ for $s = 0, 1, \ldots, m_q - 1 = q - 2$, $\hat{G}_q^{q-1}(0,1) \neq 0$, and the representation (1.73) does not contain negative powers of p.

Necessity. Let $\widehat{G}_q^{(s)}(0,1) = 0$, $s = 0, 1, \ldots, m_q - 1$. Without loss of generality, we may assume that $m_1 \leq m_2 \leq \ldots m_{br}$. It follows from the Lopatinskiĭ condition that $m_1 = 0$: otherwise, all the elements of the first row of the matrix $\mathcal{B}^{-1}(p)$ would be zero. Suppose that we have proved that $m_q = q - 1$ for $q = 1, \ldots, \alpha$. Then

$$\det \mathcal{B}^{-1}(p) = b^{11}(p) \ldots b^{\alpha\alpha}(p) \det \mathcal{B}_{\alpha\alpha}^{-1}(p)$$

$$\det \mathcal{B}(p) = b_{11}(p) \ldots b_{\alpha\alpha}(p) \det \mathcal{B}_{\alpha\alpha}(p)$$

where $\mathcal{B}_{\alpha\alpha}(p)$ [resp., $\mathcal{B}_{\alpha\alpha}^{-1}(p)$] is the matrix obtained from $\mathcal{B}(p)$ [resp., $\mathcal{B}^{-1}(p)$] by deleting the first α rows and columns. Let us prove that $m_{\alpha+1} = \alpha$. Indeed, if $m_{\alpha+1} > \alpha$, the first row of $\mathcal{B}_{\alpha\alpha}^{-1}(p)$ would consist of zeroes; if $m_{\alpha+1} < \alpha$, the first row of $\mathcal{B}_{\alpha\alpha}(p)$ would consist of zeroes. Both cases are impossible due to the Lopatinskiĭ condition. Therefore, $m_q = q - 1$ for $q = 1, \ldots, br$.

Note that the stabilization of the solution $u(x,t)$ of the Dirichlet problem for a single parabolic equation of first order in t for $n = 1$ and any stabilizing boundary data $\varphi_q(t)$ has been established in [81].

VII.1.9. Examples

a) The oblique derivative problem for the heat equation:

$$D_t u = D_1^2 u + D_2^2 u, \quad u(x,+0) = 0, \quad (a_1 D_1 + a_2 D_2)u|_{x_2=0} = \varphi(x_1,t), \quad a_2 > 0.$$

For this problem the expansion (1.13) of the function $\widehat{G}(x,p)$ contains $\ln p$, but no negative powers of p. Correspondingly, the asymptotic representation (1.45) of $u(x,t)$ contains $\ln t$. Therefore, if $\varphi(x_1,t) \in S_0$, then for the stabilization of $u(x,t)$ it is necessary and sufficient that $\int\limits_{-\infty}^{\infty} a_0(x_1)dx_1 = 0$.

b) Consider the problem:

$$D_t u = -\Delta^2 u, \quad u(x,+0) = 0, \quad u(x,t)|_{x_n=+0} = \varphi_1(x',t), \quad D_n^3 u|_{x_n=+0} = \varphi_2(x',t).$$

Here $m_1 = 0$, $m_2 = 3$, $r_1 - n + 1 < 0$ for $n \geq 2$. Hence, the component $u_1(x,t)$ stabilizes for any function $\varphi_1(x',t) \in S_0$. Calculation of $\widehat{G}_2(x,p)$ shows that $\widehat{G}_2(0,1) = 0$. Expansion (1.13) of $\widehat{G}_2(x,p)$ for $n = 3$ contains only $\ln p$. If $\varphi_q(x',t) \in S_0$, $q = 1,2$, and $\lim\limits_{T\to\infty} \left(\frac{1}{T} \int\limits_0^T \varphi_2(x',t)dt \right) = \varphi_2(x')$, then for the stabilization of $u(x,t)$ it is necessary and sufficient that $\int\limits_{-\infty}^{\infty} \varphi_2(x') \, dx' = 0$.

VII.2. Tikhonov's problem

VII.2.1. Statement of the problem. Notations. Conditions

In 1950, A.N. Tikhonov [100] posed the following problem. Consider the bounded solutions $u(x,t)$ of the one-dimensional heat equation on the half-line $x_1 > 0$,

$$D_t u = D_1^2 u, \tag{2.1}$$

that satisfy the initial condition

$$u|_{t=0} = 0 \tag{2.2}$$

and the boundary condition

$$B(D_1)u|_{x_1=0} = \sum_{\alpha_1=0}^{r} b_{\alpha_1} D_1^{\alpha_1} u|_{x_1=0} = \varphi(t). \tag{2.3}$$

One is required to find conditions on the polynomial $B(z)$ that guarantee stabilization of solutions of the problem (2.1)–(2.3) constructed for any stabilizing boundary data $\varphi(t)$. In his paper, Tikhonov solved the above problem completely and gave some physically-meaningful examples to which there correspond a problem (2.1)–(2.2) with a boundary operator $B(D_1)$ of arbitrary order.

Next we will consider Tikhonov's problem in the case of the heat equation with an arbitrary number of independent variables:

$$D_t u = \Delta u, \tag{2.4}$$

$$u|_{t=0} = 0, \tag{2.5}$$

$$B(D)u|_{x_n=0} \equiv \sum_{|\alpha'|+\alpha_n \leq r} b_{\alpha'\alpha_n} D_{x'}^{\alpha'} D_{x_n}^{\alpha_n} u|_{x_n=0} = \varphi(x't). \tag{2.6}$$

Although at the present time considerably more general results are known, we resisted the temptation to embark upon their exposition because we felt that the investigation technique can be demonstrated in the most accessible manner for the simplest (and most important) problem.

In this technique an essential role is played by a special theorem on the properties of the Poisson kernels (Poisson matrices) for the problems under consideration, which, in our opinion, is of independent interest.

Note that the boundary value problem (2.6) involves differential operators of various orders, and so as regards both its formulation and the obejctive of the study this problem differs from that discussed in the previous section.

Let us represent the boundary operator (2.6) in the form

$$B(D) = \sum_{\nu=0}^{r} B_\nu(D), \quad B_\nu(D) = \sum_{|\alpha'|+\alpha_n=\nu} b_{\alpha'\alpha_n} D_{x'}^{\alpha'} D_{x_n}^{\alpha_n}.$$

The vectors of $\mathbb{R}^{n-1}$ will be denoted by $\xi' = (\xi_1,\ldots,\xi_{n-1})$, $\xi'^2 = (\xi_1^2+\ldots+\xi_{n-1}^2)$.

Let us solve (formally, for the time being) the problem (2.4)–(2.6) by using the Laplace transformation in t and the Fourier transformation in x'. We get

$$u(x,t) = \int_0^t d\tau \int_{\mathbb{R}^{n-1}} G(x-y',t-\tau)\varphi(y',\tau)\,dy', \tag{2.7}$$

where $G(x,t)$ is the Poisson kernel of the problem under consideration:

$$G(x,t) = -i(2\pi)^{-1} \int_{\mathbb{R}^{n-1}} \exp\{ix'\cdot\xi'\}d\xi'$$

$$\times \int_{a-i\infty}^{a+i\infty} \exp\{pt - \sqrt{p+\xi'^2}x_n\}\big(B(i\xi',-\sqrt{p+\xi'^2})\big)^{-1}dp; \tag{2.8}$$

here $\sqrt{p+\xi'^2}$ is the branch with positive imaginary part of the radical and a is an arbitrary positive number.

We denote by $P_{\xi'}^-$ the half-plane $\operatorname{Re}p < -\xi'^2$ of the complex p-plane and by $\overline{P}_{\xi'}^- = P_{\xi'}^- \cup \{\operatorname{Re}p = -\xi'^2\}$ the corresponding closed half-plane, and we put $F(\xi',p) = B(i\xi',-\sqrt{p+\xi'^2})$, $F_r(\xi',p) = B_r(i\xi',-\sqrt{p+\xi'^2})$.

Next we list assumptions that guarantee the validity of (2.7) and (2.8) and take part in formulating the main results of the present section:

β_1. *For any $\xi' \in \mathbb{R}^{n-1}$ the p-zeroes of the function $F_r(\xi',p)$ lie in $\overline{P}_{\xi'}^-$;*

β_2. *For any $\xi' \in \mathbb{R}^{n-1}$ the p-zeroes of the function $F(\xi',p)$ lie in $P_{\xi'}^-$;*

Note that β_1 is the parabolicity condition for problem (2.4)–(2.6), while β_2 says that this problem satisfies the *stability condition*.

In the case of the problem (2.1)–(2.3) condition β_2 means that the zeroes of the function $B(-\sqrt{p})$ lie in the half-plane $\operatorname{Re}p < 0$. Introduce the new variable $z = -\sqrt{p}$ and write the following chain of inequalities:

$$-\frac{\pi}{2} \le \arg p \le \frac{\pi}{2}, \quad -\frac{\pi}{4} \le \arg\sqrt{p} \le \frac{\pi}{4}, \quad \frac{3}{4}\pi \le \arg(-\sqrt{p}) \le \frac{5}{4}\pi.$$

Thus, the fact that the zeroes of $B(-\sqrt{p})$ do not belong to $P_0^- = \{p : \operatorname{Re} p < 0\}$ means that the zeroes of $B(z)$ lie in the sector $\arg z \in [3/4\pi, 5/4\pi]$ of the complex z-plane. In [100] Tikhonov formulated condition β_2 as the requirement that the zeroes of $B(z)$ should not belong to the said sector. In [100] it is shown that condition β_2 is necessary and sufficient for the stabilization of the bounded solutions of problem (2.1)–(2.3), constructed for arbitrary stabilizing boundary data $\varphi(t)$.

Our argument below significantly differs from that of [100]. The reasoning there is based on techniques developed in the last decades in the theory of parabolic systems (in particular, parabolic boundary value problems) which enable one to solve the above problem for single equations and for rather general systems of equations.

VII.2.2. Lemmas

Here we prove two lemmas which explain why we are allowed to deform the integration contours in (2.8).

Introduce the functions $F^*(\zeta', p) = B(i\zeta', -\sqrt{p})$, $F_\nu^*(\zeta', p) = B_\nu(i\zeta', -\sqrt{p})$, $\nu = 0, \ldots, r$, where $\zeta' = \xi' + i\eta'$, $\eta = (\eta_1, \ldots, \eta_{n-1})$, and the sets $\Omega_{\varepsilon,k} = \{(\zeta', p) : \operatorname{Re} p \geq -k|\operatorname{Im} p|, |\eta'| \leq \varepsilon(|\zeta'| + |p|^{1/2})\}$, with ε and k positive constants.

Lemma VII.4. *Let condition β_1 be satisfied. Then there exist positive constants ε_1 and k_1 such that $F_r^*(\zeta', p)$ does not vanish in the domain $\Omega_{\varepsilon_1, k_1} \cap \{(\zeta', p) : |p| + |\zeta'|^2 > 0\}$.*

Lemma VII.5. *Let conditions β_1 and β_2 be satisfied. Then there exist positive constants ε_2, k_2, δ_0 ($\varepsilon_2 < \varepsilon_1, k_2 < k_1$) such that the inequality $|F^*(\zeta', p)| \geq \delta_0$ holds for any $(\zeta', p) \in \Omega_{\varepsilon_2, k_2}$.*

Proof: Let us present the proof of Lemma VII.4. Perform the scale transformation $\widetilde{p} = p/A^2$, $\widetilde{\zeta}' = \zeta'/A$, $A = |\zeta'| + |p|^{1/2}$ and notice that $|\widetilde{p}|^{1/2} + |\widetilde{\zeta}'| = 1$. Then the generalized homogeneous function $F_r^*(\zeta', p)$ can be rewritten in the form

$$F_r^*(\zeta', p) = A^r F_r^*(\widetilde{\zeta}', \widetilde{p}). \tag{2.9}$$

The function $F_r^*(\widetilde{\zeta}', \widetilde{p})$ is defined and continuous on the compact set $\widetilde{\Omega} = \{(\widetilde{\zeta}', \widetilde{p}) : |\widetilde{\zeta}'| + |p|^{1/2} = 1\}$. Consider the subset $\widetilde{\Omega}_1$ of Ω where $\operatorname{Re} \widetilde{p} \geq 0$, $\operatorname{Im} \zeta' \equiv \eta' = 0$. Condition β_1 implies that $F_r^*(\widetilde{\zeta}', \widetilde{p}) \neq 0$ in $\widetilde{\Omega}_1$. Since $F_r^*(\widetilde{\zeta}', \widetilde{p})$ is continuous on the compact set $\widetilde{\Omega}$, there exists a neighbourhood of $\widetilde{\Omega}_1$ in $\widetilde{\Omega}$ in which $F_r^*(\widetilde{\zeta}', \widetilde{p})$ does not vanish. In particular, this is true for the domain $\Omega_{\varepsilon_1, k_1} \cap \{(\widetilde{\zeta}', \widetilde{p}) : |\widetilde{\zeta}'| + |p|^{1/2} = 1\}$ for sufficiently small ε_1 and k_1. In conjunction with relation (2.9), this shows that $F_r^*(\widetilde{\zeta}', \widetilde{p}) \neq 0$ in $\Omega_{\varepsilon_1, k_1}$, if $|\zeta'| + |p|^{1/2} > 0$. Lemma VII.4 is proved.

Now let us prove Lemma VII.5. Using the same scale transformation, we rewrite $F^*(\zeta', p)$ as

$$F^*(\zeta', p) = A^r \sum_{\nu=0}^{r} F_\nu^*(\widetilde{\zeta}', \widetilde{p}) A^{\nu-r}. \tag{2.10}$$

By Lemma VII.4, for any positive A_0 there exists a positive $\delta_0(A_0)$ such that for $|\zeta'| + |p|^{1/2} = A_0$

$$|F_r^*(\zeta', p)| \geq \delta_0(A_0) A_0^r. \tag{2.11}$$

It follows from (2.10) and (2.11) that for a sufficiency large A_0

$$|F^*(\zeta', p)| \geq \delta_0' \tag{2.12}$$

for any $(\zeta', p) \in \Omega_{\varepsilon_1, k_1} \cap \{|\zeta'| + |p|^{1/2} \geq A_0\}$.

Now consider the continuous function $F^*(\zeta', p)$ on the compact set $\Omega_{\varepsilon_1, k_1} \cap \{|\zeta'| + |p|^{1/2} \leq A_0\}$; on the subset Ω_2 of this set where $\operatorname{Re} p \geq 0$, $\operatorname{Im} \zeta' \equiv \eta' = 0$, $F^*(\zeta', p)$ does not vanish. Hence, there exist (sufficiently small) positive constants ε_2, k_2, $\varepsilon_2 < \varepsilon_1$, $k_2 < k_1$ and δ_0'' such that for $(\zeta', p) \in \Omega_{\varepsilon_2, k_2} \cap \{|\zeta'| + |p|^{1/2} \leq A_0\}$

$$|F(\zeta', p)| \geq \delta_0''. \tag{2.13}$$

Set $\delta_0 = \min(\delta_0', \delta_0'')$. Then (2.12) and (2.13) imply the inequality $|F(\zeta', p)| \geq \delta_0$ for any $(\zeta', p) \in \Omega_{\varepsilon_2, k_2}$. Lemma VII.5 is proved.

VII.2.3. Study of the Poisson kernel

Here we present a technique for studying important properties of the Poisson kernel of the problem (2.4)–(2.6) for large time t. From these properties it follows almost immediately that conditions β_1 and β_2 guarantee stabilization of the bounded solutions $u(x, t)$ of our problem which are constructed for arbitrary stabilizing boundary data $\varphi(x', t)$. The results obtained here complement in essential manner the information on Poisson kernels given in Chapter VI.

Theorem VII.7. *Let conditions β_1 and β_2 be satisfied. Then the Poisson kernel $G(x, t)$ for the problem (2.4)–(2.6) has the following properties:*
1) $|G(x, t)| \leq C t^{-(n+1-r)/2} \exp\{-cx^2/t\}$ *for all $t \in (0, 1]$;*
2) $G(x, t)$ *has the representation* $G(x, t) = G_0(x, t) + G_1(x, t)$, *where*

$$G_0(x, t) = \left(B(0', 0)\right)^{-1} (x_n/t)(4\pi t)^{-n/2} \exp\{-x^2/4t\},$$

and $G_1(x, t)$ obeys the estimate $|G_1(x, t)| \leq C t^{-(n+2)/2} \exp\{-cx^2/t\}$ *for $t \geq 1$;*
3) *the integral $\int\limits_0^\infty dt \int |G(x, t)| dx'$ is bounded uniformly in x_n;*
4) *the function $K(x) = \int\limits_0^\infty G(x, t) dt$ is the Poisson kernel of the problem (2.6) for the Laplace equation. Morevoer, $|K(x)| \leq C(1 + x_n)|x|^{-n}$.*

Proof: We begin by proving the central fact, that is, property 2) of the Poisson kernel. To this end we write the relation

$$\left(B(i\zeta', -\sqrt{p})\right)^{-1} = \left(B(0', 0)\right)^{-1} + Q(i\zeta', -\sqrt{p}),$$

where

$$Q(i\zeta', -\sqrt{p}) = -\frac{\sum\limits_{\nu=1}^{r} B_\nu(i\zeta', -\sqrt{p})}{B(0', 0)(i\zeta', -\sqrt{p})}.$$

Then $G(x, t)$ has the representation

$$G(x, t) = \left(B(0'0)\right)^{-1}(-i)(2\pi)^{-n} \int\limits_{\mathbb{R}^{n-1}} \exp\left\{ix' \cdot \xi'\right\} d\xi'$$

$$\times \int\limits_{a-i\infty}^{a+i\infty} \exp\left\{pt - \sqrt{p + \xi'^2}\, x_n\right\} dp + G_1(x, t), \qquad (2.14)$$

where

$$G_1(x, t) = (-i)(2\pi)^{-n} \int\limits_{\mathbb{R}^{n-1}} \exp\left\{ix' \cdot \xi'\right\} d\xi'$$

$$\times \int\limits_{a-i\infty}^{a+i\infty} \exp\left\{pt - \sqrt{p + \xi'^2}\, x_n\right\} \left(Q(i\zeta', -\sqrt{p + \zeta'^2})\right) dp.$$

The first of the integrals in (2.14) can be calculated directly:

$$(-i)(2\pi)^{-n} \int\limits_{\mathbb{R}^{n-1}} \exp\left\{ix' \cdot \xi'\right\} d\xi' \int\limits_{a-i\infty}^{a+i\infty} \exp\left\{pt - \sqrt{p + \xi'^2}\, x_n\right\} dp =$$

$$(x_n/t)(4\pi t)^{-n/2} \exp\{-x^2/4t\}.$$

This is the Poisson kernel of the Dirichlet problem for the heat equation, that is, the kernel of the surface heat double-layer potential.

Now let us examine $G_1(x, t)$. First, replace p by the new integration variable $p_1 = p + \xi'^2$; by the analyticity of the function under the integral sign, the fact that it is decreasing, and the Cauchy integral theorem, we can use the same integration contour for p_1 (and actually keep the notation p). Then

$$G_1(x, t) = (-i)(2\pi)^{-n} \int\limits_{\mathbb{R}^{n-1}} \exp\left\{ix' \cdot \xi' - \xi'^2 t\right\} d\xi'$$

$$\times \int\limits_{a-i\infty}^{a+i\infty} \exp\left\{pt - \sqrt{p}\, x_n\right\} Q(i\xi', -\sqrt{p})\, dp.$$

Now perform the scale transformation $\xi' = t^{-n/2}\sigma'$, $p_2 = tp$ and introduce the variables $\tilde{x}_j = t^{-1/2} x_j$, $j = 1, \ldots, n$. In these new variables the integration with respect to p_2 will be carried out along the line $\operatorname{Re} p_2 = at$; since the function under the integral sign is analytic and decays as $|p_2| \to \infty$, one can keep the previous integration contour (and we will also preserve the notation p). Thus,

$$G_1(x,t) = (-i)(2\pi)^{-n} t^{-(n+1)/2} \int_{\mathbb{R}^{n-1}} \exp\left\{i\widetilde{x}' \cdot \sigma' - \sigma'^2\right\} d\sigma'$$

$$\times \int_{a-i\infty}^{a+i\infty} \exp\left\{p - \sqrt{p}\,\widetilde{x}_n\right\} Q(it^{-1/2}\sigma', -t^{-1/2}\sqrt{p})\, dp. \quad (2.15)$$

Here and below we use the fact that when the integration variables σ' and p run within the integration limits in (2.15), while $t \geq 1$, the following inequalities hold:

$$\left|\sum_{\nu=1}^{r} B_\nu(it^{-1/2}\sigma', -t^{-1/2}\sqrt{p})\right| \leq C \sum_{\nu=1}^{r} t^{-\nu/2}\left(|\sigma'| + |p|^{1/2}\right)^\nu, \quad (2.16)$$

$$|B(it^{-1/2}\sigma', -t^{-1/2}\sqrt{p})| \geq \delta_0. \quad (2.17)$$

Now in the integral with respect to p let us replace the contour by $l(a)$: $\operatorname{Re} p = -k_2|\operatorname{Im} p| + a$ (where k_2 is as in Lemma VII.5). That this transition is possible follows from the analyticity of the integrand in the domain $\operatorname{Re} p \geq -k_2|\operatorname{Im} p|$ and from the fact that on the contour $l(a)$ one has $p = a + iv\exp\{i(\operatorname{sgn}v)\varphi\}$, $\varphi = \arctan k_2$, $v \in (-\infty, \infty)$, $|\exp p| = \exp\{a - |v|\sin\varphi\}$.

This yields the following convenient representation for $G_1(x,t)$:

$$G_1(x,t) = -i(2\pi)^{-n} t^{-(n+1)/2} \int_{\mathbb{R}^{n-1}} \exp\left\{i\widetilde{x}' \cdot \sigma' - \sigma'^2\right\} d\sigma'$$

$$\times \int_{l(a)} \exp\left\{p - \sqrt{p}\,\widetilde{x}_n\right\} Q(it^{-1/2}\sigma', -t^{-1/2}\sqrt{p})\, dp. \quad (2.18)$$

Assuming that $|\widetilde{x}'| \geq 1$ we again perform the scale transformation $p = p_3\widetilde{x}'^2$, $\sigma' = |\widetilde{x}'|\widetilde{\sigma}'$. Here the integration with respect to p_3 is carried out along the contour $l(a|x'|^{-2})$. Again, using the analyticity of the integrand and the inequalities (2.17) one can return to integrating along $l(a)$ (and preserve the previous notation of the integration variables).

Then formula (2.18) takes the form

$$G_1(x,t) = -i(2\pi)^{-n} t^{-(n+1)/2} |\widetilde{x}'|^{n+1} \int_{\mathbb{R}^{n-1}} \exp\left\{i\widetilde{x}' \cdot \sigma'|\widetilde{x}'| - \sigma'^2\widetilde{x}'^2\right\} d\sigma'$$

$$(2.19)$$

$$\times \int_{l(a)} \exp\left\{p\widetilde{x}'^2 - \sqrt{p}\,\widetilde{x}_n|\widetilde{x}'|\right\} Q(i|\widetilde{x}'|t^{-1/2}\sigma', -|\widetilde{x}'|t^{-1/2}\sqrt{p})\, dp.$$

Next, replace integration with respect to σ_j by integration along the line in the complex s_j-plane which is parallel to the real axis, and which intercepts on the imaginary axis the segment $\eta_j = \eta_0 \operatorname{sgn} \widetilde{x}_j$, $s_j = \sigma_j + i\eta_j$, $(j = 1, \ldots, n-1)$. This again is possible thanks to the analyticity of the function under the integral sign, the fact that it is decreasing, and the Cauchy integral theorem.

It remains to estimate the obtained integral. To this end, we first estimate the function

$$\operatorname{Re}[i\widetilde{x}'(\sigma' + i\eta')|\widetilde{x}'| - (\sigma' + i\eta')^2 \widetilde{x}'^2] + a\widetilde{x}'^2$$

$$= -\eta_0 \sum_{j=1}^{n-1} |\widetilde{x}_j||\widetilde{x}'| - \sigma'^2 x'^2 + n\eta_0^2 \widetilde{x}'^2 + a\widetilde{x}'^2 \leq -\delta_1 \widetilde{x}'^2 - \sigma'^2 \qquad (2.20)$$

for η_0 and a sufficiently small; here δ_1 is a positive constant. Further, we write the equality $|s'|^2 = \sigma'^2 + (n-1)\eta_0^2$ and observe that for $p \in l(a)$ the following relations hold:

$$p = a - |v| \sin \varphi + iv \cos \varphi, \quad |p|^2 = a^2 + v^2 - 2a|v| \sin \varphi \leq a^2 + v^2;$$

$$p = |p| \exp\{i\psi\}; \quad \sqrt{p} = |p|^{1/2} \exp\{i\psi/2\}; \quad 0 \leq \frac{1}{2}|\psi| \leq (\pi/4) + (\varphi/2),$$

$$1 \geq \cos(\psi/2) \geq \cos\big((\pi/4) + (\varphi/2)\big) = (\sqrt{2}/2)(\cos(\varphi/2) - \sin(\varphi/2));$$

$$|p|^{1/2} \geq (1 - \sin \varphi)^{1/4}(a^2 + v^2)^{1/4} \geq 1/2(1 - \sin \varphi)^{1/4}(\sqrt{a} + \sqrt{v}).$$

Consequently,

$$- \operatorname{Re} \sqrt{p} = -|p|^{1/2} \cos(\psi/2)$$
$$\leq -(\sqrt{2}/4)(\cos(\varphi/2) - \sin(\varphi/2))(1 - \sin \varphi)^{1/4}(\sqrt{a} + \sqrt{v}) = -\delta_2(\sqrt{a} + \sqrt{v}).$$

Relying on the foregoing analysis (in particular, the inequalities (2.16), (2.17) and (2.20)), we obtain the following estimate for $G_1(x, t)$:

$$|G_1(x, t)| \leq C|\widetilde{x}'|^{n+1} t^{-(n+1)/2} \exp\big\{ - \delta_1 \widetilde{x}'^2 - \delta_2 \widetilde{x}_n |\widetilde{x}'|\sqrt{a}\big\} \int_{\mathbb{R}^{n-1}} \exp\{-\sigma'^2\} d\sigma'$$

$$\times \int_0^\infty \exp\big\{ - \delta_2 \widetilde{x}_n |\widetilde{x}'|\sqrt{v} - k_2 v\widetilde{x}'^2\big\} \sum_{\nu=1}^r |\widetilde{x}'|^\nu t^{-\nu/2}\big[a^2 + v^2 + \sigma'^2 + (n-1)\eta_0^2\big]^{\nu/4} dv.$$

Using the fact that the function $|\widetilde{y}'|^N \exp\{-c\widetilde{y}'^2\}$ is bounded for any $N \geq 0$ and $c \geq 0$ and bounding the function $\exp\{-\delta_2 \widetilde{x}_n |\widetilde{x}'|(\sqrt{v} + \sqrt{a})\}$ from above by 1, for $|\widetilde{x}| \geq 1$, $t \geq 1$ we have

$$|G_1(x, t)| \leq C t^{-(n+2)/2} \exp\{-cx'^2/t\}. \qquad (2.21)$$

To this point we have assumed that $|\tilde{x}'| \geq 1$. If $|\tilde{x}'| < 1$, then there is no need to perform the scale transformation $p = p_3\tilde{x}'^2$, $\sigma' = |\tilde{x}'|\tilde{\sigma}'$. Arguing in the same manner in this case we obtain the inequality

$$|G_1(x,t)| \leq Ct^{-(n+2)/2}\exp\{-c|x'|/t\} \leq Ct^{-(n+2)/2}\exp\{-cx'^2/t\}. \qquad (2.22)$$

Now let us estimate the kernel $G_1(x,t)$ as a function of x_n; we assume that $\tilde{x}_n \geq 1$. Change the integration variables in (2.19) by the formulas $\sigma_j = \tilde{x}_n\xi_j'$, $j = 1,\ldots,n-1$, $p = \tilde{x}_n'^2 p_4$ (preserving the notation p). This yields

$$G_1(x,t) = -i(2\pi)^{-n}t^{-(n+1)/2}|\tilde{x}'|^{n+1}\int\limits_{\mathbb{R}^{n-1}}\exp\left\{i\xi'\cdot\tilde{x}'x_n - \xi'^2\tilde{x}^2\right\}d\xi'$$

$$\times \int\limits_{l(a)}\exp\left\{p\tilde{x}_n^2 - \tilde{x}_n^2\sqrt{p}\right\}Q(i\tilde{x}_nt^{-1/2}\xi', -\tilde{x}_nt^{-1/2}\sqrt{p})\,dp.$$

Using the above estimates we obtain, for sufficiently small a,

$$\left|\exp\{(p-\sqrt{p})\tilde{x}_n^2\}\right| = \exp\left\{(\operatorname{Re}p - \operatorname{Re}\sqrt{p})\tilde{x}_n^2\right\}$$

$$\leq \exp\left\{(a-\delta_2\sqrt{a})\tilde{x}_n^2\right\}\exp\left\{-(|v|\sin\varphi + \delta_2\sqrt{|v|})\tilde{x}_n^2\right\} \leq$$

$$\exp\left\{-\delta_3\tilde{x}_n^2\right\}\exp\left\{-(|v|\sin\varphi + \delta_2\sqrt{|v|})\tilde{x}_n^2\right\},$$

where $\delta_3 > 0$. Consequently,

$$|G_1(x,t)| \leq C_1 t^{-(n+1)/2}\tilde{x}_n^{n+1}\exp\left\{-\delta_3\tilde{x}_n^2\right\}\int\limits_{\mathbb{R}^{n-1}}\exp\{-\xi'^2\}\,d\xi'$$

$$\times \int\limits_0^\infty \exp\left\{-(v\sin\varphi + \delta_2\sqrt{v})\tilde{x}_n^2\right\}\sum_{\nu=1}^r \tilde{x}_n^\nu t^{-\nu/2}(a^2 + v^2 + |\xi'|^4)^{\nu/4}\,dv.$$

As in the preceding case, this last estimate yields

$$|G_1(x,t)| \leq Ct^{-(n+2)/2}\exp\{-cx_n^2/t\}. \qquad (2.23)$$

Multiplying (2.21) and (2.23) (with accounting for (2.22)), we obtain the final inequality

$$|G_1(x,t)| \leq Ct^{-(n+2)/2}\exp\{-cx^2/t\}.$$

Assertion 2) of the theorem is proved.

Now we pass to the proof of assertion 1). Consider the case $t \in (0,1)$. Let $|\widetilde{x}'| \geq 1$. Perform the same transformations in the formula for $G(x,t)$ that we performed for $G_1(x,t)$. This yields (cf. (2.8))

$$G(x,t) = i(2\pi)^{-n} t^{-(n+1)/2} |\widetilde{x}'|^{n-1} \int_{\mathbb{R}^{n-1}} \exp\left\{ ix' \cdot s'|\widetilde{x}'| - s'^2 \widetilde{x}'^2 \right\} d\sigma'$$

$$\times \int_{l(a)} \exp\left\{ p\widetilde{x}'^2 - \widetilde{x}_n |\widetilde{x}'| \sqrt{p} \right\} \left(B(i|\widetilde{x}'| t^{-1/2} s', -|\widetilde{x}'| t^{-1/2} \sqrt{p}) \right)^{-1} dp.$$

Let us estimate the function $B \equiv B(i|\widetilde{x}'| t^{-1/2} s', -|\widetilde{x}'| t^{-1/2} \sqrt{p})$ from below. Set $\widetilde{A} = (|s'|^4 + |p|^2)^{1/4} t^{-1/2} |\widetilde{x}'|$ and consider two cases: $\widetilde{A} \leq A_1$ and $\widetilde{A} \geq A_1$, where A_1 is a sufficiently large positive constant which will be chosen later. In the first case B is a function defined and continuous on a compact set where it does not vanish. Therefore, there exists a positive constant $\delta(A_1)$ such that $|B| \geq \delta(A_1)$. Consequently,

$$B^{-1} \leq \delta(A_1)^{-1} \cdot 1 < \delta(A_1)^{-1} A_1^{r/4} t^{r/2} (|s'|^4 + |p|^2)^{-r/4}$$

$$= \delta(A_1)^{-1} A_1^{r/4} t^{r/2} \left[(|\sigma'|^2 + (n-1)\eta_0^2)^2 + a^2 + v^2 \right]^{-r/4}. \quad (2.24_1)$$

Now consider the second case. We have

$$|B(i|\widetilde{x}'| t^{-1/2} s', -|\widetilde{x}'| t^{-1/2} \sqrt{p})| \geq \widetilde{A}^r \left(|B_r(s'/\widetilde{A}, -\sqrt{p}/\widetilde{A})| \right.$$

$$- \sum_{\nu=0}^{r-1} \widetilde{A}^{\nu-r} |B_\nu(is'/\widetilde{A}, -\sqrt{p}/\widetilde{A})| \right) \geq (\delta_0/2)(|s'|^4 + |p|^2)^{r/4} |x'|^r t^{-r/2}, \quad (2.24_2)$$

provided A_1 is sufficiently large. Estimates (2.24_1) and (2.24_2) yield the inequality

$$(B(i|\widetilde{x}'| t^{-1/2} s', -|\widetilde{x}'| t^{-1/2} \sqrt{p}))^{-1} \leq C_0 t^{r/2} (|s'|^4 + |p|^2)^{-r/4},$$

where $C_0 = \max(\delta(A^{-1}), 2\delta_0^{-1})$, which holds for all the values of the variables under consideration.

Arguing as we did in the estimation of $G_1(x,t)$, we get the inequality

$$|G(x,t)| \leq C t^{-(n+1-r)/2} \exp\left\{ -cx'^2/t \right\},$$

which, as in the case of $G_1(x,t)$, yields the final estimate

$$|G(x,t)| \leq C t^{-(n+1-r)/2} \exp\left\{ -cx^2/t \right\}. \quad (2.25)$$

In conclusion, let us mention that there exists another way of obtaining the estimate (2.25) based on the fact that in the model case when $B(D) \equiv B_r(D)$, the estimate (2.25) has been established in the monographs [94, 108], while the general case can be studied by techniques of perturbation theory.

Next, let us establish property 3) of $G(x,t)$. For $r = 0$, $G(x,t)$ is the Poisson kernel for the Dirichlet problem, and property 3) can be verified by direct calculation.

For $r \geq 1$ we carry out the following estimation:

$$\int_0^\infty dt \int |G(x,t)|\,dx' \leq C\bigg\{ \int_0^1 t^{-(2-r)/2}dt \int_{\mathbb{R}^{n-1}} t^{-(n-1)/2} \exp\big\{ -cx'^2/t\big\}\,dx'$$

$$+ \int_1^\infty x_n t^{-3/2} \exp\big\{ -x_n^2/4t\big\}\,dt \int_{\mathbb{R}^{n-1}} t^{-(n-1)/2} \exp\big\{ -x'^2/4t\big\}\,dx'$$

$$+ \int_1^\infty t^{-3/2}dt \int_{\mathbb{R}^{n-1}} t^{-(n-1)/2} \exp\big\{ -cx'^2/t\big\}\,dx'\bigg\} \leq C.$$

Finally, let us prove assertion 4). First, note that $G(x,t)$ is an infinitely differentiable function (for $t > 0$, $x_n > 0$) and its derivatives obey the estimates

$$|D^\alpha G(x,t)| \leq C_\alpha t^{-(n+1+|\alpha|-r)/2} \exp\big\{ -cx^2/t\big\}.$$

By 3), there exists the function $K(x)$, defined as $K(x) = \int_0^\infty G(x,t)dt$. We have the following estimate of $K(x)$, which will be needed below:

$$|K(x)| = \bigg| \int_0^t G(x,t)\,dt\bigg| \leq \int_0^1 |G(x,t)|\,dt + \int_1^\infty |G_0(x,t)|\,dt + \int_1^\infty |G_1(x,t)|\,dt$$

$$\leq C\bigg(\int_0^1 t^{-(n+1-r)/2} \exp\big\{ -cx^2/t\big\}\,dt + \int_1^\infty x_n t^{-(n+2)/2} \exp\big\{ -x^2/4t\big\}\,dt$$

$$+ \int_1^\infty t^{-(n+2)/2} \exp\big\{ -cx^2/t\big\}\,dt\bigg) \leq C(1+x_n)|x|^{-n}.$$

Note that this estimate guarantees that the integral $\int K(x-y')\varphi(y')\,dy'$ converges for any bounded function $\varphi(x')$ $(x_n > 0)$.

We claim that $K(x)$ is a harmonic function for $x_n > 0$. Indeed, integrating the identity $D_t G \equiv \Delta G$ with respect to t from ε to N, we get the identity

$$G(x,N) - G(x,\varepsilon) = \Delta \int_\varepsilon^N G(x,t)dt.$$

Now let $\varepsilon \to 0$ and $N \to \infty$. By the estimates for $G(x,t)$, we get that $\Delta K \equiv 0$ (for $x_n > 0$). Further,

$$
B(D)K(x) = B(D) \int_0^\infty G(x,t)\,dt = \int_0^\infty B(D)G(x,t)\,dt
$$

$$
= \int_0^\infty (x_n/t)(4\pi t)^{-n/2} \times \exp\big\{ -x^2/4t \big\}\,dt
$$

$$
= -\frac{1}{2}\frac{\partial}{\partial x_n} \int_0^\infty \big(2\sqrt{\pi t}\,\big)^{-n} \exp\big\{ -x^2/4t \big\}\,dt = \frac{1}{2}D_n K_n(x),
$$

where $K_n(x) = -(1/(n+2)\omega_n)|x|^{-n+2}$ is the fundamental solution of the Laplace equation and ω_n is the area of the unit sphere in $\mathbb{R}^n$ [10, p. 203], $n \geq 3$; appropriately modified, the above formula is also valid for $n = 2$. Thus, $B(D)K(x)$ is the Poisson kernel of the Dirichlet problem for the Laplace equation, and consequently

$$
\lim_{x_n \to 0} \int B(D)K(x-y')\varphi(y')\,dy' = \varphi(x')
$$

for any smooth function with compact support $\varphi(x')$. Thus, we have established that $K(x)$ is the Poisson kernel for the Laplace equation in the case of the problem $B(D)u|_{x_n=0} = \varphi_0(x')$.

Remark VII.3. Without modifying the proof of Theorem VII.7, we can assume in property 1) that $t \in (0, \varepsilon_0)$, and assume in 2) that $t \geq \varepsilon_0$, where ε_0 is a fixed positive number.

Remark VII.4. Consider the problem

$$
\sum_{\alpha_n=0}^{r} b_{0'\alpha_n} D_n^{\alpha_n} u\big|_{x_n=0} = \varphi(x',t). \tag{2.6$_0$}
$$

Let us show that *the Poisson kernel of the problem (2.4)–(2.5) and (2.6$_0$) is given by the formula*

$$
G(x,t) = \big(2\sqrt{\pi t}\,\big)^{1-n} \exp\big\{ -x'^2/4t \big\}\widetilde{G}(x_n,t), \tag{2.26}
$$

where $\widetilde{G}(x,t)$ is the Poisson kernel for the problem (2.1)–(2.3).

Proof: We use formula (2.8):

$$G(x,t) = -i(2\pi)^{-n} \int\limits_{\mathbb{R}^{n-1}} \exp\{ix'\cdot\xi'\}\,d\xi'$$

$$\times \int\limits_{a-i\infty}^{a+i\infty} \exp\{pt - \sqrt{p+\xi'^2}\,x_n\}\big(B(0',-\sqrt{p+\xi'^2})\big)^{-1}dp,$$

$$= -i(2\pi)^{-1} \int\limits_{\mathbb{R}^{n-1}} \exp\{ix'\cdot\xi' - \xi'^2\}\,d\xi' \int\limits_{a-i\infty}^{a+i\infty} \exp\{pt - \sqrt{p}\,x_n\}\big(B(0',-\sqrt{p})\big)^{-1}dp,$$

$$= (2\sqrt{\pi t})^{1-n}\exp\{-x'^2/4t\}\widetilde{G}(x_n,t).$$

Note that (2.26) can also be verified directly.

VII.2.4. Stabilization theorem

First let us introduce a class of stabilizing boundary data.

Definition VII.3. *A function $\varphi(x',t)$ defined in E_+^n is said to belong to the class $L_0(E_+^n)$ if 1) $\varphi(x',t)$ is continuous and bounded in E_+^n and 2) $\varphi(x',t) \to \varphi_0(x')$ as $t \to \infty$, uniformly in $x' \in \mathbb{R}^{n-1}$.*

Theorem VII.8. *Let conditions β_1 and β_2 be satisfied. Then the solution $u(x,t)$ of the problem (2.4)–(2.6) given by (2.7) and constructed for any boundary function $\varphi(x',t) \in L_0(E_+^n)$ stabilizes, uniformly in each layer $\Pi_H^+ = \{x : x' \in \mathbb{R}^{n-1}, 0 \le x_n \le H\}$, to a harmonic function $u_0(x)$ that satisfies the boundary condition $B(D)u|_{x_n=0} = \varphi_0(x')$ and has the representation*

$$u_0(x) = \int\limits_{\mathbb{R}^{n-1}} K(x-y',x_n)\varphi_0(y')\,dy'.$$

Proof: We begin by writing the difference $u(x,t) - u_0(x)$ in the form

$$u(x,t) - u_0(x) =$$

$$\int\limits_0^t d\tau \int\limits_{\mathbb{R}^{n-1}} G(x-y',t-\tau)\varphi(y',\tau)\,dy' - \int\limits_0^\infty d\tau \int\limits_{\mathbb{R}^{n-1}} G(x-y',\tau)\varphi_0(y')\,dy'$$

$$= \left[\int\limits_0^t d\tau \int\limits_{\mathbb{R}^{n-1}} G(x-y',t-\tau)\varphi(y',\tau)\,dy' - \int\limits_0^t d\tilde{t} \int\limits_{\mathbb{R}^{n-1}} G(x-y',\tilde{t})\varphi_0(y')\,dy'\right]$$

$$- \int\limits_t^\infty d\tilde{t} \int\limits_{\mathbb{R}^{n-1}} G(x-y',\tilde{t})\varphi_0(y')\,dy' = J_1(x,t) + J_2(x,t). \tag{2.27}$$

Consider first $J_1(x,t)$; changing the variable in the subtracted term, we obtain

$$J_1(x,t) = \int_0^t d\tau \int_{\mathbb{R}^{n-1}} G(x-y',t-\tau)[\varphi(y',\tau) - \varphi_0(y')]\,dy'$$

$$= \int_0^T \int_{\mathbb{R}^{n-1}} G(x-y',t-\tau)[\varphi(y',\tau) - \varphi_0(y')]\,dy' \tag{2.28}$$

$$+ \int_T^t d\tau \int_{\mathbb{R}^{n-1}} G(x-y',t-\tau)[\varphi(y',\tau) - \varphi_0(y')]\,dy'$$

$$= J_{11}(x,t) + J_{12}(x,t).$$

By the assumptions on $\varphi(x',t)$, for any $\varepsilon > 0$ and for any $x' \in \mathbb{R}^{n-1}$ there exists a $T(\varepsilon)$ such that $|\varphi(x',t) - \varphi_0(x')| < \varepsilon$ for any $t \geq T(\varepsilon)$. Using this fact and the properties of the Poisson kernel established in Theorem VII.7, we obtain the estimate

$$|J_{12}| \leq \varepsilon \int_0^\infty d\tau \int_{\mathbb{R}^{n-1}} |G(x-y',\tau)|\,dy' \leq C\varepsilon. \tag{2.29}$$

Let us estimate J_{11}, assuming that $t - T(\varepsilon) \geq 1$. Using property 2) of the function $G(x,t-\tau)$ and denoting by M_0 a constant that bounds $|\varphi(x',t)|$ and $|\varphi_0(x')|$ from above, we get

$$|J_{11}| \leq 2M_0 C \int_0^{T(\varepsilon)} d\tau \int_{\mathbb{R}^{n-1}} (x_n + 1)(t-\tau)^{-(n+2)/2}$$

$$\times \exp\{-c|x'-y'|^2/(t-\tau) - cx_n^2/(t-\tau)\}dy' \leq C_1(1+H)M_0 \int_0^{T(\varepsilon)} (t-\tau)^{-3/2}\,d\tau$$

$$= 2C_1 M_0 (1+H)\big[(t-T(\varepsilon))^{-1/2} - t^{-1/2}\big]. \tag{2.30}$$

Finally, $J_2(x,t)$ obeys the estimate

$$|J_2| \leq M_0 \int_t^\infty d\tilde{t} \int_{\mathbb{R}^{n-1}} \tilde{t}^{-(n+2)/2} \exp\big\{-c|x'-y'|^2/\tilde{t}\big\}dy' = CM_0 t^{-1/2}. \tag{2.31}$$

Inequalities (2.29)–(2.31) imply that $u(x,t)$ stabilizes to $u_0(x)$ uniformly in Π_H^+.

Remark VII.5. In order that the theorems on well-posedness in spaces of Hölder functions discussed in Section VII.3 be applicable to the problem (2.4)–(2.6), it is required that the boundary functions $\varphi(x',t)$ be smooth and satisfy the condition of compatibility with zero at $x = 0$ up to the order dictated by the boundary operator $B(D)$. As it follows from the facts established above, these conditions have no connection whatsoever with stabilization.

Note that given a boundary datum $\varphi(x',t)$ one can define the function $\varphi_1(x',t) = \varphi(x',t)\chi_0(t)$, where $\chi_0(t)$ is a non-negative function bounded by 1, such that $\chi(x) = 0$ for $t \in [0,1]$ and $\chi(x) = 1$ for $t \geq 2$. It is readily verified that the solutions $u(x,t)$ of problem (2.4)–(2.6) constructed by means of formula (2.7) from the functions $\varphi(x',t)$ and $\varphi_1(x',t)$ both stabilize (or both do not stabilize).

VII.2.5. Examples

Let us give a number of examples taken from applied areas which illustrate and, to some extent, complement the results presented above.

Example VII.4. (The Dirichlet problem) $B(D_x) \equiv 1$, $F_0(\xi',p) \equiv F(\xi',p) \equiv 1$. Conditions β_1 and β_2 are satisfied; therefore, by Theorem VII.8, the solution of the Dirichlet problem for the heat equation represented by the heat surface double-layer potential with an arbitrary density $\varphi(x',t) \in L_0(E_+^n)$ stabilizes.

Example VII.5. (The Neumann problem) $B(D) \equiv D_n$, $F_1(\xi',p) \equiv F(\xi',p) = -\sqrt{p+\xi'^2}$. Condition β_1 holds, but β_2 does not, since the function $F(\xi',p)$ has a zero $p(\xi') = -\xi'^2$ which does not lie in the half-plane $P_{\xi'}^-$. Therefore, Theorem VII.8 does not apply. Let us show that stabilization of solutions constructed for arbitrary boundary data of class $L_0(E_+^n)$ also does not hold. The Poisson kernel of the Neumann problem is given by the formula

$$G^{(1)}(x,t) = -2(4\pi t)^{-n/2}\exp\{-x^2/4t\}. \tag{2.32}$$

Take $\varphi(x',t) \equiv 1$ as a boundary function; then the solution of the Neumann problem (the single-layer potential with unit density) admits the representation

$$u^{(1)}(x,t) = -\frac{2}{(2\sqrt{\pi})^n}\int_0^t d\tau \int_{\mathbb{R}^{n-1}} \exp\{-|x'-y'|^2/4(t-\tau)\}$$

$$\times \exp\{-x_n^2/4(t-\tau)\}(t-\tau)^{-n/2}dy' = \pi^{-1/2}\int_0^t \exp\{-x_n^2/4(t-\tau)\}2d(t-\tau)^{1/2}$$

$$= -2\pi^{-1/2}\int_0^{\sqrt{t}}\exp\{-x_n^2/4\tilde{\tau}^2\}d\tilde{\tau} \to -\infty \quad \text{as} \quad t\to\infty.$$

Note that the results of the previous subsection imply that for the solutions of the Neumann problem to stabilize it is necessary and sufficient that the limiting function $\varphi_0(x')$ be orthogonal to the function identically equal to 1 (this is true for a different class of stabilizing boundary data).

Example VII.6. (The heat exchange problem) $B(D) = \sum\limits_{j=1}^{n} b_j D_j + b_0$, $F_1(\xi', p) = i\xi' \cdot b' - b_n \sqrt{p + \xi'^2}$, $F(\xi', p) = i\xi' \cdot b' - b_n \sqrt{p + \xi'^2} + b_0$, $b' = (b_1, \dots, b_{n-1})$.

If $b_n \neq 0$, then condition β_1 is satisfied; if $\alpha = (b_0/b_n) < 0$ then condition β_2 is also satisfied. If, however, $\alpha > 0$, then condition β_2 fails: for ξ' orthogonal to b' the function $F(\xi', p)$ has a zero $p = -\xi'^2 + \alpha^2$ which does not lie in $P_{\xi'}^-$. The same holds true for $\alpha = b_0 = 0$.

Let us calculate the Poisson kernel for the heat exchange problem. By formula (2.8),

$$G^{(2)}(x, t) = -i(2\pi)^{-n} \int\limits_{\mathbb{R}^{n-1}} \exp\left\{ ix' \cdot \xi' \right\} d\xi'$$

$$\times \int\limits_{a-i\infty}^{a+i\infty} \left[\exp\left\{ pt - \sqrt{p + \xi'^2}\, x_n \right\} / \left(i\xi' \cdot b' + b_0 - b_n \sqrt{p + \xi'^2} \right) \right] dp$$

$$= i(2\pi)^{-n} \int\limits_{\mathbb{R}^{n-1}} \exp\left\{ ix' \cdot \xi' - \xi'^2 t \right\} d\xi'$$

$$\times \int\limits_{a-i\infty}^{a+i\infty} \left[\exp\left\{ pt - \sqrt{p}\, x_n \right\} / \left(i\xi' \cdot b' + b_0 - b_n \sqrt{p} \right) \right] dp.$$

Next, we need to calculate the integral

$$J(x_n, t; \xi') = (2\pi i)^{-1} \int\limits_{a-i\infty}^{a+i\infty} \left[\exp\left\{ pt - \sqrt{p}\, x_n \right\} / \left(q(\xi') - b_n \sqrt{p} \right) \right] dp. \qquad (2.33)$$

where $q(\xi') = i\xi' \cdot b' + b_0$. To this end we write

$$\left[\exp\{-\sqrt{p}\, x_n\} / (q(\xi') - b_n p) \right] = -\left[\exp\{-\sqrt{p}\, x_n\} / b_n \sqrt{p} \right]$$
$$+ \left[q(\xi') \exp\{-\sqrt{p}\, x_n\} / b_n \sqrt{p}(q(\xi') - b_n \sqrt{p}) \right]. \qquad (2.34)$$

The first term has the inverse Laplace transform $-(\sqrt{\pi + b_n})^{-1} \exp\{-x_n^2/4t\}$. Consider the function $\hat{\psi}(p) = q(\xi') \exp\{-px_n\} / (q(\xi') - b_n p)$. According to operational calculus, the corresponding inverse Laplace transform $\psi(t)$ is given by the formula

$$\psi(t) = \operatorname{res}_{b_n^{-1} q(\xi')} \left(\hat{\psi}(p) \exp(pt) \right) = -\eta(t - x_n) q(\xi') b_n^{-2} \exp\{ q(\xi') b_n^{-1}(t - x_n) \},$$

$\eta(t)$ is the Heaviside function. By Efros' theorem (on the inverse Laplace transform of a composite function) [58], the second term of (2.34) has the inverse Laplace transform

$$-(\pi t)^{-1/2}(i\xi' \cdot b' + b_0)b_n^{-2} \int_{x_n}^{\infty} \exp\{(i\xi' \cdot b' + b_n)b_n^{-1}(\tau - x_n) - \tau^2/4t\}\, d\tau.$$

Thus

$$J(x_n, t; \xi') = -(\sqrt{\pi t}\, b_n)^{-1} \exp\{-x_n^2/4t\} - (\pi t)^{1/2}(i\xi' \cdot b' + b_0)b_n^2$$
$$\times \int_{x_n}^{\infty} \exp\left\{(i\xi' \cdot b' + b_0)b_n^{-1}(\tau - x_n) - \tau^2/4t\right\} d\tau.$$

Substituting $J(x_n, t; \xi')$ in formula (2.32), we get the equality

$$G^{(2)}(x, t) = -\left(2/(4\pi t)^n b_n\right) \exp\{-x^2/4t\} + G_1^{(2)}(x, t), \tag{2.35}$$

where $G_1^{(2)}(x, t)$ is defined by the formula

$$G_1^{(2)}(x, t) = -2^{1-n}\pi^{1/2-n}b_n^{-2}t^{-1/2}\left(\sum_{j=1}^{n-1} b_j D_j + b_0\right) \int_{\mathbb{R}^{n-1}} \exp\left\{ix' \cdot \xi' - \xi'^2 t\right\} d\xi'$$
$$\times \int_{x_n}^{\infty} \exp\left\{(i\xi' \cdot b' + b_0)(\tau - x_n)b_n^{-1} - \tau^2/4t\right\} d\tau$$
$$= -2^{1-n}\pi^{1/2-n}b_n^{-2}t^{-1/2} J_2(x, t). \tag{2.36}$$

In the integral $J_2(x, t)$ we introduce the new variables $\tilde{\xi}_j = t^{1/2}\xi_j$, $j = 1, \ldots, n-1$, $\tilde{\tau} = (\tau - x_n)/2t^{1/2}$, $\tilde{x}_j = t^{-1/2}x_j$; then $J_2(x, t)$ can be rewritten as

$$J_2(x, t) = \tilde{J}_2(\tilde{x}, t) = 2t^{-(n-2)/2} \int_{\mathbb{R}^{n-1}} \exp\left\{i\tilde{\xi}' \cdot \tilde{x}' - \tilde{\xi}'^2\right\} d\tilde{\xi}'$$
$$\times \int_0^{\infty} \exp\left\{(2i\tilde{\xi}' \cdot b' + 2b_0 t^{1/2})b_n^{-1}\tilde{\tau} - (\tilde{\tau} + \tilde{x}/2)^2\right\} d\tilde{\tau}.$$

The latter integral is transformed in the standard manner: first one calculates the integral with respect to $\tilde{\xi}$ and then one transforms the integral with respect to $\tilde{\tau}$.

We obtain

$$\tilde{J}_2(\tilde{x}, t) = -2t^{-(n-2)/2}\pi^{(n-1)/2}(b_n/|b|)\left(\sum_{j=1}^{n-1} b_j D_j + b_0\right)$$

$$\times \exp\left\{ -(x^2/4t) + [(x \cdot b - 2b_0 t)^2/4|b|^2 t] \right\} \int_{(x\cdot b - 2b_0 t)/2|b|t^{-1/2}}^{\infty} \exp\left\{ -\tilde{\tau}^2 \right\} d\tilde{\tau}.$$

$$(2.37)$$

Relations (2.32)–(2.37) yield the final formula

$$G^{(2)}(x, t) = -\left(2/(4\pi t)^{n/2} b_n\right)\left[\exp\{-x^2/4t\} + 2(\sqrt{t}/|b|)\left(\sum_{j=1}^{n-1} b_j D_j + b_0\right) \right.$$

$$\left. \times \exp\left\{ -x^2/4t + [(x \cdot b - 2b_0 t)^2/4|b|^2 t] \right\} \int_{(x\cdot b - 2b_0 t)/2|b|t^{-1/2}}^{\infty} \exp\left\{ -\tilde{\tau}^2 \right\} d\tilde{\tau} \right].$$

$$(2.38)$$

Note that for $b_0 = 0$ this formula has been derived in detail (in a slightly different manner) in [60], pp. 311–317.

Recall that Theorem VII.8 guarantees stabilization of solutions of the problem under consideration when $\alpha = (b_0/b_n) < 0$. We claim that this condition is not only sufficient, but also necessary for stabilization of a solution of the heat exchange problem constructed for any boundary function $\varphi(x', t) \in L_0(E_+^n)$. Indeed, take $\varphi(x', t) \equiv 1$; the corresponding solution is given by the formula

$$u(x, t) \equiv u(x_n, t) = \int_0^t G^{(2)}(x_n, t - \tau)\, d\tau = \int_0^t G^{(2)}(x_n, \tau)\, d\tau,$$

$$G^{(2)}(x_n, t) = -(\pi t)^{-1/2} b_n^{-1} \exp\left\{ -x_n^2/4t \right\}\left[1 + 2\alpha\sqrt{t}\exp\left\{ (x_n - 2\alpha t)^2/4t \right\} \right.$$

$$\left. \times \int_{(x_n - 2\alpha t)/2\sqrt{t}}^{\infty} \exp\left\{ -\tilde{\tau}^2 \right\} d\tilde{\tau} \right].$$

$$(2.39)$$

Here we will use the asymptotic equality 7.1.23 of [98]:

$$z\exp\{z^2\}\int_z^{\infty} \exp\{-\tilde{\tau}^2\}\, d\tilde{\tau} \sim 1 + \sum_{m=1}^{\infty}(-1)^m \frac{1 \cdot 3 \cdot \ldots \cdot (2m-1)}{(2z^2)^m},$$

$$(2.40)$$

valid for $z \to \infty$, $|\arg z| < 3\pi/4$. It implies that for $\alpha < 0$ and large t

$$2\alpha\sqrt{t}\exp\left\{(x_n - 2\alpha t)^2/4t\right\} \int_{(x_n-2\alpha t)/2\sqrt{t}}^{\infty} \exp\left\{-\tilde{\tau}^2\right\}d\tilde{\tau} \sim 2\alpha t(x_n - 2\alpha t)^{-1}$$

$$-2 \quad \alpha t^2(x_n - 2\alpha t)^{-3} + \ldots \sim -1 + (1/2\alpha)\left[(1/2\alpha - x_n)\right]t^{-1} + \ldots .$$

$$(2.41)$$

From (2.39) and (2.41) it follows that the integral $\int_0^\infty G^{(2)}(x_n, \tau)\, d\tau$ converges. If, on the other hand, $\alpha > 0$, then $\displaystyle\int_{(x_n-2\alpha t)/2\sqrt{t}}^{\infty} \exp\left\{-\tilde{\tau}^2\right\}d\tilde{\tau} \to \sqrt{\pi}$ as $t \to \infty$ and the integral $\int_0^\infty G^{(2)}(x_n, \tau)\, d\tau \to -\infty$.

Example VII.7. (**Third boundary value problem**) $B(D) \equiv (D_n + b_0)$, $F_1(\xi', p) = -\sqrt{p + \xi'^2}$, $F(\xi', p) = -\sqrt{p + \xi'^2} + b_0$. This is a particular case of the heat exchange problem ($b_j = 0$, $j = 1, \ldots, n - 1$). As follows from the foregoing analysis, for stabilization of solutions of the third boundary value problem for arbitrary $\varphi(x', p) \in L_0(E_+^n)$ it is necessary and sufficient that $b_0 < 0$.

VII.2.6. Necessity

Let us start by establishing a necessary condition for stabilization of solutions of the problem (2.4), (2.5), (2.6_0).

Theorem VII.9. *If the solutions $u(x, t)$ of the problem (2.4), (2.5), (2.6_0) constructed for arbitrary boundary data $\varphi(x', t) \in L_0(E_+^n)$ stabilize, then all the p-zeroes of the function $B(0', -\sqrt{p + \xi'^{,2}})$ lie in the half-plane $\operatorname{Re} p < -\xi'^2$ of the complex p-plane, where ξ' is an arbitrary vector in $\mathbb{R}^{n-1}$ (that is, condition β_2 is satisfied).*

Proof: We need to show that there is no $\xi'_0 \in \mathbb{R}^{n-1}$ for which the function $B(0', -\sqrt{p + \xi'^2})$ has a p-zero with $\operatorname{Re} p > -\xi_0'^2$. This is equivalent to showing that $B(0', -\sqrt{z})$ has no z-zeroes with $\operatorname{Re} z \geq 0$.

Consider the solution of the problem (2.4), (2.5), (2.6_0) constructed for $\varphi(x', t) \in L_0(E_+^n)$. Then, by hypothesis, $u(x, t)$ stabilizes, and consequently is a bounded function. This bounded function has the Laplace transform $\widehat{u}(x, z)$ which is a analytic in z in the half-space $\operatorname{Re} z > 0$. Now consider the solution $u(x, t)$ of our problem constructed for a function $\varphi(x', t) \equiv \varphi(t) \in L_0(E_+^n)$. Obviously, this solution coincides with the bounded solution $v(x_n, t)$ of the problem

$$D_t v = D_n^2 v, \quad v|_{t=0} = 0, \quad B(0', D_n)v|_{x_n=0} \equiv B(D_n)v|_{x_n=0} = \varphi(t). \quad (2.42)$$

The Laplace transform $\hat{v}(x_n, t)$ of $v(x_n, t)$ solves the problem $z\hat{v} = D_n^2\hat{v}$, $B(D_n)\hat{v}|_{x_n=0} = \hat{\varphi}(z)$, whose bounded solution is given by the formula

$$\hat{v}(x_n, z) = \exp\{-\sqrt{z}\,x_n\}\hat{\varphi}(z)/B(-\sqrt{z}). \quad (2.43)$$

It immediately follows from (2.43) that $B(-\sqrt{z})$ cannot have zeroes in the half-plane $\operatorname{Re} z > 0$. Indeed, if there exists z_0, $\operatorname{Re} z_0 > 0$, such that $B(-\sqrt{z_0}) = 0$, then, taking $\varphi(t) \in L_0(E_+^n)$, $\hat{\varphi}(z_0) \neq 0$, we arrive at a contradiction with the analyticity of $\hat{v}(x_n, z)$ in the half-plane $\operatorname{Re} z > 0$ established above.

It is more difficult to prove that $B(-\sqrt{z})$ does not have any zeroes on the imaginary axis of the complex z-plane. Here is how this is done. Let $z_1 = i\gamma$, $\gamma \neq 0$ be a purely imaginary z-zero of $B(-\sqrt{z})$, of multiplicity ν. Consider the boundary function $\varphi_1(t)$ whose Laplace transform is given by the formula

$$\hat{\varphi}_1(z) = \exp\{-\sqrt{z}\} B(-\sqrt{z})(\sqrt{i\gamma} - \sqrt{z})^{-\nu} = \exp\{-\sqrt{z}\} \mathcal{R}(\sqrt{z}), \qquad (2.44)$$

where $\mathcal{R}(\zeta)$ is a polynomial, $\mathcal{R}(-\sqrt{i\gamma}) \neq 0$.

To calculate $\varphi_1(t)$, write the following operational equality:

$$(\pi t)^{-1/2} \exp\{-x_n^2/4t\} \Rightarrow \sqrt{z} \exp\{-\sqrt{z}\, x_n\}.$$

From this relation and (2.44) it follows that

$$\varphi_1(t) = -D_n \mathcal{R}(D_n)((\pi t)^{-1/2} \exp\{-x_n^2/4t\})|_{x_n=1},$$

which shows that $\varphi_1(t) \in L_0(E_+^1)$.

Now substitute the expression of $\hat{\varphi}_1(z)$ given by (2.44) in (2.43). We get $\hat{v}(x_n, z) = (\sqrt{i\gamma} - \sqrt{z})^{-\nu} \exp\{-(x_n + 1)\sqrt{z}\}$.

Thus, the solution of the problem (2.42) constructed for the boundary function $\varphi_1(t)$ is given by the formula

$$v_1(x_n, t) = (1/2\pi i) \int_{a-i\infty}^{a+i\infty} \exp\{zt - (x_n + 1)\sqrt{z}\}(\sqrt{i\gamma} - \sqrt{z})^{-\nu} dz.$$

Let us examine successively the cases $\nu = 1$ and $\nu \geq 2$.

In the case $\nu = 1$ we can repeat the calculation of the function $J(x_n, t; \xi')$ (cf. (2.33)). Here instead of x_n we need to write $x_n + 1$, and take $q(\xi') \equiv b_0 = \sqrt{i\gamma}$. This yields the following formula (cf. (2.38)):

$$v_1(x_n, t) = -(\pi t)^{-1/2} \exp\left\{ -(x_n + 1)^2/4t\right\}\left[1 + 2\sqrt{i\gamma}t \right.$$
$$\left. \times \exp\left\{(x_n + 1 - 2\sqrt{i\gamma}t)^2/4t\right\} \int_{(x_n+1-2\sqrt{i\gamma}t/)2\sqrt{t}}^{\infty} \exp\{-\tilde{\tau}^2\}\, d\tilde{\tau}\right]. \qquad (2.45)$$

The integral in (2.45) is understood as a complex curvilinear integral. Since $\exp\{-\tilde{\tau}^2\}$ is an entire function, the integral of $\exp\{-\tilde{\tau}^2\}$ does not depend on the choice of the integration path, and consequently the notation used above is meaningful. It is probably more customary to define the integral as follows:

$$(2/\sqrt{\pi}) \int_z^\infty \exp\{-\tilde{\tau}^2\}\, d\tilde{\tau} = 1 - (2/\sqrt{\pi}) \int_0^z \exp\{-\tilde{\tau}^2\}\, d\tilde{\tau}.$$

Using the readily verified equality

$$(2/\sqrt{\pi}) \int_z^\infty \exp\{-\tilde{\tau}^2\}\, d\tilde{\tau} = 2 - (2/\sqrt{\pi}) \int_{-z}^\infty \exp\{-\tilde{\tau}^2\}\, d\tilde{\tau},$$

we can rewrite $v_1(x_n, t)$ as

$$v_1(x_n, t) = -(\pi t)^{-1/2} \exp\left\{ -(x_n+1)^2/4t \right\} - 2\sqrt{i\gamma}\,\exp\left\{ i\gamma t - (x_n+1)\sqrt{i\gamma} \right\}$$

$$-\sqrt{i\gamma}\,\exp\left\{ -(x_n+1)^2 \right\}(2/\sqrt{\pi})$$

$$\times \left[\exp\left\{ (2\sqrt{i\gamma}t - x_n - 1)^2/4t \right\} \int_{(2\sqrt{i\gamma}t-x_n-1)/2\sqrt{t}}^{\infty} \exp\left\{ -\tilde{\tau}^2 \right\} d\tilde{\tau} \right].$$

$$(2.46)$$

To simplify notations, we denote the function in square brackets in (2.46) by F. Since the variable $z = \sqrt{i\gamma}t - (x_n+1)/2\sqrt{t}$ satisfies the conditions of applicability of the asymptotic equality (2.40) $(\arg \sqrt{i\gamma} = \pm\pi/4)$, (2.40) yields

$$F \sim (\sqrt{i\gamma}t - (x_n+1)/2\sqrt{t})^{-1} - (1/2)(\sqrt{i\gamma}t - (x_n+1)/2\sqrt{t})^{-3} + \ldots \quad (2.47)$$

It follows from (2.46) and (2.47) that the first and third terms in (2.45) tend to zero as $t \to \infty$, while the second term is a complex harmonic that does not have a limit for $t \to \infty$. Thus, the assumption that the function $B(-\sqrt{z})$ has a simple purely imaginary zero $z = i\gamma$, $\gamma \neq 0$, leads to a contradiction.

Next let us discuss the assumption that there is such a zero of multiplicity $\nu \geq 2$. Arguing in the same manner, by successively applying the theorem of residues at a multiple pole, the time-shift theorem for the Laplace transformation, and Efros' theorem, we obtain the following formula for $v_1(x_n, t)$:

$$v_1(x_n, t) = ((-1)^\nu/(\nu-1)!\sqrt{\pi t}) \int_{x_n+1}^{\infty} \left[(\nu-1)(\tau - x_n - 1)^{\nu-2} + \right.$$

$$\left. \sqrt{i\gamma}(\tau - x_n - 1)^{\nu-1} \right] \exp\left\{ \sqrt{i\gamma}(\tau - x_n - 1 - \tau^2)/4t \right\} d\tau. \quad (2.48)$$

By standard transformations one can derive from (2.43) a representation of $v_1(x_n, t)$ through integrals of error function type:

$$v_1(x_n, t) = ((-1)^\nu/(\nu-1)!) \exp\left\{ -(x_n+1)^2/4t \right\} \exp\left\{ (x_n + 1 - 2\sqrt{i\gamma}t)^2/4t \right\}$$

$$\times (2\sqrt{\pi}) \int_{(x_n+1-2\sqrt{i\gamma t})/2\sqrt{t}}^{\infty} (2\sqrt{t}\tau' + 2t\sqrt{i\gamma} - x_n - 1)^{\nu-2}[\nu - 1 + 2\sqrt{i\gamma}t\tau' + 2ti\gamma$$

$$- \sqrt{i\gamma}(x_n + 1)]\exp\{-\tau'^2\}\,d\tau'. \tag{2.49}$$

In the function under the integral sign isolate the group of principal terms (as $t \to \infty$),

$$2^{\nu-1}\sqrt{i\gamma}t^{\nu-1}(\tau' + \sqrt{i\gamma t})^{\nu-1} = 2^{\nu-1}\sqrt{i\gamma}\,t^{\nu-1}\sum_{j=0}^{\nu-1}\binom{\nu-1}{j}\tau'^{j}(i\gamma t)^{(\nu-1-j)/2}.$$

$$\tag{2.50}$$

We need the readily verifiable identity

$$(2/\Gamma((j+1)/2))\int_{z}^{\infty}\tau^{j}\exp\{-\tau^2\}\,d\tau$$

$$\tag{2.51}$$

$$= 1 + (-1)^j + \big(2/\Gamma((j+1)/2)\big)(-1)^{j+1}\int_{-z}^{\infty}\tau^j\exp\{-\tau^2\}\,d\tau.$$

Formulas (2.49)–(2.51) immediately imply that $v_1(x_n, t)$ tends to infinity as $t \to \infty$ at least as fast as $t^{\nu-1}$. Thus, the assumption that $B(-\sqrt{z})$ has a purely imaginary zero $i\gamma$, $\gamma \neq 0$ of multiplicity $\nu \geq 2$ leads again to a contradiction.

Finally, suppose that $B(-\sqrt{z})$ has a zero $z_0 = 0$ of multiplicity ν_0 and zeroes $z_1, \ldots, z_N$, $\operatorname{Re} z < 0$, of multiplicities $\nu_1, \ldots, \nu_N$, respectively. Take $\varphi_2(t) \equiv 1$ as a boundary function; then

$$\hat{u}_2(x_n, t) = \hat{\varphi}_2(z)\exp\{-x_n\sqrt{z}\}/B(-\sqrt{z}) \equiv \exp\{-x_n\sqrt{z}\}/z(-\sqrt{z})^{\nu_0}\mathcal{R}(-\sqrt{z}),$$

where $\mathcal{R}$ is a polynomial, $\mathcal{R}(0) \neq 0$. We compute $u_2(x_n, t)$, appealing again to Efros' theorem. In the present case $F(z) = (-1)^{\nu_0}z^{-\nu_0-1}\exp\{-x_n z\}/\mathcal{R}(-z)$. Applying the time-shift theorem and the generalized Heaviside expansion theorem, we get

$$f(t) = (-1)^{\nu_0}\eta(t - x_n)\left[\operatorname{res}_0\big(\exp\{z(t-x_n)\}z^{-\nu_0-1}/\mathcal{R}(-z)\big)+\right.$$

$$\left.\sum_{k=1}^{N}\operatorname{res}_{z_k}\big(\exp\{z(t-x_n\}z^{-\nu_0-1}/\mathcal{R}(-z)\big)\right]$$

$$= \eta(t - x_n)\left[P_{\nu_0}(t - x_n) + \sum_{k=1}^{N}P_{\nu_k-1}(t - x_n)\exp\{z_k(t - x_n)\}\right],$$

where $P_\mu(t - x_n)$ is a polynomial of degree μ.

By Efros' theorem

$$u_2(x_n, t) = (\pi t)^{-1/2} \int\limits_{x_n}^{\infty} \left[P_{\nu_0}(\tau - x_n) + \sum_{k=1}^{N} P_{\nu_k - 1}(\tau - x_n) \exp\{z_k(\tau - x_n)\} \right]$$

$$\times \exp\{-\tau^2/4t\} \, d\tau = \pi^{-1/2} \int\limits_{x_n/2\sqrt{t}}^{\infty} P_{\nu_0}(2\sqrt{t}\tau' - x_n) \exp\{-\tau'^2\} \, d\tau'$$

$$+ \pi^{-1/2} \int\limits_{x_n/2\sqrt{t}}^{\infty} \left[\sum_{k=1}^{N} P_{\nu_k - 1}(2\sqrt{t}\tau' - x_n) \exp\{z_k(2\sqrt{t}\tau' - x_n)\} \right] \exp\{-\tau'^2\} \, d\tau'.$$

Let us examine the last formula. The first term tends to infinity as $t \to \infty$ at least as fast as $\sqrt{t}$, while the second one is bounded, since

$$|2\sqrt{t}\tau' - x_n|^{\mu} \exp\{\operatorname{Re} z_k(2\sqrt{t}\tau' - x_n)\} \le C_{\mu}$$

for any non-negative μ.

Thus, $u_2(x, t) \to \infty$ as $t \to \infty$, and so the assumption that $B(-\sqrt{z})$ has a zero $z_0 = 0$ leads to a contradiction.

We see that the assumption that the solutions of the problem (2.4), (2.5), (2.6_0) constructed for any boundary function $\varphi(x', t) \in L_0(E_+^n)$ stabilize implies condition β_2.

Our next goal is to establish a necessary condition for the stabilization of solutions of problem (2.4)–(2.6).

Theorem VII.10. *Suppose that condition β_1 is satisfied. If the solutions $u(x, t)$ of problem (2.4)–(2.6), constructed for arbitrary boundary data $\varphi(x', t) \in L_0(E_+^n)$, stabilize, then all the p-zeroes of the function $B(i\xi', -\sqrt{p + \xi'^2})$ lie in the half-plane $\operatorname{Re} p \le 0$, where ξ' is an arbitrary vector of $\mathbb{R}^{n-1}$.*

Proof: Let ξ_0' be a fixed vector of $\mathbb{R}^{n-1}$. We are interested in the solutions $u(x, t)$ of the problem (2.4)–(2.6) constructed for boundary data of the form $\varphi(x', t) = \exp\{i\xi_0 \cdot x'\}\varphi(t)$, where $\varphi(t) \in L_0(E_+^1)$. It is natural to seek such solutions in the form $u(x, t) = \exp\{i\xi_0 \cdot x'\}v(x_n, t)$. Then the problem (2.4)–(2.6) for $u(x, t)$ becomes the following problem for $v(x_n, t)$:

$$D_t v = -\xi_0'^2 v + D_n^2 v; \quad v|_{t=0} = 0, \quad B(i\xi_0, D_n)v|_{x_n=0} = \varphi(t). \tag{2.52}$$

By hypothesis, $u(x, t)$ stabilizes, and consequently so does $v(x_n, t)$. From this it follows that $v(x_n, t)$ is bounded, and hence its Laplace transform, $\widehat{v}(x_n, p)$, is an analytic function in the half-plane $\operatorname{Re} p > 0$. For the function $\widehat{v}(x_n, p)$ we derive from (2.52) the boundary value problem

$$(p + \xi_0'^2)\widehat{v} = D_n^2 \widehat{v}, \quad B(i\xi_0', D_n)v|_{x_n=0} = \widehat{\varphi}(p). \tag{2.53}$$

Its solution is given by the formula

$$\widehat{v}(x_n, p) = \exp\left\{ -\sqrt{p + \xi_0'^2}\, x_n \right\} \widehat{\varphi}(p) / B(i\xi_0', -\sqrt{p + \xi_0'^2}). \qquad (2.54)$$

From the analyticity of $\widehat{v}(x_n, p)$ in p for $\operatorname{Re} p > 0$ and the arbitrariness of $\varphi(t)$ it follows that the function $B(i\xi_0', -\sqrt{p + \xi_0'^2})$ cannot have a p-zero with $\operatorname{Re} p > 0$.

VII.2.7. Discussion

In the previous subsections we gave a detailed account of the simplest version of the problem considered, using, in our opinion, elementary technique easily accesible to a wide audience. Of course, when considering more general versions of the problem, one is forced to replace direct calculations by arguments and estimations, often rather tedious. However, the schemes of the proofs and the results formulated are of a general character. The reader may convince herself that this is the case by consulting the papers of Ya.S. Kushitskiĭ and S.D. Eidel'man [54–57]. Some of these results will be presented below.

A prominent place is given to the construction of Poisson kernels for boundary value problems where the boundary operators contain derivatives of various orders, as well as to their large-time behaviour. This analysis substantially complements and sharpens the important results on Poisson kernels given in [60, 108]. When studying stabilization, a special role is played by the Poisson kernels of the Dirichlet problem; these kernels have properties which enable one to easily establish stabilization of the solutions of the Dirichlet problem constructed for a wide class of boundary data. For problems other than the Dirichlet problem, the Poisson kernels manifestly lack such properties. However, under special assumptions on the boundary operators, the difference between the Poisson kernels of the problem under consideration and the Poisson kernels of the Dirichlet problem admits an estimate that enables one to prove stabilization theorems. This consideration plays a decisive role in establishing sufficient conditions for stabilization. In the case when the boundary operators involve no derivatives with respect to the tangential variables $x_1, \ldots, x_{n-1}$, the aforementioned conditions become both necessary and sufficient. In the general case the problem of finding sharp necessary conditions remains open.

Note that in order to implement the approach described above it is necessary that the Dirichlet problem be parabolic. For a single equation parabolic in the sense of Petrovskiĭ this is indeed the case, while for an arbitrary parabolic system one should additionally require that the Dirichlet problem in question be parabolic.

Below we give a brief survey of some generalizations of the above results that admit a simple enough formulation.

VII.2.8. Heat and mass exchange equations

For simplicity, we restrict ourselves to the case of one space variable.

The problem under consideration is to find a vector-valued function $u(x,t) = \big(u_1(x,t), \ldots, u_m(x,t)\big)$ that satisfies the parabolic system

$$D_t u = AD^2 u, \tag{2.55}$$

with the null initial condition and the boundary condition

$$\mathcal{B}(D)u|_{x=0} = \varphi(t), \tag{2.56}$$

where A is a $(m \times m)$-matrix with constant elements whose eigenvalues have positive real parts (the parabolicity condition for (2.55)), and $\mathcal{B}(D)$ is a matrix operator:

$$\mathcal{B}(D) = \big(B_{qj}(D)\big)_{q,j=1}^{m} = \sum_{\alpha=0}^{r_q} b_\alpha^{qj} D^\alpha, \quad r_1 \le r_2 \le \ldots \le r_m,$$

where b_α^{qj} are constants.

The Dirichlet problem for (2.55) is parabolic (this fact has been discussed in the first chapter). This implies, by Lopatinskiĭ's theorem, that the matrix $\mathfrak{A}(\xi,p) = pI + A\xi^2$ admits the matrix factorization $\mathfrak{A}(\xi,p) = \mathfrak{A}_-(\xi,p)\mathfrak{A}_+(\xi,p)$ where the ξ-zeroes of $\det\mathfrak{A}_-(\xi,p)$ [resp, $\det\mathfrak{A}_+(\xi,p))$] lie in the lower [resp., upper] half-plane of the complex ξ-plane, $p \ne 0$. The first important assertion is

Theorem VII.11. 1) *The Poisson matrix $G^{(\mathrm{D})}(x,t)$ of the Dirichlet problem for equation (2.55) is defined by the formula*

$$G^{(\mathrm{D})}(x,t) = (1/2\pi) \int_{a+i\infty}^{a-i\infty} \exp\{pt\}\Big((1/2\pi i)\int_{\gamma_+} \exp\{i\xi x\}(\mathfrak{A}_+(\xi,p))^{-1}d\xi\Big)dp.$$

2) $G^{(\mathrm{D})}(x,t)$ *admits the estimate*

$$|G^{(\mathrm{D})}(x,t)| \le Cxt^{-3/2}\exp\{-cx^2/t\}.$$

In order to formulate conditions to be imposed on the boundary operator $\mathcal{B}$, we introduce the matrix-valued function

$$\mathcal{F}(p) = (1/2\pi i)\int_{\gamma_+} \mathcal{B}(i\xi)(\mathfrak{A}_+(\xi,p))^{-1}d\xi,$$

and denote its determinant by $F(p) = \det \mathcal{F}(p)$; consider also the matrix-valued function

$$\mathcal{F}^0(p) = (1/2\pi i)\int_{\gamma_+} \mathcal{B}^0(i\xi)(\mathfrak{A}_+(\xi,p))^{-1}d\xi,$$

where $\mathcal{B}^0(i\xi)$ is the principal part of the matrix $\mathcal{B}(i\xi)$, and put $F^0(p) = \det \mathcal{F}^0(p)$.

The main assumptions are:

β_1'. $F^0(p)$ *does not vanish for $p \neq 0$.*

β_2'. *The zeroes of the function $F(p)$ lie in the half-plane* $\operatorname{Re} p < 0$.

Note that

$$F^0(p) = \delta_0 p^{\sum\limits_{q=1}^{m} r_q},$$

therefore condition β_1' means that $\delta \neq 0$, which is precisely the parabolicity condition for the problem (2.55) - (2.56). Condition β_2' is known as the *Tikhonov's condition* (or the *stability condition*) for this problem.

Theorem VII.12. *Let conditions β_1' and β_2' be satisfied. Then the Poisson kernels $G_{qj}(x,t)$, $q,j = 1,\ldots,m$ of the boundary value problem (2.55)–(2.56) have the following properties:*

1) *For $t \in (0,1]$ there holds the estimate*

$$|G_{qj}(x,t)| \leq Ct^{-(2-r_j)/2} \exp\{-cx^2/t\}.$$

2) *One has the representation*

$$G(x,t) = G^{(0)}(x,t) + G^{(1)}(x,t),$$

where $G^{(0)}(x,t) = G^{(D)}(x,t)\mathcal{F}(0)^{-1}$, and $G^{(1)}(x,t)$ satisfies for $t \geq 1$ the following estimate:

$$|G^{(1)}(x,t)| \leq Ct^{-3/2} \exp\{-cx^2/t\}.$$

3) *The integrals $\int\limits_0^\infty |G(x,t)|\,dt$ are bounded uniformly in $x \in [0,\infty)$.*

Theorem VII.13. *Let conditions β_1' and β_2' be satisfied. Then the solutions $u(x,t)$ constructed for continuous stabilizing boundary vector-valued functions $\varphi(t)$ (i.e, such that there exists the limit $\lim\limits_{t\to\infty} \varphi(t) = \varphi^0$) have the property that there exists the limit $\lim\limits_{t\to\infty} u(x,t) = \mathcal{B}(0)^{-1}\varphi^0$.*

The following statement is also valid.

Theorem VII.14. *Let condition β_1' hold. If the solutions $u(x,t)$ constructed for continuous stabilizing boundary functions have a limit as $t \to \infty$, then condition β_2' is satisfied.*

VII.2.9. A model equation of higher order

Consider the following boundary value problem:

$$D_t u + (-1)^b(\Delta^b_{x'} + D^{2b}_n)u = 0, \tag{2.57}$$

$$u|_{t=0}, \tag{2.58}$$

$$B_q(D_{x'}, D_n)u|_{x_n=0} \equiv \sum_{|\alpha'|+\alpha_n \leq r_q} b^{(q)}_{\alpha',\alpha_n} D^{\alpha'}_{x'} D^{\alpha_n}_n u|_{x_n=0} = \varphi_q(x', t), \tag{2.59}$$

$q = 1,\ldots,m$, $r_1 \leq r_2 \leq \ldots \leq r_b$. Denote by $\omega_s = \exp\{i(\pi/2b + (s-1)\pi/b)\}$, $s = 1,\ldots,b$, the roots of order $2b$ of -1 with positive real parts. Let $(p + \xi'^{2b})^{1/2b}$ be the $2b$th root of the function $p + \xi'^{2b}$ that is positive for $p + \xi'^{2b} > 0$ and let B^0_q be the principal part of the operator B_q. Define the functions

$$F(\xi', p) = \det\left(B_q(i\xi', i\omega_s(p + \xi'^{2b})^{1/2b})\right)^b_{q,s=1},$$

$$F_0(\xi', p) = \det\left(B^0_q(i\xi', i\omega_s(p + \xi'^{2b})^{1/2b})\right)^b_{q,s=1},$$

Denote by $Q^-_{\xi'}$ the half-plane $\operatorname{Re} p < -\xi'^{2b}$ of the complex p-plane, and put $\overline{Q}^-_{\xi'} = Q^-_{\xi'} \cup \{p = -\xi'^{2b}\}$.

We are now ready to formulate the main conditions:

$\hat{\beta}'_1$. *For any $\xi' \in \mathbb{R}^{n-1}$ the p-zeroes of the function $F_0(\xi', p)$ lie in $\overline{Q}^-_{\xi'}$;*

$\hat{\beta}'_2$. *For any $\xi' \in \mathbb{R}^{n-1}$ the p-zeroes of the function $F_0(\xi', p)$ lie in $\overline{Q}_{\xi'}$;*

As usual, $\hat{\beta}_1$ is the parabolicity condition for problem (2.57)–(2.59), while $\hat{\beta}_2$ is the stability condition for this problem.

The bounded solutions of problem (2.57)–(2.59) can be represented through the Poisson kernels $G_j(x, t)$, $j = 1,\ldots,b$, by the formula

$$u(x, t) = \sum_{j=1}^b \int_0^t d\tau \int_{\mathbb{R}^{n-1}} G_j(x - y', t - \tau)\varphi(y', \tau)\, dy'.$$

Let us state the main results. Denote $\tilde{x} = t^{-1/2b}x$, and let $P^{(k)}_j(x_n)$ be polynomials of degree at most k with non-negative coefficients.

Theorem VII.15. *Let conditions $\hat{\beta}_1$ and $\hat{\beta}_2$ be satisfied. Then the Poisson kernels $G_j(x, t)$, $j = 1,\ldots,b$, of the boundary value problem (2.57)–(2.59) have the following properties:*

1)   *For $t \in (0, 1]$ and $r_j > 0$*

$$|G_j(x, t)| \leq P^{(b-1)}_j(x_n)t^{-(n-1+2b-r_j)/2b} \exp\left\{-c|\tilde{x}|^{2b/(2b+1)}\right\},$$

while for $r_j = 0$ there holds the estimate

$$|G_j(x,t)| \le x_n P_j^{(b-1)}(x_n) t^{-(n+2b)/2b} \exp\left\{ -c|\tilde{x}|^{2b/(2b-1)} \right\}.$$

2) *There holds the representation*

$$G_j(x,t) = G_j^{(0)}(x,t) + G_j^{(1)}(x,t),$$

where

$$G_j^{(0)}(x,t) = \sum_{s=1}^{b} C_{sj} D_n^{s-1} G_b^{(D)}(x,t),$$

C_{sj} *are constants,* $G_b^{(D)}(x,t)$ *is the b-th Poisson kernel of the Dirichlet problem; further, for $t \ge 1$ one has the estimates*

$$|G_j^{(0)}(x,t)| \le x_n P_j^{(b-1)}(x_n) t^{-(n+2b)/2b} \exp\left\{ -c|\tilde{x}|^{2b/(2b-1)} \right\},$$

and

$$|G_j^{(1)}(x,t)| \le P_j^{(b-1)}(x_n) t^{-(n+2b)/2b} \exp\left\{ -c|\tilde{x}|^{2b/(2b-1)} \right\}.$$

3) *The integrals $\int\limits_0^\infty dt \int\limits_{\mathbb{R}^{n-1}} |G_j(x,t)| \, dx'$, $j = 1,\ldots,b$ are uniformly (in x_n) bounded on any finite segment $[0,H]$.*

4) *The functions $K_j(x) = \int\limits_0^\infty G_j(x,t)\,dt$, $j = 1,\ldots,b$, are the Poisson kernels of the boundary value problem (2.59) for the elliptic equation $(\Delta_{x^1}^b + D_n^{2b})u = 0$.*

Theorem VII.16. *Let conditions $\hat{\beta}_1$ and $\hat{\beta}_2$ be satisfied. Then the bounded solutions $u(x,t)$ of the problem (2.57)–(2.59) constructed for boundary data $\varphi_q(x',t)$, $q = 1,\ldots,b$, of class $L_0(E_+^n)$ stabilize uniformly in each layer $\mathbb{R}^n \cap \{0 \le x_n \le H\}$ to a bounded function $u_0(x)$, which is the solution of the elliptic boundary value problem*

$$(\Delta_{x'}^b + D_n^{2b})u = 0, \quad B_q(D)u|_{x_n=0} = \varphi_q^0(x')$$

and is given by the formula

$$u_0(x) = \sum_{j=1}^{b} \int\limits_{\mathbb{R}^{n-1}} K_j(x-y')\varphi_j^0(y')dy'.$$

Finally, let us consider a special class of boundary operators which do not contain derivatives with respect to the tangential variables:

$$B_q(D_n)u|_{x_n=0} \equiv \sum_{\alpha_n=0}^{r_q} b_{0'\alpha_n}^{(q)} D_n^{\alpha_n} u|_{x_n=0} = \varphi_q(x',t), \quad q = 1,\ldots,b. \tag{2.59_0}$$

Theorem VII.17. *If the bounded solutions $u(x,t)$ of the boundary value problem (2.57), (2.58), (2.59$_0$) constructed for arbitrary boundary data $\varphi_q(x',t) \in L_0(E_+^n)$ stabilize, then condition $\hat{\beta}_2$ is satisfied.*

Note that in the case of the model parabolic equation with one space variable Tikhonov's problem has been completely solved by D.V. Proka [81]. His technique differs from those presented above.

Comments

Chapter I has an introductory character and contains basic definitions, formulations of problems, and various illustrative examples.

Chapter II is auxilliary and contains basic definitions and properties of distributions and the basic information on Sobolev spaces $\mathcal{H}^s, H^s$ necessary for the exposition. The sources for the material presented here Chapter II are the papers and monographs [1, 10, 14, 15, 92, 94, 113].

Chapter III contains known assertions on the boundedness of various operators in $\mathcal{H}^s$, H^s as well as the commutation formulas (Green formulas) (3.14)–(3.16). Together with the definitions of the spaces $\widetilde{\mathcal{H}}^s$, results of the investigations of these spaces and of the action of differential operators in them, the commutation formulas form the basis for the exposition in Chapters IV and V.

The crucial material is presented in Chapter IV, which contains a study of model parabolic boundary value problems in the spaces $\widetilde{\mathcal{H}}^s$ of ordinary functions and distributions. The theory presented is similar to the theory of elliptic equations [85, 86, 89] and is based on Lemmas III.5–III.7, III.16–III.19, and Theorem IV.2 on the solvability of parabolic systems in $\mathcal{H}^s_{++}(\mathbb{R}^{n+1})$. Theorem IV.2 is a direct analogue of Lemmas 3 and 5 of [86]. The method of raising the smoothness of generalized solutions up to the boundary relies on the compatibility conditions of initial and boundary values and makes essential use of the "doubly" anisotropic spaces $\mathcal{H}^{s,r}_{(x,t),(x',t)}$, like in the theory of elliptic equations [70, 106] and [86, 89, 13]. The chapter presents novel results that have been partially reported in [30, 31, 34].

Chapter V contains results of the investigation of boundary value problems in cylindrical domains and some of their applications. These results generalize and complement (in the framework of L_2-theory) the well-known results in the case of smooth [1, 94, 108, 120] as well as generalized [120] solutions. The results obtained for boundary value problems are generalized in a natural way to nonlocal boundary value problems (local and nonlocal conjugation problems). The conditions on the coefficients of the operators are sufficient conditions and can be considerably sharpened. The results obatined are apllicable to problems whose right-hand sides are allowed to have regular (power-like) singularities and to problems concerning the existence of limit values of weak solutions to parabolic equations and their normal derivatives on the boundary of a cylindrical domain. Some results of this chapter were published in [25, 28, 32, 33].

Chapter VI is a survey, with the relevant references given in the course of the exposition.

Chapter VII is based on publications of the authors and Ya.S. Kushitskiĭ [26, 27, 29, 54–57] and presents results on the asymptotic behavior of the solutions of model parabolic boundary value problems and their stabilization for $t \to \infty$.

Let us add that the paper [129] contains basic results of the L_p-theory, for arbitrary $p > 1$, of parabolic boundary value problems in the spaces $\mathcal{H}_p^s$ of distributions of arbitrary finite order. This work naturally develops the L_2-theory presented in Chapters IV and V of this monograph.

References

[1] Agranovich, M. S. and Vishik, M. I., *Elliptic problems with a parameter and parabolic problems of general type*, Uspekhi Mat. Nauk **19**, no. 3(117), 53–161 (1964); English transl. in Russian Math. Surveys **19** (1964). [MR **33** #415]

[2] Agranovich, M. S. and Sukhorutchenko, V. V., *The necessity of the algebraic parabolicity conditions for nonstationary problems*, Mat. Zametki **2**, no. 6, 615–625 (1967); English transl. in Math. Notes **2**, no. 6, 864-869 (1967). [MR **37** #3181]

[3] Besov, O. V., Il'in, V. P., and Nikol'skiĭ, S. M., *Integral Representations of Functions and Embedding theorems*, "Nauka", Moscow, 1975; English transl.: Vols. I and II, V. H. Winston and Sons, Washington, D.C.; Halstead Press [John Wiley and Sons], New York, 1978, 1979. [MR **55** #3776]

[4] Bergh, Jöran and Löfström, Jörgen, *Interpolation Spaces. An Introduction*, Springer-Verlag, Berlin, Heidelberg, New York, 1976. [MR **58** #2349]

[5] Berezanskiĭ, Yu. M., *Expansion in Eigenfunctions of Selfadjoint Operators*, "Naukova Dumka", Kiev, 1965; English transl.: Translations of Math. Monographs, Vol. 17, Amer. Math. Soc., Providence, R. I., 1968. [MR **36** # 5769]

[6] Bitsadze, A. V. and Samarskiĭ, A. A., *Some elementary generalizations of linear elliptic boundary value problems*, Dokl. Akad. Nauk SSSR **185**, No. 4, 739–740 (1969); English transl. in Soviet Math. Dokl. **10**, 398–400 (1969). [MR **40** #540]

[7] Bourbaki, N., *Eléments de Mathématique. XV and XVIII, Espaces Vectoriels Topologiques*, Hermann, Paris, 1953 and 1955; English transl., Springer-Verlag, Berlin, Heidelberg, New York, 1987. [MR **14,880b** and **17,1109d**]

[8] Paley, R. E. A. C. and Wiener, N., *Fourier Transforms in the Complex Domain*, Amer. Math. Soc. Colloquium Publications, 19, Amer. Math. Soc., Providence, R. I., 1934.

[9] Vishik, M. I. and Eskin, G. I., *Parabolic convolution equations in a bounded region*, Mat. Sb. **71 (113)**, 162–190 (1966); English transl. in Amer. Math. Soc. Transl. **95** no. 2 (1970). [MR **34** #6311]

[10] Vladimirov, V. S., *Generalized Functions in Mathematical Physics*, 2nd edition, "Nauka", Moscow, 1979; English transl.: "Mir", Moscow, 1979. [MR **80j:46062a,b**]

[11] Volevich, L. R., *A problem in linear programming stemming from differential equations*, Uspekhi Mat. Nauk **18**, no. 3(111), 155–162 (1963); English transl. in Russian Math. Surveys **18** (1963). [MR **28** #5248]

[12] Volevich, L. R,., *Solubility of boundary value problems for general elliptic systems*, Mat. Sb. **68(110)** 373–416 (1965); English transl. in Amer. Math. Soc. Transl. (2) **67**, 182–225 (1968). [MR **33** #418]

[13] Volevich, L. R. and Gindikin, S. G., *The method of energy estimates in the mixed problem*, Uspekhi Mat. Nauk **35**, no. 5(215), 53–210 (1980); English transl. in Russian Math. Surveys **35**, no. 5, 57-137 (1980). [MR **82c:35042**]

[14] Volevich, L. R. and Paneyakh, B. P., *Some spaces of generalized functions and embedding theorems*, Uspekhi Mat. Nauk **20**, no. 1(121), 3–74 (1965); English transl. in Russian Math. Surveys **20**, no. 1, 1–73 (1965). [MR **30** #5160]

[15] Gel'fand, I. M. and Shilov, G. E., *Generalized Functions*. Vols. 1, 2, 3, Gosud. Izdat. Fiz-Mat. Lit., Moscow, 1958, 1959; English transl.: Academic Press, New York, 1964, 1967, 1968. [MR **20** #4182; **21** #5142a]

[16] Golubov, B. I, *A method of Abel-Poisson summation of multiple Fourier series*, Mat. Zametki **27** no.1, 49–59 (1980); English transl. in Math. Notes **27**, 28–33 (1980). [MR **81a:42024**]

[17] Gorbachuk, V. I. and Gorbachuk, M. L., *Boundary Value Problems for Differential-Operator Equations*, "Naukova Dumka", Kiev, 1984; English transl.: Kluwer Academic Publishers, Dordrecht, 1991. [MR **86g:34081**]

[18] Gushchin, A. K. and Mikhaĭlov, V. P., *Boundary values in L_p, $p > 1$, of solutions of elliptic equations*, Mat. Sb. **108(150)**, 3–21 (1979); English transl. in Math. USSR Sb. **36**, 1–19 (1980). [MR **80g:35062**]

[19] Gyul'misaryan, A. G., *General boundary value problems for parabolic equations with discontinuous coefficients*, Izv. Akad. Nauk Armyan. SSR, Ser. Fiz.-Mat. **18**, no. 1, 14–33 (1965) (in Russian). [MR **33** #7688]

[20] Günter, N. M., *Potential Theory and its Application to Basic Problems of Mathematical Physics*, Gostekhizdat, Moscow, 1953; English transl.: F. Ungar, New York, 1967. [MR **16,357f**]

[21] Ditkin, V. A. and Prudnikov, A. P., *Operational Calculus*, "Vyssh. Shkola", Moscow, 1966; French transl.: "Mir", Moscow, 1979. [MR **82f:44001**]

[22] Drin', M. M. and Ivasishen, S. D., *The Green matrix of a general boundary value problem for a system with discontinuous coefficients that is parabolic in the sense of I. G. Petrovskiĭ's* , Dokl. Akad. Nauk Ukrain. SSR Ser. A, **2**, no. 11, 7–10 (1984) (in Russian). [MR **86g:35095**]

[23] Doetsch, Gustav, *Guide to the Application of Laplace and Z-Transforms*, Van Nostrand Reinhold Co., London, 1971. [Zbl 274.44001]

[24] Zhitarashu, N. V., *Schauder estimates and solvability of general boundary value problems for general parabolic systems with discontinuous coefficients*, Dokl. Akad. Nauk SSSR **169**, 511–514 (1966); English transl. in Soviet Math. Dokl. **7**, no.4, 952–956 (1969). [MR **34** #482]

[25] Zhitarashu, N. V. and Eidel'man, S. D., *A certain nonlocal parabolic boundary value problem*, Mat. Issled. **5**, no. 3(17), 83–100 (1970) (in Russian). [MR **55** #10855]

[26] Zhitarashu, N. V. and Eidel'man, S. D., *Necessary and sufficient conditions for the stability of solutions to model parabolic boundary value problems*, Dokl. Akad. Nauk SSSR **244**, no. 4, 809–813 (1979); English transl. in Soviet Math. Dokl. **20**, no.1, 110–114 (1979). [MR **80d:35019**]

[27] Zhitarashu, N. V. and Eidel'man, S. D, *Asymptotic representation and necessary and sufficient conditions for the stabilization of the solutions of model parabolic boundary value problems*, Mat. Issled. No. 58, 26–47 (1980), (in Russian). [MR **82a:35054**]

[28] Zhitarashu, N. V., *A remark on general boundary value problems for elliptic and parabolic systems of equations with discontinuous coefficients*, Mat. Issled. no. 46, 53–60 (1978) (in Russian). [MR **80e:35022**]

[29] Zhitarashu, N. V., *Asymptotic representation and stabilization of solutions of boundary value problems for systems of parabolic equations in a half space*, Mat. Issled. no. 63, 42–48 (1981) (in Russian). [MR **83f:35061**]

[30] Zhitarashu, N. V., *Theorems on the complete set of isomorphisms in the L_2-theory of generalized solutions of boundary value problems for an equation that is parabolic in I. G. Petrovskiĭ's sense*, Mat. Sb. (N.S.) **128(170)**, 451–473 (1985); English transl. in Math. USSR Sb. **56**, no. 2, 447–471 (1987). [MR **87h:35149**]

[31] Zhitarashu, N. V., *L_2-theory of the generalized solutions of general linear model parabolic boundary value problems*, Izv. Akad. Nauk SSSR Ser. Mat. **61**, no. 5, 962–993 (1987). English transl. in Math. USSR Izv. **31**, no. 2, 273–305 (1988). [MR **89c:35075**]

[32] Zhitarashu, N. V., *On the solvability of a parabolic boundary value problem in the presence of power singularities in the right-hand sides*, Mat. Issled. no. 92, 69–77 (1987) (in Russian). [MR **89e:35071**]

[33] Zhitarashu, N. V., *Boundary values of weak generalized solutions of systems of parabolic equations*, Mat. Issled. no. 106, 88–96 (1989) (in Russian). [MR **90d:35052**]

[34] Zhitarashu, N. V., *L_2-theory of generalized solutions of general linear parabolic boundary values problems*, Mat. Issled. no. 112, 104–115 (1990) (in Russian). [MR **91j:35129**]

[35] Zhitomirskiĭ, Ya. I., *Cauchy's problem for parabolic systems of linear partial differential equations with increasing coefficients*, Izv. Vyssh. Uchebn. Zaved. Mat. no.1, 55–74 (1959) (in Russian). [MR **26** #4060]

[36] Zagorskiĭ, T. Ya., *Mixed Problems for Parabolic Systems of Partial Differential Equations*, L'vov State University, L'vov, 1961 (in Russian).

[37] Ivasishen, S. D., *The correct solvability of general parabolic boundary value problems in negative Hölder spaces*, Dokl. Akad. Nauk Ukrain. SSR Ser. A, no.5, 396–400 (1977) (in Russian). [MR **58** #1650]

[38] Ivasishen, S. D., *Green's matrices of boundary value problems for systems of a general form that are parabolic in the sense of I. G. Petrovskiĭ*, I and II, Mat. Sb. (N.S.) **114(156)**, no. 1, 110–166 (1981) and no. 4, 523–565 (1981); English transl. in Math. USSR Sb. **42**, no. 1, 93–144, and no. 4, 461–498 (1982). [MR **84i:35081a,b**]

[39] Ivasishen, S. D., *Linear Parabolic Boundary Value Problems*, "Naukova Dumka", Kiev, 1987 (in Russian). [Zbl 704.35059]

[40] Ivasishen, S. D. and Lavrenchuk, V. P., *The correct solvability of general boundary value problems for parabolic systems with increasing coefficients*, Ukrain. Mat. Zh. **30**, no.1, 100–106 (1978); English transl. in Ukrainian Math. J. **30**, 75–79 (1978). [MR **58** #11924]

[41] Ivasishen, S. D. and Eidel'man, S. D., $\vec{2b}$-*parabolic systems*, Trudy. Sem. Funktsional. Anal. Vyp. 1, 3–175, 271–273, Kiev, 1963 (in Russian). [Zbl 245.35043]

[42] Ivasishen, S. D. and Eidel'man, S. D., *Parabolic equations: examples, the Cauchy problem, properties of solutions*, in: Mathematics Today '87, "Vishcha Shkola", Kiev, 74–108 (1987) (in Russian). [MR **91b:35047**]

[43] Ivasishen, S. D. and Eidel'man, S. D., *Linear parabolic boundary value problems: examples, theorems on well-posedness, applications*, in: Mathematics Today '88, "Vishcha Shkola", Kiev, 76–104, (1988) (in Russian). [MR **90d:35132**]

[44] Ivasishen, S. D., *Green's Matrices of Parabolic Boundary Value Problems*, "Naukova Dumka", Kiev, 1990 (in Russian) [Zbl. 823.35001]

[45] Ivasishen, S. D., *Integral representation and initial values of solutions of $\vec{2b}$-parabolic systems*, Ukrain. Mat. Zh. **42**, no. 4, 500–506 (1990); English transl. in Ukrainian Math. J. **42**, no. 4, 443–448 (1990). [MR **91c:35033**]

[46] Il'in, A. M., *On the fundamental solution of a parabolic equation*, Dokl. Akad. Nauk SSSR **147**, no. 1, 768–771 (1962); English transl. in Soviet Math. Dokl. **3**, 1697–1700 (1962). [MR **29** #1453]

[47] Il'in, A. M., *Parabolic equations whose coefficients do not satisfy the Dini condition*, Mat. Zametki **1**, no. 1, 71–80 (1967); English transl. in Math. Notes **1**, 46–51 (1967). [MR **34** #7972]

[48] Il'in, A. M., Kalashnikov, A. S., and Oleĭnik, O. A., *Second-order linear equations of parabolic type*, Uspekhi Mat. Nauk **17**, no. 3, 3–146 (1962); English transl. in Russian Math. Surveys **17**, no. 3, 1–146 (1962). [MR **25** #2328]

[49] Ionkin, N. I. and Moiseev, E. I. *A problem for the heat equation with two-point boundary conditions*, Differentsial'nye Uravneniya **15**, no. 7, 1284–1295 (1979); English transl., Differential Equations **15**, 915–923 (1979). [MR **80i:35087**]

[50] Kamynin, L. I., *The uniqueness of the solution of a boundary value problem with A. A. Samarskiĭ's boundary conditions for a second order parabolic*

equation, Zh. Vychisl. Mat. i Mat. Fiz. **16**, no. 6, 1480–1488 (1967) (in Russian). [MR **58** #6705]

[51] Kondrat'ev, V. A. and Eidel'man, S. D. *Positive solutions of linear partial differential equations*, Trudy Mosk. Mat. Obshch. **31**, 85–146 (1974); English transl. in Trans. Moscow Math. Soc. **31**, 81–148 (1974). [MR **51** #13427]

[52] Kruzhkov, S. N., *Estimates of the highest derivatives for the solutions of elliptic and parabolic equations with continuous coefficients*, Mat. Zametki **2**, no. 5, 549–560 (1967); English transl. in Math. Notes **2**, no. 5, 824–830 (1967). [MR **36** #4123]

[53] Krylov, N. V. and Safonov, M. V., *A property of the solutions of parabolic equations with measurable coefficients*, Izv. Akad. Nauk SSSR, Ser. Mat. **44**, no. 1, 161–175 (1980). English transl. in Math. USSR Izv. **16**, no. 1, 151–164 (1981). [MR **83c:35059**]

[54] Kushitskiĭ, Ya. S. and Eidel'man, S. D., *A multi-dimensional variant of a problem of A. N. Tikhoñov*, Dokl. Akad. Nauk SSSR **299**, no. 5, 1056–1059 (1988); English transl. in Soviet Math. Dokl. **37**, no. 2, 518–521 (1988). [MR **90d:35121**]

[55] Kushitskiĭ, Ya. S. and Eidel'man, S. D., *Stabilization of the solutions of boundary value problems for higher-order model parabolic equation* , Dokl. Akad. Nauk Ukrain. SSR Ser. A, no. 1, 19–22 (1988) (in Russian). [MR **89f:35099**]

[56] Kushitskiĭ , Ya. S. and Eidel'man, S. D., *Stabilization of the solutions of boundary value problems for the multidimensional heat equation*, Ukrain. Mat. Zh. **41**, no. 3, 327–334 (1989); English transl. in Ukrainian Math. J. **41**, no. 3, 290–296 (1989). [MR **90f:35084**]

[57] Kushitskiĭ, Ya. S. and Eidel'man, S. D., *Stabilization of solutions of boundary value problems for second-order parabolic systems*, Dokl. Akad. Nauk Ukrain. SSR Ser. A, no. 6, 23–26 (1990) (in Russian). [MR **91g:35042**]

[58] Lavrent'ev M. A. and B. V. Shabat, *Methods of the theory of functions of a complex variable*, "Nauka", Moscow, 1987 (in Russian).

[59] Ladyzhenskaya O. A, V. Ya. Rivkind, and N. N. Ural'tseva, *On classical solvability of diffraction problems*, Trudy Mat. Inst. Steklov. **92**, 116–146 (1966); English transl. in Proc. Steklov Inst. Math. **92** (1966).

[60] Ladyzhenskaya O. A, V. A. Solonnikov, and N. N. Ural'tseva, *Linear and quasilinear equations of parabolic type*, "Nauka", Moscow, 1967; English transl., Amer. Math. Soc., Providence, R. I., 1968.

[61] Lopatinskiĭ, Ya. B., *On a method of reducing boundary value problems for a system of equations of elliptic type to regular integral equations*, Ukrain. Mat. Zh. **5**, no. 2, 123–151 (1953); English transl. in Amer. Math. Soc. Transl. (2) **89**. 149–183 (1970). [MR **17,494b**]

[62] Lopatinskiĭ, Ya. B., *Factoring of a polynomial matrix*, L'vov, Nauch. Zap. Polit. Inst., Ser. Fiz.-Mat. No. 2, 3–7 (1958) (in Russian).

[63] Markushevich A. I., *Theory of Analytic Functions , Vol. II. Further Construction of the Theory*, "Nauka", Moscow, 1968 (in Russian).

[64] Matiĭchuk, M. I. and Eidel'man, S. D., *On fundamental solutions and Cauchy problem for parabolic systems whose coefficients satisfy Dini condition*, Proc. of the Seminar on Functional Analysis. Vyp. 9, Voronezh, 1967, 54–83 (in Russian).

[65] Matiĭchuk, M. I. and Eidel'man, S. D., *The Cauchy problem for parabolic systems whose coefficients have low smoothness*, Ukrain. Mat. Zh. **22**, no.1, 22–36 (1970); English transl. in Ukrainian Math. J. **22**, no. 1, 18–30 (1970). [MR **41** #8830]

[66] Mikhaĭlov, V.P., *Boundary values of the solutions of elliptic equations in domains with a smooth boundary*, Mat. Sb. (N. S.) **101(143)**, no. 2, 163–188 (1976); English transl. in Math. USSR Sb. **30**(1976), 143–166 (1978). [MR **54** #13311]

[67] Nikiforov, A. F. and Uvarov, V. B., *Special Functions of Mathematical Physics*, "Nauka", Moscow, 1978; English transl., Birkhäuser, Basel, 1988. [MR **81b:33001**]

[68] Oleĭnik, O. A., *Boundary-value problems for linear equations of elliptic and parabolic type with discontinuous coefficients*, Izv. Akad. Nauk SSSR Ser. Mat. **25**, no. 1, 3–20 (1961); English transl. in Amer. Math. Soc. Transl (2) **42**, 175–194 (1964) [MR **23** #A1136]

[69] Oleĭnik, O. A., *The uniqueness of the solution of boundary value problems and the Cauchy problem for general parabolic systems*, Dokl. Akad. Nauk SSSR **220**, no. 6, 1274–1277 (1975); English transl. in Soviet Math. Dokl. **16**, 247–251 (1975). [MR **51** #8638]

[70] Peetre, Jaak, *Another approach to elliptic boundary problems*, Comm. Pure Appl. Math. **14**, 711–731 (1961). [MR **30** #1301]

[71] Petrovskiĭ, I. G., *Selected Works. Systems of Partial Differential Equations. Algebraic Geometry*, "Nauka", Moscow, 1986 (in Russian). [MR **88f:01059**]

[72] Petrovskiĭ, I. G., *Selected Works. Differential Equations. Probability Theory*, "Nauka", Moscow, 1987 (in Russian). [MR **88m:01108**]

[73] Porper, F. O. and Eidel'man, S. D., *Two-sided estimates of the fundamental solutions of second-order parabolic equations and some applications of them*, Uspekhi Mat. Nauk **39**, no. 3(237), 107–156 (1984); English transl. in Russian Math. Surveys **39**, no.3, 119–178 (1984). [MR **86b:35078**]

[74] Porper, F. O. and Eidel'man, S. D., *Properties of weak fundamental solutions of parabolic equations with lower-order derivatives*, Dokl. Akad. Nauk SSSR **288**, no. 4, 827–831 (1986); English transl. in Soviet Math. Dokl. **33**, 789–793 (1986). [MR bf 88g:35098]

[75] Porper, F. O., *Estimates of the derivatives of the fundamental solution of a stationary parabolic divergence equation with constants that do not depend on the smoothness of the coefficients*, Dokl. Akad. Nauk SSSR **235**, no. 5,

1022–1025 (1977); English transl. in Soviet Math. Dokl. **18**(1977), 1092–1096 (1978). [MR **56** #9071]

[76] Petrushko, I. M. and Eidel'man, S. D., *Solvability of the Cauchy problem for parabolic equations of second order in classes of arbitrarily growing functions*, Ukrain. Mat. Zh. **19**, no. 1, 108–113 (1967); English transl. in Ukrainian Math. J. **19**, 93–97 (1967). [MR **34** #4718]

[77] Petrushko, I. M., *Boundary values in L_p, $p > 1$, of solutions of elliptic equations in domains with Lyapunov boundary*, Mat. Sb. (N.S.) **120(162)**, no. 4, 569–588 (1983); English transl. in Math. USSR Sb. **48**, 565–585 (1984). [MR **84h:35065**]

[78] Petrushko, I. M., *Boundary and initial conditions in L_p, $p > 1$, for solutions of parabolic equations*, Mat. Sb. (N.S.) **125(167)**, no. 4, 489–521 (1984); English transl. in Math. USSR Sb. **53**, 489–522 (1986). [MR **88a:35118**]

[79] Porper, F. O. and Eidel'man, S. D., *Asymptotic behavior of classical and generalized solutions of second-order one-dimensional parabolic equations*, Trudy Mosk. Mat. Obshch. **36**, 85–130 (1978); English transl. in Trans. Moscow Math. Soc. **36**, 83–131 (1979). [MR **80g:35015**]

[80] Porper, F. O. and Eidel'man, S. D., *Properties of solutions of second-order parabolic equations with lower-order terms*, Trudy Mosk. Mat. Obshch. **54**, 118–160 (1992); English transl. in Trans. Moscow Math. Soc., 101–137 (1993). [MR **95b:35084**]

[81] Proka, D. V., *The stabilization of the solution of a boundary value problem for a parabolic equation*, Trudy Mat. Inst. Steklov. **126**, 145–170 (1973); English transl. in Proc. Steklov Inst. Math. **126**(1973), 159–188 (1975). [MR **49** #856]

[82] Ramm, A. G., *Behavior of a solution of a boundary value problem for a hyperbolic equation for $t \to \infty$*, Izv. Vyssh. Uchebn. Zaved. Matematika, no. 1 (50), 124–138 (1966) (in Russian). [MR **33** #7674]

[83] Roĭtberg, Ya. A., *Elliptic problems with non-homogeneous boundary conditions and local increase of smoothness of generalized solutions up to the boundary*, Dokl. Akad. Nauk SSSR **157**, no. 4, 798–801 (1964); English transl. in Soviet Math. Dokl. **5**, 1034–1038 (1964). [MR **29** #2511]

[84] Roĭtberg, Ya. A., *The values on the boundary of generalized solutions of elliptic equations*, Mat. Sb. (N.S.) **86(128)**, no. 2, 248–267 (1971); English transl. in Math. USSR Sb. **15** (1971). [MR **47** #3813]

[85] Roĭtberg, Ya. A., *A theorem on the complete set of isomorphisms for systems that are elliptic in the sense of Douglis and Nirenberg*, Ukrain. Mat. Zh. **27**, no. 4, 544–548 (1975); English transl. in Ukrainian Math. J. **27**, 523–526 (1975). [MR **52** #6176]

[86] Roĭtberg, Ya. A. and Serdyuk, V. A., *Elliptic problems with a parameter in L_2-spaces of generalized functions for general systems of equations*, Preprint No. 82.30, Akad. Nauk Ukrain. SSR Inst. Mat., 1982 (in Russian). [MR **84b:35041**]

[87]   Roĭtberg, Ya. A., *On the boundary values of generalized solutions of systems that are elliptic in the sense of Douglis and Nirenberg*, Sibirsk. Mat. Zh. **18**, no. 4, 846–860 (1977); English transl. in Siberian Math. J. **18**, 600–610 (1978). [MR **58** #6649]

[88]   Roĭtberg, Ya. A., *The solvability of general boundary value problems for elliptic equations involving power singularities in their right-hand sides*, Ukrain. Mat. Zh. **20**, no. 3, 412–418 (1968); English transl. in Ukrainian Math. J. **20**, 358–362 (1968). [MR **37** #6602]

[89]   Roĭtberg, Ya. A., *Elliptic boundary value problems in the class of generalized functions*, I–IV. Preprint, Chernigov, 1990; see also *Elliptic Boundary Value Problems in the Spaces of Distributions*, Mathematics and its Applications, 384, Kluwer, Dordrecht, 1996.

[90]   Serdyuk, V. A., *General elliptic boundary value problems whose right-hand sides degenerate nonsmoothly*, Ukrain. Mat. Zh. **34**, no. 2, 237–241 (1982); English transl. in Ukrainian Math. J. **34**, 195–197 (1982). [MR **83j:35056**]

[91]   Slobodetskiĭ L. N., *The fundamental solution and the Cauchy problem for parabolic system*, Mat. Sb. (N.S.) **46(88)**, no 2, 229–258 (1958) (in Russian). [MR **22** #830]

[92]   Slobodetskiĭ, L. N., *Generalized Sobolev spaces and their applications to boundary value problems for partial differential equations*, Leningrad. Gos. Ped. Inst. Uchen. Zap. **187**, 54–112 (1958); English transl. in Amer. Math. Soc. Transl. (2) **57** (1966). [MR **34** #3075]

[93]   Solonnikov, V. A., *General boundary value problems for systems elliptic in the sense of A. Douglis-L. Nirenberg*, I and II, Izv. Akad. Nauk SSSR Ser. Mat. **28**, no. 3, 665–706 (1964) and Trudy Mat. Inst. Steklov. **92**, 233–297, (1966); English transl. in Amer. Math. Soc. Transl. (2) **56**, 193–232 (1964) and Proc. Steklov Inst. Math. **92** (1966), 269–339 (1968). [MR **35** #1952 and **35** #1953

[94]   Solonnikov, V. A., *On boundary value problems for linear parabolic systems of differential equations of general form*, Trudy Mat. Inst. Steklov. **83**, 3–163 (1965); English transl. in Proc. Steklov Inst. Math. **83** (1965). [MR **35** #1965]

[95]   Solonnikov, V. A., *Estimates in L_p of solutions of elliptic and parabolic systems*, Trudy Mat. Inst. Steklov. **102**, 137–160 (1967); English transl. in Proc. Steklov Inst. Math. **102** (1967). [MR **37** #4388]

[96]   Solonnikov, V. A., *The Green's matrices for parabolic boundary value problems*, Zap. Nauchn. Sem. Leningrad. Otdel. Mat. Inst. Steklov. (LOMI) **14**, 256–287 (1969); English transl. in Semin. Math. Steklov Mat. Inst. Leningr. **14**, 109–121 (1971) [MR **45** #5587]

[97]   Solonnikov, V. A. and Khachatryan, A. G., *Estimates for solutions of parabolic initial-boundary value problems in weighted Hölder norms*, Trudy Mat. Inst. Steklov. **147**, 147–155 (1980); English transl. in Proc. Steklov Inst. Math. **147**, 153–162 (1981). [MR **81f:35060**]

[98] Abramowitz, M. and Stegun, I. A. (eds), *Handbook of Mathematical Functions with Formulas, Graphs, and Mathematical Tables*, Reprint of the 1972 edition. A Wiley-Interscience Publication. Selected Government Publications. John Wiley and Sons, Inc., New York; National Bureau of Standards, Washington, D.C., 1984. [MR **85j:00005a**]

[99] Stepanov, V. V., *A Course on Differential Equations*, Moscow, 1959 (in Russian)

[100] Tikhonov, A. N., *On boundary conditions containing derivatives of order higher than the order of the equation*, Mat. Sbornik (N.S.) **26(98)**, no. 1, 35–56 (1950) (in Russian). [MR **11,440f**]

[101] Tikhonov, V. I. and Kul'man, N. K., *Nonlinear Filtering and Quasicoherent Reception of Signals*, "Radio i Svyaz'", Moscow, 1975 (in Russian).

[102] Triebel, Hans, *Interpolation Theory, Function Spaces, Differential Operators*, VEB Deutscher Verlag Wiss., Berlin, 1978, and North-Holland, Amsterdam, 1978. [MR **80i:46032a,b**]

[103] Fedoryuk, M. V., *Asymptotic behavior of the Green function for a pseudodifferential parabolic equation*, Differentsial'nye Uravneniya **14**, no. 7, 1296–1301 (1978); English transl. in Differential Equations **14**, 923–927 (1978). [MR **80:35018**]

[104] Fedoryuk, M. V., *Asymptotics: Integrals and Series*, Mathematical Reference Library, "Nauka", Moscow, 1987 (in Russian). [MR **89j:41045**]

[105] Fedoryuk, M. V., *The Saddle-Point Method*, "Nauka", Moscow, 1977 (in Russian). [MR **58** #22580]

[106] Hörmander, L., *Linear Partial Differential Equations*, Springer-Verlag, Berlin, and Academic Press, New York, 1963. [MR **28** #4221]

[107] Chaus, N. N., *Unique solution classes of the Cauchy problem and representation of positive definite kernels*, in: Proc. Sem. Functional Analysis (Kiev, 1968), no. 1, , 176–270, Akad. Nauk Ukrain SSR Inst. Mat., Kiev, 1968 (in Russian). [MR **41** #633]

[108] Eidel'man, S. D., *Parabolic Systems*, "Nauka", Moscow, 1964; English transl., North-Holland, Amsterdam, 1969. [MR **29** #4998]

[109] Eidel'man, S. D., *On a class of parabolic systems*, Dokl. Akad. Nauk SSSR **133**, No. 1, 40–43 (1960); English transl. in Soviet Math. Dokl. **1**, 815–818 (1960). [MR **23** #A1167]

[110] Eidel'man, S. D. and Ivasishen, S. D., *Investigation of the Green's matrix of a homogeneous parabolic boundary value problem*, Trudy Moskov. Mat. Obshch. **23**, 179–234 (1970); English transl. in Trans. Moscow Math. Soc. **23**(1970), 179–242 (1972). [MR **51** #3697]

[111] Eidel'man, S. D. and Porper, F. O., *The behavior of solutions of second order parabolic equations with dissipation*, Differentsial'nye Uravneniya **7**, no. 9, 1684–1695 (1971); English transl. in Differential Equations **7**, no. 9, 1280–1288 (1974). [MR **46** #3603]

[112] Eidel'man, S. D. and Ivasishen, S. D., *Estimates of derivatives of the Green's matrix for a parabolic boundary value problems in a half-space*, Dopovidivi Akad. Nauk Ukrain. RSR **7**, 846–850 (1966) (in Ukrainian). [MR **34** #4669]

[113] Eskin, G. I., *Boundary Value Problems for Elliptic Pseudodifferential Equations*, "Nauka", Moscow, 1973; English transl., Translations of Mathematical Monographs, 52, Amer. Math. Soc., Providence, R. I., 1981. [MR **56** #6170]

[114] Agmon, S., Douglis, A., and Nirenberg L., *Estimates near the boundary for solutions of elliptic partial differential equations satisfying general boundary conditions*, I and II, Comm. Pure Appl. Math. **12**, 623–727 (1959) and **17**, 35–92 (1964). [MR **23** #A2610] and [MR **28** #5252]

[115] Aronson, D. G., *Bounds for the fundamental solution of a parabolic equation*, Bull. Amer. Math. Soc. **73**, 890–896 (1967). [MR **36** #534]

[116] Aronson, D. G., *Non-negative solutions of linear parabolic equations*, Ann. Scuola Norm. Sup. Pisa (3) **22**, 607–694 (1968); addendum in Ann. Scuola Norm. Sup. Pisa (3) **25**, 221–228 (1971). [MR **55** #8553] and [MR **55** #8554]

[117] Friedman, A., *Partial Differential Equations of Parabolic Type*, Prentice-Hall, Englewood Cliffs, 1964. [MR **31** #6062]

[118] Hopf, E., *Über den funktionalen, insbesondere den analytischen Charakter der Lösungen elliptischer Differentialgleichungen zweiter Ordnung*, Math. Z. **34**, 194–233 (1931). [Zbl. 002.34003]

[119] Hörmander, L., *The Analysis of Linear Partial Differential Operators*, Vols. I, II (1983) and III, IV (1965), Springer-Verlag, Berlin (1983) and (1985). [MR **85g:35002a,b** and **87d:35002a,b**]

[120] Lions, J. L. and Magenes, E., *Problèmes aux Limites non Homogènes et Applications*. Vol. 2, Dunod, Paris, 1968; English transl., Springer-Verlag, New York, 1972. [MR **40** #513]

[121] Moser, J., *A Harnack inequality for parabolic differential equations*, Comm. Pure Appl. Math. **17**, no. 1, 101–134 (1964); correction in Comm. Pure Appl. Math. **20**, no. 1, 231–236 (1967). [MR **28** #2357 and **34** #3121]

[122] Moser, J., *On a pointwise estimate for parabolic differential equations*, Comm. Pure Appl. Math. **24**, no. 5, 727–740 (1971). [MR **44** #5603]

[123] Nash, J., *Continuity of solutions of parabolic and elliptic equations*, Amer. J. Math. **80**, no. 4, 931–954 (1958). [MR **20** #6592]

[124] Shirota, T., *On Cauchy problem for linear partial differential equations with variable coefficients*, Osaka Math. Journ. **9**, no. 1, 43–59 (1957). [MR **21** #3668]

[125] Täclind, Sven, *Sur les classes quasianalitiques des solutions des equations aux derivée partielles du type parabolique*, Nova Acta Soc. Sci. Upsal., IV Ser. **10**, no. 3, 1–57 (1936). [Zbl. 014.02204]

[126] Watson, N. A., *Parabolic Equations on an Infinite strip*, Monographs and Textbooks in Pure and Appl. Math., 127, M. Dekker, New York, 1988. [MR **90e:35001**]

[127] Widder, D. V., *Positive temperatures on an infinite rod*, Trans. Amer. Math. Soc. **55**, 85–95 (1944). [MR **5,203f**]

[128] Zhitarashu, N. V., *On the existance of limit boundary and initial values of weak solutions of parabolic equations*, Kishinev, Izv. Akad. Nauk Respub. Moldova Mat. no. 1, 57–65 (1993) (in Russian). [MR **94f:35059**]

[129] Zhitarashu, N. V. and Rud', I. I., *Solvability of linear parabolic boundary value problems in the anisotropic spaces* $\mathcal{H}_p^s$, $s \in \mathbb{R}$, *of Bessel potentials*, Izv. Akad. Nauk Respub. Molodova Mat. no. 3, 41–48 (1993) (in Russian). [MR **96e:35064**]

Index

82. **T. Constantinescu**: Schur Parameters, Factorization and Dilation Problems, 1996, (ISBN 3-7643-5285-X)

83. **A.B. Antonevich**: Linear Functional Equations. Operator Approach, 1995, (ISBN 3-7643-2931-9)

84. **L.A. Sakhnovich**: Integral Equations with Difference Kernels on Finite Intervals, 1996, (ISBN 3-7643-5267-1)

85/ **Y.M. Berezansky, G.F. Us, Z.G. Sheftel**: Functional Analysis, Vol. I + Vol. II, 1996,
86. Vol. I (ISBN 3-7643-5344-9), Vol. II (3-7643-5345-7)

87. **I. Gohberg, P. Lancaster, P.N. Shivakumar** (Eds): Recent Developments in Operator Theory and Its Applications. International Conference in Winnipeg, October 2–6, 1994, 1996, (ISBN 3-7643-5414-5)

88. **J. van Neerven** (Ed.): The Asymptotic Behaviour of Semigroups of Linear Operators, 1996, (ISBN 3-7643-5455-0)

89. **Y. Egorov, V. Kondratiev**: On Spectral Theory of Elliptic Operators, 1996, (ISBN 3-7643-5390-2)

90. **A. Böttcher, I. Gohberg** (Eds): Singular Integral Operators and Related Topics. Joint German-Israeli Workshop, Tel Aviv, March 1–10, 1995, 1996, (ISBN 3-7643-5466-6)

91. **A.L. Skubachevskii**: Elliptic Functional Differential Equations and Applications, 1997, (ISBN 3-7643-5404-6)

92. **A.Ya. Shklyar**: Complete Second Order Linear Differential Equations in Hilbert Spaces, 1997, (ISBN 3-7643-5377-5)

93. **Y. Egorov, B.-W. Schulze**: Pseudo-Differential Operators, Singularities, Applications, 1997, (ISBN 3-7643-5484-4)

94. **M.I. Kadets, V.M. Kadets**: Series in Banach Spaces. Conditional and Unconditional Convergence, 1997, (ISBN 3-7643-5401-1)

95. **H. Dym, V. Katsnelson, B. Fritzsche, B. Kirstein** (Eds): Topics in Interpolation Theory, 1997, (ISBN 3-7643-5723-1)

96. **D. Alpay, A. Dijksma, J. Rovnyak, H. de Snoo**: Schur Functions, Operator Colligations, and Reproducing Kernel Pontryagin Spaces, 1997, (ISBN 3-7643-5763-0)

97. **M.L. Gorbachuk / V.I. Gorbachuk**: M.G. Krein's Lectures on Entire Operators, 1997, (ISBN 3-7643-5704-5)

98. **I. Gohberg / Yu. Lyubich** (Eds): New Results in Operator Theory and Its Applications The Israel M. Glazman Memorial Volume, 1997, (ISBN 3-7643-5775-4)

99 **T. Ayerbe Toledano / T. Dominguez Benavides / G. López Acedo**: Measures of Noncompactness in Metric Fixed Point Theory, 1997, (ISBN 3-7643-5794-0)